Transportation Analysis

Springer
Berlin
Heidelberg
New York
Barcelona
Budapest
Hong Kong
London
Milan
Paris
Santa Clara
Singapore
Tokyo

Titles in this Series

L. Bianco · A. R. Odoni (Eds.)
Large Scale Computation and Information Processing in Air Traffic Control

N. H. Gartner · G. Improta (Eds.)
Urban Traffic Networks

Lucio Bianco · Paolo Toth (Eds.)

Advanced Methods in Transportation Analysis

With 146 Figures
and 49 Tables

Springer

Professor Dr. Lucio Bianco
University of Rome "Tor Vergata"
Dipartimento di Informatica, Sistemi e Produzione
and Consiglio Nationale delle Ricerche
Progetto Finalizzato Trasporti
Viale dell'Università 11
00185 Rome, Italy

Professor Dr. Paolo Toth
University of Bologna
Dipartimento di Elettronica
Informatica e Sistemistica
Viale Risorgimento 2
40136 Bologna, Italy

ISBN-13:978-3-642-85258-9

Cataloging-in-Publication Data applied for
Die Deutsche Bibliothek – CIP-Einheitsaufnahme
Advanced methods in transportation analysis: with 49 tables/Lucio Bianco; Paolo Toth (ed.). – Berlin; Heidelberg; New York; Barcelona; Budapest; Hong Kong; London; Milan; Paris; Santa Clara; Singapore; Tokyo: Springer, 1996
(Transportation analysis)
ISBN-13:978-3-642-85258-9 e-ISBN-13:978-3-642-85256-5
DOI: 10.1007/978-3-642-85256-5
NE: Bianco, Lucio [Hrsg.]

Softcover reprint of the hardcover 1st edition 1996

SPIN 10516312 43/2202-5 4 3 2 1 0 – Printed on acid-free paper

FOREWORD

This volume is a compendium of papers presented during the second TRIennal Symposium on Transportation ANalysis (**TRISTAN II**) that took place in Capri, Italy on June 23-28, 1994. The Symposium was organized by the Progetto Finalizzato Trasporti and the Istituto di Analisi dei Sistemi ed Informatica of the Italian National Research Council jointly with the Italian Operations Research Society.

The purpose of this kind of meetings is to periodically allow an exchange of views and findings by scientists in the field of transportation analysis methods and tools. Therefore, the papers presented dealt with a wide range of topics and cover the different aspects of transportation analysis.

The material contained in this book gives particular emphasis to the development of mathematical modelling and algorithms. This development is due to the evolution of digital computers and the continuous increase of the computing power. In fact the need of solving large scale problems (crew scheduling, network traffic control, pollution monitoring and control, etc...) involves in some case, thousands of variables and therefore sophisticated mathematical models and computational algorithms.

Moreover, in the last years, the need for integration of transportation planning and control with management problems has increased the complexity of these models. The selected papers attempt to depict the most up to date state of the art in this direction and can be classified in three broad categories:

1. Transportation Planning;
2. Transportation Management;
3. Vehicle Management.

The **Transportation Planning** category covers a large set of problems. More specifically the first four papers concern different traffic planning models.

Y. Chen and M. Florian, after reviewing the previous contributions made in development of models and algorithms to estimate an (O-D) matrix by using the observed flows on the links of a congested traffic network, propose a non linear bilevel programming formulation for this problem. They prove the existence of solutions under several assumptions, reformulate the problem in terms of single-level optimization, and derive the optimality conditions.

In a second paper, S. Berka and D. E. Boyce use an asymmetric static user-equilibrium traffic assignment model to generate link travel times for a large scale real-world transportation network. The implementation of the model shows the convergence to a stable solution which is close to the observed traffic conditions.

The successive paper by P. Marcotte and D. Zhu deals with a traffic assignment model where two criteria, namely time and money, are explicitly considered. The authors propose a solution procedure more efficient than the similar previous models.

The fourth paper by M.G.H. Bell et al. provides a path flow estimator on the basis of measurements of link flows. The estimator proposed is designed to produce, under the assumption of stochastic user equilibrium assignment, turning movements for the DEDALE network in Lyon.

The second set of transportation planning models address in particular the design and evaluation of urban transportation networks.

A. Di Febbraro and S. Sacone propose a discrete event model to evaluate the performances and develop control strategies of a urban intermodal network. Successively, in the paper by A. Schöbel, the problem of zone planning in public transportation systems is analyzed. The goal is to design the zones so that the zone tariff is as close as possible to the distance tariff. The developed model has been applied to the data of a regional public transportation company in Germany.

A multicriteria evaluation model to reorganize the bus transit system, is described in the paper by M. Bielli et al. The proposed method has been applied to the urban network of Parma, that is a medium size Italian city.

In the following paper by Y. Israeli and A. Ceder, the transit route design problem is considered and formulated in terms of non linear programming with mixed variables. For this problem a heuristic solution procedure is proposed.

The last two papers of this set deals the parking problem in urban areas.

S. Carrese et al. develop a model to define the parking locations on the basis of a simulation of choices of different user classes on a multi-modal network. A. Sciomachen instead, given the parking facilities, develops a prototype of a Decision Support System able to guide car drivers to a parking site and to plan, at the same time, the multimodal paths.

The last set of papers included in the first category concerns with traffic dynamic assignment models.

M. Di Gangi and V. Astarita present a dynamic network loading model which allows to simulate the effects of regulation strategies on users' behaviour. The model removes some of the drawbacks of other similar approaches proposed in the literature, and explicitly allows the en-route modification of the followed path.

The successive paper by G. Bellei and M. Bielli deals with the evaluation of potential benefits of new information systems to road users and of demand regulatory measures in congested networks. The model utilized, based on a simple link performance model embedded in a dynamic assignment procedure, has been applied to small size test networks.

A multidestination system optimal dynamic traffic assignment model with distributed parameters is presented by E. Codina and J. Barcelò. This model allows to overcome the limitations of the previous models, since it represents more adequately the evolution of traffic flows under route guidance systems.

The last paper included in the category of transportation planning is by A. Nuzzolo and E. Russo. They analyze the problem of intercity and extraurban transit assignment in the framework of the stochastic approach. Random utility models of path choice and time-space network models, explicitly considering timetables, are utilized.

The issue of **Transportation Management** is tackled in the following four papers.

The first two address the large-scale network management.

L. Grandinetti et al. present an innovative method which utilizes high performance parallel computers to solve large-scale non linear network flow problems arising in the transportation field.

Successively, N. Christofides et al. examine the problem of data management of a large-scale transportation network. They propose a new dynamic structure for the representation of the tree associated to the network, and present a scheme for its memory management.

The next two papers, both by W.J. Hurley and E.R. Petersen, are referred to the economic aspects of freight transportation.

In the first-one, the problem of prices regulation in rail freight markets is discussed. The authors analyze the efficiency of the two standard economic solutions, Marginal Cost Pricing and Ramsey Pricing, relative to Deregulated Prices. They conclude that these regulatory solutions do not enhance efficiency relative to a regime of commercial freedom. The second paper is devoted to the freight network equilibrium problem. Both the cases of carriers acting cooperatively and noncooperatively are examined. The authors show that both formulations are equivalent.

The third category of papers, denoted as **Vehicle Management**, includes nine contributions.

The first three papers address the utilization of information systems, based on the information technology, in traffic management.

The acceptance of route guidance systems by motorists is analyzed in the first paper by M. Ben-Akiva et al. In particular the impact of the main factors on the behaviour of users is investigated for a small prototypical network. The analytical results allow to identify the critical factors involved in the design of driver information systems.

Successively, again M. Ben-Akiva et al. present an on-line traffic prediction system for an interurban motorway network. It forecasts in real time, traffic conditions on the network and provides information to a traffic control centre.

In the following paper, Y. Asakura describes a simulation model which allows to evaluate the effects of parking guidance and information systems on drivers' parking choice behaviour and overall transportation system performance.

The traffic control problems in urban systems are tackled in three papers.

Firstly, A. Adamski presents different dispatching control tools for the improvement of public transit service realiability. Then, A. Ceder and I. Reshetnik propose a self-tuning system for controlling intersections in undersatured and oversatured conditions. Finally, P. Dell'Olmo and P.B. Mirchandani illustrate a model for real-time traffic coordination in a urban network. This model provides a prediction of traffic movements and evaluates the performance of any signal timing strategy.

The last three papers are devoted to some important routing and scheduling problems strictly related to applications in the transportation field.

In the first paper, W.B. Powell and D.H. Gittoes develop an approximate algorithm for the dynamic assignment problem. This problem can represent several dynamic routing and scheduling problems arising in different contexts as, for example, the fleet management for motor carriers.

Successively, P. Toth and D. Vigo examine the Vehicle Routing Problem with Backhauls and propose an effective heuristic algorithm for its solution.

The last paper of the volume, by R.S.K. Kwan and A. Wren, deals with the bus driver scheduling problem. A hybrid approach, which incorporates Genetic Algorithms, a rule-based driver duties Estimator, and an integer linear programming method, is described.

Concluding, this book is, of course, by no means a complete survey of the transportation analysis field. However, we believe that the previously illustrated papers provide additional evidence of the strides being made in advanced transportation analysis research. It is noteworthy that, to varying degrees, the methods described herein have benefited not only from advances in data processing technologies and management science, as might be expected, but also from striking recent progress in such areas as software engineering, telecommunications, and other acquisition data technologies.

Finally, the editors hope that this volume will give a further contribution to the progress of transportation science.

Lucio Bianco **Paolo Toth**

Acknowledgment

We wish to thank M. Bielli and A. Scognamiglio for their contribution to organizing the **TRISTAN II** Symposium on which this volume is based.

We are also indebted with the numerous referees for their outstanding work in the selection of the papers.

Finally we gratefully acknowledge P. Carotenuto for the coordination of technical aspects of the book, and C. Landi together with C. Petrosillo for their carefull assistance in the editorial work.

CONTENTS

O-D Demand Adjustment Problem with Congestion: Part I. Model Analysis and Optimality Conditions
Y. Chen, M. Florian 1

Generating Highway Travel Times with a Large-Scale, Asymmetric User Equilibrium Assignment Model
S. Berka, D.E. Boyce 29

An Efficient Algorithm for a Bicriterion Traffic Assignment Problem
P. Marcotte, D. Zhu 63

A Stochastic User Equilibrium (SUE) Path Flow Estimator for the Dedale Database in Lyon
M.G.H. Bell, C.M. Shield, J.J. Henry, L. Breheret 75

Modelling and Performance Analysis of Urban Transportation Networks
A. Di Febbraro, S. Sacone 93

Zone Planning in Public Transportation Systems
A. Schöbel 117

Multicriteria Evaluation Model of Public Transport Networks
M. Bielli, M. Gastaldi, P. Carotenuto 135

Multi-Objective Approach for Designing Transit Routes with Frequencies
Y. Israeli, A. Ceder 157

Relationship between Parking Location and Traffic Flows in Urban Areas
S. Carrese, S. Gori, T. Picano 183

A DSS Prototype for Urban Intermodal Path Planning with Parking Management
A. Sciomachen 215

Structure of a Dynamic Network Loading Model for the Evaluation of Control Strategies
M. Di Gangi, V. Astarita 235

Dynamic Traffic Assignment in Congested Networks
G. Bellei, M. Bielli 263

A System Optimal Traffic Assignment Model with Distributed Parameters
E. Codina, J. Barceló 299

Stochastic Assignment Models for Transit Low Frequency Services: Some Theoretical and Operative Aspects
A. Nuzzolo, F. Russo 321

A Parallel Approach to Large-Scale Nonlinear Network Optimization
L. Grandinetti, F. Guerriero, R. Musmanno 341

Data Management of Large-Scale Transportation Networks
N. Christofides, H.O. Badra, Y.M. Sharaiha, G. Salkin 359

Why Regulate Prices in Freight Transportation Markets?
W.J. Hurley, E.R. Petersen 373

Optimal Freight Transport Pricing and the Freight Network Equilibrium Problem
W.J. Hurley, E.R. Petersen 389

The Impact of Predictive Information on Guidance Efficiency: An Analytical Approach
M. Ben-Akiva, A. De Palma, I. Kaysi 413

Dynamic Traffic Prediction for Motorway Networks
M. Ben-Akiva, E. Cascetta, H. Gunn, D. Inaudi, J. Whittaker 433

A Parking Simulation Model for Evaluating Availability Information Service
Y. Asakura 457

Flexible Dispatching Control Tools in Public Transport
A. Adamski 481

Queuing Optimization of Signalized Intersections
A. Ceder, I. Reshetnik 507

A Model for Real-Time Traffic Coordination Using Simulation Based Optimization
P. Dell' Olmo, P.B. Mirchandani 525

An Approximate Labelling Algorithm for the Dynamic Assignment Problem
W.B. Powell, D.H. Gittoes 547

A Heuristic Algorithm for the Vehicle Routing Problem with Backhauls
P. Toth, D. Vigo 585

Hybrid Genetic Algorithms for Bus Driver Scheduling
R.S.K. Kwan, A. Wren 609

O-D DEMAND ADJUSTMENT PROBLEM WITH CONGESTION: PART I. MODEL ANALYSIS AND OPTIMALITY CONDITIONS

Yang Chen and Michael Florian

Centre de Recherche sur les Transports, Université de Montréal
Montréal, Québec H3C 3J7, Canada

The problem of adjusting (or estimating) an origin-destination (O-D) matrix by using observed flows on the links of a congested traffic network, which we denote *DAP*, is considered in this paper. After reviewing the previous contributions made in stating models and development solution algorithms for this problem, a nonlinear bilevel programming formulation is proposed to model the *DAP*. The existence of solutions is proved under relatively mild assumptions on the link cost functions and the property of the continuous dependence of equilibrium link flows on the demand is demonstrated under a fairly weaker condition. By using the general bilevel programming theory, the *DAP* is reformulated as a single-level like optimization problem, where the marginal function of the lower level equilibrium problem is used explicitly in a constraint. The gradient function of the implicit marginal function is derived in terms of the link cost mapping and the link proportions in an equilibrium state. Necessary optimality conditions for the *DAP* are derived based on the gradient information of the marginal function, of which the significance and application for the *DAP* are discussed as well.

1. Introduction and Review

The travel demand between origins and destinations on a transportation network is an essential component of network models used in transportation planning. It is a basic input to all network assignment methods both in static and dynamic contexts. The conventional methods of obtaining origin-destination (O-D) matrices are based on the results of surveys, which are costly, or by using various demand modeling schemes. Link counts of vehicular traffic, which are relatively easy to obtain, provide information which may be used to adjust, or estimate, known O-D matrices which are out of date. The use of the information contained in these link counts has motivated the development of various methods for adjusting O-D matrices. In a broad sense, the O-D demand adjustment problem *(DAP)* was defined by Cascetta and Nguyen [10] as that of "determining an estimate of the O-D trip demand by efficiently combining traffic count data and all other available information". Before proceeding with a brief literature review, the notation used in this paper is introduced next.

Let $R = (N, A)$ be a transportation network, where N is the set of nodes and A the set of links. Denote by I the index set of the O-D pairs of the network, $\upsilon = (\upsilon_a;\ a \in A)$ the arc flow vectors, $h=(h_k;\ k \in K_i,\ i \in I)$ the path flow vectors, where K_i refers to the set of all paths between O-D pair i, and $g = (g_i,\ i \in I) \in G$ the demand vectors (matrices) for all O-D pairs, where G is a bounded set. The average cost of transportation on arc a is denoted by $s_a(\upsilon)$ while the average cost of transportation on path k by $s_k(h)$. The cost functional of the network is then defined by $s(\upsilon)=(s_a(\upsilon);\ a \in A)$ in terms of arc flows or $s(h)=(s_k(h);\ k \in K_i,\ i \in I)$ in terms of path flows. Let $P(g)$: $G \rightarrow I \times A$ be a proportion matrix map, where the cells $p_{ia}(g)$ are the proportions which assign demand g_i to arc a.

One of the first mathematical formulations of the *DAP* is the following. It is assumed that observed link counts $\hat{\upsilon}_a$ are available for a subset of arcs $\hat{A}$, $\hat{A} \subseteq A$. The problem is to find $g^* \in G$ such that, when g^* is assigned to the network according to the proportion map $P(g^*)$, the resulting arc flows reproduce the counts on $\hat{A}$, or equivalently, to find a solution to the system of nonlinear equations:

$$\sum_{i \in I} p_{ia}(g) g_i = \hat{\upsilon}_a, \quad a \in \hat{A}, \tag{1}$$

where p_{ia} is the proportion that demand g_i is assigned to arc a.

Different assumptions made for the assignment map $P(g)$ and on the observed link counts result in a large variety of formulations and solution methods for the *DAP*. When the proportion map $P(g)$ is independent of the demand, such as in network assignment models with flow independently cost, the *DAP* may be simplified to the problem of seeking the solution of linear equations

$$\sum_{i \in I} p_{ia} g_i = \hat{\upsilon}_a, \quad a \in \hat{A}. \tag{2}$$

The systems of equations (1) or (2) are usually underspecified since the number of O-D pairs usually far exceeds the number of links with available counts. In order to obtain a unique solution some additional information is usually added in the form of a known O-D matrix $\hat{g}$, which is referred to as the "target" matrix or as the matrix "to be adjusted".

The previous contributions made for the solution of the *DAP* may be classified according to the assumptions made on the assignment map $P(g)$. If it is independent of flow, such as in (2), then these proportions are computed once, depend only on the path costs and are then used explicitly in the model formulation. The generic form of such models is

$$\min \quad F(g, \hat{g}) \tag{3}$$

$$s.t. \quad \sum_{i \in I} p_{ia} g_i = \hat{\upsilon}_a, \quad a \in \hat{A} \tag{4}$$

$$g_i \geq 0, \quad i \in I. \tag{5}$$

The objective function $F(g, \hat{g}): I \rightarrow R$ may be any suitable metric between g and $\hat{g}$. Such metrics may be motivated by assumptions made on the stochastic process which generates g or by arguments originating in information theory. Examples of such functions are

$$F(g, \hat{g}) = \frac{1}{2} \sum_{i \in I} (g_i - \hat{g}_i)^2, \tag{6}$$

$$F(g, \hat{g}) = \sum_{i \in I} g_i \ln \frac{g_i}{\hat{g}_i}, \tag{7}$$

$$F(g,\hat{g})\sum_{i\in I}\hat{g}_i \ln\frac{g_i}{\hat{g}_i}. \tag{8}$$

Van Zuylen [40], Van Zuylen and Willumsen [42], Carey et al. [9], McNeil and Hendrickson [30], Spiess [37] and Brenninger-Göthe et al. [7] have made various contributions to justify the metrics used. Cascetta and Nguyen [10] discuss the statistical foundations of various matrices used for the *DAP* problem and the corresponding assumptions.

Often, (2) does not have a feasible solution, that is, the link counts are inconsistent. In order to take into account the variability in the link counts, a more general model includes a link term in the metric used, such as

$$\min \gamma_1 F_1(g,\hat{g})+\gamma_2 F_2(\upsilon,\hat{\upsilon}) \tag{9}$$

$$s.t. \quad \sum_{i\in I} p_{ia} g_i = \upsilon_a, \ \ a\in \hat{A} \tag{10}$$

$$g_i \geq 0, \ \ i \in \mathrm{I} \tag{11}$$

In this formulation, the link flows need not be reproduced exactly and the weights γ_1, γ_2 reflect the relative "confidence" in the information contained in $\hat{g}$ and $\hat{\upsilon}$. The solution methods for (2), (3)-(5), and (9)-(11) that have been proposed depend on the functional forms adopted for $F_1(\bullet)$ and $F_2(\bullet)$. The link flow inconsistency issue and related matters are discussed in Willumsen [44], Bell [4] and Van Zuylen and Branston [41]. Jörnsten and Wallace [27] propose the use of a stochastic programming approach to deal with link flow inconsistencies.

If the assignment map *P(g)* is flow dependent, such as in a network equilibrium model which is the appropriate model for the congested traffic networks, the *DAP* is considerably more complex, since *P(g)* is determined endogenously as a mapping of the demand *g*. The early contributions made in approaching this problem were based on the assumption that the counts are obtained from an assignment which is a perfect equilibrium assignment. Nguyen [31] shows that there exist an infinity of O-D matrices which reproduce exactly the equilibrium flows and are the solution of the network equilibrium problem with variable demand:

$$\min \sum_{a\in A} \int_0^{\upsilon_a} s_a(t)dt - \sum_{i\in I} \hat{u}_i g_i \tag{12}$$

$$s.t. \quad \sum_{k\in K_i} h_k - g_i = 0, \quad i \in I \tag{13}$$

$$h_k \geqq 0, \quad k \in K_i, \quad i \in I, \tag{14}$$

$$g_i \geq 0, \quad i \in I, \tag{15}$$

where $\hat{u}_i$ are minimum path lengths between O-D pair i computed with link costs $s_a(\upsilon_a^*)$, $a \in A$, where υ_a^* are equilibrium flows. In an attempt to find the O-D matrix which is most likely from a statistical viewpoint, Jörnsten and Nguyen [26] suggest to solve the problem

$$\min \sum_{i\in I} g_i \ln g_i \tag{16}$$

$$s.t. \quad \sum_{i\in I} g_i = G', \tag{17}$$

$$g \in O(g), \tag{18}$$

where $O(g)$ denotes the set of optimal solutions of (12)- (15) and G' is an estimated or prescribed total number of trips. (16) may be replaced by other metrics, such as proposed by Gur et al. [20], which suggest the use of (6) where $\hat{g}$ is a matrix obtained from a gravity type trip distribution model.

In practice it is not likely that one can obtain a full set of link counts nor is it reasonable to assume that the link flows are sampled from a perfect network equilibrium flow pattern. An early attempt to tackle the *DAP* without assuming a complete set of observed link flows available is that of Holm et al. [24] in which the *DAP* is formulated by

$$\min \; F(g, \hat{g}) \tag{19}$$

$$s.t. \;\; \sum_{i \in I} p_{ia}(g) g_i = \hat{\upsilon}_a, \quad a \in \hat{A} \tag{20}$$

$$\sum_{i \in I} g_i = g', \tag{21}$$

$$g_i \geq 0, \quad i \in I, \tag{22}$$

where $p_{ia}(g)$ are assumed to be equilibria for given g and g' is a constant vector. The proposed method uses a heuristic assignment algorithm (see Smock [36]) to iteratively refine estimates of the proportion map until a satisfactory demand is found.

If one relaxes the assumption that the observed link counts are sampled from a perfect equilibrium pattern, the issue of seeking more appropriate mathematical model for the *DAP* may be raised. Fisk ([16], [17]) extends the model proposed by Van Zuylen and Willumsen [42] by including a variational inequality constraint in the formulation, which enforces the flow pattern to be equilibrium (see Smith [35] and Dafermos [13]). The resulting model is

$$\max \;\; \sum_{i \in I} (g_i - g_i \ln g_i) \tag{23}$$

$$s.t. \;\; \sum_{a \in A} s_a(\upsilon_a)(\upsilon_a - \upsilon_a') \leq 0, \tag{24}$$

$$\sum_{i \in I} \sum_{k \in K_i} \delta_{ak} h_k = \hat{\upsilon}_a, \quad a \in \hat{A} \tag{25}$$

$$\sum_{k \in K_i} h_k - g_i = 0, \quad i \in I \tag{26}$$

$$h \geq 0, \tag{27}$$

where υ'_a and υ_α are constrained by (25)- (27) for the same g and any feasible υ'_a. Problems like (23)-(27) are usually referred to as generalized bilevel programming problems due to the constraints (24)-(27). Fisk [16] suggests using a penalty function approach to solve the problem and discusses some possible computational difficulties which may occur.

Spiess [38] considers the *DAP* in the form of the bilevel programming problem

$$\min \frac{1}{2} \sum_{a \in \hat{A}} (\upsilon_a - \hat{\upsilon}_a)^2 \tag{28}$$

$$s.t. \quad \upsilon = \arg\min \sum_{a \in A} \int_0^{\upsilon_a} s_a(t)\, dt \tag{29}$$

$$s.t. \quad \sum_{k \in K_i} h_k - g_i = 0, \quad i \in I, \tag{30}$$

$$h_k \geq 0, \; k \in K_i, \; i \in I. \tag{31}$$

This formulation is a particular instance of the general problem

$$\min \quad \gamma_1 F_1(g, \hat{g}) + \gamma_2 F_2(\upsilon, \hat{\upsilon}) \tag{32}$$

$$s.t. \quad \sum_{i \in I} p_{ia}(g) g_i = \upsilon_a, \quad a \in \hat{A} \tag{33}$$

$$g_i \geq 0, \quad i \in I, \tag{34}$$

where $p_{ia}(g)$, $i \in I$, $a \in A$ is the assignment map corresponding to an equilibrium assignment. Spiess [38] proposes an approximate gradient approach based on the assumption that the proportion map $P(g)$ is locally constant and implements the method in an ingenious way, where the proportion map is generated implicitly by using the value of the step sizes computed in solving the inner network equilibrium problem by the linear approximation method and by using the paths generated at each iteration. This method is implemented by using the macro language of the EMME/2 [25] software package and has been applied successfully to numerous problems originating from practice.

Florian and Chen [19], [18] address a similar *DAP* model as that studied by Spiess [38] by emphasizing the analysis of the model as a bilevel programming problem (BLPP). Some analytical properties of the *DAP* are exploited explicitly by adopting the analysis for general *BLPP* (Chen [11] and Chen and Florian [12]). A penalty function approach with trust region strategy and a Gauss-Seidel type method are proposed to solve the *DAP*, where the Gauss-Seidel method has been used to solve some real traffic networks with very good results. Drissi-Kaïtouni and Lundgren [15] present a sensitivity-analysis based heuristic algorithm to solve the *DAP*, where the link flow υ is treated as a function of the demand g and the derivatives of υ with respective to g, if it exists, is calculated based on sensitivity analysis theory for variational inequalities (Tobin and Friesz [39], for example). The gradient information may be obtained by solving a set of quadratic programs and is used to construct descent directions for g at a given point. Yang et al. [45] study the *DAP* in bilevel programming setting and suggest to use a Gauss-Seidel approach similar to that used by Holm et al. [24] to solve the *DAP*. Some statistical tests are provided for the obtained numerical results. Yang et al. [46] also model the *DAP* (16)-(18), studied by Jörnsten and Nguyen [26] and Nguyen [32] before, and indicate that it is a bilevel programming formulation. It should be pointed out that (16)-(18), with the optimal solution set $O(g)$, does not have the structure of a bilevel programming problem, since there is no leader-follower (or upper-lower level) structure in that formulation at all; it is in fact a convex programming problem! Evidently, the claim of Yang et al. [46] that reduce a bilevel program to a convex program is not valid.

Generally, the *BLPP* is very difficult to solve in practice due to its inherent nonconvexity and nonsmoothness. For other contributions on the *BLPP* and applications to transportation planning problems, readers are referred to [1], [2], [3], [6], [8], [11], [12], [5], [29] and [22] etc.. Most suggested solution procedures in the literature for the congested *DAP* are various heuristics and there are no optimality conditions available to date for the *DAP* with congestion. Clearly, it will be very difficult (if it is not impossible) to develop exact solution methods for the *DAP* without having optimality conditions for the problem. In order to fully understand and to solve the *DAP,* it is necessary to analyze the *DAP* in a more general setting and derive effective optimality conditions for the problem. This paper aims at providing such an analysis and optimality conditions for the congested *DAP* by analyzing the *DAP* in a theoretical bilevel programming framework, exploiting the analytical properties of the *DAP* and, consequently, deriving a set of necessary optimality conditions. The paper is organized as follows. After the brief review, in Section 2, a general nonlinear *BLPP* formulation for the *DAP* is proposed. The existence of solutions is proved under relatively mild assumptions on the link cost functions and the property of the continuous dependence of equilibrium link flows on the demand is demonstrated under a fairly weaker condition. In Section 3, the *DAP* is reformulated as a single-level like optimization problem, where the marginal function of the lower level equilibrium problem is used explicitly in a constraint. The gradient function of the implicit marginal function is derived in terms of the link cost mapping and the link proportions in an equilibrium state. Necessary optimality conditions for the *DAP* is derived and analyzed, and is applied to a typical example. A conclusion ends the paper with highlighting the application of the obtained optimality condition to design solution methods for the *DAP*.

2. Model Analysis

Let Ω be defined by

$$\Omega = \{(g, h) \mid \Lambda h - g = 0,\ h \geq 0,\ 0 \leq g \leq \overline{g}\},$$

where Λ is the O-D/path incidence matrix and $\overline{g}$ is an upper bound vector for the demand. Then, Ω gives all possible demands and path flows which satisfy the conservation of flow for some bounded demand. Evidently, Ω possesses the following property:

Lemma 1 Ω *is nonempty, compact and convex.*

It will be convenient sometimes to consider the feasible set Ω in a different way. Let $\Omega(g)$ be the set of all path flows satisfying the conservation of flow for the specified g. Then $\Omega(g)$ defines a set-valued map, from g-space to h-space:

$$\Omega(g) = \{h \mid \Lambda h - g = 0,\ h \geqq 0\}, \qquad 0 \le g \le \bar{g}.$$

The feasible set Ω for demands and path flows may be expressed by

$$\Omega = \{(g,h) \mid h \in \Omega(g),\ 0 \le g \le \bar{g}\}.$$

As a set-valued map, $\Omega(g)$ has the following property:

Lemma 2 *The set-valued map $\Omega(g)$ is upper semi-continuous on its domain.*

Proof. Consider $\forall g' \in [0, \bar{g}]$, where $[0, \bar{g}]$ is an $|I|$-cube, and a convergent sequence $\{g^j\}$ such that $g^j \to g'$, when $j \to \infty$. Consider a sequence h^j such that $h^j \in \Omega(g^j)$. Since Ω is compact by Lemma 1, there exists a subsequence $\{h^{jk}\}$ such that $h^{jk} \to h,\ k \to \infty$. By the linear relation between g and h, it is concluded that $h' \in \Omega(g')$. Therefore, $\Omega(g)$ is upper semi-continuous or closed on its domain.
Denote by Δ the link/path incidence matrix with elements δ_{ak} where

$$\delta_{ak} = \begin{cases} 1 & \text{if link } a \text{ is in path } k \\ 0 & \text{otherwise} \end{cases} \tag{35}$$

Then the link flow υ may be determined by

$$\upsilon = \Delta h. \tag{36}$$

It is clear that a link flow pattern may be uniquely determined by a path flow pattern but the reverse is not generally true.

Now the *DAP* with congestion may be stated as follows:

Find a demand vector (matrix) g^* and a corresponding equilibrium link (path) flow pattern υ^* (h^*) such that g^* and υ^* (h^*) minimize some objective determined by the transportation planner over all feasible demands g and the corresponding equilibrium flows. Mathematically, the *DAP* may be formulated by

$$\min F(g, h) = \gamma_1 F_1(g) + \gamma_2 F_2(\upsilon) \tag{37}$$

$$s.t. \qquad s(\upsilon)(\upsilon - \upsilon') \leq 0, \quad \forall \upsilon' = \Delta h', \quad h' \in \Omega(g), \tag{38}$$

$$h \in \Omega(g), \tag{39}$$

$$0 \leq g \leq \bar{g}, \tag{40}$$

where $\upsilon = \Delta h$, the objective function $F_1(g)$ may be any suitable distance metric, such as a quadratic or an entropy form, between the estimated trip demand and a reference trip demand $\hat{g}$, and $F_2(\upsilon)$ may be any suitable distance metric between the observed traffic count $\hat{\upsilon}$, on a subset of links, and the desired equilibrium flow pattern. The coefficients γ_1 and γ_2 are predetermined weights, which reflect the different confidence in the observed traffic counts $\hat{\upsilon}$ and the reference demand $\hat{g}$.

The constraints (38)-(39), which stand for a variational inequality in terms of link flows, bring the two level problem nature into the formulation and enforce all link (path) flows to be equilibrium flow patterns (see [35], [13]). Then, the formulation (37)-(40) gives a generalized bilevel programming formulation since a special case of (38)-(39) corresponds to an optimization problem. The demand g is the decision variable for the leader (upper level) and the flows υ via h are decision variables for the follower (lower level, which faces an equilibrium assignment problem for a given demand). It is clear that to evaluate the objective function at any given demand g, one has to solve the variational inequality (38)-(39).

In order to show that the *DAP* is well defined by this bilevel programming formulation, it is necessary to take a closer look at the solution set of the variational inequality (38)-(39) parameterized by the demand g. The following theorem indicates that the solution mapping of (38)-(39) depends continuously on the demand g.

Theorem 1

If the cost mapping $s(\upsilon)$ *is continuous, then the solution mapping of the variational inequality (38)-(39) is upper semi-continuous on* g.

Proof. Denote by $\Omega^*(g)$ the set of the equilibrium path flows corresponding to the solutions of (38)-(39) for given g:

$$\Omega^*(g) = \{h \mid h \in \Omega(g),\ s(\upsilon)(\upsilon - \upsilon') \leq 0,\ \text{for } \forall \upsilon' = \Delta h',\ h' \in \Omega(g)\}.$$

Since $s(\upsilon)$ is continuous, the variational inequality (38)-(39) has at least a solution for given g. Then $\Omega^*(g)$ is nonempty. Consider $\forall \tilde{g} \in [0, \bar{g}]$ and a convergent sequence $\{g^j\}$, $g^j \rightarrow \tilde{g}$ when $j \rightarrow \infty$. Consider a sequence h^j such that $h^j \in \Omega^*(g^j) \subseteq \Omega(g^j)$ and $h^j \rightarrow \tilde{h}$. By the discussion in the proof of Lemma 2, $\tilde{h} \in \Omega(\tilde{g})$.

Suppose that $\tilde{h} \notin \Omega^*(\tilde{g})$. That is, there exists $\underline{h} \in \Omega(\tilde{g})$ and $\underline{\upsilon} = \Delta \underline{h}$ such that

$$s(\tilde{\upsilon})(\tilde{\upsilon} - \underline{\upsilon}) > 0. \tag{41}$$

A contradiction will be generated. In order to do so, we construct μ^j for each g^j by using a similar argument as that used in [14] in the following way:

$$0 \leq \mu_k^j \leq h_k^j \quad \text{if} \quad g_i^j \geq \tilde{g}_i, \quad k \in K_i, \tag{42}$$

$$-\underline{h}_k \leq \mu_k^j \leq 0 \quad \text{if} \quad g_i^j < \tilde{g}_i, \quad k \in K_i, \tag{43}$$

$$\sum_{k\in K_i} \mu_k^j = g_i^j - \tilde{g}_i \quad i \in I. \tag{44}$$

Let $\tau^j = \Delta\mu^j$ Then,

$$\underline{\upsilon} + \tau^j = \Delta\underline{h} + \Delta\mu^j,$$
$$\underline{h} + \mu^j \in \Omega\ (g^j),\ \forall j.$$

Since $h^j \in \Omega^*(g^j)$ by assumption, it follows that

$$s(\upsilon^j)(\upsilon^j - (\underline{\upsilon} + \tau^j)) \leq 0 \quad \forall j, \tag{45}$$

where $\upsilon^j = \Delta h^j$. It is not difficult to choose a particular sequence of μ^j satisfying (42)-(44) such that $\mu^j \to 0$ when $j \to \infty$ Thus, $\tau^j \to 0$ when $j \to \infty$. Taking limits for both sides of (45) and using the continuity of $s(\upsilon)$, it follows that

$$s(\tilde{\upsilon})(\tilde{\upsilon} - \underline{\upsilon}) \leq 0, \tag{46}$$

which contradicts (41). Thus, it is proved that $\tilde{h} \in \Omega^*(\tilde{g})$ and, consequently, $\Omega^*(g)$ is upper semi-continuous on its domain.

Theorem 1 may be viewed as a generalization of the sensitivity analysis results obtained in [21] and [14], where the cost mapping $s(\upsilon)$ is assumed to be separate, continuous and monotone and strongly monotone respectively, while only the continuity condition is needed in this result in a more general setting.

The existence of solutions for the *DAP* follows immediately.

Theorem 2

If the cost mapping $s(\upsilon)$ is continuous, then the problem (37)-(39) has at least one solution.

Proof. By Theorem 1 and the results in [23], the graph of $\Omega^*(g)$ is closed, which implies that the feasible region of (37)-(39) is closed.

Furthermore, by the compactness of Ω, it is compact. The theorem follows.

The discussion above is mainly concerned with the existence of solutions of the *DAP* in the bilevel programming formulation in an abstract manner. Discussions focusing on transportation engeneering viewpoint may be found in some references mentioned in Section 1.

3. Necessary Optimality Conditions of the *DAP*

In this section, the discussion is focused on some necessary optimality conditions for the *DAP* and the relationship between the O-D demand and the equilibrium link (path) flows at optimality. To this end, the bilevel *DAP* is studied in an environment of single-level like optimization problem by using the theory developed in Chen [11] and Chen and Florian [12] for the general *BLPP*. To facilitate the presentation, it is assumed that the cost mapping $s(\upsilon)$ is monotone and separable with respect to links, that is, $s(\upsilon) = (s_a(\upsilon_a))$. Hence, the variational inequality (38)-(39) is equivalent to the following optimization problem for a given demand g ([13], [35]):

$$\min \sum_{a \in A} \int_0^{\upsilon_a} s_a(t)dt \tag{47}$$

$$s.t. \quad \sum_{k \in K_i} h_k - g_i = 0, \quad i \in I \tag{48}$$

$$h \geq 0. \tag{49}$$

Thus the *DAP* may be formulated as a *BLPP:*

$$\min F(g,h) \tag{50}$$

where, for given g, h solves

$$\min f(g, h) = \sum_{a \in A} \int_0^{\upsilon_a} s_a(t)dt \tag{51}$$

$$s.t. \quad \sum_{k \in K_i} h_k - g_i = 0, \quad i \in I \tag{52}$$

$$g \geq 0, \; h \geq 0, \tag{53}$$

where, as usual, $\upsilon = \Delta h$ and the upper-bound constraints for the demand g are relaxed without loss of generality. This formulation may be interpreted as a leader-follower game where the leader (planner) wants to find the best demand to fulfill an objective (i.e. minimum difference or maximum entropy), while the follower (network user) wants to minimize the transportation costs (user optimal equilibrium). For any given demand g, the equilibrium problem (51)-(53) corresponds to a unique optimal objective value. When g varies on its domain, a function of g will be defined implicitly by the optimal values of the problem in this manner. Denote by $V(g)$ the optimal value (marginal) function of (51)-(53), defined by

$$V(g) = min\{ \sum_{a \in A} \int_0^{\upsilon_a} s_a(t)dt | h \in \Omega(g), \}$$

which is the minimum of the sum of integrals of the link cost function s_a for a given demand g. According to the properties of equilibrium assignment problem and the compactness of Ω, $V(g)$ is well defined and has finite value:

$$-\infty < V(g) < \infty \,, \qquad 0 \leq g \leq \overline{g}$$

By using the marginal function $V(g)$, the *DAP* may be reformulated as a single-level like optimization problem:

$$\min F(g, h) \tag{54}$$

$$s.t. \quad \sum_{a \in A} \int_0^{\upsilon_a} s_a(t)dt - V(g) = 0, \tag{55}$$

$$\sum_{k \in K_i} h_k - g_i = 0, \quad i \in I \tag{56}$$

$$0 \leq g \leq \bar{g}, \tag{57}$$

$$h \geq 0. \tag{58}$$

It is noted that any feasible solution to (54)-(58) must correspond to an optimal solution of an equilibrium assignment problem for that given demand, implied by the constraint (55) as well as (56) and (58). Evidently, we have

Lemma 3 *The problem (54)-(58) is equivalent to the problem (37)-(40).*

Though the formulation (54)-(58) seems to resemble that of a single level optimization problem, the *DAP* remains a bilevel one because of the implicitness of $V(g)$. In fact, to evaluate any single value of $V(g)$ for a given g, one has to solve an entire optimization problem. However, this single-level like reformulation does provide a useful mean to overcome some difficulties brought in by the two level nature of the problem. More importantly, to enable one to deal with the *DAP* by using the well developed single-level mathematical programming theory, for example, to derive optimality conditions for the *DAP*. In order to do so, the differentiability of $V(g)$ should be addressed first.

Theorem 3

In addition to the assumptions on the cost mapping $s(\upsilon)$ made earlier, if it is strictly monotone on υ, then $V(g)$ is differentiable and the gradient may be computed by

$$\nabla V(g) = (\sum_{a \in A} \delta_{ak^i} \, s_a(\upsilon_a^*(g)), \quad i \in I), \tag{59}$$

or $$\nabla V(g) = (\sum_{a\in A} P^*_{ia}(g)\, s_a(\upsilon^*_a(g)), \quad i \in I), \tag{60}$$

where υ^*_a *(g) are equilibrium link flows,* p^*_{ia} *(g) are equilibrium link proportions for a given demand g, and* k^i *are the paths with minimum costs between O-D pairs i,* $i \in I$.

Proof. Denote by $l_g(h,u,\alpha)$ the Lagrange of the equilibrium assignment problem (51)-(53):

$$l_g(h,u,\alpha) = \sum_{a\in A} \int_0^{\upsilon_a} s_a(t)dt - u^T(\Lambda h - g) - \alpha^T h,$$

with multipliers u and α. Since $s(\upsilon)$ is strictly monotone on υ, the problem (51)-(53) has a unique optimal solution, $\upsilon^*(g)$, for given g, and $\upsilon^*(g)$ can be characterized by

$$\sum_{a\in A} \delta_{ak} s_a(\upsilon^*_a(g)) - u_i - \alpha_k = 0, \quad k \in K_i, \quad i \in I \tag{61}$$

$$h^*_k(g)\alpha_k = 0, \ \alpha_k \geq 0, \quad k \in K_i, \quad i \in I, \tag{62}$$

as well as the feasibility conditions (52)-(53), where $h^*(g)$ are equilibrium path flows related to g. As it is well known, these conditions imply that

$$\sum_{a\in A} \delta_{ak} s_a(\upsilon^*_a(g)) - u_i \geq 0, \quad k \in K_i, \quad i \in I, \tag{63}$$

$$h^*_k(g)(\sum_{a\in A} \delta_{ak} s_a(\upsilon^*_a(g)) - u_i) = 0, \quad k \in K_i, \quad i \in I. \tag{64}$$

Thus, u represents the minimum cost vector between all O-D pairs of the traffic network in an equilibrium state. Consequently, the optimal u is uniquely determined by

$$u_i = \sum_{a \in A} \delta_{ak^i} s_a(\upsilon_a^*(g)) \quad (65)$$

$$= \sum_{a \in A} p_{ia}^*(g) s_a(\upsilon_a^*(g)), \quad i \in I, \quad (66)$$

where (66) holds because of the relation between link and path proportions, $p_{ia} = \Sigma_{k \in K_i} \delta_{ak} p_k$, and the fact that all path costs are equal at an equilibrium state. According to Theorem 5.1.1 in [11],

$$\nabla_g V(g) = \nabla_g l_g(h, u, \alpha) = u$$

$$= (\sum_{a \in A} \delta_{ak^i} s_a(\upsilon_a^*(g)), \quad i \in I)$$

$$= (\sum_{a \in A} p_{ia}^*(g) s_a(\upsilon_a^*(g)), \quad i \in I)$$

This completes the proof.

Theorem 3 states that the gradient of $V(g)$ can be completely characterized by the link cost mapping $s(\upsilon)$ and link or path proportions at an equilibrium state. In other words, the gradient of $V(g)$ is just the minimum cost vector for all O-D pairs for a given demand g. Then it is possible to use the gradient information to help designing solution methods for solving the *DAP*. Another significant consequence of the differentiability of $V(g)$ is that a Fritz-John type necessary optimality condition for the *DAP* may be established.

Theorem 4

If (g^*, h^*) solves the DAP (54)-(58), then there exist multipliers $\lambda \in \mathbf{R}^2$ and $u \in \mathbf{R}^{|I|}$ such that the following inequalities hold:

$$\lambda_0(\frac{\partial F}{\partial g_i}|_{g_i=g_i^*} + \frac{\partial F}{\partial h_k}|_{h_k=h_k^*}) + \lambda_1(\sum_{a \in A} \delta_{ak} s_a(\upsilon_a^*) - u_i) \geq 0, \quad (67)$$

$$\lambda_0 h_k^* \left(\frac{\partial F}{\partial g_i}\Big|_{g_i=g_i^*} + \frac{\partial F}{\partial h_k}\Big|_{h_k=h_k^*}\right) = 0, \tag{68}$$

$$\sum_{a\in A} \delta_{ak} s_a(\upsilon_a^*) - u_i \geq 0, \tag{69}$$

$$h_k^* \left(\sum_{a\in A} \delta_{ak} s_a(\upsilon_a^*) - u_i\right) = 0, \tag{70}$$

for all $k \in K_i$ *and* $i \in I$.

Proof. It is evident that the constraint (55) is equivalent to the equilibrium conditions (69)-(70) for some $u \in \mathbf{R}^{|I|}$. By Theorem 5.3.1 in [11] and Theorem 3 above, there exist multipliers $\lambda \in \mathbf{R}^2$, $\eta \in \mathbf{R}^{|I|}$, $\xi \in \mathbf{R}^{|I|}$ and $\gamma \in \mathbf{R}^{\Sigma i \in I|K_i|}$ such that

$$\lambda_0 \frac{\partial F}{\partial g_i}\Big|_{g_i=g_i^*} - \lambda_1 u_i - \eta_i - \xi_i = 0, \qquad i \in I \tag{71}$$

$$\lambda_0 \frac{\partial F}{\partial h_k}\Big|_{h_k=h_k^*} + \lambda_1 \sum_{a\in A} \delta_{ak} s_a(\upsilon_a^*) + \eta_i - \gamma_k = 0, \quad k \in K_i,\ i \in I \tag{72}$$

$$g_i^* \xi_i = 0, \ \ \xi_i \geq 0, \qquad i \in I \tag{73}$$

$$h_h^* \gamma_K = 0, \ \ \gamma_K \geq 0, \qquad k \in K_i, \quad i \in I \tag{74}$$

as well as the feasibility conditions (57)-(58).

(71)-(74) may be rewritten as

$$\lambda_0 \frac{\partial F}{\partial g_i}\Big|_{g_i=g_i^*} - \lambda_1 u_i - \eta_i \geq 0, \qquad i \in I \tag{75}$$

$$\lambda_0 \frac{\partial F}{\partial h_k}\Big|_{h_k=h_k^*} + \lambda_1 \sum_{a\in A} \delta_{ak} s_a(\upsilon_a^*) + \eta_i \geq 0, \quad k \in K_i, \quad i \in I \tag{76}$$

$$g_i^*(\lambda_0 \frac{\partial F}{\partial g_i}|_{g_i=g_i^*} - \lambda_1 u_i - \eta_i) = 0, \quad i \in I \tag{77}$$

$$h_k^*[\lambda_0 \frac{\partial F}{\partial h_k}|_{h_k=h_k^*} + \lambda_1 \sum_{a\in A} \delta_{ak} s_a(\upsilon_a^*) + \eta_i] = 0, \quad k \in K_i, \quad i \in I \tag{78}$$

It may be assumed that g_i^*, $i \in I$ is positive without loss of generality by the physical interpretation of g. Then, (77) becomes

$$\eta_i = \lambda_0 \frac{\partial F}{\partial g_i}|_{g_i=g_i^*} - \lambda_1 u_i, \quad i \in I \tag{79}$$

Substituting η_i in (76) and (78) by (79), it follows that

$$\lambda_0 (\frac{\partial F}{\partial g_i}|_{g_i=g_i^*} + \frac{\partial F}{\partial h_k}|_{h_k=h_k^*}) + \lambda_1 (\sum_{a\in A} (\delta_{ak} s_a(\upsilon_a^*) - u_i) \geq 0, \tag{80}$$

$$h_k^*[\lambda_0 (\frac{\partial F}{\partial g_i}|_{g_i=g_i^*} + \frac{\partial F}{\partial h_k}|_{h_k=h_k^*}) + \lambda_1 (\sum_{a\in A} (\delta_{ak} s_a(\upsilon_a^*) - u_i) = 0, \tag{81}$$

for all $k \in K_i$, $i \in I$. By combining (69)-(70) and (80)-(81), the theorem follows.

A little analysis is helpful to gain some insight into these necessary optimality conditions for the *DAP*. It should be pointed out that the multiplier λ_0 may not be necessarily nonzero since, generally, the nonlinear *BLPP* may not satisfy a constraint qualification at an optimal solution of the problem, see [12]. If $\lambda_0 = 0$, the conditions (67)-(68) coincide the conditions (69)-(70), which is the Wardrop's user-optimal principle. In nondegenerate case, that is, if $\lambda_0 \neq 0$, (68) reduces to

$$\frac{\partial F}{\partial g_i}|_{g_i=g_i^*} + \frac{\partial F}{\partial h_k}|_{h_k=h_k^*} = 0, \quad k \in K_i, i \in I \tag{82}$$

for any used path k, $h_k^* > 0$, $k \in K_i$, $i \in I$. The condition (82), then, implies that, at optimality, an increase in O-D demand g_i leads to a decrease in path flows $h_k > 0$, $k \in K_i$ in the same amount of variation rate, and vice versa. On the other hand, the conservation of flow constraints enforce that the demand increase will cause the flow increasing in an average sense. By taking both aspects into account, some equilibrium state between the O-D demands and corresponding positive path flows must be attained at optimal solutions. Therefore, the necessary optimality condition for the *DAP* characterizes an equilibrium state between the determination of O-D demands and selection of paths, as well as the equilibrium state between the path choices (user-optimal principle).

If the objective function $F(g, h)$ is taken to be a quadratic form:

$$F(g,h) = \frac{1}{2}[\sum_{i\in I} (g_i - \hat{g}_i)^2 + \sum_{a\in \hat{A}} (\upsilon_a - \hat{\upsilon}_a)^2],$$

the conditions (67)-(70) reduce to

$$\lambda_0(g_i^* - \hat{g}_i + \sum_{a\in \hat{A}} p_{ia}^*(\upsilon_a^* - \hat{\upsilon}_a)) + \lambda_1(\sum_{a\in A} \delta_{ak} s_a(\upsilon_a^*) - u_i) \geq 0, \qquad (83)$$

$$\lambda_0 h_k^*(g_i^* - \hat{g}_i + \sum_{a\in \hat{A}} p_{ia}^*(\upsilon_a^* - \hat{\upsilon}_a)) = 0, \qquad (84)$$

$$\sum_{a\in A} \delta_{ak} s_a(\upsilon_a^*) - u_i \geq 0, \qquad (85)$$

$$h_k^*(\sum_{a\in A} \delta_{ak} s_a(\upsilon_a^*) - u_i) = 0. \qquad (86)$$

for all $k \in K_i$ and $i \in I$. If $\lambda_0 \neq 0$, the condition (82) implies that

$$g_i^* + \sum_{a\in \hat{A}} p_{ia}^*\upsilon_a^* = \hat{g}_i + \sum_{a\in \hat{A}} p_{ia}^*\hat{\upsilon}_a, \quad i \in I. \qquad (87)$$

That is, the summation of the expected demand g_i^* and the corresponding flows υ_a^* on the observed links related to the O-D pair i is the same as that of observed ones. An extreme example which satisfies this condition is that the observed link counts $\hat{\upsilon}$ are an equilibrium pattern which can be generated by the target demand $\hat{g}$, and $g^* = \hat{g}$. Consequently, $\upsilon^* = \hat{\upsilon}$. (g^*, υ^*) is obviously a global optimal solution of the problem in this context.

Though the existence of a nonzero λ_0 may not be guaranteed generally in theory, many problems arisen from reality do possess the property. In the following, the necessary optimality condition derived above is applied to a small example with one O-D pair and four links. Nonzero values of λ_0 occur in this example.

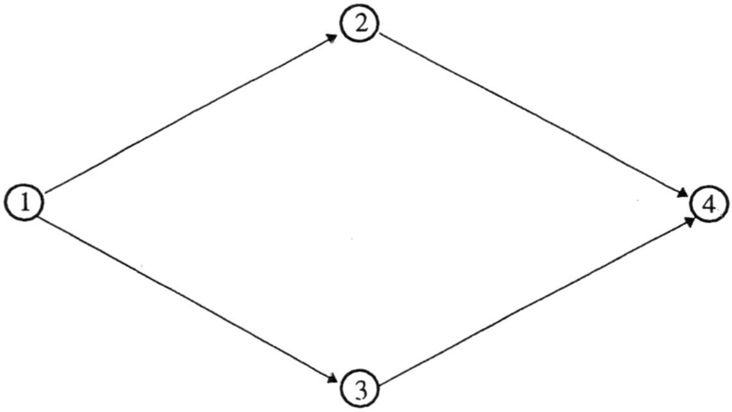

Figure 1. A Small Example

Example

$$\min \frac{1}{2}((g - \hat{g})^2 + (\upsilon_{12} - \hat{\upsilon}_{12}))^2 + (\upsilon_{13} - \hat{\upsilon}_{13})^2)$$

where υ solves

$$\min \sum_{a \in A} \int_0^{\upsilon_a} s_a(t)dt$$

$$\begin{aligned} s.t. \quad & \upsilon_{12} + \upsilon_{13} = g, \\ & \upsilon_{12} - \upsilon_{24} = 0, \\ & \upsilon_{13} - \upsilon_{34} = 0, \\ & \upsilon \geq 0,\ g \geq 0, \end{aligned}$$

where the cost mapping is given by

$$\begin{aligned} c_{12}(\upsilon_{12}) &= 10\ \upsilon_{12} \\ c_{24}(\upsilon_{24}) &= \upsilon_{24} + 10, \\ c_{13}(\upsilon_{13}) &= \upsilon_{13} + 50, \\ c_{34}(\upsilon_{34}) &= 10\ \upsilon_{34}, \end{aligned}$$

and $\hat{\upsilon}_{12}$ and $\hat{\upsilon}_{13}$ are observed link flows. The optimal solution can be determined by

$$\begin{aligned} g^* &= (2/3)\ \hat{g} + (1/3)(\hat{\upsilon}_{12}+\hat{\upsilon}_{13}), \\ \upsilon_{12}^* = \upsilon_{12}^* &= (g^*/2) + 20/11, \\ \upsilon_{13}^* = \upsilon_{34}^* &= (g^*/2) - 20/11, \end{aligned}$$

and the optimality condition is satisfied at these solutions for any observed link flows with $\lambda_0 \neq 0$, provided $\hat{\upsilon}_{12} - \hat{\upsilon}_{13} = 10/11$.

4. Conclusions

In this paper, the *DAP* on congested network is analyzed in a *BLPP* framework from theoretical aspects. The existence of solutions and sensitivity analysis of the *DAP* are conducted under mild conditions on the link cost mapping. The optimality condition for the *DAP* is first stated in the literature to the authors' knowledge. One should be aware that, since the *DAP* is an inherently nonconvex problem, the derived optimality condition may be only a stationary condition for the *DAP* in general. However, from the experience of the authors on the solution methods for the *DAP*, it does provide useful and effective stopping criterion for solution procedures to solve the problem. The algorithmic issue will be discussed in detail in Part II of this paper.

References

[1] Aiyoshi, E. and Shimizu K., "A Solution method for the static constrained Stackelberg problem via penalty functions", *IEEE Trans. Auto. Control* AC-29 (1984), 1111-1114.

[2] Bard, J.F., "Convex two-level optimization", *Mathematical Programming* 40 (1988), 15-27.

[3] Bard, J.F. and J.E. Falk, "An explicit solution to the multilevel programming problem", *Computers and Operations Research* 9 (1982), 77-100.

[4] Bell, M. G. H., "The estimation of an origin-destination matrix from traffic counts", *Transportation Science* 17 (1983), 198-217.

[5] Ben-Ayed, O., D.E. Boyce and C.E. Blair, "A General Bilevel Linear Programming Formulation of the Network Design Problem", *Transportation Research B* 22 (1988), 311-318.

[6] Bialas, W.F. and M.H. Karwan, "Two-level Linear Programming", *Management Science* 30 (1984), 1004-1020.

[7] Brenninger-Göthe, M. K.O. Jörnsten and J.T. Lundgren, "Estimation of origin-destination matrices from counts using multiobjective programming formulations", *Transportation Research B* 23 (1989), 257-269.

[8] Candler, W. and R. Townsley, "A Linear Two-level Programming Problem", *Computers and Operations Research* 9 (1982), 59-76.

[9] Carey, M., C. Hendrickson and K. Siddarthan, "A method for estimation of origin/destination trip matrices", *Transportation Science* 15 (1981), 32-49.

[10] Cascetta, E. and S. Nguyen, "A Unified Framework for Estimating or Updating Origin-Destination Matrices from Traffic Counts", *Transportation Research B* 22 (1988), 437 455.

[11] Chen, Y., "Bilevel Programming Problems: Analysis, algorithms and applications", Ph.D. dissertation, Université de Montréal, Montréal, Canada (1993).

[12] Chen, Y. and M. Florian, "The Nonlinear Bilevel Programming: Formulations, Regularity and Optimality Conditions", to appear in *Optimization*, Publication CRT-794, Université de Montréal, Montréal, Canada (revised version, 1993).

[13] Dafermos, S., "Traffic equilibrium and variational inequalities", *Transportation Science* 14 (1980), 42-54.

[14] Dafermos, S. and A. Nagurney, "Sensitivity Analysis for the Asymmetric Network Equilibrium Problem", *Mathematical Programming* 28 (1984), 174-184.

[15] Drissi-Kaïtouni, O. and J. Lundgren, "Bilevel Origin-destination Matrix Estimation Using a Descent Approach", *Technique Report*, LiTH-MAT-R-1992-49, Linköping University, Linköping, Sweden (1992).

[16] Fisk, C.S., "On Combining Maximum Entropy Trip Matrix Estimation with User Optimal Assignment", *Transportation Research B* 22 (1988), 69-73.

[17] Fisk, C.S., "Trip Matrix Estimation from Link Traffic Counts: The Congested Network Case", *Transportation Research B* 23 (1989), 331-336.

[18] Florian, M and Y. Chen, "A coordinate descent method for bilevel O-D matrix adjustment problem", to appear in *International Transactions in Operational Research* (1994).

[19] Florian, M and Y. Chen, "A Bilevel Programming Approach to Estimating O-D Matrix by Traffic counts", *Publication CRT-750, Université de Montréal, Montréal,* Canada (1991).

[20] Gur, Y.J., M. Turnquist, M. Schneider, L. LeBlanc and D.Kurth, "Estimation of an Origin-Destination Trip Table Based on Observed Link Volumes and Turning Movements" - Volume 1., *Technical Report*, FHWA, U.S. Department of Transportation, Washington, D.C., RD-801034, (1980).

[21] Hall, M. A., "Properties of the Equilibrium State in Transportation Networks", *Transportation Science* 12 (1978), 208-216.

[22] Hansen, P., B. Jaumard and G. Savard, "New Branching and Bounding Rules for Linear Bilevel Programming", *SIAM J. Sci. Stat. Comput.* 13 (1992),

[23] Hogan, W., "Point-to-set Maps in Mathematical Programming", *SIAM Review* 15 (1973), 591-603.

[24] Holm, J., T. Jenson, S.K. Nielsen, A. Christensen, B. Johnsen and G. Ronby, "Calibrating Traffic Models on Traffic Census Results Only", *Traffic Engineering and Control* 17 (1976), 137-140.

[25] INRO Consultants Inc., "EMME/2 Users' Manual", 735 pp, Montréal, Québec, Canada (1990).

[26] Jörnsten, K. and S. Nguyen, "On the estimation of a trip matrix from network data", *Publication No. 153, CRT, Université de Montréal, Montréal,* Canada (1979).

[27] Jörnsten, K. and S. Wallace, "Overcoming the (apparent) problems of inconsistency in origin-destination matrix estimations", *Transportation Sciences* 27 (1993), 374-380.

[28] Maher, M.J., "Inferences on Trip Matrices from Observations on Link Volumes: A Bayesian Statistical Approach", *Transportation Research* 17B (1983), 435-447.

[29] Marcotte, P., "Network Design Problem with Congestion Effects: A Case of Bilevel Programming", *Mathematical Programming* 34 (1986), 142-162.

[30] McNeil, S. and C. Hendrickson, "A regression formulation of the matrix estimation problem", *Transportation Science* 19 (1985), 278-292.

[31] Nguyen, S., "Estimating an O-D Matrix from Network Data: A Network Equilibrium Approach", *Publication No. 60, CRT, Université de Montréal, Montréal, Canada* (1977).

[32] Nguyen, S., "Modeles de distribution spatiale tenant compte des itineraires", *INFOR* (1983)

[33] Nguyen, S., "Estimating Origin-Destination Matrices from Observed Flows", in *Transportation Planning Models* (Eds. M. Florian) (1984), 363-380.

[34] Nguyen, S. and C. Dupuis, "An Efficient Method for Computing Traffic Equilibria in Networks with Asymmetric Transportation Costs", *Transportation Science* 18 (1984), 185-232.

[35] Smith, M.J., "Existence, Uniqueness and Stability of Traffic Equilibria", *Transportation Research B* 13 (1979), 295-304.

[36] Smock, R., "An iterative Assignment approach to capacity restraint on arterial networks", *HRB, Bulletin 347* (1962).

[37] Spiess, H., "A Maximum Likelihood Model for Estimating Origin-Destination Matrices", *Transportation Research* 21B, 395-412 (1987).

[38] Spiess, H., "A Gradient Approach for the O-D Matrix Adjustment Problem", *Publication No.693, CRT, Université de Montréal, Montréal, Canada* (1990)

[39] Tobin, R. and T.Y. Friesz, "Sensitivity Analysis for Equilibrium Network Flow", *Transportation Science* 22 (1988), 242-250.

[40] Van Zuylen, J.H., "The information minimizing method: Validity and applicability to transport planning", in *New Developments in Modeling Travel Demand and Urban Systems* (Edited by Jansen G. R. H., et al). Saxon, Farnborough (1978).

[41] Van Zuylen, J.H. and D.M. Branston, "Consistent link flow estimation from counts", *Transportation Research* 16B (1982), 473-476.

[42] Van Zuylen, J.H. and L.G. Willumsen, "The Most Likely Trip Matrix Estimated from Traffic Counts", *Transportation Research B* 14 (1980), 281-293.

[43] Wardrop, J.G., "Some Theoretical Aspects of Road Traffic Research", *Proceedings of the Insititute of Civil Engineers* Part II (1952), 325-378.

[44] Willumsen, L.G., "Simplified transport models based on traffic counts", *Transportation Planning* 10 (1981), 257-278.

[45] Yang, H., T. Sasaki, Y. Iida and Y. Asakura, "Estimation Origin-destination Matrices from Link Traffic Counts on Congested Networks", *Transportation Research* 26B (1992), 417-434.

[46] Yang, H., Y. Iida and T. Sasaki, "The Equilibrium-based Origin-destination Matrix Estimation Problem", *Transportation Research* 28B (1994), 23-33.

GENERATING HIGHWAY TRAVEL TIMES WITH A LARGE-SCALE, ASYMMETRIC USER EQUILIBRIUM ASSIGNMENT MODEL

Stanislaw Berka and David E. Boyce

Urban Transportation Center, University of Illinois at Chicago
1033 West Van Buren Street, Suite 700 South
Chicago, Illinois 60607, USA

An asymmetric, static user-equilibrium traffic assignment model with traffic-engineering based turning-specific link travel time functions is used to generate highway travel time functions for an 800 square km network. Specialized link delay functions for all intersection and roadway types occurring in the study network are adapted from the literature or developed. Conditions for the existence and uniqueness of the solution based on the theory of variational inequalities are checked. The network coding process and the implementation of the model are described. Results are presented and discussed. The model is found to converge to a stable solution and the solution obtained is close to the observed traffic conditions.

1. Introduction

The ultimate objective of the study summarized here and described in detail in [4], is to generate link travel times for a large-scale real-world transportation network for ADVANCE, a dynamic route guidance system described in [6]. Link travel time estimates are required by time-of-day periods. The method selected for this task is the static asymmetric user equilibrium assignment model. The model is asymmetric because of the interactions among flows and delays within each intersection. In the literature known to the authors, the only documented models of this type were applied to small and medium-size networks; see [7], [21] and [27].

This study utilizes a realistic network with all the variety of intersection configurations and types of transportation facilities including freeways occurring in the test area, the area in which the route guidance system is being planned. The reasons that there is no study of this kind documented in the literature may be several. First, the cost of obtaining the data and computing the results for this kind of model may be too high compared with the benefits. For the implementation of a route guidance system, however, detailed travel time estimates are needed by turning movement, which justifies the higher cost. The set of link travel times is the initial basis for planning reasonable routes. Although such systems are designed to obtain travel time data from participating vehicles, obtaining travel time data in this way may require the system to operate a very long time, even with several thousand participating vehicles [17]. Therefore, there is a need for initial travel time estimates.

Another reason may be the current discrepancy between the realistic link delay functions applied in the model and the solution properties of the user equilibrium assignment problem. Theoretical studies indicate that sufficient conditions for existence and uniqueness of the equilibrium solution may be violated for this type of function. However, computational studies suggest that the solution obtained is unique [22]. The third possible reason is the lack of input data required for this kind of model, especially detailed intersection geometry data. The cost of collecting and coding this data for this study was relatively high. The advantage of applying the model to a network served by a route guidance system is that once the system operates, it should be relatively easy to obtain measured travel time data to validate the model.

The original contributions of this study can be outlined as follows.

1. Adapting delay functions from the literature, or developing new functions when none is available, for each type of road facility and each type of intersection occurring in the ADVANCE test area.

2. Developing a methodology for collecting large amounts of detailed transportation network data, designing the database, coding and testing the data.
3. Extending existing codes for an asymmetric route choice model to solve a problem for a large network. The test area, located mainly north and northwest of Chicago, covers approximately 300 square miles (800 square km) and consists of nearly 8000 links and more than 2500 nodes.
4. Solving the model by time of day period and generating the travel time estimates to be used within the ADVANCE Project.

Throughout the paper, turning movements are described as LT, TH and RT, which stand for left-turn, through and right-turn movements. For convenience, references to the 1985 Highway Capacity Manual [32] are denoted as 1985 HCM.

The paper begins by presenting the background of the asymmetric user equilibrium assignment problem with the conditions of existence and uniqueness of the solution. The next section presents the turning-movement specific travel time functions used in the study. Then, the study network, trip tables and the process of creating them are briefly described, followed by a discussion of the implementation of the model. The last section includes some conclusions and observations on the results obtained.

2. Problem Statement and Background

Route choice models have a long tradition and large literature. A good overview and classification of the models can be found in [27] or [23]. Route choice problems can be classified according to different criteria: route choice principle; time period; and source of travel demand. The model described in this paper is a static, deterministic, user equilibrium assignment (UE) with fixed demand. The optimization formulation of this problem is presented below, followed by the variational inequalities formulation for the asymmetric case.

Consider a network consisting of L links. The vector

$$\mathbf{f} = (f_1,\ldots,f_L) \tag{1}$$

represents the link flow pattern in the network, where f_a is the flow of vehicles on link a expressed in vehicles/hour. The cost function

$$\mathbf{c} = (c_1(\mathbf{f}),\ldots,c_L(\mathbf{f})) \tag{2}$$

associates with any given flow pattern a link cost pattern, where c_a is the travel cost function of link a. The UE assignment problem consists of assigning a given trip pattern to the network, so that Wardrop's principle is satisfied: drivers choose their routes so that the travel time on all routes that are actually used between each origin and destination are equal and not more than the travel time on any unused route [35]. The trip pattern is defined by the number of trips per hour from each origin zone to each destination zone T_{ij}; in this study, $T=(T_{11},\ldots,T_{ij},\ldots,T_{IJ})$ is given as a fixed matrix.

If each cost function depends on the flow on the same link only, the static UE problem with fixed demand can be stated as follows [3]:

$$\text{minimize} \sum_a \int_0^{f_a} c_a(x)dx \tag{3}$$

$$\text{subject to} \sum_{r\in R_{ij}} h_r = T_{ij} \quad \forall i,j \tag{4}$$

$$h_r \geq 0 \quad \forall r \in R \tag{5}$$

where

$$f_a = \sum_{r\in R} \delta_{ar} h_r \quad \forall a \in L \tag{6}$$

and h_r is the flow on route r, R_{ij} is the set of routes from origin i to destination j, T_{ij} is the number of trips per hour from origin zone i to destination zone j and d_{ar} equals 1 if link a belongs to route r, and 0 otherwise.

Problem (3)-(6) is called diagonal because the Jacobian matrix (7) of the delay function c,

$$J = \left[\frac{\partial c_i}{\partial f_j}\right], \quad i = 1,\ldots, L;\ j = 1,\ldots, L \tag{7}$$

is diagonal. The Karush-Kuhn-Tucker necessary optimality conditions for this problem are equivalent to Wardrop's principle commonly used within the UE assignment literature as the assumption about drivers' route choice behavior.

In this study a version of problem (3)-(6) with an asymmetric Jacobian matrix (7) is considered. The asymmetric case has been studied rather extensively. Abdulaal and LeBlanc [1] proved that the mathematical programming formulation for the problem does not exist if matrix is asymmetric. However, Smith [29] formulated the problem as a system of inequalities, which is a variational inequality problem (VIP), as pointed out by Dafermos [8]. The variational inequality formulation of the problem and these conditions are presented next.

The variational inequality formulation of a UE assignment can be stated as:

$$\text{find } \mathbf{f}^* \text{ such that } \quad \mathbf{c}(\mathbf{f}^*)\cdot(\mathbf{f}-\mathbf{f}^*) \geq 0 \quad \forall \mathbf{f} \in F \tag{8}$$

where F is a set of feasible link flows satisfying constraints (4)-(6). Note that due to (4)-(6) and the fact that all T_{ij} are finite, F is closed and bounded. For alternative variational inequality formulations of the UE problem see [13]. The conditions for the existence and uniqueness of the solution of (8) were partially obtained by Smith [29], but later completed and extended by Dafermos [8] and Dafermos and Nagurney [10],[11], who applied the theory of VIP originally developed for infinite-dimensional problems in physics.

A sufficient condition for problem (8) to have a solution is that the cost function c is continuous and F is a compact convex set. As mentioned above the F is bounded and closed so it is compact. A sufficient condition for the solution to be unique is that the cost function c is strictly monotone, that is:

$$[\mathbf{c}(\mathbf{f})-\mathbf{c}(\mathbf{f}')]\cdot(\mathbf{f}-\mathbf{f}')>0 \quad \forall \mathbf{f},\mathbf{f}' \in F \tag{9}$$

The condition of strict monotonicity is satisfied if and only if the Jacobian (7) of cost function c is positive definite [16]. Matrix J is positive definite if it satisfies the following conditions (see, for example, [31]):

$$\frac{\partial c_i(\mathbf{f})}{\partial f_i}>0 \quad \forall i \tag{10}$$

$$\frac{\partial c_i(\mathbf{f})}{\partial f_i}>\frac{1}{2}\sum_{i\neq j}\left|\frac{\partial c_i(\mathbf{f})}{\partial f_j}+\frac{\partial c_j(\mathbf{f})}{\partial f_i}\right| \quad \forall i \tag{11}$$

Condition (10) means that the cost on each link is strictly increasing with the flow on that link. The second condition is satisfied when the link cost depends mainly on the flow on the same link. Note that in the standard UE assignment when

$$\mathbf{c}(\mathbf{f})=\left[c_1(f_1),\ldots,c_L(f_L)\right] \tag{12}$$

condition (11) is satisfied by definition.

In the case of a transportation network, the Jacobian matrix usually has a block-diagonal structure:

$$\mathbf{J}=\begin{bmatrix} \mathbf{J}_1\ldots\ldots 0\ldots\ldots 0 \\ 0\ldots\ldots\mathbf{J}_k\ldots\ldots 0 \\ 0\ldots\ldots 0\ldots\ldots\mathbf{J}_l \end{bmatrix} \tag{13}$$

where J_k is the Jacobian matrix of the derivatives of travel costs for intersection k and I is the total number of intersections in the network. The reason that the Jacobian has this structure is that the delay at an intersection usually depends only on the flows approaching the same intersection. The travel time functions used in this study generate a Jacobian that is block-diagonal. Smith [30] proved that in this case c is strictly monotone if and only if c_k corresponding to a single intersection is strictly monotone. However, according to Meneguzzer et al [21], conditions (11) are not guaranteed to be satisfied by the class of cost functions applied in this study. The question arises whether the sufficient conditions for the uniqueness of the solution are too strict. Computational results obtained by Meneguzzer et al [21] suggest that a unique solution does exist. The question of weaker sufficient conditions remains open.

To solve problem (8), the so-called diagonalization algorithm or nonlinear Jacobi is used (see, for example, [21] or [16]). The algorithm consists of solving a sequence of nonlinear approximations of the problem, called diagonalized problem, each of which is a nonlinear optimization problem and is solved using the Frank-Wolfe algorithm [15]. The diagonalized problems are created by fixing the non-diagonal flows at the level from last iteration, so that within each diagonalized problem each link delay is a function of the flow on the same link only. Dafermos [9] proved that the diagonalization algorithm converges if the Jacobian matrix is positive definite and loosely speaking "not too asymmetric". These two conditions are only sufficient conditions, so the algorithm may still converge even though they are not satisfied, as appears to be the case in our study. Section 6 presents the implementation of the algorithm.

3. Link Travel Time Functions

This section presents mathematical functions used within an assignment model to estimate link travel times for given flow rates. The choice of the delay functions involves several criteria: (a) the desired mathematical properties of the function to satisfy the conditions for a unique solution of the assignment model; (b) the cost and limited availability of road data; (c) computational effort required by the model and (d) the desired accuracy of the travel time estimates generated by the model. One of the goals of this study was to model travel times by turning movement. Analytical functions are preferred to regression-based models, because the former generate reasonable estimates over a much wider range of input flows and other parameters. Criteria (a) and (b) above exclude many highly detailed traffic engineering models. Criterion (c) excludes simulation models, which are suitable only for small networks.

Delay functions selected for this study can be classified by road type and intersection type. First, delay functions adapted from literature for arterials and collectors are presented (Section 3.1). Then link delay functions partially developed for freeway and tollway related facilities (Section 3.2) are presented. Table 2 presents the classification of links by facility type. Table 3 presents a breakdown of intersections by the type of control.

The delay functions are generally based on the Highway Capacity Manual [32], [33]. Some of these models are extended to obtain better estimates (e.g. the model for left-turn capacity at signalized intersections); new models were developed for freeways and freeway-related facilities since those proposed in HCM cannot be used for a wide range of traffic conditions.

Time-dependent delay functions [19] are used to incorporate queuing effects at intersections in oversaturation conditions. These functions are defined for any non-negative volume-to-capacity ratio; for high volumes they are asymptotic to the deterministic model that best describes the delay in heavily oversaturated conditions. See Figure 4 for an example of a time-dependent delay function. These functions are also continuous and strictly increasing, which is required by the sufficient conditions for the existence and uniqueness of the solution of the model stated in Section 2. Some of these functions were adapted from the literature (for example, for signalized intersection or for freeways), while others were derived within this study using the transformation technique developed by Kimber and Hollis (for example for priority intersections and all-way-stop intersections).

The first term of any link travel time function is cruise time, which is the time needed to traverse the road segment when no delay occurs at downstream intersections. The second term is the delay experienced by the vehicle at the downstream intersection and is called intersection delay. Models for each of these two terms require usually two stages: (a) the analysis of the capacity under prevailing traffic conditions, and (b) the calculation of the delay.

In all models presented in this paper, capacities, saturation flows and volumes are expressed in actual vehicles per hour (vph) unless stated otherwise. The implementation of the delay functions presented in the following sections is described in Section 6.

3.1 Arterials and Collectors

This section presents capacity and delay functions for arterial and collector links. Delay formulas for unsignalized intersections were derived as a part of this study. No time-dependent delay formula for this type of intersection was known to the authors. Other models were adapted from the literature. For purpose of intersection delay analysis, intersections are classified into several categories according to the type of intersection control, the intersection layout and the approach geometry. Approaches to signalized intersections are assigned a category according to lane designation and intersection geometry. However, actual lane use may be different than the lane designation. An example is an approach with the left lane designated as a LT/TH lane. In the case when the LT movement flow exceeds or is close to LT capacity, the TH drivers try to use the right lane to minimize their delay, and eventually the left lane functions as an exclusive LT lane. The actual lane usage is determined based on the lane flow analysis which estimates flows by turning movement on each approach lane. The analysis is based on the assumption that drivers choose their lane to minimize their delay.

3.1.1 Cruise Time

Cruise time is determined based on the length of the link and its free flow travel time which, according to 1985 HCM, is also a function of the link length. The capacity of an arterial/collector link is practically always higher than the downstream intersection capacity and is not involved in the cruise time estimation.

3.1.2 Signalized Intersections

Signalized intersection models can be used for intersection design, in which the problem may consist of selecting the type of control, geometry of the facility or phasing plan (for signalized intersections). The literature on this topic is rather large. Some models optimize all of the above factors (for example [18]; [26]), while others concentrate on cycle length and green split optimization (see [5], [36], [37]). See also [27]. However, in this study the intersection analysis is applied to generate intersection delay estimates.

Delay models for signalized intersections consist of three modules: (a) saturation flow analysis; (b) signal timing procedure; (c) the delay function itself. Because of mutual dependencies between the inputs and outputs of the modules (e.g. saturation flows are inputs for the signal setting module, while its output is an input to the saturation flow module), an iterative procedure is required to obtain consistent results. This procedure is described in Section 6. The following sections describe each of the above modules of signalized intersection analysis. The effect of the progression on the vehicle delays is a rather important factor. However, the procedure applied in this study does not account for progression which remains a topic for future research.

Saturation Flow - Saturation flow is the capacity of the approach, lane or lane group under prevailing traffic conditions if the approach received all of the green time of the signal. It is used together with the length of the actual green time to calculate the capacity of a link. The procedure chosen for this study is a synthesis of the 1985 HCM and Australian method (see [2]). Conceptually, saturation flow can be defined as:

$$S_a = w(\mathbf{S}_i, \mathbf{f}_i, \mathbf{g}, \mathbf{p}) \quad \forall a \in L \tag{14}$$

where S_a is the saturation flow of link a, S_i is a vector of saturation flows discharging into intersection i, f_i is a vector of vehicle flows discharging into intersection i, g is a vector of signal timing parameters at the intersection and p is a vector of other parameters including the intersection geometry.

The relationship between the saturation flow of a permissive LT movement and the opposing flow is the most complex of all turning movements. Figure 1 shows an example of such a relationship. In the figure, f^{LT} is the LT flow, g/C is the green time-to-cycle length ratio and f is the total flow on subject approach. Computationally, the function w is an iterative procedure rather than a closed form function. To start the procedure, some default values of saturation flows and signal setting parameters are assumed.

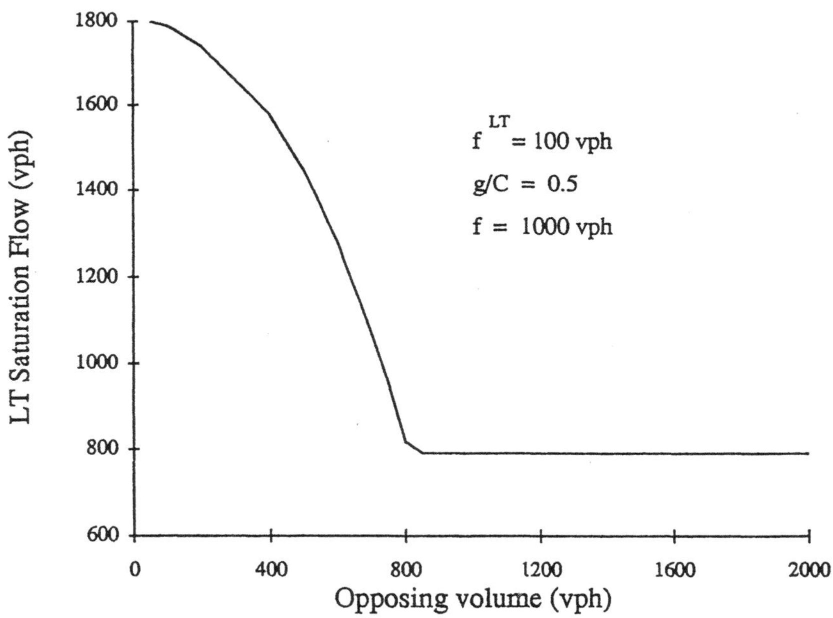

Figure 1. Saturation Flow of the Permissive LT Movement as a Function of the Opposing Volume

Signal Timing - The main parameters describing how the traffic signals operate are: phase pattern, cycle length and green split. The phase pattern defines which movements are allowed in each part of the cycle. In this study it is fixed for each intersection class defined in terms of approach geometry and the number and designation of approach lanes. The cycle length and the green split are modeled using the Webster procedure [36] described briefly below. This procedure is intended to determine signal settings similar to those used for real intersections, rather than optimizing these settings. Moreover, the majority of the signals in the study area are actuated signals, which adjust the settings according to existing traffic patterns. Thus, the procedure described below should imitate those actuated signal controllers, which are designed to minimize total intersection delay. While the procedure generates capacity maximizing signal settings, it is applied as an approximation of minimum delay settings. The procedure is also computationally attractive.

The concept of the procedure consists of assigning green time for each turning movement according to its volume-to-capacity ratio. More precisely, for each phase a critical movement is considered, that is one which requires the longest green time. The procedure consists of a sequence of equations, which in general form can be defined as follows:

$$C = z(y) \tag{15}$$

$$g_p = u_p(y, C) \quad p \in P \tag{16}$$

where C is the cycle length, y is a vector of volume-to-saturation flow ratios y^i, where $y^i = f^i/S^i$ for all movements j at intersection, g_p is the time assigned for phase p and P is set of all phases.

In the case of a four-leg intersection like the one in Figure 2 and a two phase pattern, relationships (15) and (16) take the following form:

$$C = \frac{A}{1 - y^{NS} - y^{EW}} \quad (17)$$

$$g_i = (C - b)\frac{y^i}{y^{NS} + y^{EW}}, \quad i = \{NS, EW\} \quad (18)$$

where A and b are parameters.

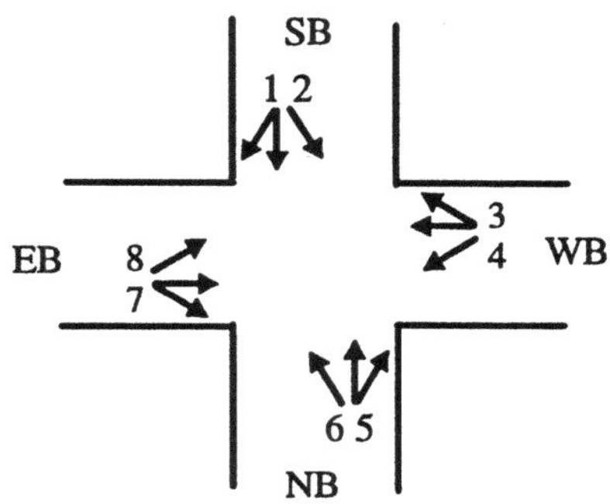

Figure 2. Four-Leg Intersection Used in the Example

Because of the mutual dependency of the saturation flow procedure and signal setting procedure, both need to be recomputed iteratively 3 to 4 times to obtain consistent results. An example of the relationship between the green time for one approach and the volume on the cross street at a four-leg intersection is shown in Figure 3.

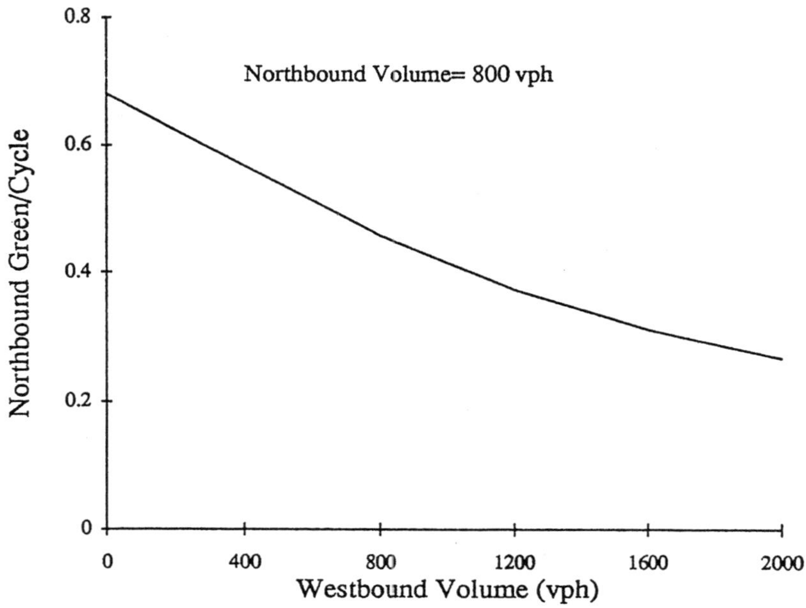

Figure 3. Green Split for One Approach as a Function of the Volume on the Cross

Delay - The delay function is a relationship between the flow and the delay experienced by the vehicle at the intersection. For a signalized intersection, the delay d can be expressed in general as follows:

$$d = d(f, S, g) \tag{19}$$

The specific delay function applied in the study has the following form:

$$d_a = \begin{cases} \dfrac{C(1-u)^2}{2(1-ux)} & for\, x < 0.5 \\ \dfrac{C(1-u)^2}{2(1-ux)} + 900T\left[x-1+\sqrt{(x-1)^2+\dfrac{8(x-0.5)}{KT}}\right] & for\, 0.5 \le x < 1 \\ \dfrac{C(1-u)}{2} + 900T\left[x-1+\sqrt{(x-1)^2+\dfrac{8(x-0.5)}{KT}}\right] & for\, x \ge 1 \end{cases} \tag{20}$$

where d_a is the average delay per vehicle for link a (sec/vehicle), C is the signal cycle length (sec), u= g/C is the green split, g is the green time (sec), x= f/K is the volume-to-capacity ratio, and T is the duration of the flow (hrs). Capacity K is determined from the following relationship:

$$k = \frac{g}{C} S \tag{21}$$

where S is the saturation flow. Figure 4 shows an example of function (20).

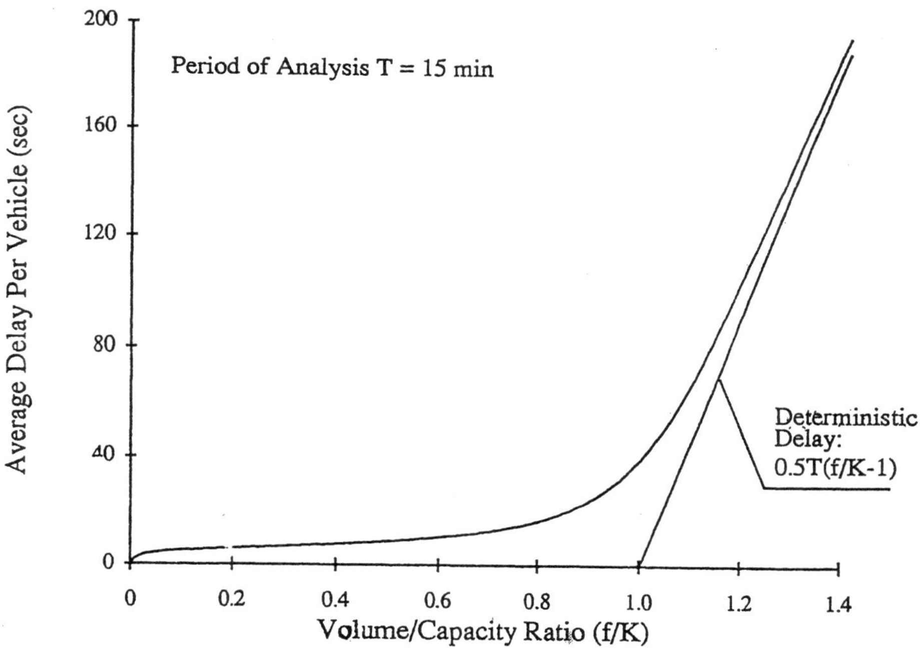

Figure 4. Time Dependent Delay Formula

The first term, called *uniform delay*, was originally developed by Webster [36]. It reflects the average delay experienced by drivers in *undersaturation conditions*, that is when the arrival flow does not exceed capacity. In oversaturation conditions, $x = 1$ is used in the uniform delay term. The second term of function (20) is called overflow delay. It reflects the delay experienced by the vehicles when the flow rate is close to the capacity or exceeds it. Temporary overflow of the intersection occurs also when the average arrival rate is lower than the capacity, due to a random character of the arrival pattern. The earliest delay functions (for example, [36]) were based on the steady-state model and were defined only for undersaturation conditions.

3.1.3 Unsignalized Intersections

Delay models for unsignalized intersections, whether major/minor priority intersections or all-way-stop controlled intersections, are much simpler than models for signalized intersections. In all cases, first the capacity is estimated, followed by the delay based on the flow-to-capacity ratio. The delay functions are similar the second term of delay functions for signalized intersection. Only a few notes are given here concerning the models for unsignalized intersections.

For priority intersections, the 1985 HCM capacity model is applied. The HCM capacity procedure is computationally attractive and generates rather realistic results in a wide range of traffic conditions. For a comparison of alternative capacity models, see [34]. For delay estimation, the time-dependent formula developed by Kimber and Hollis [19] is applied.

The capacity of all-way-stop controlled intersections is modeled using the analytical iterative procedure developed by Richardson [25]. An analytical procedure is preferred over a regression-based one because of the wide range of traffic patterns occurring during the equilibration process. Regression-based models reported in literature (for example, [20]), are validated only within a narrow range of traffic flows. A time-dependent delay function was developed as part of the study using the technique by Kimber and Hollis [19].

3.2 Freeway Related Facilities

Several kinds of freeway-related facilities occur within the test area: basic freeway segments, ramps and ramp-freeway junctions, weaving sections and toll plazas. All of them are modeled using appropriate capacity models and time-dependent delay functions similar to the delay function for signalized intersections (only the second term). Most of the capacity models are based on the 1985 HCM. Some of them, however, had to be developed (for example, freeway-ramp merge capacity model). Existing models were either too complex for use within the UE assignment problem or were not suitable for oversaturated conditions. These models are described in detail in [4].

4. Test Area and Network

The study area depicted in Figure 5 is located in northwest suburbs of Chicago and covers about 300 square miles (800 square km). To develop the test network, detailed intersection control, layout and some traffic data were collected in the field and merged with topology data from a regional planning network created by the Chicago Area Transportation Study (CATS). The process of coding and testing the network was made possible by using the Network Display Tool [24]. The network was then extensively tested and debugged for coding errors. Tables 1, 2 and 3 show the size and various characteristics of the test network.

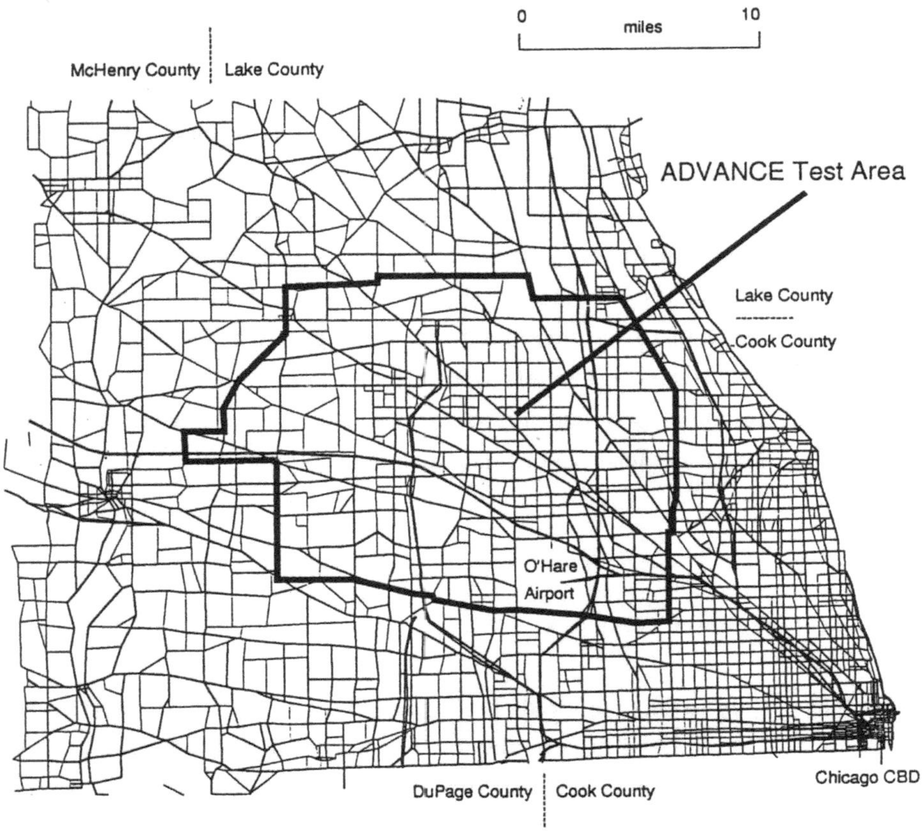

Figure 5. ADVANCE Test Area Located in North and Northwest Suburbs of Chicago

In a conventional assignment model, the network is defined so that each intersection is represented as a single node and each road segment is represented as a single approach link (precisely a two-directional segment is represented as two links, one in each direction). This network representation defines travel times in terms of approach. In this study a classical turning movement representation of intersection is applied. This representation is described below for the sake of completeness.

Number of Links	Number of Nodes	Number of Zones
7850	2552	447

Table 1. Size of the Study Network

Type of Facility	Number of Links
Arterial	2715
Collector	1346
Tollway/Freeway	197
Freeway Ramp	202
Toll Plaza	14
Freeway Weaving Section	11
Total of Actual Links	4485
Local Connector Links	874
Centroid Connectors	2491
Total	7850

Table 2. Links by Facility Type

Type of Control	Number of Intersections
Signalized	699
Major/minor priority	821
All-way-stop	60
Freeway-ramp merge	99
Non-intersection nodes	873
Total	2552

Table 3. Intersections by Type of Control

The approach consists of defining a special network representation so that each turning movement is represented by a separate link called an intersection link. More precisely, an approach node is defined for each approach to the intersection, and the number of intersection links originating from this node equals the number of turning movements. Similarly, an exit node is defined for each exit from the intersection. For a typical four-leg intersection with two-way approaches without turning restrictions, four approach nodes, four exit nodes and twelve intersection links are required in this representation (see Figure 6).

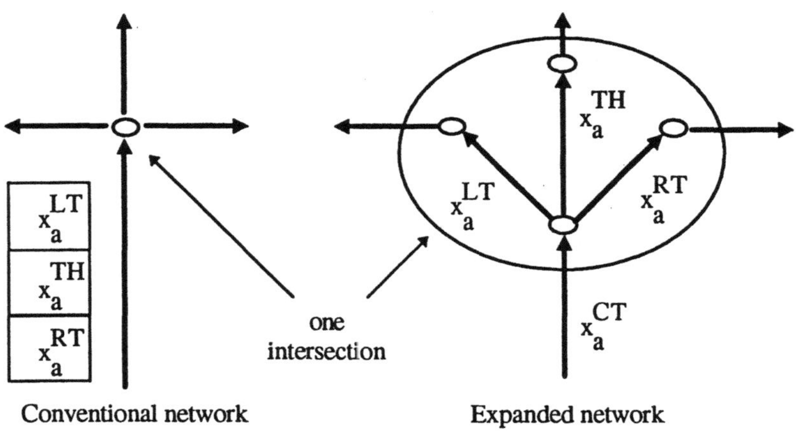

Figure 6. Expanded Intersection Representation

Networks of this type were used in different variations for detailed intersection analysis e.g. by Fisk [12] and Florian and Nguyen [14]. The disadvantage of this approach is that the number of links and nodes in a detailed network is much higher than in a conventional network. Also, to create a detailed network representation, a special procedure is required. This kind of network builder is described by Sharaf-Eldien [28] and Meneguzzer et al [21].

During the computations, some operations need to be performed on an intersection-by-intersection basis; therefore, the detailed network representation needs to retain the original intersection numbers either explicitly or by means of appropriately defined link and node labels. In this study the latter method is adopted from [21] (p. 51). The disadvantage of this method is that for a network with say 5-digit node labels, the node labels in the detailed network have 10 digits, while the number of the nodes require only 5 or 6 digits. The program that creates this expanded network from the conventional network, called a network builder, was adapted in this study from an existing code [21]. As part of this study, the network builder was revised to accept not only two-way but also one-way approaches and exits to/from intersection. Also the interface was coded to convert data collected in the field to unavailable data (e.g. presence of turning arrows). The code was revised to process the centroid connectors connected to intersections not only to dummy nodes. Finally, the algorithm of creating expanded intersection representation was generalized to process intersections with any number of legs.

5. Travel Demand

Travel demand data for this study are based on CATS estimates for 1990. The process of creating trip tables from the CATS data is described in [38]. There are two steps in this process. The first step consists of time-of-day factorization of the original 24-hour estimates. The factors are based on travel surveys performed by CATS tabulated to yield departure rates by 30-minute intervals. According to these data, five time-of-day periods are determined such that, within each period, the travel demand fluctuates as little as possible. The time-of-day periods chosen are shown in Table 4. The factors obtained in this way are applied to generate five time-of-day matrices.

The second step consists of the analysis of trips beginning or ending or both outside the test area. First, the standard UE assignment problem is solved for the Chicago region which includes the test area. Routes generated are analyzed and external zones are created on the boundary of the test area to represent flows entering and exiting the test area. Each of the final matrices consists of the internal matrix and external trip matrix (see Table 4).

	Period	Time-of-Day	Internal Trips	Total Trips
1	Night	12pm-6am	7,839	19,439
2	AM Peak	6am-9am	83,708	184,185
3	Midday	9am-4pm	88,543	170,573
4	PM Peak	4pm-6pm	98,532	203,278
5	Evening	6pm-12pm	81,245	146,092

Table 4. Total Number of Trips per Hour

6. Model Solution

The implementation of the model is based on [21]. As part of this study the code was tuned and modified to operate efficiently for large size networks. Also, the code was enhanced to model accurately one-way approaches or exits to/from intersections as well as multi-leg intersections. The golden section subroutine performing the line search step was replaced with the bisection procedure. In this way the computationally-demanding objective function evaluation could be eliminated. Several other modifications improved the usability of the code: (a) reorganizing the program input into a link data file, trip data file and zone file; (b) computation of various statistics; (c) selective saving of the routes generated and (d) saving periodically the status of the computation which can be used to restart the computation from the point where it is terminated. Figure 7 shows the simplified structure of the algorithm. The algorithm is also described briefly in Section 2.

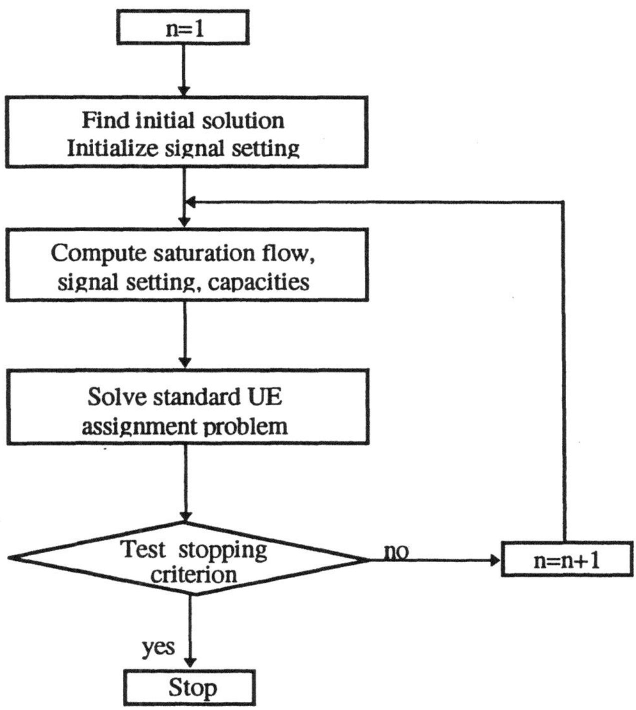

Figure 7. Implementation of Delay Models within an Asymmetric Assignment Algorithm

The algorithm starts with the computation of initial estimates of link flows by turning movement. For these flows, capacities and other input parameters of delay functions are computed. Because of the mutual dependencies between the different parameters described in the section on delay functions, obtaining consistent values of these parameters requires iterative recomputation of the intersection analysis procedure not shown the figure. Capacities and other parameters of the delay functions obtained in this way are used to solve the standard UE assignment model. The resulting link flows are then used to reevaluate intersection capacities. The algorithm continues until the convergence condition is satisfied.

7. Computational Results

This section presents results of the solution of the asymmetric UE assignment model. The model was implemented in Fortran 77. The computations were performed on a Sun 330 workstation (with 16 MB of memory), a Sparc 2 workstation (24 MB), and a IPX workstation (8 MB). As each Frank-Wolfe iteration required about 30-40 minutes of real time, one diagonalization iteration consisting of 2 Frank-Wolfe iterations required about 1.5 hours. The computational time for the intersection analysis is about 6% of the total computational time.

Results were evaluated using several global performance measures as well as by checking the reasonableness of the routes generated in the last diagonalization iteration. Table 5 presents the results obtained for the five time-of-day periods.

	Time Period	Distance (miles)	Travel Time (minutes)	Speed (mph)	v/c Ratio (-)	Links Converged (%)	Diagonal. Iterations
1	Night	9.8	14.0	41.9	0.18	95.8	3
2	AM Peak	11.2	31.6	21.2	0.78	95.9	15
3	Midday	10.7	24.0	25.5	0.70	96.2	20
4	PM Peak	11.1	32.6	20.4	0.82	95.8	16
5	Evening	9.5	18.9	30.3	0.62	98.0	19

Table 5. Network Performance Measures for the Asymmetric UE Assignment Problem

The speed shown in the table is the space-mean speed defined as ratio of total distance to total travel time; time is the average travel time and distance is the average trip distance for trips within the test area. The value of the convergence criterion is shown in the last column. The convergence criterion is the percent of the links in the network whose flow differs by no more than 10% from its flow in the previous diagonalization iteration. All five problems were solved to convergence, which was set at the level of 95% of all links.

Figure 8 shows how the convergence criterion behaves during the equilibration process. It can be seen that the model converges non-monotonically and that at the end of the equilibration process the convergence criterion behaves non-smoothly. For the Night period it can be seen that for an uncongested network, the model converges quite rapidly. This fact is also known from the literature. However, there is no indication that the more heavily the network is loaded, the slower is the convergence. The reason that the model converges relatively fast for highly congested time-of-day periods (like PM Peak) may be that high congestion and corresponding high travel times drive the system strongly toward the equilibrium solution.

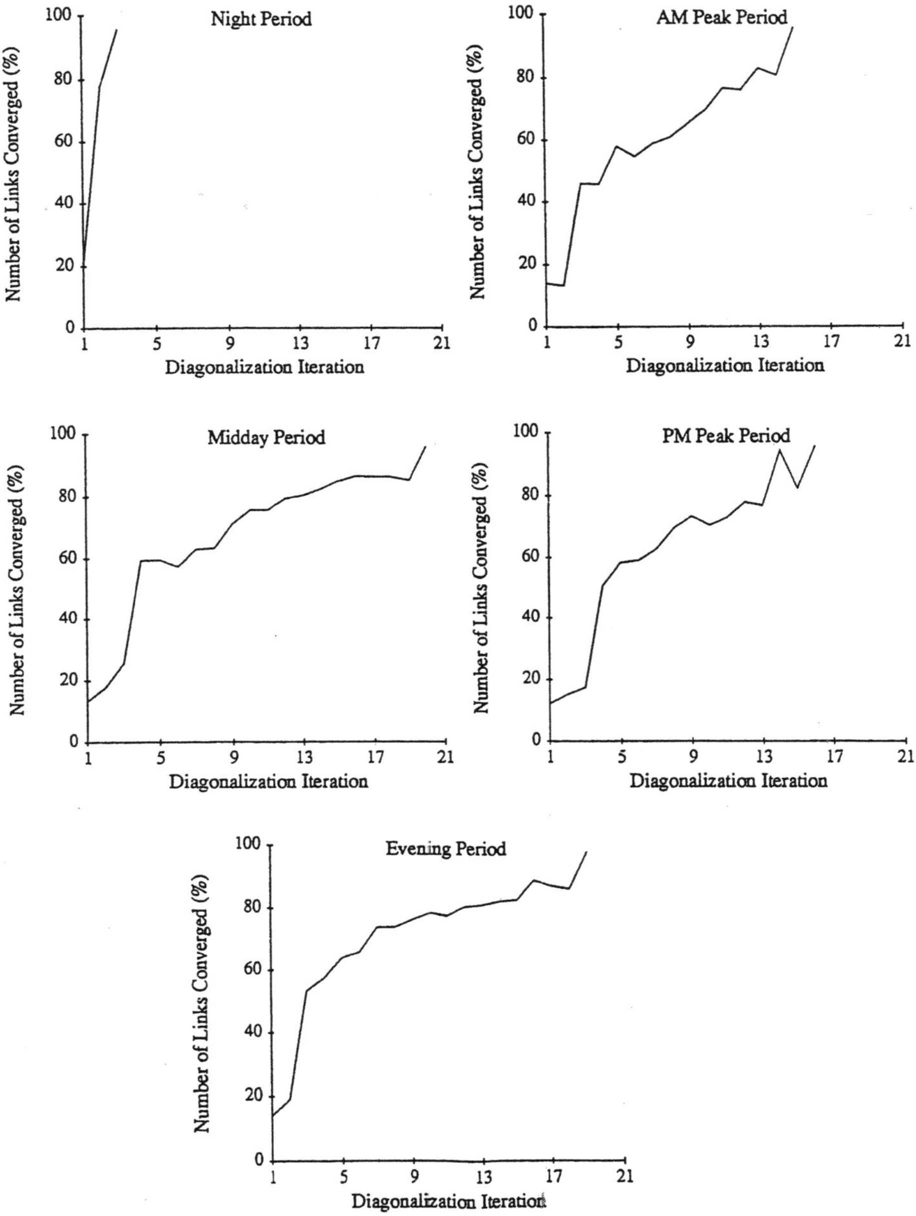

Figure 8. Converge Criterion Value as a Function of the Diagonalization Iteration

Another measure of network performance presented in Table 6 is the space-mean speed based on the last diagonalization iteration for each time-of-day period, computed separately for four functional classes. According to the table, the speeds for all road classes are inversely related to the congestion level.

	Time Period	Collectors	Arterials	Freeways
1	Night	36.2	39.0	62.7
2	AM Peak	13.5	18.9	45.7
3	Midday	18.6	23.0	47.9
4	PM Peak	14.4	17.4	45.4
5	Evening	23.3	28.2	50.1

Table 6. Travel Speed by Facility Type (mph)

Figure 9 presents a sample of routes generated by the model as an auxiliary solution during the equilibration process in each iteration of the Frank-Wolfe algorithm. The routes presented are generated in the last diagonalization iteration for one origin-destination pair. The thick outer line represents the boundary of the test area. Two rectangles represent the origin (thinner line) and destination (heavier line) zones. The two routes use the same links to a large extent, so that in the figure they overlap. Since delays are by turning movements, the routes do not include too many turns, which is frequently the characteristic of the routes generated by conventional assignment models. Depicted routes are found as auxiliary solutions to the subproblem generated in each iteration of the Frank-Wolfe algorithm. It is noted here that the uniqueness of route flows is not guaranteed by the deterministic UE assignment problem.

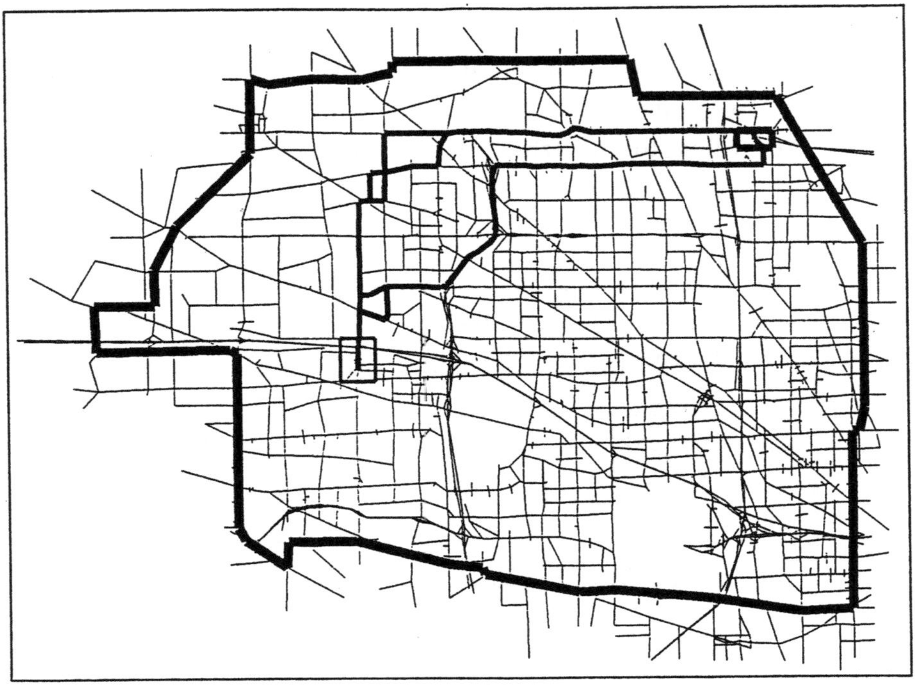

Figure 9. Routes Generated by the Model for One Origin/Destination Pair

8. Conclusions

The purpose of the study was to generate link travel time estimates in the transportation network for use as initial estimates for the ADVANCE, a dynamic route guidance system being implemented in Chicago suburbs. In general the estimates generated are satisfactory, even though certain link travel times are somewhat unrealistic. Moreover, the estimates for different turning movements at the same intersection, and within some proximity of it, are consistent. Average travel speeds are similar to actual speeds according to traffic engineers familiar with the test area. In the future, it would be desirable to perform a comparison between computed flows and observed counts.

However, it would be difficult to identify the source of any differences since the trip tables may be inaccurate. Validation of the link travel time functions is envisioned in the future as well.

Three shortcomings of the model should be mentioned. First, in our model the intersection analysis is performed on an isolated intersection basis, so that the same flows are assigned for all intersection approaches along a route from an origin to a destination. In the real world, the flow passing through one intersection is never higher than its capacity and the flow at the next intersection should reflect that. This shortcoming should be avoided and is a topic for future research. Treating intersections as isolated also does not account for progression effects along arterials, leading to an overestimation of delay. The third issue concerns the validity of route choice behavior based on the travel time only. Roadway class (freeway vs. arterial vs. collector) is not taken into account in our model, and may be a significant factor in route choice behavior as well.

Acknowledgments

The research on which this paper is based was performed for the ADVANCE Project. The financial support of the Illinois Department of Transportation and the Federal Highway Administration is gratefully acknowledged. We thank Professor Andrew Tarko for his important contributions to this research especially in its traffic engineering aspects. Also, we acknowledge the support of the National Center for Supercomputing Applications for some of computer resources utilized in the research.

References

[1] Abdulaal, M., and LeBlanc, L.J. (1979) "Methods for combining modal split and equilibrium assignment models", *Transportation Science*, 13, 292-314.

[2] Akcelik, R. (1988) "Capacity of a Shared Lane", *Australian Road Research Board Proceedings*, 14(2), 228-241.

[3] Beckmann, M., McGuire, C.B., and Winsten, C.B. (1956) "Studies in the Economics of Transportation", *Yale University Press*, New Haven, CT.

[4] Berka, S., Boyce, D.E., Raj, J., Ran, B., Tarko, A., and Zhang, Y. (1994) "A Large-Scale Route Choice Model with Realistic Link Delay Functions for Generating Highway Travel Times", *Report to the Illinois Department of Transportation*, Urban Transportation Center, University of Illinois, Chicago.

[5] Boyce, D.E., Chen, H.K., Rouphail, N.M., and Sen, A. (1989a) "Subregional Route Choice Models with Link Travel Times Reflecting Intersection Flows", *Final Report to the Illinois Department of Transportation*, Urban Transportation Center, University of Illinois, Chicago.

[6] Boyce, D.E., Kirson, A.M. and Schofer, J.L. (1994) "ADVANCE: The Illinois Dynamic Navigation and Route Guidance Demonstration Program", *Chapter 11 in Advanced Technology in Transport: IVHS and ATT*, ed. Ian Catling, Artech House, Boston.

[7] Boyce, D.E., Meneguzzer, C., Rouphail, N., Sen, A. and Lauritzen, T. (1989b) "*A User-optimal Route Choice Model with Asymmetric Cost Functions Incorporating Intersection-related Travel Times*", *Final Report to the Illinois Department of Transportation*, Urban Transportation Center, University of Illinois, Chicago.

[8] Dafermos, S. (1980) "Traffic Equilibrium and Variational Inequalities", *Transportation Science*, 14, 42-54.

[9] Dafermos, S. (1982) "Relaxation Algorithms for the General Asymmetric Traffic Equilibrium Problem", *Transportation Science*, 16, 231-240.

[10] Dafermos, S., and Nagurney, A. (1984a) "Sensitivity Analysis for the Asymmetric Network Equilibrium Problem", *Mathematical Programming*, 28, 174-184.

[11] Dafermos, S., and Nagurney, A. (1984b) "Stability and Sensitivity Analysis for the General Network Equilibrium-travel Choice Model", *Proceedings of the Ninth International Symposium on Transportation and Traffic Theory*, J. Volmuller and R. Hamerslag (eds.), Utrecht, The Netherlands, 217-231.

[12] Fisk, C. (1978) "A Transportation Planning Model for Detailed Intersection Analysis", *Canadian Journal of Civil Engineering*, 5, 18-25.

[13] Fisk, C.S., and Boyce, D.E. (1983) "Alternative Variational Inequality Formulations of the Network Equilibrium-Travel Choice Problem", *Transportation Science*, 17, 454-463.

[14] Florian, M., and Nguyen, S. (1976) *An Application and Validation of Equilibrium Trip Assignment Methods*, Transportation Science, 10, 374-390.

[15] Frank, M., and Wolfe, P. (1956) "An Algorithm for Quadratic Programming", *Naval Research Logistics Quarterly*, 3, 95-110.

[16] Harker, P.T., and Pang, J-S. (1990) "Finite-dimensional Variational Inequality and Nonlinear Complimentarity Problems: A Survey of Theory", *Algorithms and Applications*, Mathematical Programming, 48, 161-220.

[17] Hicks, J.E., Boyce, D.E., and Sen, A. (1992) "Static Network Equilibrium Models and Analyses for the Design of Dynamic Route Guidance System", *Technical Report to the Illinois Department of Transportation*, Urban Transportation Center, University of Illinois, Chicago.

[18] Improta, G., and Candarella, G.E. (1984) "Control system design for an individual signalized junction", *Transportation Research*, 18B, 147-167.

[19] Kimber, R.M., and Hollis, E.M. (1979) "Traffic Queues and Delays at Road Junctions", *Transport and Road Research Laboratory Report*, 909, Crowthorne, Berkshire.

[20] Kyte, M. (1991) "Interim Materials on Unsignalized Intersection Capacity", *Transportation Research Circular*, 373.

[21] Meneguzzer, C., Boyce, D.E., Rouphail, N., and Sen, A. (1990) "Implementation and Evaluation of an Asymmetric Equilibrium Route Choice Model Incorporating Intersection-Related Travel Time", *Report to the Illinois Department of Transportation*, Urban Transportation Center, University of Illinois, Chicago.

[22] Nagurney, A. (1984) "Comparative Tests of Multimodal Traffic Equilibrium Methods", *Transportation Research*, 18B, 469-485.

[23] Patriksson, M. (1994) "The Traffic Assignment Problem: Models and Methods", *VSP*, Utrecht, The Netherlands.

[24] Ramakrishnan, R., Berka, S., and Hicks, J. (1994) "Network Display Tool: An Object Oriented Visualization Method for Intelligent Vehicle-Highway Systems", *Report to the Illinois Department of Transportation*, Urban Transportation Center, University of Illinois, Chicago.

[25] Richardson, A.J. (1987) "A Delay Model for Multiway Stop-sign Intersections", *Transportation Research Record*, 1112, 107-114.

[26] Rouphail, N.M., and Radwan, A.E. (1989) "Simultaneous Optimization of Signal Settings and Left Turn Treatments", *paper presented at the 69th Annual Meeting of the Transportation Research Board*, Washington D.C.

[27] Said, K., Rouphail, N.M., Boyce, D.E., and Sen, A. (1993) "Optimization of Traffic Signal Controls within Equilibrium Route Choice Models", *Report to the Illinois Department of Transportation, Urban Transportation Center, University of Illinois*, Chicago.

[28] Sharaf-Eldien, O.M. (1988) "Hierarchical Traffic Flow Optimization for Congested Networks", *Ph.D. thesis, University of Maryland.* College Park.

[29] Smith, M.J. (1979) "The existence, uniqueness and stability of traffic equilibria", *Transportation Research*, 13B, 295-304.

[30] Smith, M.J. (1982) "Junction interactions and monotonicity in traffic assignment", *Transportation Research*, 16B, 1-3.

[31] Stewart, G.W. (1973) "Introduction to Matrix Computation", *Academic Press*, New York.

[32] Transportation Research Board (1985) "Highway Capacity Manual", *Special Report 209*, Washington D.C.

[33] Transportation Research Board (1993) "Basic Freeway Section", *Revised Chapter 3 of the 1985 Highway Capacity Manual.*

[34] Troutbeck, R. (1991) "Recent Developments in the Theory of Unsignalized Intersections", *Report of the Queensland University of Technology*, Brisbane.

[35] Wardrop, J.G. (1952) "Some Theoretical Aspects of Road Traffic Research", *Proceedings of the Institution of Civil Engineers*, Part II, 1, 325-378.

[36] Webster, F.V. (1958) "Traffic Signal Settings, Road Research Technical Paper", 39, *Her Majesty's Stationary Office*, London.

[37] Webster, F.V., and Cobbe, B.M. (1966) "Traffic Signals", *Road Research Laboratory Technical Paper*, 56, Her Majesty's Stationary Office, London.

[38] Zhang, Y., Hicks, J., Boyce, D.E. (1994) "Trip Data Fusion in ADVANCE", *ADVANCE Working Paper Series Number. 43*, Urban Transportation Center, University of Illinois, Chicago.

AN EFFICIENT ALGORITHM FOR A BICRITERION TRAFFIC ASSIGNMENT PROBLEM

Patrice MARCOTTE *

Daoli ZHU

Centre de Recherche sur les Transports - Université de Montréal
Montréal, Québec H3C 3J7, Canada

We consider a bicriterion traffic assignment model where the network users have different values of time. We present an infinite-dimensional formulation of this model that lends itself to an extremely simple and efficient algorithm that takes into account the network structure of the problem.

1. Introduction

Traffic assignment procedures, with fixed or elastic demand, are frequently based on Wardrop's assumption that, at equilibrium, the cost of any used path is minimal among all used or unused paths. In this context, the term "cost" is used as a synonym for "generalized transportation cost", i.e. a combination of several criteria: actual cost, travel time, waiting time, average speed, etc. In this paper, having in mind the problem of pricing of the network arcs, we consider explicitly two criteria, namely time and money. Following Leurent [6], Dafermos [2] and Dial [3], the preferences of the users are represented by a density function $h(\alpha)$ that yields information pertaining to the "number" of people whose value of time equals. Our model is similar to that of Dafermos and Leurent. However, our analysis is simpler and yields a solution procedure that is more efficient and obviates path enumeration.

---

* Research supported by NSERC grant A5789

2. The Mathematical Model and its Solution Algorithm

For the sake of simplicity, we will assume that demand is fixed and that the function $h(\alpha)$ that characterizes the distribution of time values α is common to all origin-destination pairs. These restrictions can easily be relaxed. More specially, define

$G=(N,A)$:	A network with node set N and arc set A
FS(i):	The forward star associated with node $i \in N$
BS(i):	The backward star associated with node $i \in N$
α:	The value of time (state variable)
k:	A commodity index (origin-destination pair index)
K:	The set of origin-destination couples
d^k:	The demand for commodity (origin-destination) k
$h(\alpha)$:	A probability density function over the values of time. The number of people having values of time in the interval $[\alpha_1,\alpha_2]$ is given by $d^k\int_{\alpha_1}^{\alpha_2} h(\alpha)d\alpha$. We have: $\int_0^\infty h(\alpha)d\alpha = 1$ and we assume: $\int_0^\infty h(\alpha)/\alpha\, d\alpha < \infty$.
$x_a^k(\alpha)$:	The flow density along arc a with respect to the value of time α.
$X_a(\alpha)$:	The total flow density of class α along arc a : $X_a(\alpha)=\sum_k x_a^k(\alpha)$
$x(\alpha)$:	The vector of functions $x_\alpha^k(\alpha), \alpha \in A, k \in K$
X_a:	The total flow along arc a . One has: $X_a = \sum_{k\in K} \int_0^\infty x_a^k(\alpha)d\alpha = \int_0^\infty X_a(\alpha)d\alpha$
G_a:	The fixed monetary cost of travel along arc a
$C_a(X_a)$:	The delay (congestion) function associated with arc a ; the function C_a is continuous, positive and nondecreasing.

The feasible set of this problem corresponds to the flow density functions lying in the set Ω, where Ω is a subset of the set of nonnegative square-integrable vector functions (commodity flows) that satisfy the flow conservation equations, for all time values α and all commodity indices k, i.e.:

$$\sum_{a\in FS(i)} x_a^k(\alpha) - \sum_{a\in BS(i)} x_a^k(\alpha) = \begin{cases} d^k h(\alpha) & \text{if} \quad i = \text{origin of } k \\ -d^k h(\alpha) & \text{if} \quad i = \text{destination of } k \\ 0 & \text{otherwise} . \end{cases} \qquad (1)$$

In our assignment model, equilibrium is reached when, for every origin-destination indexed by k, and for (almost) every possible value of time α, the flow density vector $\{x_a^k(\alpha)\}_{a\in A}$ is assigned to a shortest path with respect to k and α, given the congestion levels $C_a(X_a)$ on the network, i.e.:

$$\sum_{a\in A} \sum_{k\in K} (x_a^k(\alpha) - y_a^k(\alpha))(C_a(X_a) + G_a/\alpha) \leq 0 \qquad (2)$$

for all feasible flow pattern $\{y_a^k(\alpha)\}_{a\in A}$ associated with the index k and time value α.

In Appendix I, we show that equilibrium flows can be characterized as feasible flows satisfying the infinite dimentional variational inequality:

$$\text{(EQ)} \qquad \sum_{k\in K} \sum_{a\in A} \int_0^\infty (x_a^k(\alpha))(C_a(X_a) + G_a/\alpha)d\alpha \leq 0 \qquad (3)$$

for all feasible flow pattern $y_a^k(\alpha), a \in A, k \in K$.

As the congestion functions are arc-separable, this variational problem can be reformulated (see Appendix II) as the convex minimization program:

$$\text{(CONV)} \qquad \min_{x\in\Omega} J(x) = \sum_{a\in A} \left\{ \int_0^{X_a} C_a(u)du + G_a \int_0^\infty \frac{X_a(\alpha)}{\alpha} d\alpha \right\}. \qquad (4)$$

This infinite dimensional problem can be solved efficiently using Frank and Wolfe's linear approximation algorithm, without resorting to numerical discretization or path enumeration. The convergence of the method follows from a result established by Auslender and Brodeau [1] (see also Dunn [4]).

For a current solution vector x and corresponding total flow vector X, the linear subproblem resulting from the linearization of the objective with respect to the flow variables is:

$$\min_{y\in\Omega} \sum_{k\in K} \sum_{a\in A} \left(C_a(X_a) + G_a/\alpha\right) y_a^k(\alpha) \qquad \forall \alpha \geq 0. \tag{5}$$

This problem is separable by commodity. For a given commodity k and value of time α its solution can be found by computing the shortest path corresponding to k, based on the current arc costs $C_a(X_a) + G_a/\alpha$. Actually, all these shortest paths can be computed in one pass by solving a parametric shortest path problem with respect to the parameter α. This can be performed efficiently using a variant of the network simplex method. The solution to this problem will take the form:

$$y_a^k(\alpha) = h(\alpha) d^k \tag{6}$$

if arc α lies on a the computed shortest path corresponding to commodity k and value of time parameter α, and $y_a^k(\alpha) = 0$ otherwise. Let $Y_a(\alpha) = \sum_{k\in K} y_a^k(\alpha)$ and $Y_a = \int_0^\infty Y_\alpha(\alpha)d\alpha$ be the associated total arc flow. The next iterate is then determined by solving the one-dimensional problem

$$\min_{t\in[0,1]} \sum_{a\in A} \int_0^{X_a + t(Y_a - X_a)} C_a(u)du + G_a \int_0^\infty \frac{1}{\alpha}\left[X_a(\alpha) + t(Y_a(\alpha) - X_a(\alpha))\right]d\alpha \tag{7}$$
$$= \phi(t) + (1-t)Q + tR$$

where

$$Q=\sum_{a\in A} G_a\int_0^\infty X_a(\alpha)/\alpha\, d\alpha \quad \text{and} \quad R=\sum_{a\in A} G_a\int_\alpha^\infty Y(\alpha)/\alpha\, d\alpha . \tag{8}$$

Now let

$$t^* \in \arg\min_{t\in[0,1]} \phi(t)+t(R-Q) \tag{9}$$

and update

$$X_a^+ = (1-t^*)X_a + t^*Y_a . \tag{10}$$

It is remarkable that the above update can be realized without explicit reference to the commodity flow variables, exactly as for the standard traffic assignment problem. One must only be careful to keep track of the integrals $\int X_a(\alpha)/\alpha$ and $\int Y_a(\alpha)/\alpha$.

Indeed:

$$\begin{aligned} Q^+ &= \sum_{a\in A} G_a\int_0^\infty X_a^+(\alpha)/\alpha\, d\alpha \\ &= \sum_{a\in A} (1-t^*)G_a\int_0^\infty X_a(\alpha)/\alpha\, d\alpha + t^*G_a\int_0^\infty Y_a(\alpha)/\alpha\, d\alpha \\ &= (1-t^*)Q + t^*R. \end{aligned} \tag{11}$$

One can obtain a closed form formula for the above integrals only if the function $h(\alpha)/\alpha$ admits a closed form primitive. If this is not the case, an approximate but closed form expression can be derived from polynomial quadratures for $h(\alpha)$.

It follows from the previous discussion that the above bicriterion traffic assignment problem can be solved in a similar fashion to the standard problem, with a parametric shortest path problem being substituted for the standard shortest path problem. Furthermore one can enhance the basic model in several ways. For instance:

1. If the Lebesgue-Stieltjes integral $\int dH(\alpha)$ is used in place of the

Lebesgue integral $\int h(\alpha)$, the model subsumes the multiclass model where only finitely α - values are considered.

2. The model can be generalized to take into account asymmetric congestion effects, if the functions $C_a(X)$ are dependent on the entire total flow vector. In this situation, there would not be an equivalent convex optimization formulation of our problem, and we would use a restriction strategy such as Lawphongpanich and Hearn's [5] restricted simplicial decomposition to solve it.
3. If no two paths associated with a given commodity have identical monetary costs, then it can be proved that the path-flow functions are unique, with the possible exception of a set with measure zero. This is not the case for the standard single-class model, where the arc flows are unique (under suitable assumptions) but the decomposition into path flows is *not*. Moreover, in the multiclass model, path flows can easily be recovered from the sole knowledge of equilibrium arc costs. These uniqueness relationships between path and arc flows hold without any monotonicity or convexity assumptions on the congestion functions C_a. If the uniqueness condition holds, the algorithm can be easily extended to solve variational inequalities, by substituting to the "natural" objective of the optimization problem the gap function (see Marcotte [7]) associated with the variational inequality. In both cases, it can be proved that the rate of convergence of this extended Frank-Wolfe algorithm is *geometric*, a surprising result.

A seemingly innocuous extension of the multiclass model would be to consider two or more flow-independent criteria. However we can show that this situation leads to difficult problems that are *always* asymmetric and nonmonotone. Work is currently underway to see whether, in this case, the extended Frank-Wolfe algorithm is still valid.

We conclude with a numerical illustration on a simple example inspired from the "Braess paradox" network. The network topology and the value of time function are illustrated in Figure 1, and the exogenous data given in Table 1. The total demand from node 1 to node 4 has been set to 6 units of flow. Table 2 illustrates the convergence of the arc flows to their equilibrium values. Notice that the extremal flow vector

y also converges to the equilibrium, albeit at a slower pace. Figure 2 gives the equilibrium flow densities along each arc.
The behavior of the algorithm is quite similar to what could be observed on the single-class Braess paradox problem, for which convergence is geometric, given that equilibrium path flows are all positive. This is atypical of the Frank-Wolfe algorithm, whose convergence cannot be expected to be better than sublinear for large-scale problems.

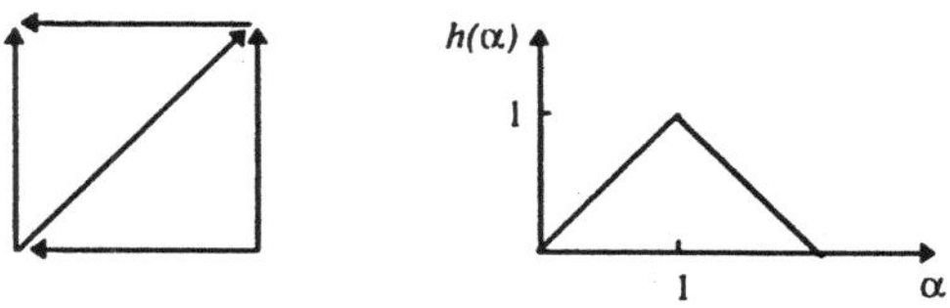

Figure 1. Network topology and VOT density.

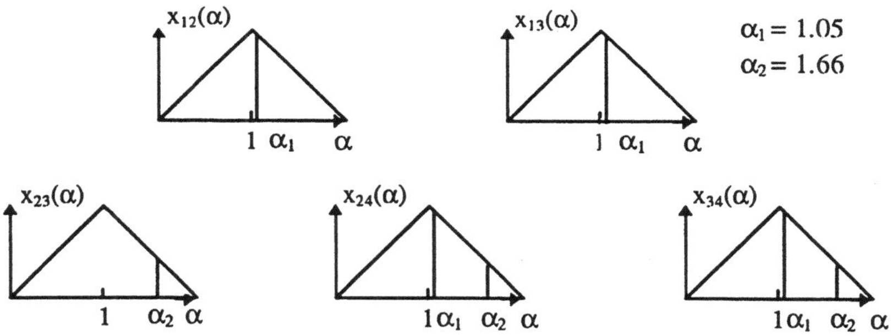

Figure 2. Equilibrium arc-flow densiry functions.

arcs:	1-2	1-3	2-3	2-4	3-4
delay function:	10x	50 + x	10 + x	50 + x	10x
G_a:	100	90	10	100	100

Table 1. Exogenous data.

iteration		arc 1-2	arc 1-3	arc 2-3	arc 2-4	arc 3-4
1	X:	6.00000	0.00000	6.00000	0.00000	6.00000
	Y:	0.00000	6.00000	0.00000	0.00000	6.00000
2	X:	1.52284	4.47715	1.52284	0.00000	6.00000
	Y:	5.87631	0.12369	0.00000	5.97631	0.12369
3	X:	3.13266	2.86734	0.95973	2.17294	3.82706
	Y:	1.43172	4.56828	0.00000	1.43172	4.56828
4	X:	1.92045	4.07955	0.27576	1.64469	4.35531
	Y:	5.58180	0.41820	0.00000	5.58180	0.41820
5	X:	2.60862	3.39138	0.22393	2.38469	3.61531
	Y:	3.55351	2.44649	0.33760	3.21591	2.78409
6	X:	3.12049	2.87951	0.28551	2.83499	3.16501
	Y:	0.00000	6.00000	0.00000	0.00000	6.00000
7	X:	2.71402	3.28598	0.24832	2.46570	3.53430
	Y:	2.38555	3.61445	0.89214	1.49341	4.50659
8	X:	2.58165	3.41834	0.50776	2.07389	3.92611
	Y:	4.62834	1.37166	0.00000	4.62834	1.37166
9	X:	2.77794	3.22206	0.45907	2.31887	3.68113
	Y:	2.96092	3.03908	0.00111	2.95981	3.04019
10	X:	2.89472	3.10528	0.16680	2.72792	3.27208
	Y:	0.00000	6.00000	0.00000	0.00000	6.00000
20	X:	2.72173	3.27827	0.35748	2.36425	3.63575
	Y:	3.03223	2.96777	0.15852	2.87371	3.12629
30	X:	2.72430	3.27570	0.34687	2.37743	3.62257
	Y:	2.92814	3.07186	0.23070	2.69743	3.30257
40	X:	2.73527	3.26473	0.34056	2.39471	3.60529
	Y:	2.73657	3.26343	0.33405	2.40252	3.59748
50	X:	2.73507	3.26493	0.33947	2.39560	3.60440
	Y:	2.73186	3.26814	0.34008	2.39178	3.60822
60	X:	2.73476	3.26524	0.33934	2.39543	3.60457
	Y:	2.73516	3.26484	0.33910	2.39606	3.60394
70	X:	2.73478	3.26521	0.33932	2.39546	3.60454
	Y:	2.73479	3.26521	0.33931	2.39548	3.60452
80	X:	2.73478	3.26521	0.33932	2.39546	3.60454
	Y:	2.73478	3.26522	0.33932	2.39545	3.60455
90	X:	2.73478	3.26521	0.33932	2.39546	3.60454
	Y:	2.73478	3.26521	0.33932	2.39546	3.60454

Table 2. Convergence of arc-flow values.

References

[1] Auslender, A. and Brodeau, F., "Convergence d'un algorithme de Frank et Wolfe appliqué à un problème de contrôle", *RIRO* **2** (1968) 3-12.

[2] Dafermos, S., "A multicriteria route-mode choice traffic equilibrium model", unpublished manuscript, *Lefschetz Center for Dynamical Systems*, June 1981.

[3] Dial, R.B., "T2: A multicriteria equilibrium traffic assignment model. Theory and algorithms", unpublished manuscript, *Volpe National Transportation Systems Center*, July 1993.

[4] Dunn, J.C., "Gradient-Related Constrained Minimization Algorithms in Function Spaces: Convergence Properties and Computational Implications", Large Scale Optimization: State of the Art, W.W. Hager et al. eds., *Kluwer Academic Publishers* (1994) 95-114.

[5] Lawphongpanich, S. and Hearn, D.W., "Simplicial decomposition of the asymmetric traffic assignment problem", *Transportation Research* B 18B (1984) 123-133.

[6] Leurent, F., "Cost versus time equilibrium over a network", *EJOR 71* (1993) 205-221.

[7] Marcotte, P., "A new algorithm for solving variational inequalities, with application to the traffic assignment problem", *Mathematical Programming* 33, 3 (1985) 339-351.

APPENDIX I

In this appendix we show that the equilibrium conditions (EQ) are equivalent to the variational inequality (VI), in the sense that any solution of (VI) satisfies (EQ), almost everywhere, and vice versa.

Let x be any solution of (VI) and assume, by contradiction, that (EQ) is not satisfied almost everywhere, i.e. that there exists a subset *I* of the nonnegative axis such that $\mu(I)$ (the measure of the set *I*) is positive and

$$\sum_{a \in A} (x_a(\alpha) - y_a(\alpha))(C_a(X_a) + G_a / \alpha) > 0 \tag{1}$$

for some (measurable) function y in Ω for all α in *I*.

Define a function $\bar{y}$ in the following way:

$$\bar{y}(\alpha) = \begin{cases} y(\alpha) & \text{if } \alpha \in I \\ x(\alpha) & \text{otherwise} . \end{cases} \tag{2}$$

We then have:

$$\sum_{a \in A} \int_0^\infty (x_a(\alpha) - \bar{y}(\alpha))(C_a(X_\alpha) + G_a / \alpha) d\alpha > 0, \tag{3}$$

which contradicts our assumption that x is a solution of (VI).

Conversely, we trivially have that any solution of (EQ) satisfies (VI).

APPENDIX II

In this appendix we prove that the variational inequality (VI) and the convex program (CONV) are equivalent.

As earlier in the paper, let us consider the functional

$$J(x)=\sum_{a\in A}\left\{\int_0^{X_a} C_a(u)du+G_a\int_0^{\infty}(X_a(\alpha)/\alpha)d\alpha\right\}. \qquad \textbf{(4)}$$

We have, for any vector function ϕ with ϕ_a in $L^2(0,\infty)$:

$$\frac{1}{\theta}(J(x+\theta\phi)-J(x))=\frac{1}{\theta}\left(\sum_{a\in A}\left\{\int_{X_a}^{X_a+\theta\Phi_a} C_a(u)du+\int_0^{\infty}\theta\frac{\phi_a(\alpha)}{\alpha}d\alpha\right\}\right), \qquad \textbf{(5)}$$

where, following our earlier convention, $\Phi_a=\int_0^{\infty}\phi_a(\alpha)d\alpha$.

Taking the limit as ϕ goes to zero, there comes:

$$\begin{aligned}\sum_{a\in A}\left\{C_a(X_a)\Phi_a+\int_0^{\infty}(G_a/\alpha)\phi_a(\alpha)d\alpha\right\} &= \sum_{a\in A}\left\langle C_a(X_a)+G_a/\alpha,\phi_a\right\rangle\\ &= \left\langle C(X)+G/\alpha,\phi\right\rangle, \qquad \textbf{(6)}\end{aligned}$$

where $\langle .,.\rangle$ represents the inner product in $L^2(0,\infty)$, and $C(X)=\{C_a(X_a)\}_{a\in A}$, $\quad G=\{G_a\}_{a\in A}, \phi=\{\phi_a\}_{a\in A}$.

Since this development holds for any function ϕ in $L^2(\Omega)$, we can conclude that J is Gâteaux-differentiable, and that its derivative is given by

$$J'(x)=C(X)+G/\alpha. \qquad \textbf{(7)}$$

Now, we observe that the variational inequality (VI) corresponds to the first-order optimality condition of the mathematical program:

$$\min_{x\in\Omega} J(x). \qquad \textbf{(8)}$$

Since, by assumption, the congestion functions C_a are convex, the latter is a convex optimization problem.

A STOCHASTIC USER EQUILIBRIUM (SUE) PATH FLOW ESTIMATOR FOR THE DEDALE DATABASE IN LYON

Michael G. H. Bell, Caroline M. Shield,

Transport Operations Research Group,
Department of Civil Engineering, Claremont Tower,
University of Newcastle upon Type, NE1 7RU, UK

Jean-Jacques Henry,

Centre D'Etude Et De Recherche Toulouse,
2 Avenue Edouard Belin BP 4025, Toulouse F - 31055, FRANCE

Laurent Breheret

SODIT, Technoparc Bât 6 - Voie n°.5 - Labège Innopole - BP 533,
31674 LABEGE Cedex FRANCE

The paper sets out a path flow estimator suitable for the DEDALE database in Lyon, where traffic flow measurements are available at 6-minute intervals from 71 Traffic Data Collection Units (permanent counting stations). The estimator assumes that traffic assigns itself to paths according to the logit route choice model and that congestion leading to delay only forms on a link when its capacity is reached. An equivalent convex programming problem is formulated and an iterative solution procedure is set out. The estimation of the dispersion factor in the logit model is discussed, and a column generation method is proposed to avoid the need for path enumeration. A number of propositions are proved.

1. Introduction

Most Advanced Transport Telematics (ATT) systems require up-to-date information about the state of the road network. However, each system collects only a limited range of data. As an example, urban traffic control and EURO-SCOUT dynamic route guidance collect data from different sources (vehicle detectors and floating vehicles) but both require information about the state of the network (travel times and traffic flows). Moreover, ATT systems generally interact with each other, so there may be a requirement for management to resolve conflicts. The need for a *transport supervisor* has therefore been perceived, particularly within the DRIVE program. Such supervisors are being installed in Turin, Southampton and Lyon.

The function of the supervisor is to monitor the state of the transport system as a whole, to facilitate the exchange of data between ATT systems and, where problems arise, to manage transport in an appropriate way. Typically the ATT systems would be linked by a ring, as illustrated in Fig. 1 which relates to the Salerno pilot project (Di Taranto et al., 1993).

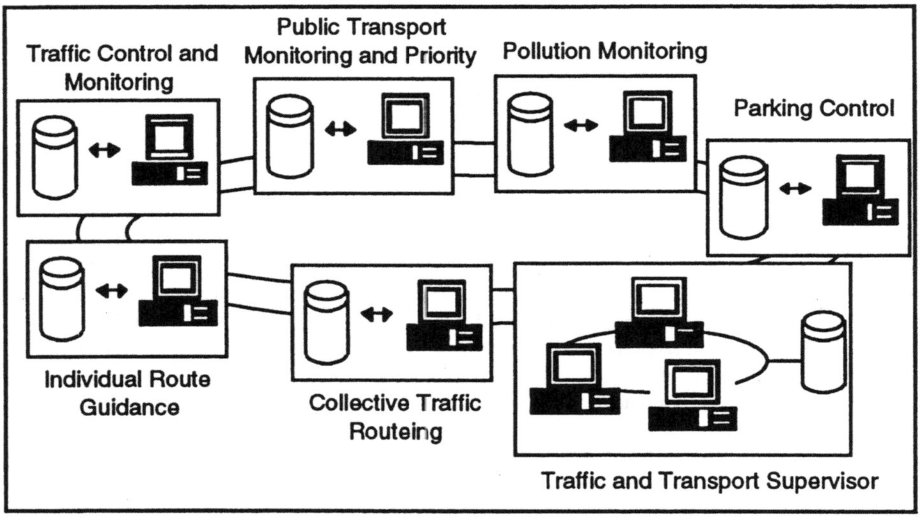

Figure 1. System architecture for the Salerno pilot project.

A form of network information that is potentially of use to a wide range of ATT systems is the current set of path flows. Knowledge of this would allow traffic signal coordination to be improved. Knowledge of the link flows and turning movements, which may be derived from the path flows, would allow the control at individual junctions to be improved. Path flows are not, however, directly observable. While link flows may be readily measured, manually or automatically, they provide only indirect information about path flows. In order to relate link to path flows, assumptions must be made about the nature of route choice (referred to as *traffic assignment*). In uncongested networks, it is possible to separate the route choice and path flow estimation problems; in congested networks the two problems are not separable. The path flow estimator presented here is particularly suited to congested networks.

There is an extensive literature on the related problem of estimating Origin-Destination (O-D) flows from traffic counts, most of it relating to uncongested networks (see Section 2). O-D flows are an aggregated form of path flows. Conventionally O-D flows are assumed to be fixed in some sense whereas path flows vary according to network conditions. It is possible to estimate path flows given O-D flows by using an appropriate traffic assignment model. However, a traffic assignment model is also required to estimate O-D flows from measurements of link flows. If the starting point is measurements of link flows, it is more direct to estimate path flows directly, which is the approach adopted in this paper.

2. Review of O-D Estimation

There are two fundamentally different approaches to O-D estimation. The static approach assumes steady state conditions in the network, and is therefore more suitable for longer time horizons and larger networks. The dynamic approach uses variation in the traffic flows in order to identify input-output coefficients, leading to recursive estimators that are more suitable for shorter time horizons and smaller networks. While O-D estimators tend to be either static or dynamic, it is possible to conceive of hybrid estimators which use time series data to identify local input-output coefficients dynamically and aggregate data to estimate O-D patterns at the network level. The broad applicability of the static, dynamic and hybrid approaches is illustrated in Table 1.

Time horizon / Type of urban network	Long term	Medium term	Short term
Small with no route choice	Static	Dynamic	Dynamic
Large with no congestion	Static	Hybrid	Hybrid
Large with congestion	Static / Bilevel programming		

Table 1. Applicability of static and dynamic estimators

When networks offer no route choice or are uncongested, the O-D estimation and the traffic assignment sub-problems may be separated, as noted in Section 1. In this case, the relationship between the O-D matrix and the link flows may be represented by a set of linear equations which underlie the following O-D estimators: the maximum entropy and minimum information estimators proposed by Van Zuylen and Willumsen (1980); the generalised least squares estimator proposed by Cascetta (1984) and Bell (1991a); the Bayesian estimator, proposed by Maher (1993), which is identical to the generalised least squares estimator; and a class of maximum likelihood estimators proposed by Spiess (1987).

When networks are congested, the estimation of O-D flows and assignment can no longer be separated. This has led to the proposed use of bilevel programming, whereby the O-D estimation and traffic assignment sub-problems are solved in sequence (see Florian et al., 1991; Yang et al., 1992). A more satisfactory approach makes use of linear programming (see Sherali et al., 1994) to estimate user optimal path flows, which may then be aggregated to yield an O-D matrix.

For junctions or small networks, time series statistical techniques may be applied to identify input-output coefficients. For small networks where there is no route choice, recursive estimation procedures are appropriate (see Cremer and Keller, 1987; Nihan and Davis, 1987; Bell,

1991b). As the size of the network increases, the time lags between inputs and outputs become more variable because of platoon dispersion within the network. This in turn makes the coefficients more difficult to identify. For larger networks which either offer no route choice or are uncongested, a hybrid static-dynamic approach, as formulated by Keller and Ploss (1987), looks attractive. Input-output coefficients for junctions are determined dynamically. These are then used as additional constraints in the static estimation of O-D flows.

The ODIN project of DRIVE I, studied the applicability of a number of promising static and dynamic O-D estimators to a range of network types (urban and inter-urban, with and without route choice) and produced preliminary recommendations. ODIN also produced a number of methodological advances, notably an event based O-D estimator suitable for very small networks (reported in Bell et al, 1991) and ways to allow for travel time variation in recursive estimation suitable for medium sized networks without route choice (reported in Bell, 1991b).

When short term prediction is required for larger networks, time dependencies must be taken into account. Approaches suggested so far have assumed proportional assignment and are therefore most appropriate for uncongested networks (see Ashok and Ben-Akiva, 1993).

3. A Path Flow Estimator

The estimator proposed in this paper is designed to produce on-line estimates of current link flows and turning movements for the DEDALE network in Lyon from 6-minute flow data emanating from 71 Traffic Data Collection Units (permanent counting stations). The approach adopted is to estimate path flows under the assumption of Stochastic User Equilibrium (SUE) assignment. Steady state network conditions are assumed. Each link is assumed to have a constant travel time and a fixed capacity, thereby simplifying the network data required to two parameters per link. When capacity is reached a queue can form leading to delay. The resulting right-angle link cost functions are particularly suitable for congested networks, and facilitate the estimation of delay (the difference between undelayed and actual travel times). The division of trips between alternative paths connecting any O-D pair is on the basis of the logit model, where the path costs consist of travel time and queuing time. The SUE assignment model adopted here is described in Bell (1995). The logit model parameter governing the degree of dispersion across alternative paths can be estimated on-line. A column generation method is proposed to avoid the specification of all paths.

4. Traffic Assignment

User equilibrium (UE) assignment has, since Wardrop (1952), played a dominating role in the literature on traffic assignment for congested networks. Underlying UE assignment are the assumptions that for each O-D pair every link has a cost known to all drivers, all used paths have minimum cost and all paths that have greater than minimum cost are not used.

In urban networks, the most significant component of delay arises at junctions because here capacity is least. This realisation focuses attention on queuing behaviour at junctions. Thus many traffic simulation models concentrate on queuing behaviour and make simplifying assumptions about link travel times. Early versions of the CONTRAM traffic simulator (Leonard et al, 1978) assumed vertical queuing and constant link travel times. The DEDALE estimator takes the same approach and assumes additionally steady state network conditions. This means that traffic on each link is processed at a constant rate, and a queue can only exist on links operating at capacity.

Smith (1987) has proved the Thompson and Payne (1975) result that a UE assignment can be found for networks of the kind assumed here by solving a linear programming problem. The total travel time in the network, excluding queuing time, is minimised subject to link capacity and non-negativity constraints.

The dual variables (the Lagrange multipliers associated with the capacity constraints) are equal to the UE delays (see Bell, 1995, for a proof).

ATT is increasing the number of possibilities for informing drivers of congestion and how to avoid it. These developments are serving to highlight the fact that drivers generally lack perfect information about network conditions, and that therefore one assumption behind UE assignment is violated significantly. This is leading to a resurgence of interest in SUE assignment (see Bell, 1993a).

The form of SUE assignment considered in this paper assumes that vehicles are assigned to routes according to the logit model. Fisk (1980) has shown how such an assignment is the solution to a particular convex programming problem formed from the Smith (1987) linear programming problem by adding a term to the objective function. Bell (1995) has demonstrated that the dual variables of the Fisk problem are still interpretable as equilibrium delays (this time SUE delays rather than UE delays).

5. Notation

Let:

c_i = the cost of link i

t_i = undelayed travel time for link i

d_i = delay for link i

v_i = the flow on link i

s_i = capacity for link i

m_i = Lagrange multiplier for link i

h_j = the flow on path j

c_j = the cost of path j

r_k = the count at station k

l_k = Lagrange multiplier for station k

Vector notation is used to represent sets of the above variables. Thus v represents $[v_1, v_2, v_3, \ldots]$.

6. Representation of Link Costs

The cost for link i, c_i, has two components; the (constant) undelayed travel time, t_i, and the delay due to queuing, d_i. On the one hand, if $d_i>0$ then $v_i=s_i$. On the other hand, if $v_i<s_i$ then $d_i=0$.

Thus for each link i

$$c_i = t_i + d_i \qquad (1)$$

The link delay d_i is determined by the demand rather than by any interactions at the junction. In reality, delay arises before the capacity of a link is reached. Where demand exceeds capacity, the simplifying cost assumptions made here are most accurate.

7. Relationship between Path and Link Flows

Let **A** be the link-path incidence matrix with elements a_{ij} equal to 1 if link i lies on path j and 0 otherwise. The relationship between the path flows and the link flows may then be expressed as

$$v = Ah \qquad (2)$$

Further, let **B** be the incidence matrix relating paths to counting stations with elements b_{kj} equal to 1 if path j passes counting station k and 0 otherwise. The relationship between the counts and the path flows may then be expressed as

$$r = Bh \qquad (3)$$

8. Stochastic User Equilibrium

Definition (SUE assignment): A stochastic user equilibrium (SUE) is achieved when the allocation of trips between alternative paths conforms to the following logit model

$$\ln(h_j/h_{j'}) = -\alpha(c_j - c_{j'}) + (l_j - l_{j'}) \quad (4)$$

where j and j' are alternative paths connecting the same O-D couple, c_j is the cost of path j, l_j is the sum of the Lagrangian multipliers l_k along path j and $\alpha > 0$ is a given parameter.

The term $(l_j - l_{j'})$ represents the relative intrinsic attractiveness of path j relative to path j'. It is required in order to obtain consistency with the traffic counts.

Consider the following problem

P_1: Minimise $\Sigma_j h_j (\ln h_j - 1) + \alpha \mathbf{t}^T \mathbf{A} \mathbf{h}$ with respect to $\mathbf{h}$ subject to $\mathbf{r} = \mathbf{B}\mathbf{h}$, $\mathbf{s} \geq \mathbf{A}\mathbf{h}$, $\mathbf{h} \geq 0$

The first term in the objective function has the effect of spreading the assignment. The following Lagrangian equation can be formed

$$\mathcal{L} = \Sigma_j h_j(\ln h_j - 1) + \alpha\, t^T v + l^T(r - Bh) + m^T(s - v) \quad (5)$$

where $\mathbf{v} = \mathbf{A}\mathbf{h}$. The Kuhn-Tucker conditions are

$$\ln h_j + (\alpha\, t - m)^T a_j - l^T b_j = 0 \quad (6)$$

and

$$r_k = \Sigma_j b_{kj} h_j \quad (7)$$

$$s_i - v_i \geq 0 \quad (8)$$

$$m_i \leq 0 \quad (9)$$

$$m_i(s_i - v_i) = 0 \quad (10)$$

are satisfied at the optimum. The form of (6) ensures that $h_j > 0$. Equation (6) reduces to

$$\ln h_j = -\alpha\, t_j + m_j + l_j \quad (11)$$

where l_j is the sum of counting station Lagrange multipliers l_k along path j, t_j is the undelayed path travel time and m_j is the sum of link capacity Lagrange multipliers m_i also along path j.

Proposition 1 proves that, for any link, $m_i = -\alpha d_i$ is necessary and sufficient for SUE.

Proposition 1 (SUE delays for link constraints): P_1 yields a SUE assignment if and only if the Lagrange multipliers $\mathbf{m}$ associated with the flow constraints $\mathbf{s} \geq \mathbf{v} = \mathbf{Ah}$ are equal to $-\alpha\mathbf{d}$, where $\mathbf{d}$ is the vector of SUE link delays.

Proof:

(i) Suppose $v_i = s_i$. There are then two possibilities. The first is that the demand for link i is exactly s_i. Nothing would be gained, in terms of a reduction in the objective function value, by shifting trips from paths not using i to paths using i. Therefore, a small increase in s_i leads to $v_i < s_i$, so refer to case (ii). The second possibility is that the demand for link i exceeds s_i, so the constraint binds. Link i is therefore the bottleneck on path j. A small increase in s_i would allow the objective function to be reduced by shifting some trips from paths not using link i to paths using link i. To minimise the objective function, the shift will be made to minimise the objective function value. Suppose that a small shift of δ trips is made from path j' to path j. Ignoring terms in δ^2 and higher powers of δ, $\ln(h+\delta) = \ln h + \delta/h$, so the change in objective function value (say F) is given by

$$dF = \delta \ln(h_j/h_{j'}) + \delta\alpha(t_j - t_{j'}) + \delta(l_j - l_{j'}) \tag{12}$$

Hence $m_i = \ln(h_j/h_{j'}) + \alpha(t_j - t_{j'}) + (l_j - l_{j'})$. But according to SUE, trips are allocated to paths according to the logit model as in (4). As link i is the bottleneck on path j, and path j' has no active bottleneck after the shift, we have

$$\ln(h_j/h_{j'}) = -\alpha(t_j + d_i - t_{j'}) + (l_j - l_{j'}) \quad (13)$$

from which we deduce that $m_i = -ad_i$ is necessary for SUE.

(ii) Suppose $v_i < s_i$. By the complimentary slackness conditions (8) to (10) $m_i = 0$. Also $d_i = 0$ as link i is operating below capacity.

Thus $m_i = -\alpha d_i$ is necessary and sufficient for SUE.

There is an issue relating to the uniqueness of the equilibrium delays. The objective function of P_1 can be shown to be strictly convex in path flows, so the path flows are uniquely defined. The link flows are therefore uniquely defined by (2). This implies that the set of links for which $m_i = d_i = 0$ is also uniquely defined. Proposition 2 proves that the linear independency of the capacity constraints is necessary and sufficient for unique equilibrium delays.

Proposition 2 (uniqueness of SUE delays): The SUE delays are unique if and only if all the constraints that bind are linearly independent.

Proof:

Drop the links (rows) for which $m_i = d_i = 0$ from **A** to produce **A'**. Writing (6) in matrix notation

$$h^T = \exp(-\alpha t^T A + m^T A' + l^T B) = \exp(-\alpha t^T A + \phi^T D) \quad (14)$$

where $\phi^T = [m^T, l^T]$ and $D^T = [A'^T, B^T]$. Thus

$$dh/d\phi = HD^T \quad (15)$$

where H is a diagonal matrix whose jth diagonal element is equal to h_j.

Let $\theta^T = [s'^T, r^T]$, where s' is a vector of capacity constraints that bind at the SUE solution. By the Chain Rule of differentiation

$$d\theta/d\phi = (d\theta/dh)(dh/d\phi) = DHD^T \quad (16)$$

Suppose ϕ is a solution to

$$\theta = D\exp(-\alpha A^T t + D^T \phi) \tag{17}$$

Consider ϕ' close to ϕ. Then by a first order approximation

$$DHD^T(\phi - \phi') = \theta - D\exp(-\phi A^T t + D^T \phi') \tag{18}$$

Suppose now that ϕ' is also a solution to (17) so that the right hand side of (18) is zero. Equation (18) implies that $\phi = \phi'$ if and only if DHD^T is non-singular. This is only the case if the rows of D are linearly independent, which in turn is only the case if all the constraints are linearly independent. Conversely, if all the constraints are linearly independent, DHD^T is non-singular and $\phi = \phi'$ if $\theta = D\exp(-\alpha A^T t + D^T \phi')$.

9. Solution by Iterative Balancing

A SUE assignment for a steady state store-and-forward network of the kind described in this paper can be readily found by solving an iterative balancing problem. By rearrangement of (11) we obtain

$$\begin{aligned} h_j &= \exp(-\alpha t_j + m_j + l_j) \\ &= \exp(-\alpha t_j)\, \Pi_{i \text{ on } j} M_i\, \Pi_{k \text{ on } j} L_k \end{aligned} \tag{19}$$

where $L_k = \exp(l_k)$ is a factor for counting station k, $M_i = \exp(m_i)$ is a factor for link i, t_j is the undelayed travel time for path j, and m_j and l_j are the sums of m_i and l_k respectively along path j. Factor L_k is calculated so that (3) holds, while factor M_i is calculated so that (8) to (10) hold. The following algorithm finds the values of the Lagrange multipliers satisfying the Kuhn-Tucker conditions.

Algorithm A_0 (to find the SUE assignment)

Step 1 (initialisation)

$M_i = 1$ for all links i

$L_k = 1$ for all counting stations k

Step 2 (iteration)

Repeat the following until convergence

For each i calculate

$$\beta = s_i \,/\, \Sigma_j\, a_{ij} \exp(-\alpha t_j)\, \Pi_{i \text{ on } j}\, M_i\, \Pi_{k \text{ on } j}\, L_k$$

$$M_i = \min[1, \beta\, M_i]$$

For each k calculate

$$\beta = r_k \,/\, \Sigma_j\, b_{kj} \exp(-\alpha t_j)\, \Pi_{i \text{ on } j}\, M_i\, \Pi_{k \text{ on } j}\, L_k$$

$$L_k = \beta\, L_k$$

Step 3 (output link flows and delays)

For each j calculate

$$h_j = \exp(-\alpha t_j)\, \Pi_{i \text{ on } j}\, M_i\, \Pi_{k \text{ on } j}\, L_k$$

For each i calculate

$$v_i = \Sigma_j\, a_{ij}\, h_j$$

$$d_i = -(\ln M_i)/\alpha$$

Note that the algorithm ensures that $M_i \leq 1$ so that $m_i \leq 0$.

Proposition 3 (convergence of iterative balancing): The above algorithm converges to the solution of P_1 provided a feasible solution exists.

Proof:

Consider the Lagrangian equation given in (5). The Saddlepoint Theorem says that at the optimum this is minimised with respect to the primal variables (link flows v) and maximised with respect to the dual variables (the Lagrange multipliers l and m).

The algorithm maximises $\mathcal{L}$ with respect to the dual variables. If $v_i>s_i$ then $\mathcal{L}$ is increased by reducing m_i. However, this also reduces v_i, so m_i should be reduced until $v_i=s_i$. If $v_i<s_i$ then $\mathcal{L}$ is increased by increasing m_i until either $v_i=s_i$ or $m_i=0$. Likewise, if $b_k^T h>r_k$ (where b_k is the kth row of B and $b_k^T h$ is therefore the sum of the path flows passing counting station k) then $\mathcal{L}$ is increased by reducing l_k until $b_k^T h=r_k$. If $b_k^T h<r_k$ then $\mathcal{L}$ is increased by increasing l_k until $b_k^T h=r_k$. As these changes to m_i and l_k exactly describe Step 2 of the algorithm, each iteration results in an increase in $\mathcal{L}$. Iterations continue until the Kuhn-Tucker conditions are satisfied. When the Kuhn-Tucker conditions are satisfied, no further increases in $\mathcal{L}$ are possible and P_1 is solved.

The requirements for uniqueness of the balancing factors M_i and L_k have been set out in Proposition 2.

10. Determination of α and Column Generation

Parameter α determines the sensitivity of assignment to path cost. As α increases, the importance of the second part of the objective function increases. In the limit, the assignment tends to UE. As α tends to zero, the driver preference for lesser cost paths disappears. As α increases, $t^T v$ reduces monotonically (proof of this is along the lines of Erlander et al., 1979). Thus a suitable way to estimate α would be to sample the flows on a sub-set of links. From the sampled link flows and knowledge of the undelayed link travel times, a sampled partial total undelayed travel time can be calculated. Parameter α can then be determined so that sampled and fitted partial totals agree.

Algorithm A_0 assumed that the paths have been enumerated before hand. In complex networks, this is a chore to be avoided. It is possible, however, to generate paths in an iterative way using an appropriate shortest path algorithm, such as Dijkstra's algorithm, as is done in the Frank-Wolfe approach to solving the UE problem. This is referred to as *column generation* because the columns of the link-path incidence matrix A are generated as the iterations progress (see Bazaraa et al, 1990). This leads to the following algorithm.

Algorithm A_1 (to find the SUE assignment by column generation)
Step 1 (initialisation)
Set d = 0 (set SUE delays to zero)
Set A to have no columns (no paths currently exist)

Step 2 (build shortest path trees)
Set c = t + d
Build fastest path trees using c
If no new paths generated, stop
Add new paths to A

Step 3 (determine SUE delays)
Run A_0
Output d, h and v
Return to Step 2

In the first few iterations, there may be insufficient capacity in the paths generated to cater for the demand. This would result in infinite delays at the bottlenecks. Although A_0 cannot converge under these conditions, it will indicate which links are the bottlenecks after a few iterations through the presence of vanishing multipliers M_i. For those links, delay can be set to high values before returning to Step 2, thereby causing new paths to be generated that avoid the bottlenecks.

11. Conclusions

The paper sets out a path flow estimator suitable for congested urban networks, such as that covered by the DEDALE database in Lyon, where traffic flow measurements are available at 6-minute intervals from 71 Traffic Data Collection Units (permanent counting stations). The estimator assumes that traffic assigns itself to paths according to the logit route choice model and that congestion leading to delay forms on a link when its capacity is reached. An equivalent convex programming problem is formulated and an iterative solution procedure is set out. Link delays are treated as dependent variables. The estimation of the dispersion factor in the logit model is discussed, and a column generation method is proposed to avoid the need for path enumeration.

A number of propositions are proved. Proposition 1 establishes the relationship between the capacity constraint Lagrange multipliers and SUE delays. Proposition 2 asserts that, while path flows and link flows will be unique, SUE delays will only be unique if the capacity constraints are linearly independent. Proposition 3 asserts that the iterative balancing method may be used to solve for the Lagrange multipliers and therefore for path flows, link flows and SUE delays.

The next step is to test the estimator against a more conventional approach to path flow estimation.

Acknowledgements

The authors are grateful to the DRIVE programme of the European Commission for funding the research reported in this paper.

References

[1] Ashok, K. & Ben-Akiva, M.E. (1993) "Dynamic origin-destination matrix estimation and prediction for real-time traffic management systems", *Proceedings of the 12th International Symposium on the Theory of Traffic Flow and Transportation*, Elsevier: Amsterdam, 465-484.

[2] Bazaraa, M.S., Jarvis, J.J. & Sherali, H.D. (1990) "Linear Programming and Network Flows", *John Wiley*: New York.

[3] Bell, M.G.H. (1991a) "The estimation of generalised least squares by constrained generalised least squares", *Transportation Research*, Vol. 25B, 13-22.

[4] Bell, M.G.H. (1991b) "The real-time estimation of origin-destination flows in the presence of platoon dispersion", *Transportation Research*, Vol. 25B, 115-125.

[5] Bell, M.G.H., Inaudi, D., Lange, J. & Maher, M. (1991) "Techniques for the dynamic estimation of O-D matrices in traffic networks", *Proceedings of the DRIVE Conference*, Brussels, 1040-1056.

[6] Bell, M.G.H., Lam, H.W.K., Ploss, G. & Inaudi, D. (1993a) "Stochastic user equilibrium assignment and iterative balancing", *Proceedings of the 12th International Symposium on Transportation and Traffic Theory*, Berkeley, July.

[7] Bell, M.G.H. (1995) "Stochastic user equilibrium assignment in networks with queues", *Transportation Research B*. In Press.

[8] Cascetta, E. (1984) "Estimation of trip matrices from traffic counts and survey data", *Transportation Research*, Vol. 18B, 289-299.

[9] Cremer, M. & Keller, H. (1987) "A new class of dynamic methods for the identification of origin-destination flows", *Transportation Research*, 21B, 117-132.

[10] Di Taranto, C., Inaudi, D., Verlardi, V. (1993) "Traffic monitoring and control: the ELASIS pilot project in Salerno", *Workshop on Integration Problems in Urban Transportation Planning and Management Systems*, Capri, October.

[11] Erlander, S., Nguyen, S. & Stewart, N.F. (1979) "On the calibration of the combined distribution and assignment model", *Transportation Research*, Vol. 13B, 259-267.

[12] Fisk, C. (1980) "Some developments in equilibrium traffic assignment", *Transportation Research*, Vol. 14B, 243-255.

[13] Florian, M. & Chen, Y. (1991) "A bilevel programming approach to estimating O-D matrix by traffic counts", *Report CRT-750*, Centre de recherche sur les transports, Montreal.

[14] Leonard, D.R., Tough, J.B. & Baguley, P.C. (1978) "CONTRAM - A traffic assignment model for predicting flows and queues during peak periods", *TRRL Laboratory* Report LR841.

[15] Maher, M.J. (1983) "Inferences on trip matrices from observations on link volumes: a Bayesian statistical approach", *Transportation Research*, Vol. 17B, 435-447.

[16] Nihan, N.L. & Davis, G.A. (1987) "Recursive estimation of origin-destination matrices from input/output counts", *Transportation Research*, Vol. 21B, 149-163.

[17] Sherali, H.D., Sivanandan, R. & Hobeika, A.G. (1994) "A linear programming approach for synthesising origin-destination trip tables from link traffic volumes", *Transportation Research*, In Press.

[18] Smith, M.J. (1987) "Traffic control and traffic assignment in a signal-controlled network with queuing", *Proceedings of the 10th Symposium on Transportation and Traffic Theory*, MIT, July, 61-77.

[19] Spiess, H. (1987) "A maximum likelihood model for estimating origin-destination matrices", *Transportation Research*, Vol.21B, 395-412.

[20] Thompson, W.A. & Payne, H.J. (1975) "Traffic assignment on transportation networks with capacity constraints and queuing" Paper presented at the *47th National ORSA/TIMS North American Meeting*.

[21] Van Zuylen, H.J. & Willumsen, L.G. (1980) "The most likely trip matrix estimated from traffic counts" *Transportation Research*, Vol. 14B, 281-293.

[22] Wardrop, J.G. (1952) "Some theoretical aspects of road traffic research" *Proceedings of the Institution of Civil Engineers II*(1), 325-378.

[23] Yang, H., Sasaki, T., Iida, Y. & Asakura, Y. (1992) "Estimation of origin-destination matrices from link traffic counts on congested networks", *Transportation Research*, Vol. 26B, 417-434.

MODELLING AND PERFORMANCE ANALYSIS OF URBAN TRANSPORTATION NETWORKS

Angela Di Febbraro and Simona Sacone

Department of Communications, Computer, and System Sciences, University of Genova - Via Opera Pia 13, 16145 Genova, Italy

A discrete event approach is proposed to evaluate the performance and develop control strategies for a urban intermodal transportation network. The discrete event model presented integrates public and private means of transport. Some disturbances are included to model the stochastic nature of the system. Based on such a model, a urban traffic simulator (INTRANET) has been designed on purpose. INTRANET includes two major modules. The first one, the Traffic Simulation Kernel, allows to study the dynamic behaviour of the transportation system, analyzing its performances and applying control strategies to optimize them. The second one, the Passenger Information Service, gives the system users at any time updated information about the different intermodal paths between any pair of origin/destination nodes. A case study relevant to the transportation network of an Italian city is dealt with. Related experimental results, showing the effectiveness of the proposed control strategy, are presented and discussed.

1. Introduction

In recent years, the integration between public and private transport services, with the aim of increasing the 'attractiveness' of public means of transport with respect to private ones, has become a research topic of great interest. Such an integration results in realizing intermodal transportation. Intermodal transportation can be defined as the serial use of different modes of transport to move passengers and/or freight from a place to another [1], [2], [3], [4]. In this paper, the problem of integrating passenger transport services in a urban area is dealt with. The realization of efficient intermodal transportation systems is especially intended to lead people not to use their cars to move in the cities, which would result in both an optimization of the travelling times and a decrease in air pollution.

A transportation system consists essentially of two major components: the *demand* system, which represents the needs and the behaviour of the users, and the *supply* system, which includes all what concerns the service production from the infrastructures to the planning rules of the system [5]. Some demand aspects usually depend on supply characteristics and vice versa. Then, a model of a transportation system can be seen as the integration of the models of demand and supply, and their interaction [6], [7]. This paper deals with a supply model relevant to an intermodal transportation network, proposed with the aim of analyzing the performances of the system and designing control strategies to optimize them. The network realizes the integration among urban public transport services themselves and private traffic. The proposed model can be viewed as an extension of the one presented in [8]. The system is represented as an oriented graph, in which the nodes are stations where it is possible to switch from a mode of transport to another one.

For the peculiar characteristics that an intermodal transportation network presents, it seems suitable to model it as a discrete event dynamic system. In fact, discrete event modelling is suitable for those systems in which significant changes in their state only occur at discrete time instants. Discrete event systems are asynchronous and modular, and often include control strategies and communication systems. Ruling communication and control strategies, it is possible to make the event flow satisfy the project requirements [9], [10]. For the transportation system under study, a peculiar discrete event model is proposed. Due to the stochastic characteristics of the transportation system considered, some disturbances are also included. The behaviour of the discrete event system modelling the transportation network is studied by means of a simulation tool on purpose designed, which is the kernel of the urban traffic simulation program INTRANET (INtegrated TRAnsportation NETwork).

INTRANET is designed to perform two major functions, almost independent of each other. The first objective is the validation of integrated timetables for the different modes of transport, so as to consider the various transport services to be parts of a whole intermodal transportation system. In this sense, INTRANET can evaluate the performance of the system, supporting the user in modifying adaptively a given timetable until reaching the desired performance. Different works can be found in the literature regarding scheduling problems in urban networks (see for instance [11],[12]).

The second objective pursued in designing INTRANET is to give the users of the intermodal transportation network some real-time updated information about the state of the network itself. The optimal situation to realize is that in which a user can choose, when approaching the transportation network, the best path to reach his destination. Each user could enter the information system in the stations of the network, or, in a more advanced realization, even from his car. In particular, if a user approaches the network by car, he can evaluate where it is possible (and convenient) to leave his car and carry on with his journey by public means of transport. To this end, some intermodal stations have a parking area associated with. More specifically, the available information regard essentially : i) the travelling times along all the possible paths, intermodal or not, between any pair of nodes in the network; ii) the availability of places in the parking areas; iii) the occurrence of irregular traffic conditions in some part of the network. Such an on-line information system could lead to a fair distribution of the traffic on the different (multimodal) paths and modes of transport, resulting in an improvement of the system performance. An evaluation tool which presents some analogy with INTRANET, but simulating individual vehicle movement in urban networks to study alternative information supply and traffic control strategies, has been described in [13].

The paper has the following structure. Next section is devoted to the description of the discrete event model designed for the intermodal transportation network considered. Section 3 describes the main modules of the simulation program INTRANET. The Traffic Simulation Kernel is essentially divided into two modules: the Discrete Event Model, described in Section 2, and the Control Module. A heuristic control strategy, based on the measure of the delays of the means of transport, is proposed with the objective of guaranteeing the regularity of the transport service, i.e., the fulfilment of the timetables for the considered modes of transport. Then, a case study relevant to the city of Genova is presented. Some conclusions end the paper, also giving some insight into possible further developments of the presented work.

2. A Discrete Event Model for an Intermodal Transportation Network

This section is devoted to the description of the model proposed for an intermodal transportation network. First, the general characteristics of the model are described, and its main components defined. Then, the discrete events selected to represent the functioning of the network are introduced. On the whole, this section is a description of the module mentioned in the Introduction as D.E.M.. The considered intermodal transportation network is modelled as an oriented graph, whose fundamental elements are nodes, macronodes, and links. A *node* is a station for a single mode of transport, and it can only exist as a part of a macronode. A *macronode* is an intermodal station, that is, a place where people can change mode of transport, or simply enter the transportation system. Therefore, a macronode is composed of one or more nodes. A *link* is a unidirectional path which connects two macronodes and is devoted to a single mode of transport. There are special links, named *inner links*, which connect two nodes in a macronode and can be traversed only on foot.

It is assumed that there are three different kinds of public means of transport, which are underground, railway, and bus. In this framework, the links which connect the N macronodes in the network are divided into three classes: a) bus links (*B-links*); b) railway links (*R-links*); c) underground links (*U-links*). For each macronode n, n=1, .., N, the types of its input and output links (U, R, or B) are specified. Each macronode can have at most six links in common with another macronode, as shown in Fig. 1.

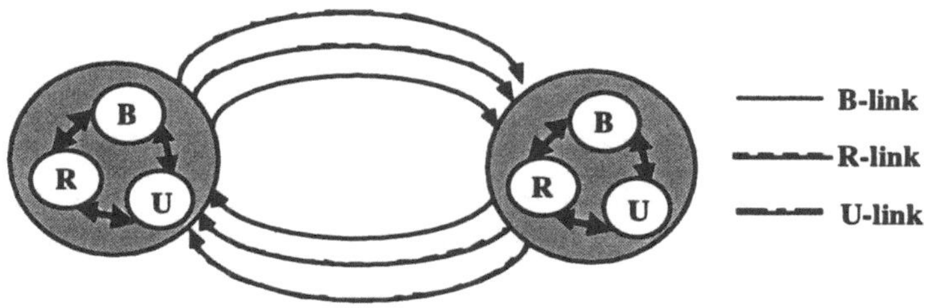

Figure 1. A generic pair of macronodes.

Macronodes n, n=1,...,N-1, represent the real intermodal stations of the transportation system under study, while the last macronode, macronode N, is a special macronode, i.e., a 'virtual' intermodal station which represents the outer world with respect to the network. In other words, each macronode which connects the urban area with its surroundings has a link towards macronode N. The presence of such a virtual macronode allows to model the considered transportation system as a closed network. This yields a major advantage consisting in the possibility of applying particular storage methods for closed graphs, which reduce sensibly the computational load. Moreover, the presence of macronode N also allows to model the movement of those means of transport which go, or come from outside the considered urban transportation network.

Users are supposed to approach the urban integrated transportation system by any means of transport. To model the arrival of passengers by private means of transport, parking areas are considered to be located near some macronodes of the network, where passengers can leave their means of transport, if room is available. The existence of a parking area near a macronode and the availability of place in it play an important role in modelling the integration between public and private means of transport.

2.1. Nodes, Macronodes, Links, and Means of Transport

In this section, the main entities of the system to be modelled are described. An *entity* is a component of the system which requires explicit representation in the model [14]. In this case, the entities of the model consist of those objects defined before as network entities (macronodes, nodes, and links) plus the means of transport. With each class of entities, static quantities (*parameters*) and dynamic quantities (*state variables*) are associated.

Nodes. A node is a part of a macronode which models a station for a single mode of transport. The parameters associated with a node are :

a) a numerical identifier;
b) the mode of transport to which it refers;
c) its capacity (the maximum number of vehicles that can be in the node);
d) the minimum stopping time of the means of transport;
e) its typology (station or terminus) with respect to the different lines which include the node. This parameter is necessary since the terminal node has the following features: i) its capacity is typically larger than that of an ordinary station; ii) vehicles can be parked; iii) means of transport can change their line numbers;
f) the time necessary for a vehicle to reach the node from the nearest depot.

The state variables of a node are:

α) the number of active transport means in it;
β) the number of out-of-service transport means in it. In each railway and underground station a place to recover broken vehicles exists. In particular, each node has a dedicated platform where the broken vehicle can stop and passengers can leave the vehicle to proceed with their travel in the network.

Macronodes. The parameters associated with a macronode, which is the entity modelling an intermodal station, are:

a) a numerical identifier;

b) the modes of transport of the nodes in it;

c) the existence of a parking place close to the station;

d) the travelling times of the inner links; obviously, these times depend on the structure of the intermodal station.

The state variable associated with a macronode is the availability of parking places in the area close to it, if such an area exists.

Links. A link, i.e., the physical connection between two macronodes, is characterized by the following parameters:

a) the mode of transport to which it refers;

b) the pair of macronodes which it connects;

c) its length;

d) its minimum travelling time;

e) its timetabling travelling time;

f) its capacity, i.e., the number of means of transport which can travel along the link (only for R and U-links). The capacity is a parameter for railway and underground links, since in these cases it cannot vary during simulation, whereas it is a state variable for B-links, where the real capacity is strongly dependent on the traffic conditions.

The state variables of a link are:

α) the number of vehicles travelling along it;

β) the real travelling time, which can differ from the minimum and the timetabling times because of the stochastic disturbances and the occurrence of some particular event;

γ) its capacity, only for B-links (see point (f) above).

Means of transport. Each means of transport is characterized by an alphanumerical parameter also indicating the relevant mode of transport. The state variables of a vehicle are:

α) the number of line;

β) the state of the vehicle, which can be one of the following : i) travelling on a link; ii) stopped in a node with closed doors (passengers cannot leave the vehicle); iii) stopped in a node (passengers can leave the vehicle); iv) ready to leave a node; v) waiting for entering a node; vi) waiting for leaving a node; vii) broken in a node; viii) broken on a link; ix) inactive (no more present in the network);

γ) timetabling leaving time; this time is significant only when the means of transport is stopped in a node.

Besides to the main static and dynamic quantities associated with the fundamental entities, the model comprehends a further set of parameters and state variables used to describe particular behaviours of the considered transportation system. For example, there exist some variables which concern the synchronization of the vehicle movement in the network. These synchronization mechanisms are quite necessary since nodes and links are limited resources shared by the means of transport. Furthermore, there are variables used to memorize breakdowns of vehicles, in order to introduce these aspects in the analysis of the system behaviour.

2.2. Events and State Transitions

An *event* can be defined as an instantaneous occurrence which may produce changes in the system state. The events that describe the evolution of the considered urban transportation network, and then rule the functioning of the INTRANET simulation tool, are divided into the following two classes:

i) events that describe the nominal traffic conditions;

ii) events that represent unpredictable conditions affecting the system.

The events belonging to the first class are:

i1) arrival of a means of transport at the end of a link;

i2) beginning of a stopping time;

i3) end of a stopping time;

i4) departure of a means of transport from a node through a link;

i5) car parking available;

i6) car parking not available;
i7) simulation end.

The second class consists of the events:

ii1) beginning of a failure;
ii2) end of a severe failure in a node;
ii3) end of a light failure in a node;
ii4) end of a severe failure in a B-link;
ii5) end of a light failure in a B-link;
ii6) end of a severe failure in a R-link or in a U-link;
ii7) end of a light failure in a R-link or in a U-link;
ii8) beginning of a B-link congestion;
ii9) end of a B-link congestion.

Each type of event is associated with an operational procedure which performs the state transitions resulting from the event occurrence. How simulation proceeds on the basis of the event occurrences is explained in next section.

3. INTRANET : A Urban Traffic Simulator

In this section, the structure of the INTRANET urban traffic simulator is described. It consists of different modules, as shown in the figure below.

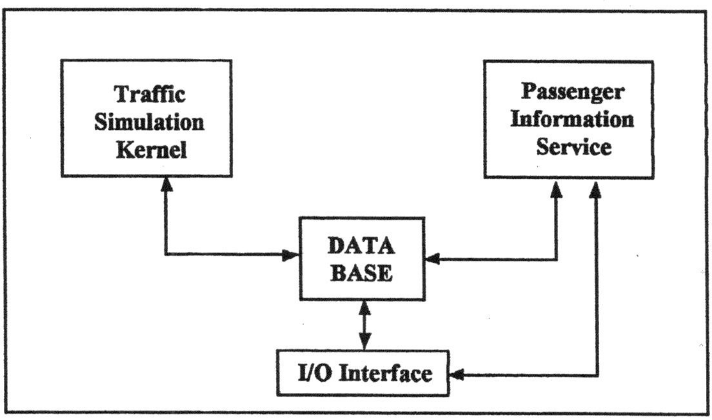

Figure 2. Structure of the INTRANET simulator.

For each module among the I/O Interface, the Traffic Simulation Kernel (T.S.K.), and the Passenger Information Service (P.I.S.), a detailed description is given in the relevant subsections. These three modules communicate by means of the data base, which contains both static and dynamic information. The static data consist of the inputs required by the simulation tool, as detailed in the following subsection. The dynamic information regard the state of the network, which is updated at each event occurrence.

3.1. The Input/Output Interface

As regards inputs, the I/O interface accepts data in a given format and makes up the static information of the data base to be used by the main blocks of INTRANET, i.e., the T.S.K. and the P.I.S.. The inputs required by the simulator are the following:

a) the topology of the network and the timetables for the three modes of transport;
b) the parameters of the probability distributions of the stochastic variables modelling the disturbances introduced in the model of the intermodal transportation network to fit its stochastic nature, which are :
    b1) perturbation of the timetabling leaving time of a transport means from a node; it consists in a time which is added to the timetabling leaving time, according to a Gaussian probability distribution;
    b2) perturbation of the timetabling travelling time of a link; again, this is a time which is added to the timetabling travelling time, according to a Gaussian probability distribution;
    b3) time interval between the arrival of a means of transport at the end of a link and the beginning of its stopping time in the node which the link enters;
    b4) time interval between the end of a stopping time of a means of transport in a node and its departure from a node;
    b5) duration of the different kinds of failures;
    b6) time interval between two subsequent failures.

Variables (b3)-(b6) are associated with different uniform probability distributions. Such distributions are generated by means of the inverse transform technique, while the Gaussian distributions are obtained by means of an approximation made using the convolution method [14].

In addition, the I/O interface manages the user's requests to the P.I.S., as mentioned above. As for outputs, the I/O interface gets from the data base information to make available the outputs of each simulation session. In particular, the output of a simulation run consists of:

a) a simulation trace; such a trace is essentially a list of all the events occurred during simulation, indicating the type and the occurrence time of each event;
b) statistics about:
   - link travelling times;
   - stopping times for each mode and means of transport at each node;
   - delays for each mode of transport, for each line, and for each means of transport;
   - duration of failures for each node and for each link.

The output statistics above can be used to analyze and optimize the performance of the transportation system. The choice of the performance indices to evaluate can be made according to the particular purposes of the simulation analysis. In this paper, the objective is to guarantee the fulfilment of the timetables for the three different modes of transport. This can be achieved by monitoring the delays of the means of transport and implementing control strategies to be applied whenever a delay is recorded.

3.2. The Traffic Simulation Kernel

The Traffic Simulation Kernel consists of two fundamental blocks, which are the Discrete Event Model, and the Control Module.

The Discrete Event Model, which represents the dynamic behaviour of the transportation network under study, has been described in section 2. The evolution of the system is reproduced by means of discrete event simulation, which can be viewed as the modelling over time of a system in which changes in the state only occur at discrete time instants, i.e., those istants when events occur. The mechanism to make simulation advance correctly, with all the events occurring in chronological order, is based on the *active event list* . Such a list is initialized before starting a simulation session, including those events which can be scheduled when the session starts. During the simulation, the active event list is examined and updated, by processing and eliminating, time after time, the event which occupies the head of the list [14].

Each type of event is associated with an operational procedure which, once invoked, modifies the state variables of the model in dependence of the event occurred. Such procedure has to insert in the active event list the events which will occur as a consequence of the current event, and cancel those events which the current event prevents from occurring. However, there are events which are not scheduled in dependence of any other event. These are the events gathered in class (ii) in subsection 2.2., i.e., those events which represent unpredictable conditions affecting the system. In fact, such events occur according to statistical distributions, a-priori given as inputs to the simulator. Whenever any nonrecurrent situation happens, the Control Module is invoked. This module implements a control strategy to try to recover the regular functioning of the system, as explained in next section.

3.2.1. Heuristic Solution of the Control Problem

In such a transportation system as the one described in this paper, i.e., when one neglects all the modelling aspects relevant to the movement of passengers, a sensible control objective is that of guaranteeing the regularity of transport services. This can be achieved by designing suitable control strategies in order to fulfil the timetable. As mentioned above, the model presented in this paper only represents the transportation supply. Then, it is supposed that the demand characteristics have been taken into account in the procedure to determine the timetables given as inputs to the simulation tool. The timetables for the different modes of transport vary during the day just to follow the demand dynamics.

The proposed criterion to analyze the performance of the intermodal transportation system is based on the measure of the delays of the different means of transport. The relevant control problem consists in minimizing the shift from the timetable of any means of transport in the network, i.e., it takes on the form of a tracking problem [15]. In this framework, the timetable for each means of transport is made up of the arrival and leaving times from each node during the whole simulation run, which usually has a one-day duration. At each event occurrence, the delay of the involved means of transport is computed as the difference between the timetabling time for that event occurrence, and the current simulation time.

The proposed control action consists in reducing the travelling times on the links and the stopping times at the nodes. In fact, there are minimum values for the travelling and the stopping times, in general less than the timetabling times, and a-priori determined according to the characteristics of both road and means of transport. Such minimum times impose upperbounds on the control action.

It is apparent that the statement and the solution of the control problem relevant to the minimization of the delays of all the means of transport moving in the network is complicated by the interactions among the means of transport themselves. Just to evaluate the complexity of such a problem, we consider its simplest version, i.e., when only one means of transport moves in the network. In this case, the control problem can be stated as follows.

Control problem. Find the optimal control action that minimizes

$$J = \sum_{i=1}^{N} |x_i - \bar{x}_i| \tag{1}$$

subject to:

$$x_i = x_{i-1} + \bar{x}_i - \bar{x}_{i-1} - u_{i-1} + \xi_i \quad i=1,..,N \tag{2}$$

$$x_i \geq 0 \qquad i=1,..,N \tag{3}$$

$$0 \leq u_i \leq u_{imax} \qquad i=1,..,N \tag{4}$$

where:

- N is the number of events relevant to the considered means of transport during a simulation run;
- x_i is the state variable representing the occurrence time instant of the i-th event;
- $\bar{x}_i$ is the (a-priori known) timetabling occurrence time of the i-th event;
- u_i is the control variable;
- u_{imax} is the upperbound for the control variable;
- ξ_i is a stochastic component due to the disturbances included in the model.

Note that constraints (2) represent the state equations of the system. The occurrence time of the i-th event, x_i, depends, as well as on the values of control and noise, on the occurrence time of the (i-1)-th event and on the difference between the timetabling occurrences times of the (i-1)-th and i-th events, $(\bar{x}_i - \bar{x}_{i-1})$. Such a difference can represent a timetabling stopping time at a node or a timetabling travelling time on a link. It is assumed that the state of the system is measured exactly. Constraints (3) represent non-negativity conditions for the state variable x_i, whereas constraints (4) define the admissible region for the control variables.

Note that ξ_i can depend on all the stochastic variables mentioned in the previous section. As some of such variables follow uniform probability distributions, whereas the others have Gaussian probability distributions, it is impossible to evaluate the probability density function of the noise ξ_i. This fact complicates the solution of the control problem. In effect, if we knew the probability density function of the noise, the above simple version of the tracking problem could be solved via the application of Dynamic Programming, as it would take on the form of a N-stage optimal control problem [15].

In this framework, we propose a simple heuristic control strategy to solve the overall control problem, that is, to make all the means of transport which move in the network fulfil the timetable. Such a heuristic policy consists in making the means of transport which are late traverse the links in the minimum travelling times until they achieve again the fulfilment of the timetable. This 'acceleration' is feasible since the timetabling travelling times of the links are usually longer than the minimum ones, so as to make it possible to regain some short delay. All the same, the timetabling leaving time from a node usually allows a stopping time longer than the minimum one. So, also the stopping times can be reduced until reaching the minimum values, in order to reduce the delay of the means of transport.

3.2.2. A Case Study

In order to analyze the behaviour of the transportation system described, simulation experiments have been performed on different intermodal networks. In this section, an applicative example relevant to the transportation network of the city of Genova is reported. As only a portion of the designed underground network has been realized so far, we model a more complex underground network than the real one, obtained by extending it according to the project lines. On the other hand, we consider only those bus stations which are significant for their position in the real Genova transportation network, and are near to railway and/or underground stations. As a result, the considered transportation network, as depicted in Fig. 3, is made up of 22 intermodal stations (macronodes), each of which consisting of at least two single-mode stations (nodes). In particular, the network consists of 6 underground, 17 railway, and 21 bus stations, and 102 links connecting the macronodes.

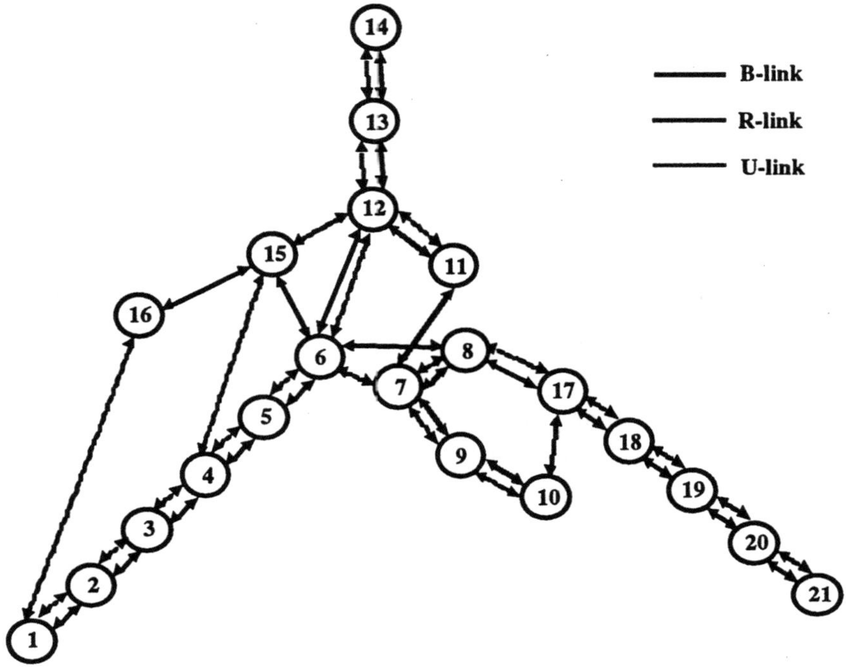

Figure 3. The considered intermodal transportation network.

The movement of the means of transport takes place according to 4 underground lines, 16 railway lines, and 48 bus lines. The mean speed is 12 km/h for buses, 30 km/h for railway, and 27 km/h for underground. Some data relevant to the network have been kindly offered by the local transportation company, the Azienda Municipalizzata Trasporti.

Many simulation experiments have been run on a Sun Sparc 10 Station, and statistics about vehicle, line, and mode delays, as well as the other outputs mentioned in subsection 3.1, have been gathered. The simulation experiments had identical initial conditions and statistically independent and identically distributed measures. The simulation results have been validated by computing the confidence interval for the output statistics of the stochastic variables.

Significant comparisons have been made between the behaviours of the transportation system with and without the application of control actions. In Fig. 4 the comparison between the mean bus delays during the day without control (plain line) and under the application of the control strategy (dashed line) is reported. The same kind of comparison is made in Fig. 5 and 6 for railway and underground, respectively. Note that the scale is different for each diagram so as to make the variations in the delay behaviours as evident as possible.

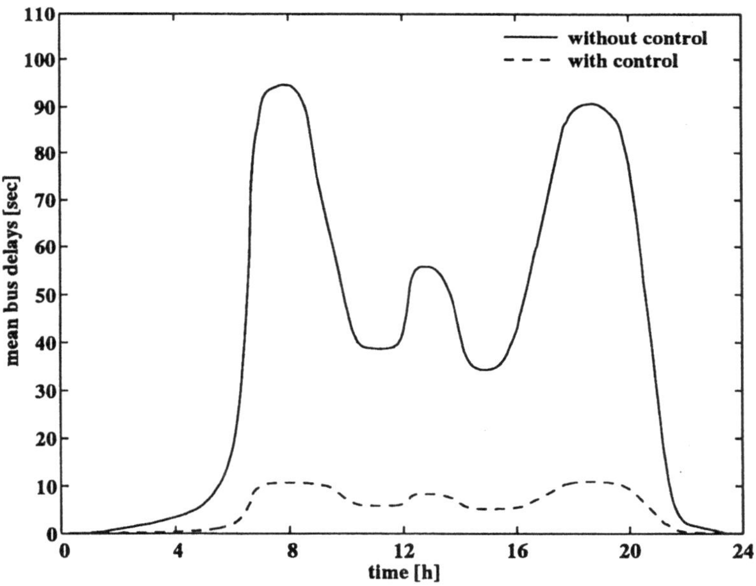

Figure 4. Mean bus delays with and without control action.

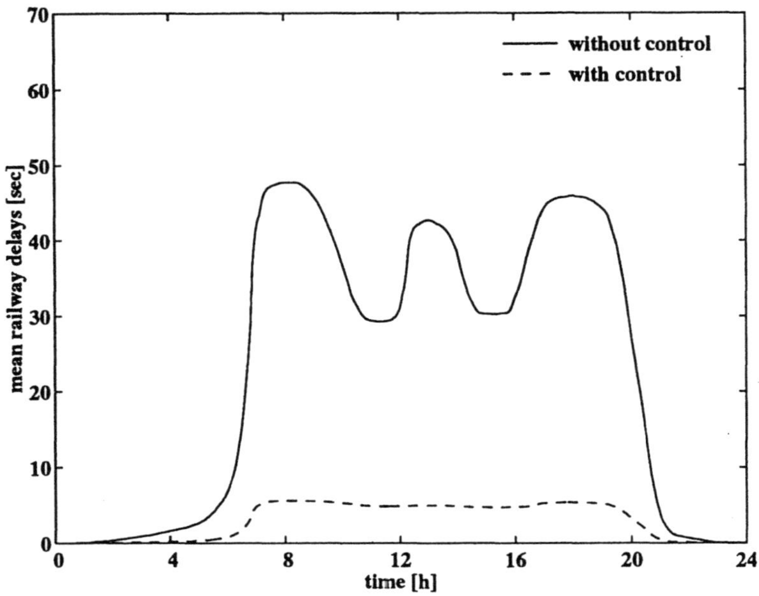

Figure 5. Mean railway delays with and without control action.

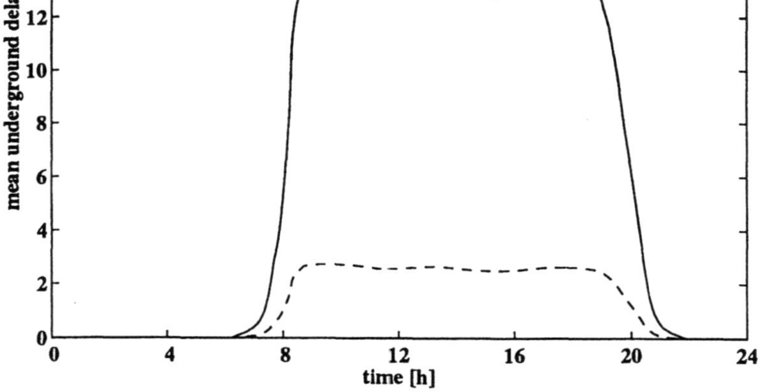

Figure 6. Mean underground delays with and without control action.

Fig. 7 reports in the same diagram the mean delays of the three modes of transport: the plain curve represents the bus delays, the dashed one stands for the railway delays and the dash-dotted one for the underground delays. This figure puts into evidence the differences between the behaviours of the three modes of transport and, in particular, the fact that bus delays are always significantly greater than the other ones.

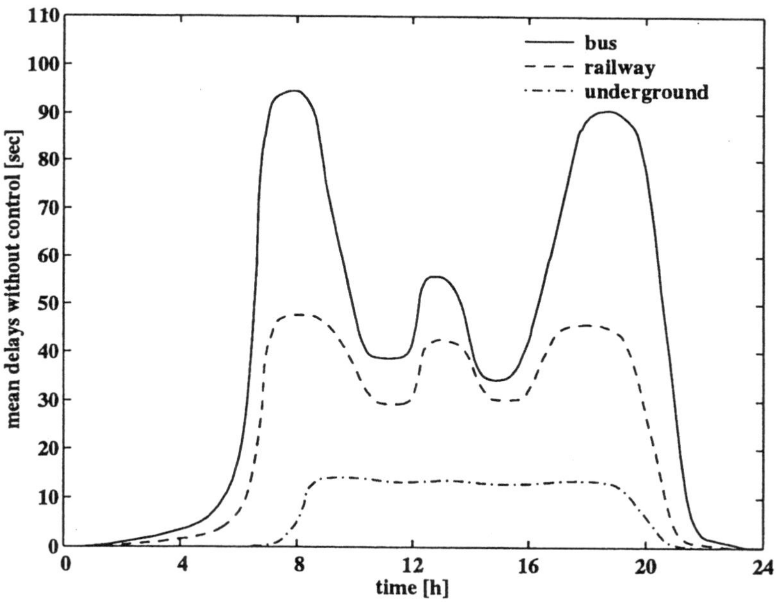

Figure 7. Mean delays of the three modes of transport.

The delays of the three modes of transport under the application of the first control strategy are shown in Fig. 8. It should be noted that the proportions among the delays of the three modes, as well as their 'shape', are similar to those shown in Fig. 7, but the control action reduces the values of such delays by a factor 10 for bus and railway, and 7 for underground.

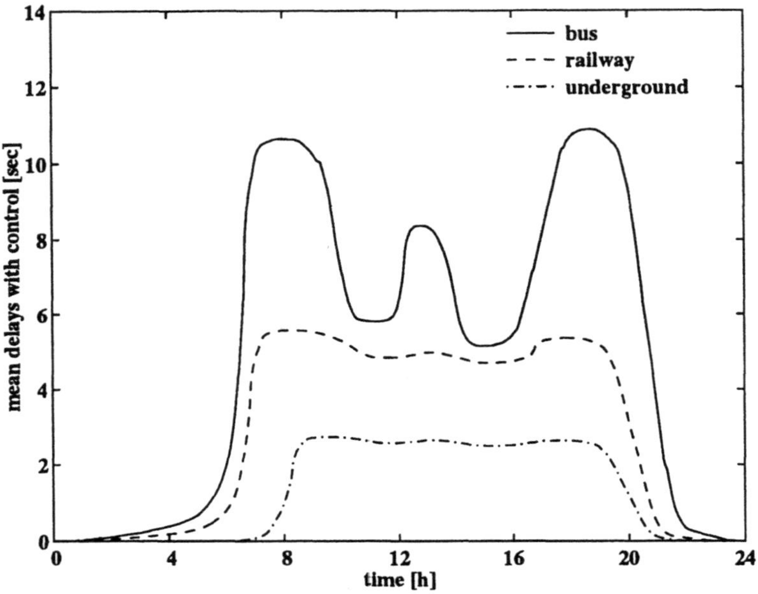

Figure 8. Mean delays of the three modes of transport with the application of the control strategy.

3.3. The Passenger Information Service

The Passenger Information Service (P.I.S.) performs a major function of the INTRANET simulator, which is to give the users of the intermodal transportation network some real-time updated information about the state of the network itself. The information offered by the P.I.S. regard essentially:

i) the travelling times along all the possible paths, intermodal or not, between any pair of nodes in the network;
ii) the availability of places in the various parking areas;
iii) the occurrence of irregular traffic conditions in some part of the network.

Such an on-line information system could lead to a fair distribution of the traffic on the different (multimodal) paths and modes of transport, resulting in an improvement of the system performance.

All the paths, intermodal or not, between a couple of origin/destination nodes in the network are a-priori determined. The real travelling times of the links between two macronodes are updated in real-time depending on the state of the network. Whenever the P.I.S. receives via the I/O interface a request from a user who needs to move from a macronode to another macronode of the network, it provides information about the current best multimodal paths between the two macronodes. A passenger's request is composed of the following data:

i) the origin macronode and the destination macronode;

ii) the modes of transport that the user is willing to use;

iii) the maximum travelling time t_{max} which the user is willing to spend to reach its destination;

iv) the number k of alternative paths requested.

The P.I.S. satisfies the users' requests using an ad-hoc modified version of the *Dijkstra algorithm* [16]. This algorithm takes into account the fact that the considered network is represented as a graph in which two nodes can be connected by more than a single link (in particular there could be at most three links connecting two macronodes in each direction). Furthermore, another feature of the method used is the cost associated with each link; the cost associated with a link l of mode m is a function of the following three quantities:

i) the real travelling time on link l;

ii) the travelling time on the inner link connecting, in the macronode origin of link l, the station of the mode of transport used to reach such a macronode and the station of mode m; this time is zero, if a change of mode in the current macronode is not required;

iii) the expected waiting time for means of transport of mode m in the relevant station of the origin macronode of link l.

The times in (ii) are known and constant, while those indicated in (i) and (iii) are updated in real time, depending on the state of the network.

The degrees of freedom offered to the user by the P.I.S. can result in different problems. In particular, if the user chooses to eliminate one or two modes of transport from his path, the graph could become not connected. For this reason, a suitable procedure to remove from the graph possibly not connected macronodes has been implemented.

The introduction of such a procedure is necessary because the Dijkstra algorithm works only for connected graphs.

Another complication arises from the fact that the user can require to know the k shortest multimodal paths to reach its destination. In the event of k>1, once the shortest path between the origin and destination macronodes is determined, the P.I.S. passes to a second phase of its work. It consists in finding the shortest (k-1) alternative paths besides to the best path already found. In doing so, the method proposed by Yen [17] is applied.

The output of the P.I.S. is a list containing the k shortest paths with total travelling time less than t_{max} ordered by increasing travelling times. This list also indicates the changes of mode of transport to be made to follow each path.

4. Conclusions

The issues of modelling, simulation, and control of an intermodal urban transportation network have been addressed in this paper. In the oriented graph which represents the transportation system, nodes can model single-mode stations or intermodal stations. The considered transportation network, integrating public and private means of transport, has been modelled as a discrete event system, whose behaviour has been studied by means of an on purpose designed discrete event simulator, called INTRANET.

INTRANET allows to study the dynamic behaviour of the transportation system, analyzing its performance and applying control strategies to guarantee the regularity of the transport service, i.e., the fulfilment of the timetables. In turn, its Passenger Information Service provides the system users with updated information about the state of the transportation network. Some experimental results relevant to a case study, showing the effectiveness of the control strategy proposed to guarantee the fulfilment of the timetables, have been presented.

In an advanced configuration, the described urban traffic simulator can be considered as a module of a more complex structure, in which an optimization module is connected to the simulator. Such a module could find, in a deterministic framework, the optimal timetables for the three modes of transport to be passed as inputs to the simulator.

References

[1] Kondratowicz, L.J. (1990) "Simulation Methodology for Intermodal Freight Transportation Terminals", *Simulation*, 55, 49-58.

[2] Guelat, J., M.Florian, and T.G.Crainic (1990) "Multimode Multiproduct Network Assignment Model for Strategic Planning of Freight Flows", *Transportation Science*, 24, 25-39.

[3] Fernandez, E., J. De Cea, M. Florian, and E. Cabrera (1994) "Network equilibrium models with combined modes", *Transportation Science*, 28, 182-192.

[4] Gedeon, C., M. Florian, and T.G. Crainic (1993) "Determining origin-destination matrices and optimal multiproduct flows for freight transportation over multimodal networks", *Transportation Research, Part B*, 27B, 351-368.

[5] Kanafani, A. (1983), "Transportation Demand Analysis", *McGraw-Hill*, New York.

[6] Ben Akiva, M., and H.Lerman (1987) "Discrete Choice Analysis", *MIT Press*, Cambridge, MA.

[7] Cascetta, E., and G.E.Cantarella (1993) "Modelling dynamics in transportation networks: State of the art and future developments", *Simulation Practice and Theory*, 65-91.

[8] Di Febbraro, A., V. Recagno, and S. Sacone (1995) "INTRANET : a new Simulation Tool for Intermodal Transportation Systems", to appear in *Simulation Practice and Theory*.

[9] Ho, Y.C., and X.R. Cao (1991) "Perturbation Analysis of Discrete Event Dynamic Systems", *Kluwer Academic*, Boston, MA.

[10] Cassandras, C. G. (1993) "Discrete Event Systems", *Irwin and Aksen Ass.*, Boston, MA.

[11] Viegas, J. (1990) "Time and space scheduling of public transport networks and the evaluation of quality of service", *Proceedings of the Fifth International Workshop on Computer-Aided Scheduling of Public Transport*, Montreal, Canada, 97-111.

[12] Grega, W. (1993) "Decomposition approach to the public transport scheduling problem", *Automatica*, 29, 745-750.

[13] Jayakrishnan, R., H.S. Mahmassani, and Ta-Yin Hu (1994) "An evaluation tool for advanced traffic information and management systems in urban networks", *Transportation Research, Part C*, 2C, 129-147.

[14] Banks, J., and J.S. Carson (1984), "Discrete-Event System Simulation", *Prentice-Hall*, Englewoods Cliffs, NJ.

[15] Bertsekas, D.P. (1987), "Dynamic Programming",*Prentice-Hall*, Englewoods Cliffs, NJ.

[16] Christofides, N. (1975) "Graph Theory" *Academic Press*, London, UK.

[17] Yen, J.Y. (1971) "Finding the k-shortest, loopless paths in a network", *Man. Sci.*, 17, 712.

ZONE PLANNING IN PUBLIC TRANSPORTATION SYSTEMS

Anita Schöbel*

Fachbereich Mathematik, Universität Kaiserslautern, Germany

In this paper, the optimal solutions of the fair zone design problem in linear station networks are studied. The fair zone design problem has been introduced in [3]. A zone in a station network is a set of stations which are treated as a unit as far as the fares for the passengers are concerned: The zone tariff is only dependent on the starting and ending zones of their travel. A fair zoning is one where the zone tariff is as close as possible to the distance tariff which relates the fare to the actual distance of a customer trip. In particular, the goal of a fair zone design is to design zones in such a way, that that neither the public transportation company nor the customers will have major disadvantages in the transition from distance tariff to zone tariff.
Three different objective functions to model that goal are introduced. Some results and an integer programming formulation for the problem in linear station networks will be given. The paper is concluded by a brief review of heuristics for the general problem and their application to the data of a regional public transportation company in Germany.

1. Introduction

We consider a public transportation network modeled by a directed graph G=(V,E) (the *station graph*). The nodes $i \in V$ represent the stations and an edge $e = (i, j) \in E$ indicates that the two stations i and j are connected by at least one direct line between i and j. Each node pair i,j has associated with it a nonnegative value d_{ij} representing the fare a customer has to pay when travelling from station i to station j. If only the prices for the edges $e \in E$ are known, we can calculate the values d_{ij}

*partially supported by a grant of the Deutsche Forschungsgemeinschaft and grant ERBCHRXCT930087 of the European HC\&M Programme as the smallest ticket prices for travelling from i to j. This can for instance be done by the algorithm of Floyd and Warshall [2,6].

In the case of *distance tariffs* we assume that the ticket price d_{ij} for going from station i to station j in the public transportation network is proportional to the distance between these two stations. Although this type of tariff is a fair one, it is not very convenient: In order to find out a specific ticket price, the distance information for all pairs of stations needs to be available. This is feasible if the number of stations is small. If the number of stations grows, this tariff system is too complex and not transparent enough for the customers.

Therefore a lot of public transportation companies consider the introduction of *zone tariffs* to simplify their tariff systems. In this case the set of nodes V is partitioned into node sets V_1, V_2, ..., V_m, which are called *zones*. For travel between two zones V_k and V_l one fixed tariff is charged, which is the same for all stations $i \in V_k$ and for all stations $j \in V_l$. I.e. the zone tariff is only dependent on the starting and ending zones of a customers' trip. Obviously, the zone tariff is highly depending on the choice of the zones V_1, V_2, ..., V_m and of the new prices between these zones.

The *fair zone design problem* now is to define the new tariff zones in such a way, that the changes in ticket price for the customers are as small as possible. That means, we try to avoid major disadvantages in the transition from distance tariff to zone tariff both for the customers and for the transportation company. Notice that the zone design problem may start as well with a tariff system (d_{ij}) which is different from the distance tariff described in that section.

The zone design problem was first mentioned in [3] and in [4] when a regional public transportation company in Germany wanted to design fair zones. This application will be described briefly in the last section. The following denotations will be used in that paper:

From a graph theoretical point of view, zoning is a contraction of the station graph G = (V, E). To model the zone design problem, we interpret each of the zones as a node in the complete graph G'=(V',E'), i.e. $V' = \{ V_1, V_2, \ldots, V_m \}$ represents the partition of V and E' = {(V_k,

V_l): k, l = 1, ..., m}. For the edges $(V_k, V_l) \in E'$ the edge weights d'_{kl} (called the *zone costs*) represent the prices for travelling from zone V_k to zone V_l. (Notice that we include loops (V_k, V_k) in G'. They are necessary to model the zone price for travelling within the same zone.) G' is called the corresponding *zone network* to the station graph G. The resulting zone tariff for travelling from station $i \in V_k$ to station $j \in V_l$ can now be defined as follows:

$$z_{ij} = \begin{cases} d'(V_k, V_l) & \text{if } i \neq j \\ 0 & \text{if } i = j \end{cases}$$

In this paper we report on some results dealing with the zone design problem. Fairness measures for zone tariffs and some elementary results will be discussed in Section 2. In Section 3 an integer programming formulation for simple linear station networks will be given. The paper is concluded by a short review of heuristics for the general problem and an application for a public transportation company in Germany.

2. Fairness Measures

The goal of the fair zone design problem is to get a zone tariff, which is as close as possible to the distance tariff (or to any other kind of tariff system which was used before). To measure the fairness of a zone tariff, we look at the differences between the old tariffs d_{ij} and the new zone tariffs z_{ij}. There are several possibilities to define the fairness of a zone tariff, depending on the station network G and the corresponding zone network G':

$$b_{max}(G,G') := \max_{i,j\in V} |d_{ij} - z_{ij}|$$

$$b_1(G,G') := \frac{1}{2} \sum_{i,j\in V} |d_{ij} - z_{ij}|$$

$$b_2(G,G') := \frac{1}{2} \sum_{i,j\in V} (d_{ij} - z_{ij})^2$$

The first objective function b_{max} models the fact, that the greatest deviation of ticket prices in the two different tariffs should be as small as possible. Its goal is to avoid that single customers have a large change in their ticket prices. The second objective, b_1 often is suitable for finding a fair transition between the old and the new tariff system, buth may yield serious disadvantages for single customers. The same properties hold for the objective function b_2 which is differentiable and may therefore be preferable to objective b_1.The factor 1/2 in b_1 and b_2 is needed to avoid that the differences (d_{ij}-z_{ij}) are counted twice for each pair of stations i and j. It also should be mentioned that deviations in price increases and decreases are treated equally, such that the model reflects both the interests of the customers and of the transportation company.

To see how to get the new zone tariffs z_{ij} and the values of the three different objective functions we look at the following example. Here the zones and the zone costs for travelling between two zones are already given:

<u>Example 1:</u> Let G be the graph of Figure 1 (a) with the zone partitioning V_1 = {1,6}, V_2 = {2,3}, and V_3 = {4,5}. If we choose the new zone costs $d'(V_k , V_l)$ as the minimum ticket price between the zones V_k and V_l in the old tariff system we get as the corresponding zone network the complete graph with three nodes and cost values $d'(V_k , V_l)$ as shown in Figure 2 (b).

We get :

$$D = \begin{pmatrix} 0 & 1 & 2 & 3 & 4 & 5 \\ 1 & 0 & 1 & 2 & 3 & 4 \\ 2 & 1 & 0 & 1 & 2 & 3 \\ 3 & 2 & 1 & 0 & 1 & 2 \\ 4 & 3 & 2 & 1 & 0 & 1 \\ 5 & 4 & 3 & 2 & 1 & 0 \end{pmatrix} \qquad Z = \begin{pmatrix} 0 & 1 & 1 & 1 & 1 & 5 \\ 1 & 0 & 1 & 1 & 1 & 1 \\ 1 & 1 & 0 & 1 & 1 & 1 \\ 1 & 1 & 1 & 0 & 1 & 1 \\ 1 & 1 & 1 & 1 & 0 & 1 \\ 5 & 1 & 1 & 1 & 1 & 0 \end{pmatrix}$$

and

$$D - Z = \begin{pmatrix} 0 & 0 & 1 & 2 & 3 & 0 \\ 0 & 0 & 0 & 1 & 2 & 3 \\ 1 & 0 & 0 & 0 & 1 & 2 \\ 2 & 1 & 0 & 0 & 0 & 1 \\ 3 & 2 & 1 & 0 & 0 & 0 \\ 0 & 3 & 2 & 1 & 0 & 0 \end{pmatrix}$$

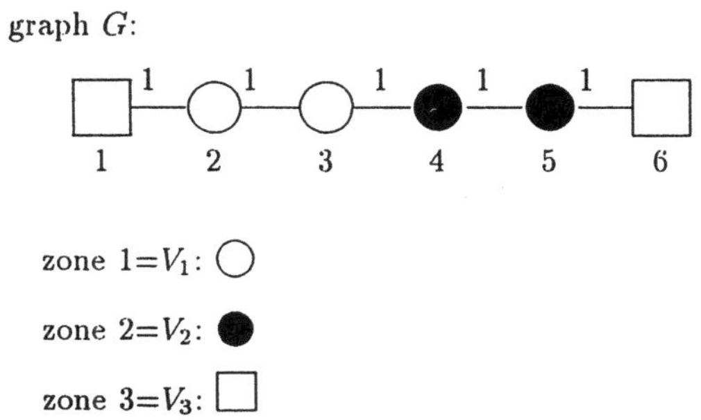

Figure 1 (a). Station graph

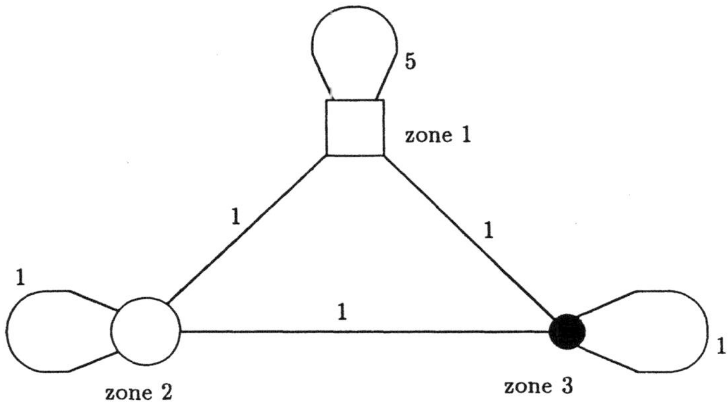

Figure 1 (b). Corresponding zone network

If the zones $V_1, V_2, \ldots, V_m$ are given, it is easy to get zone costs d'_{kl} which minimize b_{max}, b_1 and b_2.

Theorem 1: *Let $V_1, V_2, \ldots, V_m$ be a zone partitioning of G. In order to minimize $b_{max}(G,G')$, $b_1(G,G')$, and $b_2(G,G')$ we choose as new zone prices between V_k and V_l:*

a) $$d'_{max}(V_k,V_l) := \frac{1}{2}\left(\max_{i\in V_k, j\in V_l} d_{ij} + \min_{i\in V_k, j\in V_l, i\neq j} d_{ij}\right)$$

b) $$d'_1(V_k,V_l) := \operatorname{median}_{i\in V_k, j\in V_l, i\neq j} d_{ij}$$

c) $$d'_2(V_k,V_l) := \begin{cases} \dfrac{\sum_{i\in V_k, j\in V_l} d_{ij}}{|V_k||V_l|} & \text{for } l \neq k \\[2ex] \dfrac{\sum_{i\in V_k, j\in V_l} d_{ij}}{|V_k|(|V_k|-1)} & \text{for } l = k \end{cases}$$

The proof of that theorem can be found in [3,4]. Part c) is known in Statistics as the Theorem of Steiner (see e.g. [5]).

Corollary 1: *Let $V_1, \ldots, V_m$ be a zone partitioning of G and let $c'(V_l, V_k)$ be any zone cost. Then*

a) $$b_{max}(G,G') \geq \frac{1}{2} \max_{l,k=1,\ldots,L,|V_l||V_k|>1} \left\{ \max_{i\in V_l, j\in V_k} d_{ij} - \min_{i\in V_l, j\in V_k, i\neq j} d_{ij} \right\}$$

b) *Let*

$$V_{kl}^{+} := \left\{ d_{ij} : i \in V_k, j \in V_l, d_{ij} \geq \operatorname{median}_{i\in V_l, j\in V_k} d_{ij} \right\} \quad \textit{and}$$

$$V_{kl}^{-} := \left\{ d_{ij} : i \in V_k, j \in V_l, d_{ij} \leq \operatorname{median}_{i\in V_l, j\in V_k} d_{ij} \right\} \quad \textit{then}$$

$$b_1(G,G') \geq \frac{1}{2} \sum_{l,k=1,\ldots,m} \left(\sum_{d\in V_{kl}^{+}} d_{ij} - \sum_{d\in V_{kl}^{-}} d_{ij} \right)$$

c) $$b_2(G,G') \geq \frac{1}{2} \sum_{l,k=1,\ldots,m} \mathit{Variance}\left\{ d_{ij} : i \in V_l, j \in V_k, i \neq j \right\}$$

All inequalities hold with equality if the zone costs are chosen according to Theorem 1 resprectively.

In the next section we will deal with the objective function b_{max}. For this the following Corollary is helpful. Let

$$C_{max}(V_k, V_l) := \max_{i\in V_k, j\in V_l} \left| d_{ij} - z_{ij} \right|$$

be the costs betwen the zones V_k and V_l with respect to the objective function b_{max}. It is clear that b_{max} can be calculated as the maximum of all values $C_{max}(V_k, V_l)$ with $k, l \in \{1,2, \ldots, m\}$.

Corollary 2: *Let $V_1, V_2, \ldots, V_m$ be a zone partitioning of G and $d'(V_k, V_l)$ be any zone pricing. Then the following inequality holds:*

$$C_{max}(V_k, V_l) \geq \frac{1}{2} \left(\max_{i \in V_k, j \in V_l} d_{ij} - \min_{i \in V_k, j \in V_l, i \neq j} d_{ij} \right)$$

This inequality holds with equality, if new zone costs are chosen according to Theorem 1.

3. Linear Station Networks

In that section we are concerned with the question, if the optimal zones are always connected in the original station graph.

Example 2: Here, a station graph with eight stations is given. We are looking for the best zoning with two zones (with respect to b_{max}). The unique optimal solution is shown in Figure 2. As it can be seen, the outer zone is not connected in the original graph. Nevertheless this case often appears, for example when ring zones are used as a tariff system in cities.

Example 2 shows, that in general we cannot expect the optimal zones to be connected in the original graph. But in the case of linear graphs it is possible to get a connectivity result. Therefore we consider linear station networks (see Figure 3) in which the stations are numbered from 1 to n, such that 1 is the first station and n is the last one. For the distance tariff we get $d_{ij} = \sum_{k=i}^{j-1} d_{k\,k+1}$.

Some notations for zonings in linear networks should be introduced: A zone V_a in a linear graph is *connected*, if and only if there exists a node k such that $V_a = \{k, k+1, \ldots, k+|V_a|-1\}$. A zone which is not connected is called *unconnected*.

Two zones V_a, V_b are called *separated* if there is a k such that $i \leq k$ for all $i \in V_a$ and $j > k$ for all $j \in V_b$. The *diameter* of a zone V_a is defined as $\mathrm{diam}(V_a) := \max_{i,j \in V_a} d_{ij}$.

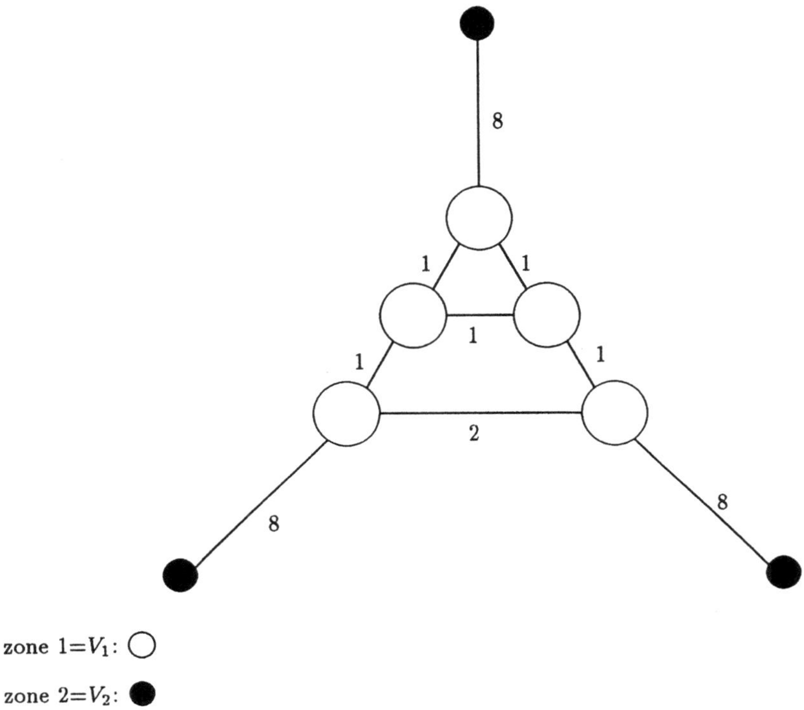

Figure 2 Example of ring zones

Lemma 1 *Let V' be a zone partitioning of a linear graph. For all separated zones V_a, V_b Î V' we have that*

1. $C_{max}(V_a, V_b) \leq \frac{1}{2}(diam(V_a)$ *for all zones* $V_a \in V'$
2. $C_{max}(V_a, V_b) = \frac{1}{2}(diam(V_a) + diam(V_a))$
3. *For all zones* $V_a \in V'$ *with* $\{1, n\} \not\subset V_a$ *there exists a zone* V_b, $V_b \in V'$, $V_b \neq V_a$, *such that* $C_{max}(V_a, V_a) \leq C_{max}(V_a, V_b)$ *i.e. the costs* $C_{max}(V_a, V_a)$ *do not influence the value of* b_{max} *for (nearly) all zones in V'.*

The proofs of the first two statements follow from Corollary 1, the proof of the last part is as follows: Because $\{1,n\} \not\subset V_a$ it is possible to choose V_b such that $diam(V_a) < diam(V_a \cup V_b)$. Then $diam(V_a \cup V_b)$ - $diam(V_a) \geq \min_{i \in V_a, j \in V_b} d_{ij}$ and we can conclude that

$$C_{max}(V_a, V_b) = \frac{1}{2}(diam(V_a \cup V_b) - \min_{i \in V_a, j \in V_b} d_{ij})$$
$$\geq \frac{1}{2}(diam(V_a) - \min_{\substack{i \in V_a, j \in V_a \\ i \neq j}} d_{ij}) = C_{max}(V_a, V_a)$$

Theorem 2 *Let V = {1, 2,..., n}, G=(V, E) a linear graph and* d_{ij} *the distance tariff between stations i and j. Then for the objective function* b_{max} *there always exists an optimal zone partitioning V' = {*V_1, V_2, ..., V_m*} with one of the following properties:*

1. *Either* ***all*** *zones* V_k *are connnected,*
2. *or there is* ***one*** *zone* V_a *with* $\{1, n\} \subseteq V_a$ *and* ***all other*** *zones are connected*

Proof: Consider a zone partitioning V' with a zone $V_1 \in V'$, such that V_1 is not connected and $\{1,n\} \subseteq V_1$ does not hold. We will show how to get a new zone partitioning V^{new} in which one more zone is connected without increasing the objective value. We define V_1^{new} as the smallest connected zone which contains V_1 and for all other zones $V_a \in V'$ we set $V_a^{new} := V_a \setminus V_1^{new}$. The new zone partitioning $V^{new} := \{V_a^{new} : V_a \in V'\}$ consists of zones which remain connected if they are connected in V'. Because V_1^{new} is connected in the new zone partitioning, we have more connected zones in V^{new} than we have in V'. (Zones which get empty are treated as connected zones.) To show that the objective value does not increase, it is sufficient to consider $C_{max}(V_1^{new}, V_a^{new})$ for all zones $V_a^{new} \in V^{new}$ as V_1^{new} is the only zone which becomes bigger.

$C_{max}(V_1^{new}, V_1^{new}) \geq C_{max}(V_1, V_1)$, but with Lemma 1 we know that these costs do not effect the objective function. For all other zones V_a we have $C_{max}(V_1^{new}, V_a^{new}) \leq C_{max}(V_1, V_a)$ which completes the proof.

As the next theorem shows, it is possible to sharpen the result of Theorem 2.

Theorem 3 *In any linear graph $G = (V, E)$ we can find an optimal zone partitioning in which there is at most one zone V_a not connected, and for that zone we have $V_a = \{1, n\}$.*

Proof: Suppose V' is a optimal zone partitioning of the graph G. Because of Theorem 2, there can be at most one unconnected zone V_1. That zone fulfills $\{1,n\} \subseteq V_1$. Similar to the proof of Theorem 2, it is possible to construct a new zone partitioning V^{new}, which contains $V_1 = \{1,n\}$ as the only unconnected zone. To prove Theorem 3, we show that the objective value b_{max} of V^{new} is equal or better than that of V'. This is done in two steps: First, it can be seen that zone V_1 cannot contain two neighboured stations i and i+1 for any $i \in \{1,2,...,n-1\}$. The second step then proves that the number of connected components of zone V_1 can be at most two.

<u>Example 3:</u> We consider two linear graphs with zone partitionings of two and three zones. (See Figure 3 to illustrate that example.) In G_1 the optimal solution consists of one zone $V_1 = \{1,n\}$ which is not connected and one connected zone V_2. In the optimal solution of the graph G_2 all zones are connected.

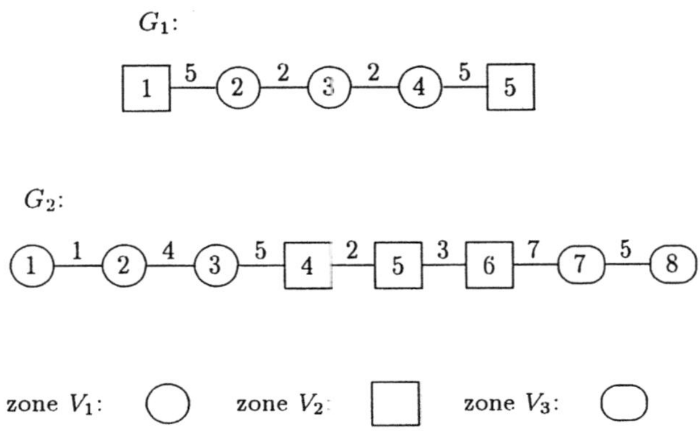

Figure 3. Optimal zonings in linear graphs

If we require, that all zones must be connected, then we get the *connected zone design problem*. We now will show that it is possible to solve the general problem in linear graphs by solving two connected zone design problems. For that we need one more lemma:

Lemma 2: *Let $G = (V,E)$ be a linear graph, $V_1 = \{1,n\}$ and $V' = \{V_2, V_3, \ldots, V_{m-1}\}$ be a zone partitioning of connected zones on $V \setminus \{1,n\}$. Then:*

1. $C_{max}(V_1, V_1) = 0$
2. $\max\{C_{max}(V_1, V_k): k=2,..m\} = \frac{1}{2}\max\{d_{2n}-d_{12}, d_{1\,n-1}-d_{n-1\,n}\}$ *that means the costs caused by V_1 are not influenced by the partition on $V \setminus \{1,n\}$.*

The proof of that lemma is similar to the proofs befor and can be found in [4].

To solve the general zone design problem in linear station networks with the help of an algorithm for the connected zone design problem, we get the following algorithm:

Algorithm

- **Given:** algorithm for the connected zone design problem in linear station graphs.
- **Input:** linear station network with weights $d_{i\ i+1}$, m= planned number of zones.
- **Output**: optimal zone partitioning $V' := \{V_1, V_2, \ldots, V_m\}$

1. Calculate the optimal zoning for the connected problem with objective value W_1
2. Calculate $a_1 := \frac{1}{2} (d_{2n} - d_{12})$ and $a_2 := \frac{1}{2} (d_{1\ n-1} - d_{n-1\ n})$
3. If $a_1 \leq W_1$ and $a_2 \leq W_1$ then goto 4. If not stop: Zoning of step 1 is optimal.
4. Remove stations 1 and n and calculate the optimal zoning with connected zones in the reduced network. Let W_2 be its objective value.
5. If $W_1 \leq W_2$ then stop: Zoning of step 1 is optimal, else $W_2 \leq W_1$ stop: Zoning of step 4 together with zone $V=\{1,n\}$ is optimal with objective value $W_3 := \max\{a_1, a_2, W_2\}$.

The correctness of that algorithm follows from Lemma 2.

To solve the problem in linear station networks quickly by any integer programming algorithm, we give a maxlinear integer programming formulation with decision variables h_{kl} for the connected zone design problem with n stations and m zones in linear graphs. For that we look at the edges (which should be numbered from 1 to n-1) of the linear graph and define:

$$h_{kl} := \begin{cases} 1 \text{ if edge k is in zone } V_l \\ 0 \text{ otherwise} \end{cases} \qquad c_{kl}^{ij} := \begin{cases} d_{kk+1} \text{ if } l \in \{i, j\} \\ 0 \text{ otherwise} \end{cases}$$

$$h := (h_{11,} h_{12}, \ldots, h_{1m}, h_{21,} h_{22}, \ldots, h_{2m}, \ldots\ldots, h_{n-1n})$$

$$c^{ij} := (c_{11}^{ij}, c_{12}^{ij}, \ldots, c_{1m}^{ij}, c_{21}^{ij}, c_{22}^{ij}, \ldots, c_{2m}^{ij}, \ldots\ldots, c_{n-1m}^{ij})$$

$$\min \{ \max \{c_{ijh} : i < j\}$$

s.t.

$$\sum_{j=1}^{n} h_{ij} \leq 1 \text{ for all } i$$

$$\sum_{i=1}^{n-1} \sum_{j=1}^{m} h_{ij} \geq n - m$$

$$h_{ij} + h_{kj} - h_{i+1j} \leq 1$$

$$\text{for all } j \in \{1, \ldots, m\},\ i,k \in \{1, \ldots, n-1\} \text{ with } i < j$$

$$h_{ik} - h_{i+1k} + h_{i+1l} \leq 1$$

$$\text{for all } k,l \in \{1, \ldots, m\} \text{ and for all } i \in \{1, \ldots, n-1\}$$

$$h_{ij} \in \{0,1\} \text{ for all } i \in \{1, \ldots, n-1\},\ j \in \{1, \ldots, m\}$$

The first restriction makes sure that each edge belongs to at least one zone. If restrictions 1 and 2 do hold together, the correct number of edges is left out. The third restriction is necessary to get connected zones. Because of restriction 4, edges between two zones are left out. Restrictions 1,2, and 4 together produce a solution in which each station belongs to exactly one zone.

4. Heuristics and Application

Greedy heuristics (in general graphs) start with a zone partitioning in which each zone corresponds to a single station. In each step two zones are combined to a new one until the planned number of zones is attained. If a value function f is available which gives the increase of the objective function when combining two zones X and Y we get the following algorithm.

Algorithm

- **Given:** f(X,Y) value function for combining zones X and Y
- **Input:** Graph G with distance tariff d_{ij}, and a number m of planned zones
- **Output**: Zone partitioning $V' := \{V_1, V_2, \ldots, V_m\}$

1. $V_i := \{i\}$ for all stations i, $V' := \{V_i : i \hat{I} V\}$
2. Choose $X,Y \in V'$ such that f(X,Y) is minimum
3. Define $NEW := X \setminus Y$, $V' := V' \setminus \{X,Y\} \cup \{NEW\}$, and update the distances of the new zones
4. If $|V'|=m$ then STOP. Else goto step 2

To get a good algorithm it is necessary to find a good value function f. As an example we show the value function f which gives a minimal increase in the objective function b_{max} in each iteration:

$$f(X,Y) = \max\{C_{max}(X \cup Y, X \cup Y), \max_{C \in V' \setminus \{X,Y\}} C_{max}(C, X \cup Y)\}$$

For b_1 and b_2 similar value functions can be derived. The resulting greedy-heuristics G_{max}, G_1, and G_2 and the single linkage (SL) and complete linkage (CL) algorithm (see e.g. [1]) were applied to the data from a regional public transportation company in Germany serving 345 stations. We compared the results for 10,11,...,25 zones, see [3]. The following tables show a partial list of our results. The objective values $b_{max}(G,G')$, $b_1(G,G')$, and $b_2(G,G')$ of zone designs were computed by algorithms G_{max},G_1,G_2,SL, and CL. The results for m=10,18,25 zones are given. The objective values b_1 and b_2 are scaled to the number of possible tours. All values are in Deutschmark per monthy ticket.

G_{max}	b_{max}	b_1	b_2
m= 10	16	4.04	25.49
18	10	2.89	13.07
25	9	2.22	7.93

SL	b_{max}	b_1	b_2
m= 10	24	7.88	94.00
18	23	6.46	68.08
25	21	4.19	30.68

G_{max}	b_{max}	b_1	b_2
m= 10	21	3.81	27.62
18	16	2.31	11.00
25	16	1.88	7.96

CL	b_{max}	b_1	b_2
m= 10	20	4.39	30.24
18	12	3.14	15.59
25	12	2.42	9.31

G_{max}	b_{max}	b_1	b_2
m= 10	26	7.10	82.58
18	26	6.12	67.47
25	26	5.74	63.02

As indicated by that partial list of results it turned out that G_{max} and G_1 produced the best results with respect to all three objective functions.

References

[1] B.S. Duran and P.L. Odell. Cluster Analysis- A Survey. Lecture Notes in Economics and Mathematical Systems 100. Springer, Berlin-Heidelberg-New York, 1962

[2] R.W. Floyd. Algorithm 97: Shortest path. Communications of the ACM, 5:345, 1962

[3] H.W. Hamacher and A. Schöbel. On fair zone designs in public transportation. Lecture Notes in Economics and Mathematical Systems 430, 8-22, 1995

[4] A. Schöbel, cited as A. Schumacher. Methoden der kombinatorischen Optimierung in der Tarifplanung im öffentlichen Personennahverkehr. Master's thesis, Universität Kaiserslautern, 1994

[5] K. Sarkadi and I. Vincze. Mathematical Methods of Statistical Quality Control. Academic Press, New York-London, 1974

[6] S. Warshall. A theorem on boolean matrices. Journal of the ACM, 9, 1962.

MULTICRITERIA EVALUATION MODEL OF PUBLIC TRANSPORT NETWORKS

Maurizio Bielli and Massimo Gastaldi

Istituto di Analisi dei Sistemi ed Informatica
National Research Council, Rome, Italy

Pasquale Carotenuto

Progetto Finalizzato Trasporti 2
National Research Council, Rome, Italy

This paper focuses on the use of multicriteria methods as a tool for adequate decision-making in urban transportation management aiming at reorganising bus transit system. In particular, it presents the main problems faced in the topic of multicriteria evaluation models, taking into account the different objectives and impacts, considering appropriate performance indicators related to environment, safety, efficiency, quality, etc. This analysis has been performed on the urban network of Parma, a medium size Italian city, assuming as reference situation the existing public transit network and proposing three alternative networks.

1. Introduction

Since the life quality in urban centers strictly depends on the efficiency of urban public transit, this topic has recently been the subject of some studies. The demographic and urban growth with the related increase of industrial, commercial and service activities, determine a growing mobility demand not supported by well-organized public transit supply. In fact, old networks have been updated without considering the system complexity, by adding and overlapping new links to the old ones not taking into account the high cost of utilized human and physical resources and the system functionality for the users. Thus, the necessity to support policy makers in choosing optimal decisions in order to plan and increase urban public transit is one of the main aims in the management of urban system. In fact, the interest of researchers is now to collect information on the planning problem, to define the various choice strategies which may be adopted, to specify in an operational way the planning objectives of decision-makers, to develop an evaluation procedure appropriate both for the decision problem and for the available information, to present optimal or at least reasonable solutions and to test the choice actually made by the decision-makers with respect to partial outcomes of the plan during the process of implementation.

In this paper two multicriteria evaluation methodologies (Utility Value Analysis [1] and the Electre method [2]) are applied on an Italian real network and the obtained results are presented and compared. These analysis have been performed on the urban network of Parma, assuming as reference situation the existing public transit network and proposing three alternative networks. The first one is planned with respect to the demand whereas the remaining two alternatives depend on the physical characteristic of the proposed networks.

2. Project Evaluation

To design an efficient transportation system on the basis of possible alternatives, it is necessary to individuate the following steps:

1) Analysis of existing situation (survey and evaluation of present network called network 0) and definition of alternative project solutions (formulation of realizable alternative projects able to replace the present one: network 1, network 2, network 3).
2) Objectives individuation: reorganization of urban bus transit network with high values of performance, evaluated considering appropriate performance indicators. In this way the demand is static, in the sense that its variations could be considered negligible both during the required time for the application of the project and for the repercussions on the system (growing demand due to the project realization, O-D changes, etc.).

3) Definition of the mathematical model representing the studied transportation system.
4) Evaluation of alternative projects: simulation of the transportation system on the basis of the available alternatives.
5) Optimal project selection: comparison among alternative projects through the evaluation of indicators that best describe the overall level of performance of transit bus system in order to select the optimal project and so the optimal network. For this purpose, the Excel worksheet using Utility Value Analysis and Electre methodologies has been used.

3. Simulation Procedure

Bus transportation demand is given by the number of users utilizing the bus service in a prefixed time interval. Since we must verify the characteristics of the networks with the higher level of demand, the peak-hour interval has been considered. The transit bus network has been defined considering the following three phases: area delimitation, area division (zones) and graph extraction.

In the first phase, we delimit the geographical area containing the effects produced by the available projects aiming at reorganizing the urban bus transit system. Thus, we considered as area of study the city center and the immediate suburbs. The external areas have been considered only for their interconnections with the studied area. The aim of the second phase is that of identifying with a point, called centroid, the starting and ending points of pedestrian movements in a sub-area (zone). We delimit 100 zones in accordance with the census area of Parma facilitating the sample extraction for the demand estimation. In the last phase, we determine the significative positions of users for the definition of graph nodes and links. The nodes can represent a real physical point in the area of study (i.e. crossing point) or the condition of the user in the transportation system (i.e. step arrival, the time instant when the user really utilizes the bus service), and the links can represent an effective connection or a wait time.

Now, it is possible to define the alternative networks (projects). The first one (network 1) is demand-oriented; it considers the user needs utilizing an appropriate O-D matrix and creating suitable paths following the required movements. The second and the third network are developed considering the physical characteristics of the proposed networks: in the former we used a circular scheme completed with a set of crossing lines assuring the existence of overlapping points for interconnections. The latter used a scheme with half and radial rings.

Once the service supply has been defined, it is necessary to describe the path choice model [3]. We utilized a behavioural model; the user choices occur in two different time instants: before leaving and during the travel. In the first case, the user utilizes his information on alternative paths (consumptive choice), whereas in the second case he can vary the first choice during the travel in accordance with additional information or with contingent events (adaptive choice). The former choice is based on a compensatory model: the user evaluates the attribute utilities of available alternatives in order to perform his choice. The latter is based on casual events during the travel; in this way the utilized path can be considered as an aleatory variable.

Moreover, two hypotheses have been assumed:

a) since the time cumulated distance of the lines useful to the user is lower than 12 minutes, we can assume the randomness of the user stop arrival instant.

b) the randomness of the vehicle stop arrival instant.

Considering these two hypotheses, the adaptive choice will be referred only to the choice of the vehicle on a line belonging to a prefixed user useful set.

4. Performance Indicators

Evaluation represents an integral part of service management with respect to the design of new services or the improvement of an existing one. Thus, to provide an efficient performance and quality evaluation of traffic/transport services, it is necessary to consider many other typical social, political and geographical factors in addition to the basic technological-economic factors.

A large body of literature has emerged dealing with a wide set of measures (indexes, variations, ratios, etc.) of performance for transportation services [4][5]. This great availability of performance measures was followed by a number of research projects designed to reduce the set of measures to a manageable size. There is a general agreement that main impacts fall into three categories: efficiency, effectiveness and quality.

In classic economic terms, a production process is efficient if it is not possible to augment any output without increasing any input or if it is not possible to decrease any input without augmenting any other input to obtain a prefixed output. However, a great number of efficiency measures appears to be inappropriate for choosing among transportation options. A further complication is introduced by an ambiguity relating to the definition of system output. In fact, in transportation service perspective, output can be measured in terms of the amount of services available (e.g. vehicle-km., vehicle-hours, etc.) or in terms of the amount of services used (e.g. passenger-km. etc.). In general, efficiency measures indicate the performance of the system in producing service at minimum cost and are related to quantities of output produced. Cost per vehicle-hour is a basic measure of the efficiency of a service provided, which reflects both the cost of inputs such as labor and fuel and the quantity of inputs used to produce the service. Another typical indicator of transit system efficiency is the vehicle-hours per employee. The number and cost of accidents are also an efficiency measure because accidents represent an unproductive use of resources.

The effectiveness performance indicators measure service (the proportion of the service area's households that have access to the system). The percentage of households served is a basic indication of the coverage of the service area provided by a transit system. In fact, it typically serves most but not all the households in the political jurisdiction or service area within which it operates, depending on the availability and density of transit stops. For these reasons, measures normally used to test system efficiency combine efficiency with effectiveness elements; in this way, it is possible to analyze how the system really serves the users besides to investigate how the system utilizes input to produce service. Effectiveness measures indicate how well a transit system works and relate the quantity of output consumed with respect to the number of passengers. In particular, the cost per passenger indicates the average total operating cost for providing service to a passenger and the revenue per passenger is calculated by dividing system revenue by total passengers carried. From the passenger's point of view, the most effective system is one that requires no transfers. From the system's perspective, it is not possible to tailor the service so that everyone gets a one-vehicle ride from origin to destination.

Methodology for measurement of quality in the service sector is a controversial topic. Quality evaluation for an urban transportation system means comparing the considered service level with the competitor ones.

Moreover, urban and metropolitan areas are very complex and decision making is characterized by a large number of conflicting goals and objectives postulated by various groups; then it is necessary to take into account the decision maker perspective.

Table 1 shows the set of indicators utilized in numerical results to evaluate the available alternatives for the improvement of bus transit service in Parma. All of them are referred to the same time unit, the peak-hour, and calculated utilizing vehicle capacity (116 per vehicle), number of users per line, frequency, length of the line and number of vehicles per line.

Performance indicators	**Symbol**		
Vehicle number	Vn	▼	E
Vehicle capacity per Km/Vehicle number	CKM/Vn	▲	E
Vehicle number/Frequency	Vn/F	▼	E
Users	U	▲	e
Users/Vehicle capacity per Km	U/CKM	▲	E
Users/Network extension	U/L	▲	e
Users/Vehicle number	U/Vn	▲	E
Users per Km	UKM	▲	e
Users per Km/Vehicle capacity per Km	UKM/CKM	▲	E
Users per Km/Network extension	UKM/L	▲	e
Users per Km/Vehicle number	UKM/Vn	▲	E
Number of stops/Number of links	Ns/Nlk	▲	e
Number of links	Nlk	▼	E
Number of lines	Nln	▼	E
Average number of transfers	Nt	▼	Q
Average time of wait	Tw	▼	Q
Average time of travel	Tt	▼	Q
Average time of moving	Tm	▼	Q
Number of pedestrian relations	Npr	▼	e
Number of pedestrians	Np	▼	e
Crowd index (passengers ≤ 20)	CI1	▲	Q
Crowd index (20 < passengers < 75)	CI2	▲	Q
Crowd index (passengers ≥ 75)	CI3	▼	Q
Equivalent pollution index	EPI	▼	Q

Effectiveness indicators	E
Efficacy indicators	e
Quality indicators	Q

The best value is the lowest one	▼
The best value is the highest one	▲

Table 1. Performance Indicators

For the efficiency indicators we assume as output unit the vehicle capacity per km and the number of vehicle per km and as resources the line frequency and the minimum number of vehicle per line. For the effectiveness indicators we assume the number of users per km as output unit. Notice that close to typical technical indicators, some environmental and crowd indices more interesting from the user point of view have been considered [6]. EPI represents the network environmental impact calculated with respect to the number of bus/km. CI1, CI2 and CI3 represent three different travel condition for the user; CI1 is the user percentage that have the possibility to be seated, CI2 represents a compromising level between firm and user giving a measure of an acceptable level of service provided by the firm and utilized by the user whereas CI3 is a discomfort measure for the user since the travel is performed with an average number of passenger >75.

5. Multicriteria Analysis and Numerical Results

The problem of decision-making with multiple criteria has been the subject of many different sciences, i.e. operation research, decision theory or regional and transportation planning [7][8][9][10]. For this reason, a lot of different evaluation and decision techniques that could be called multicriteria analysis exist. Common for all is that they are able to deal with two or more criteria measured in different units. Evaluation is basically a comparison of alternatives; this is a multistage process which encompasses the establishment of evaluation criteria, the determination of individual values and the weighted up of the different impacts. An evaluation framework is the set of evaluation tools and components used in the evaluation process. First of all the evaluation method may be fixed with the corresponding evaluation criteria and impact values. Within multicriteria decision analysis different criteria and impacts measured with monetary or non-monetary values are considered in aggregate or disaggregate form. Several different techniques can be applied according to the type of criteria or impacts and to their aggregation, as dominance analysis, compatibility analysis, cost-effectiveness and utility analysis.

In a Multicriteria evaluation, the alternatives are compared with respect to a set of criteria corresponding to the policy objectives represented by performance indicators. The first major step in the evaluation process is the definition of the impact matrix [11]; the impacts e_{ij} related to the individual criteria c_j $(j=1,2,\ldots,J)$ and the individual alternatives a_i $(i=1,2,\ldots,I)$ then establish the impact matrix.

Performance Indicators	Weight user	Weight firm	Network 0	Network 1	Network 2	Network 3
Vn	7	10	81.000	70.000	78.000	73.000
CKM/Vn	7	10	891.931	876.456	933.852	857.238
Vn/F	7	5	0.533	0.507	0.534	0.490
U	5	10	21547.000	21569.000	22134.000	21085.000
U/CKM	5	5	0.298	0.352	0.304	0.337
U/L	7	5	131.818	162.033	137.427	146.918
U/Vn	10	10	266.012	308.129	283.769	288.836
UKM	8	8	43931.635	45222.435	42523.580	43200.480
UKM/CKM	5	10	0.608	0.705	0.850	0.690
UKM/L	7	5	268.761	324.700	264.023	301.017
UKM/Vn	7	8	542.366	617.463	545.174	591.787
Ns/Nlk	7	7	0.152	0.161	0.153	0.159
Nlk	7	7	3016.000	2843.000	3006.000	2892.000
Nln	10	5	22.000	18.000	20.000	20.000
Nt	10	8	0.690	0.740	0.910	0.690
Tw	10	9	651.920	717.680	790.320	740.540
Tt	10	7	1385.440	1457.890	1523.740	1471.050
Tm	10	10	2163.030	2213.710	2311.040	2258.480
Npr	10	8	3.000	2.000	3.000	2.000
Np	8	8	44.000	28.000	60.000	28.000
CI1	5	5	0.079	0.062	0.073	0.068
CI2	10	5	0.486	0.452	0.426	0.393
CI3	10	5	0.435	0.486	0.501	0.539
EPI	10	5	3079.550	2777.380	3194.860	2945.610

Table 2. Criterion weights and impact matrix

A weight must be designed for each criterion reflecting the normative judgement of the decision-maker. Thus, the evaluation can be summarized in the following five steps:

1) Project alternatives individuation.
2) Objectives or criteria individuation.
3) Calculation of impacts matrix.
4) Transformation of impact measures in utility measures.
5) Optimal project selection.

The project alternatives are incompatible in the sense that the realization of each alternative precludes the realization of other ones. A group of indicators has been defined for each criterion weighted on the basis of its importance. The relative impact matrix is shown in Table 2. In the simulation two decision makers have been considered: the firm manager, more interested to the network technical characteristics and the user.

In order to perform the multicriteria analysis we should consider a value function for each criterion (performance indicator), transforming the initial impact in a score easier to interpret. For this purpose, we used a value function f_j in the following forms:

a) when utility grows while impact decreases (i.e. cost case):

$$u_{ij} = \frac{\min(\mathbf{e}_i)}{e_{ij}}$$

for each alternative a_i related to criterion c_j;

b) when utility grows while impact increases (i.e. benefit case):

$$u_{ij} = \frac{e_{ij}}{\max(\mathbf{e}_i)}$$

for each alternative a_i related to criterion c_j.

Performance Indicators	Rel. Weight user	Rel. Weight firm	Network 0	Network 1	Network 2	Network 3
Vn	0.036	0.057	0.864	**1.000**	0.897	0.959
CKM/Vn	0.036	0.057	0.955	0.939	**1.000**	0.918
Vn/F	0.036	0.029	0.919	0.966	0.918	**1.000**
U	0.026	0.057	0.973	0.974	**1.000**	0.953
U/CKM	0.026	0.029	0.847	**1.000**	0.864	0.957
U/L	0.036	0.029	0.814	**1.000**	0.848	0.907
U/Vn	0.052	0.057	0.863	**1.000**	0.921	0.937
UKM	0.042	0.046	0.971	**1.000**	0.940	0.955
UKM/CKM	0.026	0.057	0.715	0.829	**1.000**	0.812
UKM/L	0.036	0.029	0.828	**1.000**	0.813	0.927
UKM/Vn	0.036	0.046	0.878	**1.000**	0.883	0.958
Ns/Nlk	0.036	0.040	0.944	**1.000**	0.950	0.988
Nlk	0.036	0.040	0.943	**1.000**	0.946	0.983
Nln	0.052	0.029	0.818	**1.000**	0.900	0.900
Nt	0.052	0.046	**1.000**	0.932	0.758	**1.000**
Tw	0.052	0.051	**1.000**	0.908	0.825	0.880
Tt	0.052	0.040	**1.000**	0.950	0.909	0.942
Tm	0.052	0.057	**1.000**	0.977	0.936	0.958
Npr	0.052	0.046	0.667	**1.000**	0.667	**1.000**
Np	0.042	0.046	0.636	**1.000**	0.467	**1.000**
CI1	0.026	0.029	**1.000**	0.785	0.924	0.861
CI2	0.052	0.029	**1.000**	0.930	0.877	0.809
CI3	0.052	0.029	**1.000**	0.895	0.868	0.807
EPI	0.052	0.029	0.902	**1.000**	0.869	0.943
	1.000	1.000				

Table 3. Criterion relative weights and utilities

Table 3 shows the relative weights and the matrix of performance indicators re-evaluated through the utility functions. From an efficiency point of view, notice that no network is dominated by another one.

By examining this matrix we should obtain the indications necessary to choose the best alternative. On this subject, the current literature proposes different approaches; in our numerical results, we apply the Utility Value Analysis (UVA) and the Electre method by using the worksheet Excel. The stages of a Quantitative Utility Value Analysis are presented in Figure 1.

The Utility Value Analysis (UVA), a Multicriteria Analysis technique which aggregates the multiple criteria, is here applied. For each alternative a_i $(i=1,2,......,I)$ and criterion c_j $(j=1,2,......,J)$ the criterion-related impact e_{ij} has to be determined. These impacts e_{ij} are measured in various units (i.e. money, time, etc.). In order to make them comparable they are firstly transformed into a single unit utilizing the value function f_j $(j=1,2,......,J)$ determined for each individual criterion c_j. These functions transform the initial impact e_{ij} into the so called criterion utilities u_{ij} interpreted as goal-achievement scores. The form of value function depends on the preferences of the decision-makers.

In a second stage the individual criteria are weighted assigning a weight w_j to each criterion c_j; these weights reflect the relative importance of the criterion (the weights sum is usually equal to one). Similarly to the value function, the weights assigned to the criteria depend on the preferences of the decision makers. The weighted transformed criterion utility then reflects the partial utility (Z_{ij}).

Finally, these partial utilities have to be aggregated for each alternative. Due to the fact that preferential independence of the criteria are assumed, the aggregation can be performed by simple sum. Thus the overall utility (N_i) of alternative a_i is the sum of the weighted and transformed impacts. Where an alternative is superior to all others, no further analysis is required.

The above mentioned Tables 1-3 allow to follow the different stages to which performance indicators must undergo through the Excel worksheet in order to highlight the best solution among the proposed ones. Table 4 and 5, showing the final values of utilized indicators considering firm manager and user point of view, clearly indicates that the best alternative is Network 1, that is the demand-oriented Network presenting the higher aggregated utility index as shown in Figure 2.

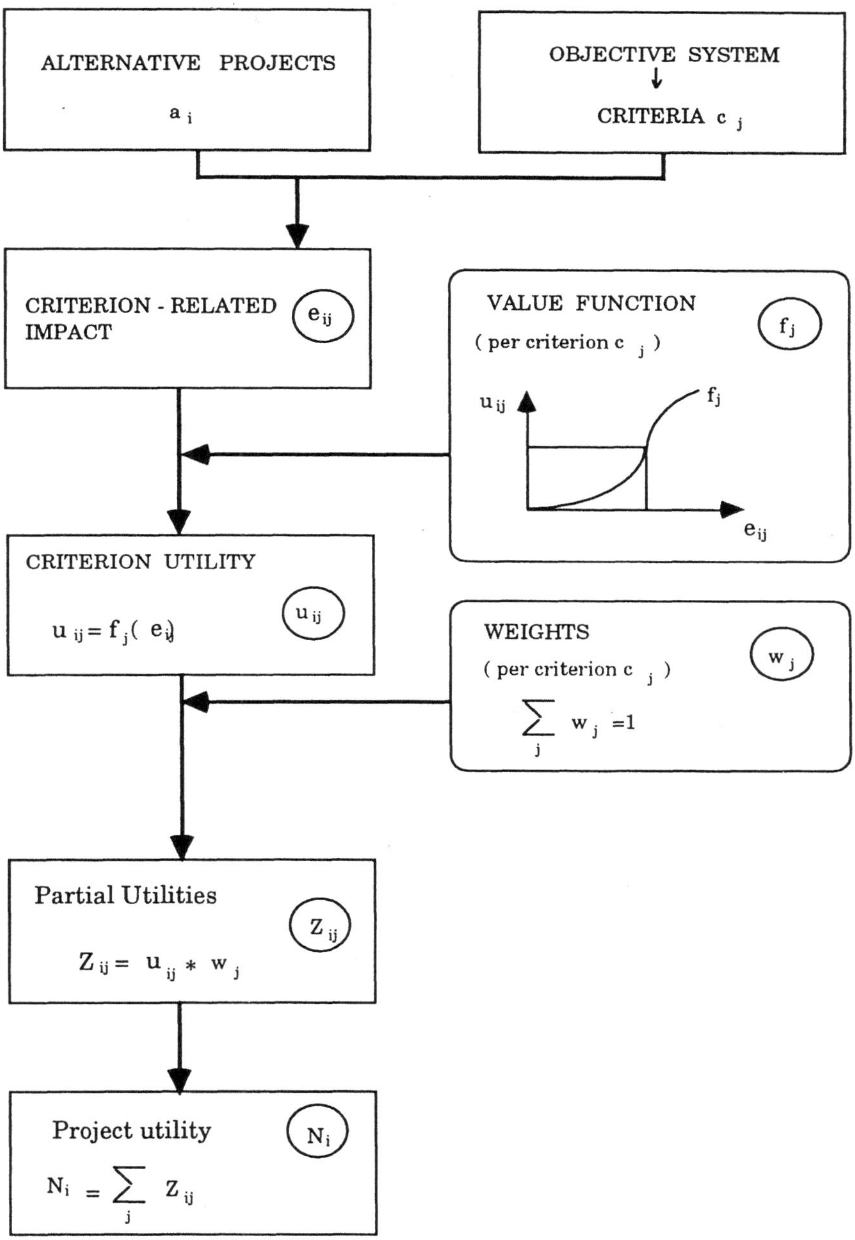

Figure 1. Stages of Quantitative Utility value analysis

Performance Indicators	Network 0	Network 1	Network 2	Network 3
Vn	0.049	0.057	0.051	0.055
CKM/Vn	0.055	0.054	0.057	0.052
Vn/F	0.026	0.028	0.026	0.029
U	0.056	0.056	0.057	0.054
U/CKM	0.024	0.029	0.025	0.027
U/L	0.023	0.029	0.024	0.026
U/Vn	0.049	0.057	0.053	0.054
UKM	0.044	0.046	0.043	0.044
UKM/CKM	0.041	0.047	0.057	0.046
UKM/L	0.024	0.029	0.023	0.026
UKM/Vn	0.040	0.046	0.040	0.044
Ns/Nlk	0.038	0.040	0.038	0.040
Nlk	0.038	0.040	0.038	0.039
Nln	0.023	0.029	0.026	0.026
Nt	0.046	0.043	0.035	0.046
Tw	0.051	0.047	0.042	0.045
Tt	0.040	0.038	0.036	0.038
Tm	0.057	0.056	0.053	0.055
Npr	0.030	0.046	0.030	0.046
Np	0.029	0.046	0.021	0.046
CI1	0.029	0.022	0.026	0.025
CI2	0.029	0.027	0.025	0.023
CI3	0.029	0.026	0.025	0.023
EPI	0.026	0.029	0.025	0.027
	0.896	**0.962**	**0.878**	**0.934**

Table 4. Partial and project utilities (firm)

Performance Indicators	Network 0	Network 1	Network 2	Network 3
Vn	0.032	0.036	0.033	0.035
CKM/Vn	0.035	0.034	0.036	0.033
Vn/F	0.034	0.035	0.033	0.036
U	0.025	0.025	0.026	0.025
U/CKM	0.022	0.026	0.022	0.025
U/L	0.030	0.036	0.031	0.033
U/Vn	0.045	0.052	0.048	0.049
UKM	0.040	0.042	0.039	0.040
UKM/CKM	0.019	0.022	0.026	0.021
UKM/L	0.030	0.036	0.030	0.034
UKM/Vn	0.032	0.036	0.032	0.035
Ns/Nlk	0.034	0.036	0.035	0.036
Nlk	0.034	0.036	0.034	0.036
Nln	0.043	0.052	0.047	0.047
Nt	0.052	0.049	0.039	0.052
Tw	0.052	0.047	0.043	0.046
Tt	0.052	0.049	0.047	0.049
Tm	0.052	0.051	0.049	0.050
Npr	0.035	0.052	0.035	0.052
Np	0.027	0.042	0.019	0.042
CI1	0.026	0.020	0.024	0.022
CI2	0.052	0.048	0.046	0.042
CI3	0.052	0.047	0.045	0.042
EPI	0.047	0.052	0.045	0.049
	0.901	**0.965**	**0.866**	**0.931**

Table 5. Partial and project utilities (user)

Notice that two of the proposed alternatives are better than the existing one. Moreover, although some networks give the higher values for some specific criteria, they are not the best as a whole as shown in table 3.

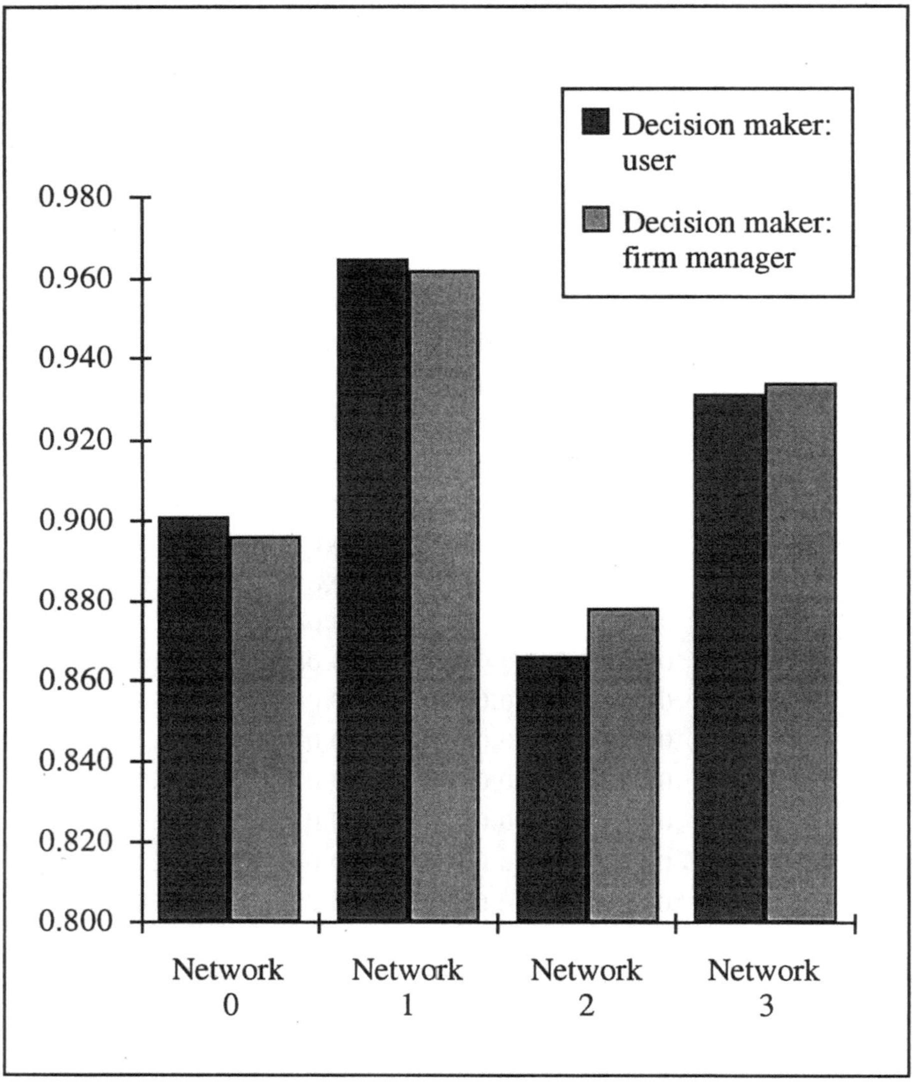

Figure 2. Project Utlities

Finally, the presented results have been compared with those obtained applying the well known multicriteria method Electre (fig. 3) This comparison demonstrates that the two considered methods provide similar results. In fact, Network 1 represents the best solution in both analysis although UVA method is based on calculation of utility values whereas Electre is based on the computation of global concordance and discordance indices (fig. 4-5). The first ones represent the difference between the extent to which the considered network dominates all others and the extent to which all other plans dominate the network under investigation. A positive value (network 1) indicates the importance of network in the selection procedure; a zero or next to zero value (network 3) implies a neutral position and a negative value (network 0 and network 2) indicates that those networks may be excluded from the selection [2]. In a similar way it is possible to define the global discordance indices (a network has higher chance of being accepted the lower index becomes).

As a result of the uncertainty implicit in the preference scores, a sensitivity analysis is necessary before definite conclusions can be drawn. The aim of such analysis is to check the stability of the better solution (network 1) with respect to a shift in the preference weights. The obtained results show the subtantial stability of networks ranking and how, changing preference weights, project utilities and concordance indeces are affected. In particular, with reference to a significative indicator (Nt - Average number of transfers) fig. 6 -7 show project utilities and concordance index variation with respect to different preference weights (1-10) both for Electre and UVA analysis. Notice that when all indicators weights are equal to 1 stability of networks ranking continues to hold.

The authors are currently working on comparing their results with other multicriteria techniques, on the possibility of adding new performance indicators to consider relevant topic in urban transport evaluation (i.e. pollutant emissions, urban trip quality, etc.).

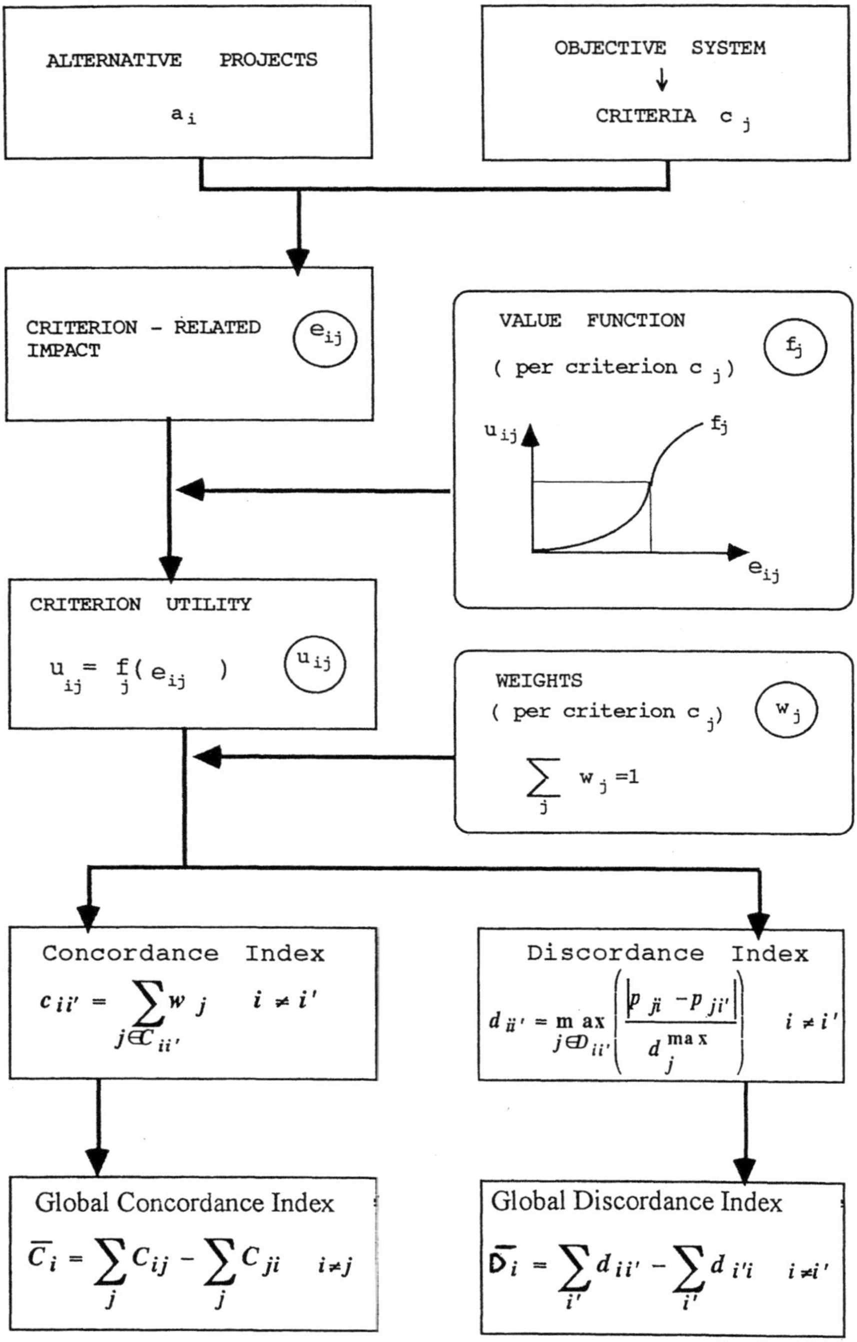

Figure 3. Stages of Electre method

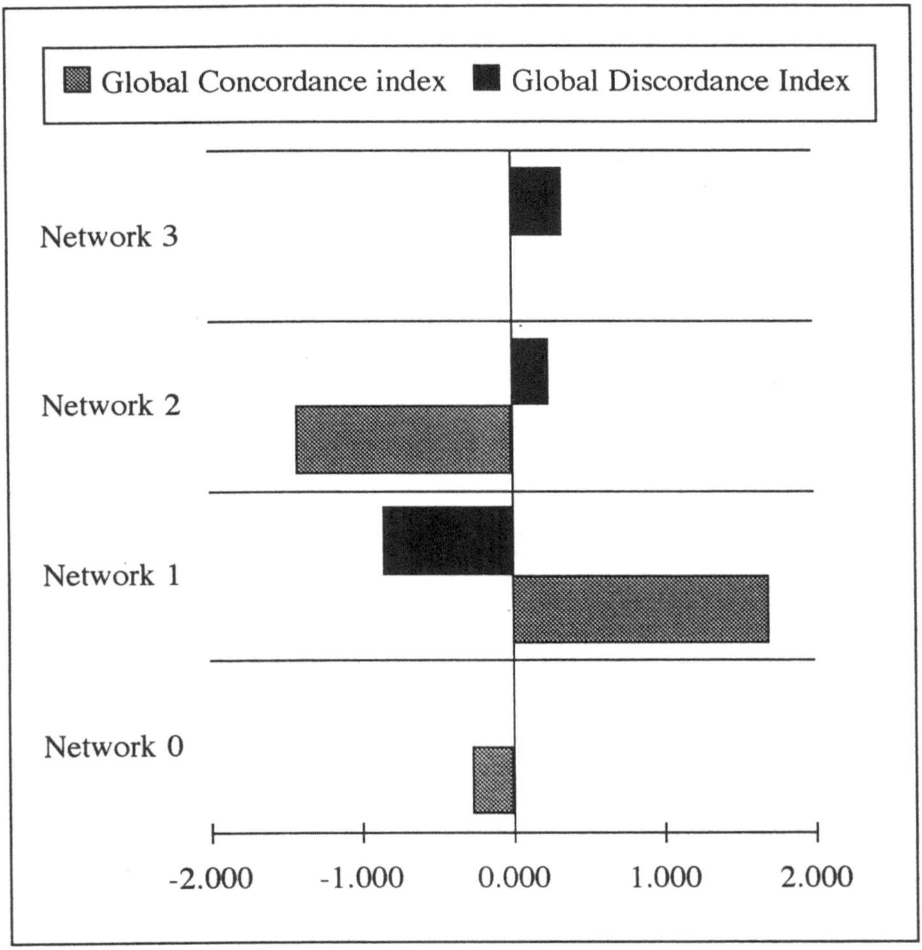

Figure 4. Results of Electre method (firm)

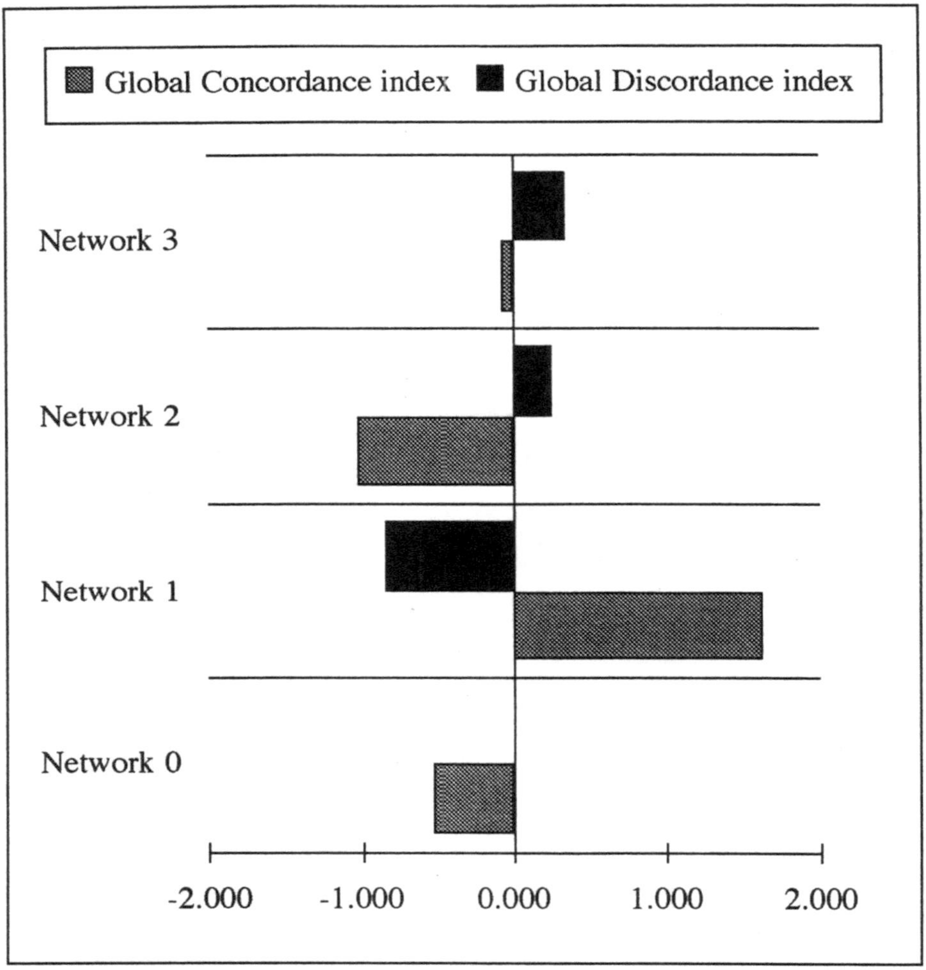

Figure 5. Results of Electre method (user)

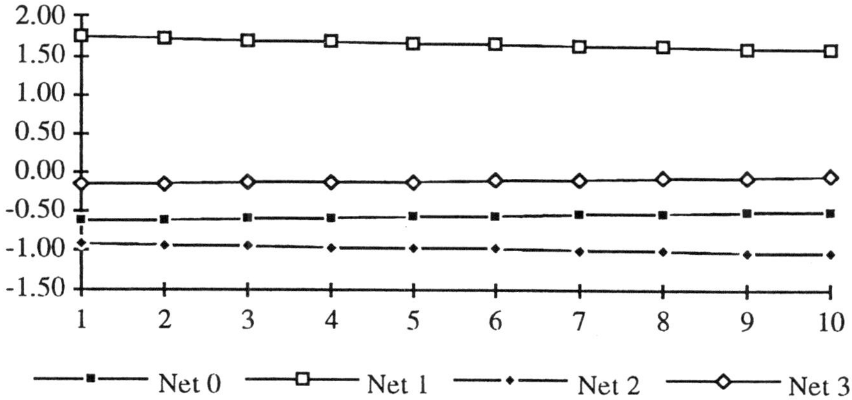

Figure 6. Sensitivity analysis (Electre, Average number of transfer, Nt)

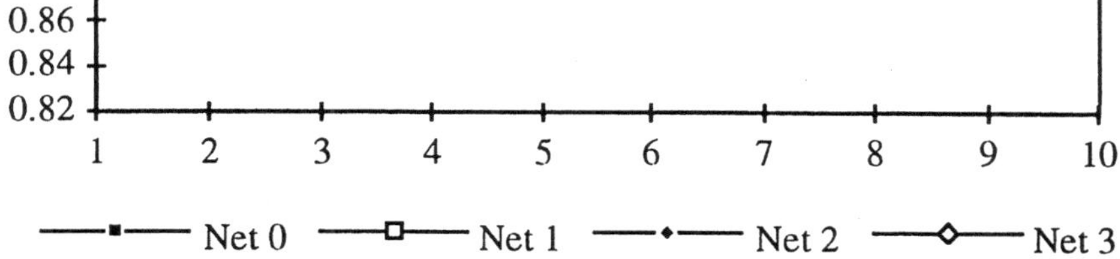

Figure 7. Sensitivity analysis (UVA, Average number of transfer, Nt)

References

[1] DRIVE PROJECT V1036, "EVA Manual - Evaluation Process for Road Transport Informatics", 1991.

[2] Nijkamp P. and van Delft A., "Multicriteria analysis and regional decision-making" *in Studies in applied regional science, Vol. 8, Martinus Nijhoff Social Sciences Division*, Leiden, the Netherlands, 1977.

[3] Cascetta E. and Nuzzolo A., "Un modello di equilibrio domanda-offerta per la simulazione dei sistemi di trasporto nelle aree urbane di medie dimensione", *Proceedings of V Conference of PFT-CNR*, Napoli, Vol. 1, 1, 141-176, 1988.

[4] Henderson G., Kwong P. and Adkins H., "Regularity indexes for evaluating transit performance", *TRB Record* 1297, 1991.

[5] Bielli M., Gastaldi M. and Sica F., "Performance Evaluation in the Management of Urban Transportation Services", *Proceedings of 1st Meeting of the EURO Working Group on Urban Traffic and Transportation*, Landshut, Germany, 1992.

[6] Rostirolla P., "Tecniche di aiuto alla decisione per la V.I.A.: un'analisi comparata su di un caso studio", *Proceedings of Second Conference of PFT2-CNR*, Vol. 1, 157-172, 1995.

[7] Bielli M., Carotenuto P. and Gastaldi M., "Multicriteria evaluation of integrated urban traffic management system", *Proceedings of 2nd Meeting of the EURO Working Group on Urban Traffic and Transportation*, Paris, France, 337-348, 1993.

[8] Nijkamp P. and Voogd H., "Conservazione e sviluppo: la valutazione nella pianificazione fisica", *Girard L.F.* (ed.), F. Angeli Editore.

[9] Voogd H., "Multicriteria Evaluation for Urban and Regional Planning", *Pergamon*, London, 1984.

[10] Rietveld P. "Multiple Objective Decision Methods and Regional Planning", *North-Holland*, Amsterdam, 1980.

[11] Bobinger R., Flowerdew R., Hammond T., Himanen A. and Keller H. "Context and framework of Drive transportation evaluation", *Proceedings of Drive Conference, Brussels, Elsevier*, Amsterdam, 389-412, 1991.

MULTI-OBJECTIVE APPROACH FOR DESIGNING TRANSIT ROUTES WITH FREQUENCIES

Yechezkel Israeli and Avishai Ceder

Transportation Research Institute, Civil Engineering Department, Technion-Israel Institute of Technology, Haifa, ISRAEL

Transit route design (TRD is considered the most complex and cumbersome problem across network route allocation problems. The wide range of the TRD's characteristics creates difficulties to formulate the problem uniquely. At the same time, the TRD's complexity of the NP-hard type creates combinatorial problems. The TRD problem is formulated as non-linear programming with mixed variables (continuous and integer). This formulation cannot be solved via known mathematical programming approaches and packages. This research provides a new and efficient approach to solve the TRD while dealing with both its complexity and its practical issues. The approach used has an impact on three components involved: the operator, the user, and the considered authority. The objectives of these three components do not always coincide. From the operator viewpoint, the system should minimize its expenses while, from the user perspective, the system should maximize its level-of-service. This trade-off situation creates this work's optimization framework. the multi-objective programming technique that was applied has not been used, to our knowledge, for solving the TRD problem. In fact, due to the problem complexity, the ordinary mathematical programming methods cannot be used in this technique, and therefore, a new approach is provided. This new approach is heuristic in nature and divided into two phases: (a) generation of finite sets of alternative efficient non-inferior solutions; and (b) evaluation and selection of the various solutions using multi-objective preference techniques for discrete variables "compromise programming" destination pairs of the network can be created by "column generation" procedure in former stages. This approach enables to solve the complex TRD problem. It combines mathematical programming with decision-making methods, using search and enumeration processes while performing the optimization. Thus, it is possible to encounter relatively large-scale problems (networks) which cannot be solved by other techniques.

1. Introduction

Transit route design (TRD) is considered the most complex and cumbersome problem across network route allocation problems. The wide range of the TRD's characteristics creates difficulties to formulate the problem uniquely. At the same time, the TRD's complexity of the NP-hard type creates combinatorial problems. The TRD problem is formulated as non-linear programming with mixed variables (continuous and integer). This formulation cannot be solved via known mathematical programming approaches and packages.

The importance of the improvement of an existing TRD solution is that with a relatively low investment, the whole transit operation can be significantly improved. This research provides a new and efficient approach to solve the TRD while dealing with both its complexity and its practical issues. The approach used has an impact on three components involved: the operator, the user, and the considered authority.

The objectives of these three components do not always coincide. From the operator viewpoint, the system should minimize its expenses while, from the user perspective, the system should maximize its level-of-service. This trade-off situation creates this work's optimization framework.

The optimization criteria of the operator-user-authority components consider, simultaneously, the formulated combination of: route design; timetabling (frequencies); and vehicle scheduling. The methodology used is unique in comparison to other methods, due to the consideration of daily operational elements (passenger counts, passenger load profiles, operational fleet size, etc.). This methodology enables to produce route changes and route design even for the short-range planning.

The mathematical formulation of this work is based on four objective functions -- each to be minimized. However, it is impossible to treat all functions simultaneously, and hence, multi-objective programming is being used. This multi-objective programming technique was not used, to our knowledge, for solving the TRD problem. In fact, due to the problem complexity, the ordinary mathematical programming methods cannot be used in this technique, and therefore, a new approach is provided. This new approach is heuristic in nature and divided into seven modules (see Israeli [1]): (i) creating large set of feasible routes; (ii) creating feasible transfer paths obeying a given level-of-service criterion; the outcome of these two modules which use "column generation" procedure is a Set Covering matrix; (iii) solving the Set Covering Problem (SCP): selection of a set of routes and transfers enabling connectivity among all the network nodes with minimum length of demand paths: (iv) demand assignment while considering the various origin-destination paths. This part determines the vehicle frequencies and the timetables; (v) derivation of an estimate for the required fleet size while deciding the vehicle scheduling based on a given timetable; (vi) generation of finite sets of alternative efficient non-inferior solutions; and (vii) evaluation and selection of the various solutions using multi-objective preference techniques for discrete variables ("compromise programming" procedure).

This research deals with the last two modules and their relation to the former three modules (SCP, assignment procedure and fleet size estimation). The research provides a general procedure and algorithm, based on a given covering matrix which generates "promising alternatives" (sets) for the multi-objective solution. The relationship between the different modules is as follows: while the solution of the SCP is based on the costs which are derived from the assignment procedure, the assignment procedure itself is based on the solution of the SCP. This yields an iterative procedure, the outcome of which is the "alternative generation process."

The suggested approach enables to solve the complex TRD problem. It combines mathematical programming with decision-making methods, using search and enumeration processes while performing the optimization. Thus, it is possible to encounter relatively large-scale problems (networks) with the possibility to interact with the solution method during intermediate steps. The final results of the analysis provide transit routes with frequencies (timetabling) and the framework for vehicle scheduling.

2. Problem Identification

2.1 Research Framework and Formulation

The transit planning process, aimed at efficient transport of origin-destination transit riders, includes four basic components performed in sequence: (a) Network Route Design; (b) Setting Timetables (Frequencies Calculation); (c) Scheduling Vehicles to Trips; and (d) Assignment of Drivers. In order for this process to be cost-effective and efficient, it should embody a compromise between passenger comfort and cost of service. For example, a good match between bus supply and passenger demand occurs when bus schedules are constructed so that the observed passenger demand is accommodated while the number of vehicles used is minimized.

Whereas most of the research and computer programming concern the last two components (see analysis of Ceder and Wilson [2]), few researchers have studied the interrelationship between the scheduling components and the network design element. The interrelationship exists in two directions: (i) each set of routes yields, based on the demand, a different set of frequencies and timetables, and ultimately, the required fleet size; and (ii) the operational cost derived from the scheduling components and the passenger level of service affect the search for the optimal route design while relying on a compromise between the operator and the user.

The approach presented in the present paper considers the first three components simultaneously: route design, timetabling (frequencies), and vehicle scheduling. It combines the philosophy of the mathematical programming approaches with decision-making techniques, in order to allow the user to select from a number of alternatives.

The mathematical formulation is based on four objective functions which incorporate the various operational and user perspectives:

$$\min\ Z_1 = PH$$
$$\min\ Z_2 = WH$$
$$\min\ Z_3 = EH \qquad (2.1)$$
$$\min\ Z_4 = FS$$

where:

PH= Total Passenger Hours between all origin-destination pairs (defined as passengers' riding time in a bus on an hourly basis. It measures how much time is spent by passengers on buses between the two nodes);

WH=Total Waiting Time between all origin-destination pairs (defined as the amount of time on a bus on an hourly basis. It measures how much time is spent by passengers at bus stops (nodes);

EH= Total Empty Space Hours on all routes of the network (defined as the unused seats in a bus on an hourly basis. Empty Space Hours measure to what capacity buses are used);

FS= Fleet Size (number of buses needed to provide all trips along the chosen set of routes).

The complete formulation, including the constraints of the network design problem, can be found in Israeli [1]); Ceder and Israeli [3]). The nature of the overall formulation is non-linear (non-linear and mixed integer programming). Its analog problem is the generalized network design problem described by Magnanti and Wong [4]), with an NP-hard computational complexity. Thus, conventional approaches are incapable of providing a solution even with a relatively high degree of simplification.

2.2 Characteristics and Application of the Multi-Objective Programming

Out of general mathematical formulation, a need appears for the minimizing of the four objective functions : Z_i, i=1,...,4. In fact, due to the conflict between the objective functions, it is impossible to arrive at an ideal solution incorporating a simultaneous minimization of all Zi, i=1, ...,4. This conflict exists in all problems of transferring demands in networks in both the level of transportation planning (Current et al., [5]) and in public transport planning (Janarthanan and Schneider [6]). The fault lies in the undefinable nature of a reasonable objective function, including all factors of relevance which can be commonly accepted (Steenbrink [7]). Since the four objective functions Z_1 up to Z_4 are complete and expressed in different units, it is complicated and perhaps undesirable to combine them into one unit scale without an accompanying information loss. Thus, to guarantee their separateness, we need to treat the problem as one in a multi-criteria mode. This problem entails a trade-off between the four objective functions. Hence, there can exist no optimal solution, but a variety of a compromise solution between the objective functions.

The choice of the "best" compromise can only be facilitated by the establishment of various solutions, themselves created by due process. Fricker and Shanteau [8] stress that, beyond the mathematical selection process, there needs to exist a various options set-up, not a single optimal option, to be of genuine assistance to the decision-makers in their comprehension and subsequent selection. Such decision-making tools which apply multi-objective programming in general networks are described by Ignizio [9].

In a transportation system, as there seem to be few models of multi-objective planning. Current and Min [10], while surveying such studies, noted a recent increasing usage of the multi-objective approach in planning transport problems. They determine this to the development of heuristic techniques for networks and computer hardware advancement facilitating the ability to address all angles of the multi-aspect problem.

Earlier techniques (dating from the mid-seventies) tended to consider evaluating alternatives instead of creating them.

From a review of relevant studies (see Israeli [1]), we realize that none combine a multi-objective function approach which successfully addresses all aspects of the problems of public transport route planning in its processes to the total definition of the question. Some studies describe a process of creating alternatives lacking evaluation and the usage of any multi-objective programming models ([8], [11], and [12]) .Other studies used multi-objective programming techniques. Some merely created alternatives ([13], [5]), while others evaluated their results ([14], [6], and [15]). Teodorovic and Krcmar-Nozic [16]) deal with air routes, including calculation of frequencies for the given routes. The uniqueness of this study is in its treatment of more than two objective functions (namely, three), in the process of creating alternatives. The research was based on a heuristic method to produce random creation of possible solutions, and solving a mini-max problem between the various alternatives, in order to choose the appropriate one.

The multi-objective programming problem can be classified into two differing alternative characteristic types: discrete problems and continuous problems. The discrete type problems are based on a number of alternatives from which one is preferred. However, the continuous mode problems require a model entailing decision variables, constraints and objective functions for creating suggested alternatives. Such variables may have any value from a given successive value structure (Cohon [17]).

The two stages in resolving the multi-objective programming problem are:

1) building the non-inferior solutions (the utilization curve); 2) choosing solutions from within that curve; in short, recognizing the "compromised set." The non-inferior solutions'set -- NIS -- expresses the collection of points in the objective functions' space in such a way that any improvement in one single objective function could only be achieved with a simultaneous damaging or reduction of, at least, one of the other objective functions. Thus, x^* will belong to NIS if another solution will not yield $x \varepsilon X$, so that :

$$Z_i(x) \leq Z_i(x^*) \qquad , i = 1,2,\ldots,p$$

with at least one inequality, while $Z_i(x)$ marks the value of the objective function in the point of x. The set of points belonging to the NIS is a subset of the feasible solution set x.

The choice of multi-objective technique demanded appraising the two stages: creating efficient solutions and choosing the compromise solutions whilst evaluating them. The problem when analyzed is of non-linear nature (concave), containing integer variables. These include the complex form of NP-hard. Such characteristics prevent the use of mathematical programming techniques inherent in small dimensional problems. It is therefore impossible to use known multi-objective programming techniques which deal with problems of a continuous nature (see the analysis of Teodorovic and Krcmar-Nozic [16]). The methods employed for these types of problems are based on a full mathematical programming; both for finding the utilization curve (NIS) and for the process of choosing the compromise - solution. Most of the existing models according to the available techniques tend to fit linear problems. Only recently has attention been paid to the resolution of non-linear problems (see [10], 18], and [19]). The treatment of multi-objective problems containing integers started later. Presently, only a few techniques exist (see [20], [21], and [22]). Consequently, Tzeng and Shiau [13] in their study used a simplified mode within which a simulating technique derived different values for the binaric variables.

Proceeding from this problem, it is impossible to construct a set of continuous NIS, thus a heuristic method for the creation of efficient solutions must be maintained. In this method, an absolute number of alternative solutions is created. These will be an approximation to the continuous set of the NIS. Its evaluation will be made with reference techniques to discrete multi-objective problems. This method fits both Janarthanan and Schneider's [6] and Teodorovic and Krcmar-Nozic's [16] approach. These studies define the problem of public transport evaluation as a discrete one.

Relating to the analysis of Teodorovic and Krcmar-Nozic's research (see Israeli [1] and Israeli and Ceder [23]), it was decided in our research to use a different heuristic process in order to generate a large set of efficiency points which can estimate the continuous utility curve NIS. Examining the different multi-objective techniques yields the use of the "compromise programming" technique.

3. Research Approach

3.1 Column Generation Approach

As described in section 1, a heuristic modular method is provided in order to overcome the mathematical complexity of the problem. The entire process is based on the first two modules which use a "column generation" approach to define the covering matrix of all possible travel paths in the network.

The process is based on a mapped network of routes and transfer paths. This incorporates within the network guaranteed connectivity under the route length criteria. It includes the deviation percentage allowed, using the projected minimum travel time (private cars) and a maximal of expected bus route transfers. This network is described comprehensively in Israeli [1]. However, the large set of routes and transfers is likely to contain many overlapping segments (of routes and transfers). An overlapped segment is one that serves O-D pairs that are wholly served by other routes and/or transfers. An overlapped route comprises segments which are all totally overlapping. The latter is treated as follows:

The system creates minimal set(s) of routes and their related transfers, such that connectivity between nodes is maintained and their total deviation from the shortest path is minimized. The problem is defined as a Set Covering Problem (SCP) which is hard to solve (see Minieka [24]; Syslo et al., [25]). The SCP can determine the minimal set of routes from the matrix of the feasible paths (routes and transfers). In this matrix, defined by $A = \{a^r{}_{ij}, a^{tr}{}_{ij}\}$, each row represents either a feasible route or a transfer. The "1" in the matrix is inserted whenever an O-D demand can be transported by the route or transfer, and "O" otherwise. (Terms will be defined hereafter).

The word "covering" refers here to at least one column with "1" in each row. The transfers are combined columns in the SCP matrix and therefore increase the complexity of the problem. No solution appears for this problem in the literature. The mathematical formulation of the SCP-transfer problem results in a non-linear programming with integer variables. Thus, if we let c_r be the cost of a direct route and c_{tr} the cost of a transfer (each one is referred to in a single column), then the formulation is:

$$\min\left[\sum_{r\in R} c_r x_r + \sum_{tr\in TR} c_{tr} \prod_{r:tr} x_r\right] \tag{3.1}$$

s.t.

$$\sum_{r\in R} a_{ij}^r x_r + \sum_{tr\in TR} a_{ij}^{tr} \prod_{r:tr} x_r \geq 1 \qquad \forall i, j \in N \tag{3.2}$$

$$Xr = \begin{cases} 1 & route \quad r \in R \quad in \quad the \quad solution \\ 0 & otherwise \end{cases} \tag{3.3}$$

when $a_{ij}^{tr} = \{0,1\} \quad \forall tr \in TR, \quad a_{ij}^r = \{0,1\} \quad \forall r \in R$

Note that in the covering matrix A, a link exists for origin-destination pair via a direct route (rεR) or transfer (TrεTR), and:

$$\sum_{r\in R} a_{ij}^r + \sum_{tr\in TR} a_{ij}^{tr} \geq 1 \tag{3.4}$$

The notation r:tr signifies that the route r consistently shows in the transfer path tr. The costs c_r and c_{tr} at this stage, before assigning the demand, describe the lengths of the travel paths: direct routes and via transfers. Hence, the solution of the covering problem will yield the travel path network (direct and indirect), with the minimum travel times.

An algorithm developed for the above formulation (which appears in Israeli [1] and to be published elsewhere), has been tested with a random network. The results showed its high efficiency under the criteria of accuracy and running time.

3.2 The Entire Iterative Process

At this stage, the real costs of matrix A's columns (routes and transfers) are not known. These costs are similar to the optimization criteria which were defined in (2.1), but associated with each column. The costs can result from the demand assignment procedure which, by itself, is based on the minimal set of routes, i.e., the SCP solution. That is why it is impossible to solve the SCP to minimum real costs, but to minimum travel time.

Thus, at this stage, commences the iterative process of constructing efficient alternatives for resolving the multi-objective problem. This comprises three main stages:

Stage 1: Producing a minimal set of direct/indirect travel paths (routes/transfers) affirming the connectivity of the network. The problem to be solved is known as SCP (Set Covering Problem).

Stage 2: The execution of the assignment process over the covering set which was created in the former stage. The outcome of the stage is: bus frequencies of the set of routes, passenger load profiles, demand assignment across the set of routes, and the optimization parameters PH, WH, EH. At the same time, the minimim fleet size required to meet the demand as well as to satisfy the determined frequency on each route is estimated.

The method used for evaluating the fleet size is based on the Deficit Function theory proposed by Ceder and Stern [26]. The final outcome of this stage is the value FS which represents the operational component's minimum required fleet size.

Stage 3: A deletion of the detrimental variables (routes/transfers) to the value of the solution acceptance of a new (diminished) set of travel paths (routes and transfers) within the network, which results in a diminished matrix A, and back to stage 1. The process executed in stage 3 guarantees two main issues: (1) there will always be a cover to the diminished matrix A; (2) previous alternatives will not be repeated.

Once the criteria reach the desired number of alternatives, the alternative creation process results in the multi-objective problem being resolved, up to the minimizing of Z_1 up to Z_4.

This paper deals mainly with the relationship between stages 1 and 3, and in resolving the preceding multi-objective problem. The second stage (demand assignment) is not detailed herein; for more information see Israeli [1].

3.3 Alternative Generation Process

The described process generates a set of feasible and efficient points in the field of the four objective functions. Such a point is typical for a feasible alternative solution which represents a set of routes. These points are an approximation of the efficient points, yet unknown, which result from solving the continuous problem. This method of generation results from the approach of "compromised programming" chosen for evaluating the results and choosing from them (see 3.4). The criteria at each stage were to reach the minimal distance (metric value) from the present ideal solution (from the minimal theoretical point of the four objective functions at the same time).

The process separates into two problems: master problem and sub-problem.

The master problem deals with incremental progress from one solution to the other in the generation process. The problem deals with building the covering matrix A of the possible routes and transfers, and solving it in such manner that different solutions will be created (new sets of routes transferring the network demands). The initial point is the covering sets which results from the first covering problem (described in (3.2)), for which the values of the objective functions are calculated according to the assignment process and the estimation of the operating vehicle fleet.

At this stage, the ideal solution is defined anew, as well as the set with the minimal distance to the ideal point. The variable most detrimental to the set's metric value cost is deleted from this set. Recovering matrix A is effected, and the process repeats itself. The minimal distance is dynamic, since the ideal point can be changed during the iterations.

The sub-problem termed SDP (Set Deletion Problem) deals with producing feasible solutions for the covering problem SCP, of the master problem. The question deals with two elements: 1) If a variable column is deleted in the master problem (or a combination of columns) from the covering matrix A, there could arise a situation where A will have no covering. In such a case, a column should be returned (or a partial combination of columns), to the matrix; 2) In the covering process for the master problem, it is essential not to enter an endless loop; this means to receive already existing sets (solutions). The SDP problem therefore, is dealing with a minimal deletion of columns from matrix A in a certain iteration so that by solving the SCP problem within the master problem (in a diminished A matrix), solutions will not be reiterated, and there will always exist coverage for A. The SDP problem is formulated as a set covering problem (non-linear), and proper resolution techniques are suggested. For clarification, it is emphasized that a few equivalent terms have been used, and have the same meaning: solution alternatives; sets (of routes); feasible solutions; feasible points (in the four objective functions' space).

3.3.1 The Master Problem

The algorithm described examines in each stage the sets of feasible points $\{(Z_1, Z_2, Z_3, Z_4)\}$ whilst generating a new solution which can guarantee approaching the ideal point. If such solutions exist, the outcome will comprise all sets of efficient solutions (efficient points in the four dimensional space). The algorithm's principal steps are described below. For the more detailed steps, the reader is referred to Israeli [1] and Israeli and Ceder [23].

define:

CAN = a group of all sets of routes (solutions of SCP) which are examined and are candidates for changing in the alternative generation process;

SOL = a group of all sets of routes which were selected to solve the multi-objective programming;

L_s^k = metric distance from the power of s of the set $k \varepsilon CAN$ (see definition of Section 3.4).

Algorithm's principal steps:

<u>Step 1</u>: A group of candidate sets:

Solve the SCP of the initial matrix A. On the solution's set(s), execute the demand assignment process and calculate the estimator of the minimum fleet size. A feasible point(s) (Z_1, Z_2, Z_3, Z_4) was created. Group the sets in CAN and in SOL.

Step 2: Finding the ideal point in the present iteration

The sets in CAN yielded

$$Z_i^* = \min \quad Z_i \qquad ,i = 1,\ldots,4$$

and define point $(Z_1^*, Z_2^*, Z_3^*, Z_4^*)$

Step 3: Metric values:

For all sets in CAN, calculate the metric values L_S from the ideal point (see formulas in section 3.4).

Step 4: Choosing the candidate set:

From the sets in CAN choose k' which keeps a minimal "potential metric distance" from the ideal solution, which means:

$$L_s^{k'} + \Delta L_s^{k'} = \min_{k \in can} (L_s^k + \Delta L_s^k) \tag{3.5}$$

ΔL_s^k shows the metric distance difference $L_s^j - L_s^k$ when the set j has been created last as a result of candidate set k in former stages. (The set which has not been generated will be marked $\Delta L_s^k = 0$). This sign marks the production potential of metric distances chosen by set k, and its aim is to prevent the degeneration of the problem. Without this sign, the candidate set can be reselected in repetitive steps even if it produces "bad" sets.

Step 5: Execution of the candidate set:

Select the column of a singular route about to be deleted from the covering matrix: Select from the set the column of a singular route r' which delineates

$$\hat{L}_s^{r'} = \max_r \quad L_s^r \tag{3.6}$$

where $\hat{L}_s^{r'}$ is the metric value of the route (after the assignment process) from the set's ideal point for a covering unit, meaning the metric value divided by the number of pairs "i,j" (rows in the covering matrix), those served by this route and the resultant transfers.

The set's ideal point is: (min PH_r, min WH_r, min PH_r, min Fs_r) and the index r related to the same optimization criteria previously defined, but which results from the load profile of route r only. FS_r is calculated by the portion usage of route r from the total FS.

Step 6: Deletion of columns from covering matrix A:

Present the column(s) of the route(s) selected in step 5 as columns in SDP matrix (see section 3.3.2).

Step 7: Examining SDP solution:

If SDP performs a feasible solution, go to step 8.

Otherwise, if the candidate set is still in CAN, replace the columns which were deleted and go to step 6. Otherwise -- process is terminated.

Step 8: Solving SCP in diminished matrix A:

Delete from matrix A the columns of routes r which are the solution of SDP and the columns of transfers tr associated with them. Solve SCP to diminished matrix A. A definite solution will be obtained.

Step 9: Obtaining another alternative:

In the set obtained in step 8 (the solution of SCP), the demand assignment procedure will be executed, and hence the calculation of the minimum fleet size. These yield a new point (Z_1, Z_2, Z_3, Z_4) in the four-dimensional space. Add the new set to CAN and to SOL.

Process terminating is due to several criteria (like number of required sets -- see [1], [23]). It will be noted that in each iteration of the described algorithm, SOL will certainly increase while CAN might (not necessarily) decrease or even receive the value CAN = ϕ, The number of alternatives to be generated depends on the density of matrix A and the bounds selected.

3.3.2 The Sub-Problem: SDP

The SDP checks the columns about to be deleted from matrix A in the master problem. A different matrix is defined whose columns are those deleted from sets in former iterations and from the present candidate, and its rows are the fitting sets. The dimensions of the matrix increase as the iteration progresses within the master problem, so that in each iteration a row must be added for marking the examined set, and a column may be added or replaced. Thus, the covering problem is solved, resulting in a minimal amount of columns to be rejected from A, so that there will be no repetition sets (which exist in SOL). The SDP problem has definite covering, and according to the chosen columns, two kinds of solutions emerge: 1) A feasible solution. -- The chosen columns are deleted from matrix A, and A has coverage; 2) An infeasible solution. -- The chosen columns will not permit a covering for A. The second case is derived from a possible combination of columns in SDP that exclusively cover one entire row in A (meaning the values of 1 are derived only for these columns, on at least one row, while for the rest of the columns, the values are 0). Such a combination, designated "A Unique Combination," if entirely rejected from A will result in 0 values at least on one row, and as a result will avoid coverages. The formulation and solution of SDP attempt to prevent the entrance of the "unique combination" to the solution. Lacking a choice, it would be possible to perform in the algorithm of the master problem, a backtrack for transferring the rejected route columns, as described in step 7 of the master problem.

Define $E = \{\{e_{ij}\}, \{e_{if}\}\}$ the matrix to solve the SDP (its binary parameters are explained hereafter).

The mathematical formulation for SDP is a set-covering problem of matrix E which is formulated as non-linear programming with integer variables:

$$\min\left[\sum_{j\in J} c_j x_j + \sum_{f\in F} c_f \underset{j:f}{\pi} x_j\right] \qquad (3.7)$$

s.t.

$$\sum_{j\in J} e_{ij} x_j \geq 1 \qquad \forall i \in I \qquad (3.8)$$

$$x_j = \begin{cases} 1 & \text{if column } j\in J \text{ is in the solution} \\ 0 & \text{otherwise} \end{cases} \qquad (3.9)$$

while j∈J is a single route column and f∈F is a "unique combination" column and exists

$$e_{ij} = \begin{cases} 1 & \text{if route } i\in I \text{ exists in set } j\in J \\ 0 & \text{otherwise} \end{cases} \qquad (3.10)$$

$$e_{if} = 0, \quad i\in I, \; f\in F \qquad (3.11)$$

$$\sum_{j\in J} e_{ij} + \sum_{f\in F} e_{if} = \sum_{j\in J} e_{ij} \geq 1 \qquad \forall i \in I \qquad (3.12)$$

The marking j:f equivalent to $j \mid j\cap f\neq\phi$ means that column j∈J takes part in the "unique combination" f∈F.

The column costs are

$$c_j = \begin{cases} -\infty & \text{if column } j\in J \text{ of the selected set chosen for deletion} \\ 1 & \text{otherwise} \end{cases} \qquad (3.13)$$

$$c_f = +\infty \quad , f\in F \qquad (3.14)$$

Another policy of choosing column costs is described in [1] and [23]. It considers the existence of columns belonging to the selected set in some other sets of the matrix.

Solution techniques to solve the SDP contain: mathematical approach, trial and error approach, heuristic approach, and complete enumeration (see [1], [23]).

3.4 Evaluation and Selection of Alternatives

The evaluation of the generated alternatives and the selection of "the best" is performed by the method of "compromised programming." This procedure, developed by Zeleny [27, [28], corresponds mainly to the solution of linear multi-objective problems (continuous), while a derivative variation of this method is used for solving discrete problems. One of the greatest benefits of this method is that it decreases the number of points to be checked in the Non-Inferior Solution set (NIS) and in concentrating on the section that lies on the NIS curve which provides the compromise solutions. In this paper, the general method will be explained first (the continuous), and following, its derivative for a discrete case which is utilized for this study. The compromise programming method -- defines the solutions which are the nearest to the ideal solution by calculating the index of the geometrical distance. These solutions are called the "compromise solutions," and they establish the "compromise set". The ideal solution is the one which yields all the objective functions to an optimum value simultaneously. Usually this is impossible; otherwise there would have been no conflict in the objective functions.

The ideal solution is brought about as a theoretical point in the objective functions' space

$$Z^* = (Z_1^*, Z_2^*, \ldots, Z_p^*)$$

which results from solving a single objective optimum problem separately for each function.

$$\begin{aligned}&\min\ (Z_1(x))\\&\text{s.t.}\\&x \in X\\&i=1,2,\dots,p\end{aligned} \tag{3.15}$$

The measure used for evaluating the proximity between the NIS solutions and the ideal point is based on a family of metric distances, Ls, with the power of s value, as defined:

$$L_s = \left[\sum_{i=1}^{p}\left[\frac{Z_i(x)-Z_i^*}{Z_i^M - Z_i^*}\right]^s\right]^{1/s} \tag{3.16}$$

when:

$$Z_i^M = \max_{x \in X} \quad Z_i(x) \qquad ,i=1,\dots,p \tag{3.17}$$

The vector x* of the solutions is the one to resolve the minimization problem:

$$\begin{aligned}&\min \quad L_s(x) = L_s(x_s^*)\\&s.t.\\&x \in X\end{aligned} \tag{3.18}$$

The compromise set is a set of the calculated compromise solutions (3.18) for each o <s<∞. In actual fact, usually only the values s = 1,2, ∞ are used (Goichoechea et al. [29]), and the decision-maker must decide which one. Yu [30] demonstrated that solving the problem will always produce an efficient point $(Z_i(x_s^*), i=1,\dots,p)$ on NIS to 1<s< ∞ , while for s = ∞ there will always be at least one efficient point.

Application of the compromise-programming method for solving discrete problems is according to Duckstein and Opricovic's [31] method. The continuous type formulas are changed into discrete ones as follows, while in the study, the value p=4 is considered (four objective functions). From the process of alternative generation, a final number of feasible points (Z_1, Z_2, Z_3, Z_4) are attained, and can be ordered in a pay-off table. Hence, the values are calculated as follows:

$$Z^*_i = \min_{k \in SOL} Z_i^k \qquad ,i = 1,\ldots,4 \tag{3.19}$$

$$Z^M_i = \max_{k \in SOL} Z_i^k \qquad ,i = 1,\ldots,4 \tag{3.20}$$

$$L_s^k = \left[\sum_{i=1}^{4} \left[\frac{Z_i^k - Z_i^*}{Z_i^M - Z_i^*} \right]^s \right]^{1/s} \qquad \forall \quad k \in SOL, \quad 1 \le s < \infty \tag{3.21}$$

The compromise solutions are those which determine minimal distances from the ideal solutions to each given s:

$$L_s^* = \min_{k \in SOL} L_s^k \qquad \forall \quad 1 \le s < \infty \tag{3.22}$$

In the research, values $s = 1,2, \infty$ have been considered.

4. Numerical Experience

The idea behind the entire heuristic process of alternative generation is to provide most of the efficient points during the first stages of the process. Executing more stages yields the production of more inefficient points. The sufficient number of alternatives is due to the structure of the matrix; otherwise, all the alternatives will be provided as in a complete enumeration procedure.

A detailed comparison between different initial conditions of the algorithm was performed (see Israeli [1], Israeli and Ceder [23]). The conclusion is that the algorithm's initial condition ((3.5) in step 4 of the master problem), is very effective.

As to testing the algorithm with different size networks, the heuristic approach's results were compared to those provided from a complete enumeration. Random networks were produced of 4-15 nodes with different densities and with a various number of terminals (and their location in the network). Each category contains 80 networks, from each 12 solution alternatives (defining SOL) were provided (if this was possible). Table 1 presents the results due to the criterion of min L_S (weighted by different s values). "Total accuracy level" denotes the number of times (in percents) the solution provided (from one of the 12 alternatives) was equal to the minimal value. "Solution location distribution" denotes the number of times the minimal solution was obtained in the first set. "Max-min relative error" is an index which examines the alternatives in SOL which do not provide the optimal solution. The error is calculated as follows: from group SOL, which comprises the 12 alternatives that do not contain the optimal solution, the alternative with the minimal L_S was selected. Then the relative error δ from the global minimum is calculated. The alternative with the "max min δ" is selected from all the groups {SOL}, which means maximal relative error. The "portion of efficient solutions" provides the portion (in percents) of efficient solutions (NIS) in the generated alternatives out of the total number of efficient solutions derived from the complete enumeration.

Approach No.	Total Acc. level	Solution location distribution 0-25%	25-50%	50-75%	75-100%	max-min δ	portion of efficient solutions
4-6	100	84	16	0	0	0.0	100
7-9	91	64	17	9	1	8.4	94
10-12	86	54	15	12	5	12.6	88
13-15	82	50	18	7	7	18.1	80
average	90	63	17	7	3	9.8	90.5

Table 1. Efficiency of Heuristic Algorithm in Achieving Minimal L_S

The table shows an "average accuracy level" of 90%. This value decreases while the network dimension increases due to the portion between the 12 alternatives and the increasing number of enumerated alternatives (which might yield efficient alternatives not contained in the solution). The "max-min relative error" shows a relatively small value even for the worst alternative, and a total average of 9.8%. The "portion of efficient solutions" denotes decreasing values due to the increase in problem size ensuing from the same reason as explained earlier. The worst case shows a relatively high percentage -- 80% of efficient solutions.

5. Summary and Conclusions

The heuristic approach described in this paper provides: (a) generation of finite sets of alternatives efficient non-inferior solutions; (b) evaluation and selection of the various solutions using multi-objective preference techniques for discrete variables ("compromise programming" procedure).

The "alternative generation process" is based on a given "travel paths matrix" which can be mapped by several techniques of "column generation" in previous stages. This matrix consists of all possible paths (direct routes and via transfers), subject to certain constraints, versus all origin-destination pairs of the network.

The iterative procedure is constructed in a modular form, successively confronting the NP-hard type complexity problem, and its formulation as non-linear (concave) programming with mixed variables (continuous and integer). Such a formulation cannot be solved via known mathematical programming approaches and packages.

The proposed approach is compared with a mathematical programming optimal solution for small networks, and by enumeration optimal solution for medium size networks. The two comparison measures related to that are the quality of solution and computational time. These tests show that the heuristics used are efficient in an overall perspective.

The proposed approach is believed to be a useful toolset for the following applications:

a) Optimal design for a new transit network;
b) Optimal design for expansion or curtailment of an existing transit network;
c) A network assessment of the performance of an existing transit network from the aspects of: (I) Operator efficiency; (ii) Passenger level-of - service;
d) Sensitivity analysis of transit network performance for a variety of system parameters (such as different bus fleet size, different level of service,changes in passenger demand, changes in frequencies, changes in traveltime and more).

References

[1] Israeli, Y., "Transit Route and Scheduling Design at the Network Level", *Doctoral Dissertation, Civil Engineering Department, Technion-Israel Institute of Technology*, Israel, 1992.

[2] Ceder, A., Wilson, N., "Bus Network Design.", *Trans. Res.*, Vol. 2O8B, No. 4, pp. 331-334. 1986.

[3] Ceder, A., Israeli, Y., "Scheduling Considerations in Designing Transit Routes at the Network Level.", *Lecture Notes in Economics Mathematical System*, No. 386, Proc. of the 5th Int. Workshop on Computer-Aided Transit Scheduling, Montreal, Canada, August 1990, by M. Descrochers and J.M. Rousseau (Eds.), Springer-Verlag, pp. 113-136, 1992.

[4] Magnanti, T.L., Wong, R.L., "Network Design and Transportation Planning: Models and Algorithms", *Trans. Sci.*, Vol. 18, No. 1, February, pp. 1-55, 1984.

[5] Current, J., Revelle, C.S., Cohon, J.L., "The Median Shortest Path Problem: A Multi-objective Approach to Analyze Cost vs. Accessibility in the Design of Transportation Networks", *Transportation Science*, Vol. 21, No. 3, August, pp. 188-197, 1987.

[6] Janarthanan, N., Schneider, J., "Multi-Criteria Evaluation of Alternative Transit System Designs", *Trans. Res. Record*, No. 1O64, pp. 26-34, 1986.

[7] Steenbrink, P.A., "Optimization of Transport Networks", Offset drukherij N.V., Nederlandse Spoorwegen, Utrecht, Holland/John Wiley & Sons Ltd., 1974.

[8] Fricker, J.D., Shanteau, R.M., "Improved Service Strategies for Small City Transit", *Trans. Res. Record*, No. 1O51, pp. 3O-34, 1986.

[9] Ignizio, J.P., "An Approach to the Modeling and Analysis of Multi-objective Generalized Networks", *European Journal of Operational Research*, Vol. 12, pp. 357-361, 1983.

[10] Current, J., Min, H., "Multi-Objective Design of Transportation Network: Taxonomy and Annotation", *European Journal of Operational Research*, Vol. 26, pp. 187-2O1, 1986.

[11] Pogun, G., Satir, A., "Alternative Bus Scheduling Policies for an Exclusive Bus Lane", *Trans. Res.*, Vol. 2OA, No. 6, pp. 437-446, 1986.

[12] Tadi, R.R., Khasnabiss, S.,Opiela, K.S., "A Methodology for Evaluating Bus Service Cutback Programs", *Transportation Quarterly*, Vol. 4O, No. 2, April, pp. 243-261, 1986.

[13] Tzeng, G.H., Shiau, T.A., "Multiple Objective Programming for Bus Operations: A Case Study for Taipei City.", *Trans. Res.*, Vol. 22B, No. 3, pp. 195-2O6, 1988.

[14] Lee, S., Moore, L., "Multi-Criteria School Busing Models", *Management Science*, Vol. 23, pp. 7O3-715, 1977.

[15] Flynn, J., Ratick, S., "A Multi-objective Hierarchical Model for the Essential Air Services Program", *Trans. Science*, Vol. 22, No. 2, May, pp. 139-147, 1988.

[16] Teodorovic, D., Krcmar-Nozic, E., "Multi-Criteria Model to Determine Flight Frequencies on an Airline Network under Competitive Conditions", *Trans. Sci.*, Vol. 23, No. 1, Feb., pp. 14-25, 1989.

[17] Cohon, J.L., "Multi-Objective Programming and Planning", *Academic Press*, New York, 1978.

[18] Fandel, G., Spronk, J. (Eds.), "Multiple Criteria Decision Methods and Application", *Selected Reading of the 1st International Summer School, Acireale, Sicily, September 1983." Springer, Berlin*, 1985.

[19] Hwang, C., Masud, A., "Multiple Objective Decision-Making Methods and Applications", *Springer*, N.Y, 1979.

[20] Lazimy, R. "Interactive Relaxation Method for a Broad Class of Integer & Continuous Non-Linear Multiple Criteria Problems.", *J. Math. Anal. Appl.*, Vol. 116, pp. 553-573, 1986.

[21] Rasmussen, L.M., "Zero-One Programming with Multiple Criteria", *European Journal of Operational Research*, Vol. 26, pp. 83-95, 1986.

[22] Gabbani, D., Magazine, M., "An Interactive Heuristic Approach for Multi-Objective Integer Programming Problems", *Journal of Operations Research Society*, Vol. 37, pp. 285-291, 1986.

[23] Israeli, Y., Ceder, A., "Multi-Objective Approach for Solving the Transit Network Design Problem" *Paper to be published in Trans. Science.*

[24] Minieka, E., "Optimization Algorithms for Networks and Graphs", *Marcel Dekker Inc.*, N.Y., pp. 181-234, 1978.

[25] Syslo, M.M., Deo, N., Kowalit, J.S., "Discrete Optimization Algorithms", *Prentice-Hall Inc.*, New Jersey, Chapter 2, pp. 176-211, 1983.

[26] Ceder, A., Stern, H. (1981), "Deficit Function Bus Scheduling with Deadheading Trip Insertion for Fleet Size Reduction", *Trans. Science*, Vol. 15, No. 4, Nov., pp. 338-363.

[27] Zeleny, M., "Compromise Programming: Multiple Criteria Decision-Making," *J.L. Cohon and M. Zeleny (Eds.). University of S. Carolina Press, Columbia*, 1973.

[28] Zeleny, M., "A Concept of Compromise Solutions and the Method of the Displaced Ideal", *Computers and Operations Research*, Vol. 1, No. 4, pp. 479-496, 1974.

[29] Goicoechea, A., Hansen, D.R., Duckstein, L., "Multi-Objective Decision Analysis with Engineering and Business Applications", *John Wiley & Sons, N.Y.*, 1982.

[30] Yu, P., "A Class of Decisions for Group Decision Problems", *Management Science*, Vol. 19, No. 936, 1973.

[31] Duckstein, L., Opricovic, S. (1980), "Multi-Objective Optimization in River Basin Development", *Water Research*, Vol. 16, No. 1, pp. 14-20.

RELATIONSHIP BETWEEN PARKING LOCATION AND TRAFFIC FLOWS IN URBAN AREAS

Stefano Carrese, Stefano Gori and Tommaso Picano

Dipartimento di Scienze dell'Ingegneria Civile
Terza Università degli Studi di Roma
Via Corrado Segre, 60 - 00146 Rome, Italy

In the last twenty years, most studies on parking demand distribution have been analyzing user-choice criteria, and after determining the utilization level of proposed parking facilities, they have identified the optimal pattern of the parking plan, based on the optimization of only two parameters: parking cost and walking distance. The most common method used to search for optimal allocation criteria, was a verification of different scenarios for parking demand satisfaction.
Today, the proposed model develops this subject of research, simulating a car volume equilibrium on the network. It makes it possible to simulate the choices of different user classes on a multi-modal network. Each class is characterized by a typical value of the utility associated to every transportation alternative. The model deals with the problem of simulating a transportation system, operating for every time interval in which the day is divided, with as many demand matrices O/D as the user classes that it intends modeling. In order to assess its performance, the model was applied to the town of Viterbo, the first one for which a usable data base was available. However, since a realistic demand function was not available, only a fixed demand assignment has been possible.

1. Introduction

In the last twenty years, most studies on parking demand distribution have been analyzing users choice criteria, and after determining the utilization level of proposed parking facilities, they have identified the optimal pattern of the parking plan, based on the optimization of only two parameters: parking cost and walking distance. The most common method used to search for optimal allocation criteria, was a verification of different scenarios for parking demand satisfaction.

The primary criterion adopted for establishing parking position, is the one of maximum utilization and consequent investment cost-effectiveness, that is maximum personal utility for the user himself. Most studies refer to walking distance and parking cost as main variables in different forms and, although these studies provide for different models, results always show that the best position for facilities is where attraction is higher. Consequently user-choice criterion is the minimization of a disutility function, e. g. generalized cost function or walking times total sum.

From a city planning point of view, in monocentric structured towns (which is the case for most ancient Italian towns) the focus of attraction lays in the center, because of the presence of commercial and administrative activities as well as tourist sites.

In most large European urban areas, more recent parking policies sustain the use of public transport and try to reduce private traffic through the application of fares and law measures. In particular the main goals are:

- to reduce traffic congestion in the historical center, penalizing the use of private cars through limitations on on-street and long-time parking;
- to avoid the increase of trips in the areas near the urban centers, through the use of Park & Ride;
- to develop public transport competitiveness, increasing the level of service (bus lane, traffic management, signal pre-emption systems, etc.)
- to satisfy parking needs of residents and customers.

The idea of creating a model capable of dealing with parking problems in urban areas was a consequence of the exceedingly sectorial nature of the approaches found in the state of the art.

2. Classification of Existing Models

The parameters characterizing most of the problems of parking location are the following:

- a single (usually public) decision-maker, which controls the spatial distribution of the service;
- a demand, based on some user-categories, moving to places where service is provided;
- an evaluation criterion for service distribution, which usually stands on some measure of user-benefit;
- costs of parking installation and management with budget constraints.

The general problem can be divided into two sub-problems, which provides the first important distinction among existing models:

- the allocation sub-problem: that is the definition of the trips scheme between user house and location of service;
- the sub-problem of parking location. That is the choice of where parking space will be established.

Obviously the two sub-problems are interrelated and cannot be solved separately (fig. 1)

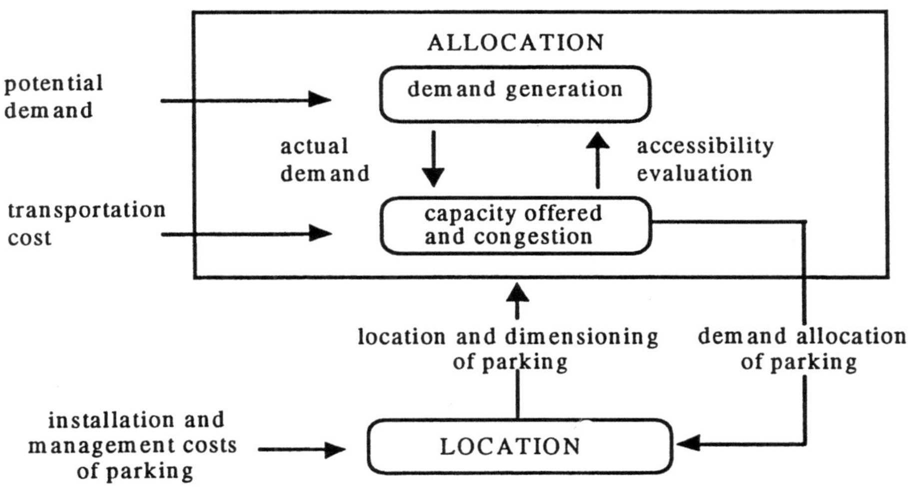

Figure 1. Relationship between allocation and location sub problems

In the allocation phase, the main decision-makers are the users (demand). Decision steps are:

- comparison of transportation cost and capacity offered by different parking and consequent evaluation of accessibility;
- choice of facility to reach, and relative trips scheme.

In the location phase, there is just one decision-maker, usually an authority in charge of the following operation:

- comparison between location and dimensions of existing parking with their allocated demand;
- cost-benefit evaluation of the effective state of the system and of its possible modification;
- generation of a new plan of location and dimensions of parking, including new ones, change of management strategy or dismissing of old ones.

A classification of existing models has been conducted according to the following criteria:

- optimal location of parking lots;
- management of attraction related to parking supply, which conditions the flow pattern on the road network;
- strategies of fare management and stay limitation.

Two main classes of models have been identified:

I) location models, based on procedures of the Operational Research and Regional Sciences;

II) allocation models, which represent parking user-behavior.

The first class consists of models that, after defining some supply and demand points on a network, choose the positions of service points in order to optimize an objective function, subject to spatial and economic constraints (Odoni & Larson, 1981). Among the others, the models frequently used in parking studies are:

- the *p-center* models, which minimize the maximum user-cost and are mainly used for emergency services location, such as ambulances and hospitals. These models make it possible to locate *p* supply points on a network in order to minimize the maximum distance between demand and supply points, so that they can favor more disadvantaged demand points;

- the *p-median* models, which minimize the average user-cost and as a result are more applicable to parking problems. In fact these models make it possible to locate *p* supply points on a network in order to minimize the average distance between demand and supply points. Problems which have been solved through these algorithms, deal with the location of non-emergency facilities, such as schools, transit terminals and public administration offices;
- the *requirement* models deal with the solution of problems in which some service efficiency standards are *a priori* fixed, and the location and the number of facilities, necessary to reach the fixed level of service, are found. These models are applicable to p-center and p-median problems, and extend their functional capability.

Until now these models have had a low degree of development as to parking problems because of the difficulty in describing particular features of parking demand, as fare and law strategies, and their interdependence with traffic mobility. However an application of the p-median problem (Perl & Ho, 1990) takes into account the elastic nature of demand, and the location criterion is the maximization of consumer surplus,: different demand functions represent user behavior in the choice of more convenient service.

Referring to the second class, many more experiments in parking simulation have been conducted, using allocation models.

3. Allocation Models for Demand Distribution

Three main categories of models for parking allocation can be detected:

1) *behavioral* models, that provide the optimum parking plan among all the possible ones, through an objective function representing the user choice;
2) *gravity* models, providing the optimal plan through a model of gravitational type;

3) *logit-based* models, simulating user-behavior through the modal split.

Models belonging to the first class represent user-criteria in choosing parking and differ by the calculation method: some use equilibrium assignment (Beinborn 84, Gray 72, Lablanche 72); other authors use linear programming (Ellis 72, Whitlock 73); others use separated utility function (Austin 73).

All of them analyze user-behavior through a disutility function, and so identify underutilized or insufficient parking. In this class the final choice for a parking location is made after the comparison of the disutility function values assumed by each scenario. This function changes expression depending on the author; this means a different sensitivity toward selected parameters, although the variables generally taken into account for demand distribution in different parking lots, are mainly the following:

C_k Parking cost in the k lot;

d_{kj} Walking distance between parking k and destination j;

s_k Capacity of parking k.

These models, for a real representation of the phenomenon of parking distribution, are first calibrated on the basis of existing situations through statistical regression, then they are applied in the feasibility study of the parking lots proposed.

For example Ellis presents a disutility function subject to two constraints:

$$\begin{aligned}
&\text{Min} \sum_j \sum_k \sum_q \sum_d Z(j,k,q,d) \times X(j,k,q,d) \\
&\text{s.t.:} \sum_j \sum_q \sum_d X(j,k,q,d) \leq s(a,k) \qquad \forall\, k \\
&\sum_k X(j,k,q,d) = T(j,q,d) \qquad \forall\, (j,q,d)
\end{aligned} \tag{1}$$

where:

j= index of the final destination zone;

a= index of the arrival period;

d= index of the departing period ($d>=a$);

$X(j,k,q,d)$ = number of users arrived in period a, departing in period d, belonging to group q, destined in zone j and parked in k;

$Z(j,k,q,d)$ = disutility for every user;

$s(a,k)$ = supply (number of spaces) available in period a in parking k;

$T(j,q,d)$ = number of users arrived in period a, departing in period d, belonging to group q, destined in zone j.

Then in order to make the disutility function explicit, three different kinds of relationship between costs and walking distances, based on user-sensitivity, are considered and calibrated by observed data.

The use of d_{kj} as the main variable brings all these models to locate parking near destinations but this criterion doesn't consider the possible negative effects of congestion due to concentration of parking lots.

The major contribution of these models is their capability of simulating user-behavior, when looking for a free space near his/her destination. In addition they can represent "how" and "how much" user-choice is influenced by in-force rules and fare measures in urban area. In this respect, the probabilistic approach used by Austin (73) must be outlined for his capability to describe user-behavior in parking and of defining the value assigned to walking time from parking to destination, according to trip purpose and stop duration. In Beinborn's model (84), the analysis of in-force rules and fare measures and their impact on user-distribution across parking lots is really interesting: what is also considered is the phenomenon of illegal stops, in relation to the following variables:

m_k Expected fine for illegal stop in parking k;

P_{kt} Probability that a user pays a fine for the illegal stop of duration t in parking k. It depends on the frequency of police controls;

$m_{kt}=P_{kt}*m_k$ Expected fine for the illegal stop of duration t in parking k.

In the second class of models (Bates 72, Pratt 84), the choice of the optimum parking plan is obtained by a comparison of zones characterized either by the lack or the surplus of supply versus demand. In other words after an appropriate zone split, the values of requested and available parking spaces can be calculated and every zone is characterized by a state of lack or surplus in parking spaces. Repeating the procedure for all the feasible parking plans, a complete framework of all possible situations of every zone can be obtained.

Bates, for example, considers that parking demand in zone i (P_i) will be equal to the sum of the expected (O-D matrix) user-demand in zone i (D_i), plus parking demand parked in zone i coming from another zone j (A_i), minus parking demand in zone i satisfied in other zones j ($\sum_j A_{ij}$).

In formula:
$$P_i = D_i + A_i - \sum_{j=1}^{n} A_{ij} \tag{2}$$

where
$$A_i = \sum_{j=1}^{n} A_{ji} - A_{ii} \tag{3}$$

and
$$A_{ij} = R_i \left[S_j f_{ij} \Big/ \sum S_j f_{ij} \right] \tag{4}$$

where:

A_{ij} = number of vehicles that would like to park in i, but go and park in j;

S_j = parking supply in zone j;

R_i = residual demand of zone i = $D_i - S_i$;

f_{ij} = distribution factor between zone i and j, proportional to the inverse of their distance.

It must be underlined that this model is valid only when parking supply in the whole town exceeds demand.

In the third class of models, called logit-based (Bifulco 93, Gattuso 92, Hunt & Teply 93), a general representation of a transport system is provided and then different parking strategies are simulated. The main goal of these models is the identification of a homogeneous class of parking users through the random utility theory. In fact a nested logit model is used to calculate the share of private traffic that chooses a specific parking lot according to time or cost minimization criteria.

In fact, in urban centers with high levels of congestion it is almost impossible to satisfy parking demand near their destinations, because of network capacity constraints and the lack of available space for building parking lots.

It is necessary to remember that an increase in parking supply can alter the current equilibrium producing an increase in parking demand (driven demand) that can reduce the efficiency of the whole plan (Papola 93). The increase in supply can take the mobility back to a congestion condition similar to a no-project situation. As it is commonly recognized the actions causing change in the supply, influence demand. They cause substantial variations in the existing equilibria between various modal-splits and transportation alternatives. They also lead to undesired modifications of the mobility that influence competitiveness, performance and the level of service of the whole transportation system.

In Gattuso's model, information required for demand representation is really specific. Although they regard only private users, the knowledge of the following data is requested: destination zone, interval during which the trip is made, access gate to urban center, trip purpose and the desired transfer opportunity between parking and destination. The specification of the model is further outlined by a logit modal-split model, which does not consider every user for every transportation mode in the area, but only of car users that have to choose a trip mode between parking lot and destination. In fact the use of a generic parking facility means a user total cost equal to the sum of the following terms: the fare (constant value), the access cost (variable with the parking load factor), the parking approach cost (constant value, independent from car volumes on the network) and the transfer cost from parking to destination (constant value, depending on the "as the crow flies" distance). For the representation of changing access cost, the model executes a phenomenon simulation, based on discreet time intervals progress, which describes the evolution of parking load, that is the number of vehicles inside the parking in the n-th temporal interval. Lastly this model can be applied only to the verification of alternative parking plans which have to be designed out of the model. It presents a strongly operative characteristic, which moreover appears to be too specific for a good representation of a multi-modal context (present in most case studies).

In Bifulco's model the context area is the complete transport network. The consequent demand and supply description considers also the information necessary to satisfy transit simulation. He uses a probabilistic model for modal split too, but the aim is the definition of volumes for every user-category and for every transport mode. As to supply modeling, every parking lot is represented by a link cost function characterized by:

- a fixed cost, due to possible stop limitation and parking fare structure;
- a variable cost, that considers the search time for a free space;
- a cost of access links to parking, that has a constant time depending on the type of parking link considered (on-street and off-street).

In this model, assignment procedure aims at calculating link volumes in the complete network and to assess different mode alternatives, even integration of them (Park & Ride is explicitly considered). Finally Bifulco's model, through its multi-user and multi-class representation of the transport system, better interprets the need to analyze mobility as a whole system. In particular, due to the powerful attraction of the parking, all those measures that change and influence parking capacity condition car volumes on network links. Thus, although this model seems a complete tool for the actual mobility representation in an urban area, a heuristic algorithm, whose convergence skill is elsewhere described, has been used due to its possible application even to simplified cases.

The original model proposed hereunder, is a simulation tool, capable to overcoming the limits of the previously described models: it integrates and improves them, outlining some interesting roles of parking analysis in mobility management in urban areas. Major contributions are the refined representation of transport supply (Bifulco 93), the algorithmic effectiveness (Gattuso 92) and utilization of most the significant variables (Ellis 72, Beinborn 84).

4. A Model for Parking Location with Mobility Verification

In order to consider two macro-effects on traffic flows, in the location test phase of an off-street parking lot an innovative methodology is proposed (Carrese, Gori 93).

The first effect to be considered in the hypothesis of parking demand (steady) lower than the supply, is the phenomenon of traffic flow redistribution in the area under study with possible vehicular congestion. Here local accessibility to the parking lot has to be considered, while maintaining the level of service in the access links over a predetermined lower bound (e.g. level of service not worse than the no-project situation).

The second effect can be noted when parking demand exceeds supply: the construction of new parking lots enables an increase in potential demand. The relaxation of the steady demand constraint increases the number of degrees of freedom, extending the flow redistribution problem to the determination of the increasing demand.

In both cases, a new parking lot is considered as an attraction element (Avarello Carrese 92), like a school or a shopping center. In practice, in a weak regime of illegal parking control, the experience shows an increasing demand due to the construction of a new parking lot.

In order to consider these phenomena, a new model, based on a typical multi-modal assignment procedure, has been built. The model is characterized by an original representation of the auto-network and a new *ad hoc* parking time-volume function for every kind of parking link.

Demand modeling

The proposed model considers a car volume equilibrium on the network. It makes it possible to simulate the choices of different user-classes on a multi-modal network. Each class is characterized by a typical value of the utility associated to every transportation alternative. The model deals with the problem of simulating a transportation system, operating for every time interval in which the day is divided, with as many origin/destination (O/D) matrices as the user-classes that it intends modeling .

The user-classes are not only characterized by trip purpose, but mostly by the stop duration after the trip and before taking the possible following trip. This stop time influences the utility perceived by the users in a decisive way, especially regarding vehicular alternatives because it conditions the value of parking cost, dealt with by the explicit modeling.

We assume the origin/destination matrices for several user-classes and the fact that they contain complete information about demand on all transportation modes. The share of demand resorting to the different transportation modes available on every O/D pair is calculated by the assignment to the network, according to the total performance of the transportation system.

Supply modeling

In order to define transport supply, a detailed description of the multi-modal individual and public networks is required. In particular the following items have to be specified:

- main road network
- centroid-network connections
- parking links (on-street and off-street)
- pedestrian links
- transit links
- transfer links
- waiting links
- parking-transit network links
- transit network-centroid links

The functions representing the different links described assume different patterns that can be generally expressed as the sum of a constant component and a variable one depending on the flows ($f_{l,m}$ with l = links and m = modes):

$$C_{l,m} = C^0_{l,m} + a_{lm} * C^v(f_{lm}). \qquad (5)$$

For better explanations of these type of functions reference should be made to other studies (Branston 76, Matsoukis 80, Bifulco 93). Here attention is paid to the modeling of the parking phenomena on-street and in lots.

The *ad hoc* developed parking time-volume functions, f_{p1}, f_{p2}, used in the simulation model, are shown in Figure 2. It can be seen in f_{p1} that for low volume of parking users, everyone takes t_0 minutes for parking. For increasing volumes, parking time slowly increases because more time is needed to find a free place (usually more time means longer walking distance, too) $t=t_0+t_{v/c}$.

When parking volume is equal to the capacity C_1, the user takes t_c minutes to park; this value is obtained by adding the value of the in-force fine M, multiplied by the probability to be fined P to the default value t_0 . In formula:

$$t_c = t_0 + (M * P) /r, \qquad (6)$$

where r is the conversion rate Lit/mins.

The model, through the parking function f_{p2}, is able to represent the construction of a toll parking lot by a step F (fare), starting from the default value, t_0. A user of a toll parking then takes t_f minutes to park;

$$t_f = t_0 + F/r, \qquad (7)$$

The parking capacity is shown as a step in C_2. If the new parking lot is to be considered free of toll, the parking function is like f_{p1} but it just presents a different value of C_1 .

In the same figure a time-volume function f_{tv}, for a secondary link is shown for comparison.

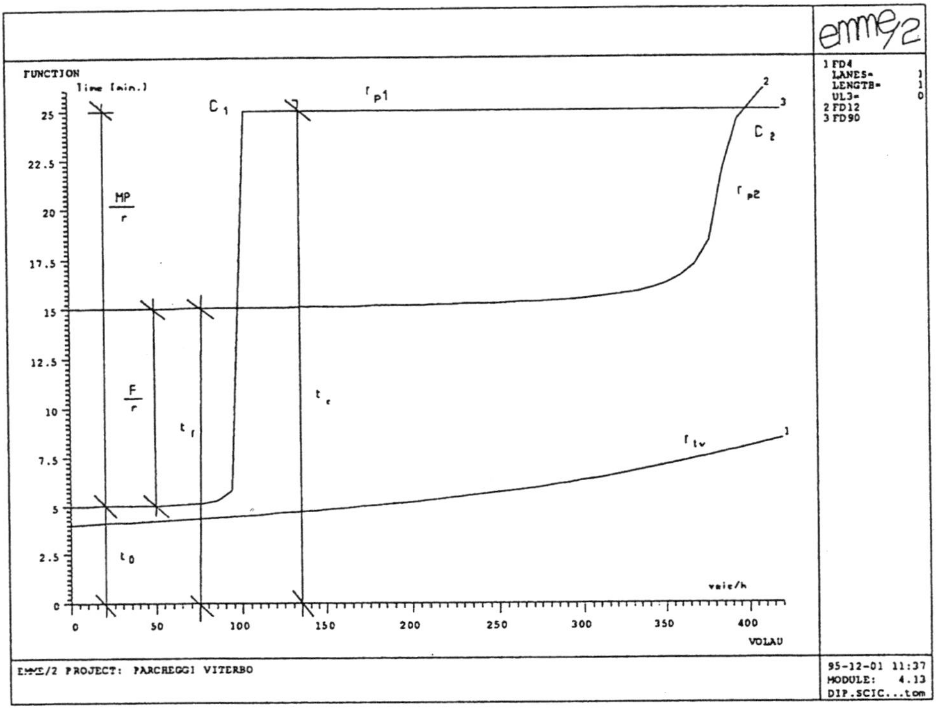

Figure 2 Parking time-volume functions

Multi-modal equilibrium for parking modeling

Supply and demand, outlined above, are inputs of the multi-modal equilibrium assignment model, which consists of:

- a *transit assignment model*, based on the concept of optimal strategies (Spiess and Florian 89). In this model the traveler's choice set consists of all feasible strategies. It is clear that, due to the asymmetry inherent to the transit trip, this model can only be solved efficiently by assigning demands from all origins to a single destination. A non-negative time (or cost) is usually used to quantify each trip component, with the exception of "waiting for a vehicle"; this is quantified by using the statistical distribution of waiting times for the arrival of the first vehicle of a given transit line at a given stop. The model may be solved by an algorithm which consists of two parts. In a first pass, from the destination node to all origins, the optimal strategy and the expected total travel times from each node to the destination node are computed. In a second pass, from all origins to the destination, the demand is assigned to the network according to the optimal strategy.

- a *variable demand auto-assignment model*. This model computes the equilibrium auto-demand, link flows and travel times by the Partial Linear Approximation Method - PLAM - (Florian 86), which is an improvement of the Linear Approximation Method - LAM - (Frank and Wolfe 56) used for Fixed Demand. Given a current feasible solution, the LAM finds a descent direction by solving a sub problem where the objective function is a linear approximation of the nonlinear objective function of the problem to be solved. Moreover, in the PLAM the algorithm may be implemented by requiring only evaluations of the inverse demand function. As only the auto-demand function is required, the inverse demand is obtained numerically, by using the secant method. Given a current solution, the PLAM finds a descent direction, by solving a sub-problem where the delay functions are linearized, but the inverse demand functions are not.
- a *multi-class user-equilibrium assignment*. In the general case different categories of travelers on a given link $a \in A$ experience different cost functions. Let $m \in M$ denote the set of distinct user classes. For each user class *m* the cost on link *a* depends on the class volume $v_a^{m'}$ of all classes *M*, i.e. $s_a^m(v_a^{m'} | m' \in M)$. The problem is equivalent to an equilibrium assignment with generalized costs and corresponds to a variational inequality problem, which, in general, has no unique solution and also cannot be transformed into a convex optimization problem. A simpler form of the multi-class assignment can be implemented. It is assumed that the cost of a link *a* perceived by a user of class *m* can be written as:

$$s_a^m(v_a) = s_a(v_a) + b_a^m \qquad a \in A, \quad m \in M \tag{8}$$

This implies that the different user classes are all subject to the same congestion effect, but each user-class perceives a different constant bias b_a^m. It is easy to show that this simplified multi-class user-equilibrium assignment problem is equivalent to a convex optimization problem of the form:

$$\text{Min} \sum_{a \in A} \int_0^{v_a} s_a(v)\, dv + \sum_{m \in M} \sum_{a \in A} v_a^m b_a^m \tag{9}$$

subject to the usual conservation of flow constraints.

When the auto-demand function is not available, it will be possible to use the fixed demand auto-assignment, simulating only the above mentioned first effect of car volumes redistribution. In fact, in the following model application to a small size town, only the fixed demand auto-assignment will be performed, due to the lack of information about demand function.

If the demand function is a mode choice function, as in the variable demand model, the transit demand can be computed as difference between the total and the auto-matrices, and assigned to the transit network. Frequently the transit impedances are computed once before the auto-assignment and are kept constant during the assignment, even though the transit time functions used have the auto-times as parameters. It is also possible to end the auto-assignment after a chosen number of iterations, recompute the transit impedances, and then readjust the auto-matrix to reflect the changes in the demand function; this may be done by obtaining the expression of the inverse of the demand function analytically (or by numerical procedure) and then recomputing the impedance values. The variable demand auto-assignment is then resumed with the effect of the adjusted transit impedances accounted for. It is possible to carry out these steps automatically, and to set up a bi-modal equilibrium assignment with variable transit times.

If the demand function is more general than a mode choice function, a similar approach may be used to set up more complex multi-modal equilibrium assignment models. This approach is valid since the travel time (impedance) variables of the demand function are separable by O-D pair. From an operational point of view the multi-modal equilibrium model proposed, representing explicitly the parking phenomenon, gives interesting and reliable results. Then it is true that the equilibrium uniqueness, is only ensured under particular conditions (Daganzo 93, Bifulco 93) for one mode assignment.

If a multi-modal model is considered, the uniqueness of the equilibrium solution is verified only under less realistic assumptions. For a multi-modal model, with for example only one user-class, the generalized cost of two transportation modes on the same O/D pair must be the same. Besides, it is necessary that the contribution of the congestion cost (transit on board time) on road links to the generalized cost is the same for all modes. This situation is not realistic because it has been proven that the contributions of cost concerning different transportation modes are weighted differently in the determination of the generalized transportation cost. Furthermore, in multi-modal models, the generalized volume on links, is not the algebraic sum of every mode volume.

Further it can be noted that network equilibrium models traditionally rely on the assumption of continuous travel cost function. However, congestion pricing schemes make use of discontinuous step function tolls and the existence of network equilibria in the presence of discontinuities has been explored. In particular it has been shown that when such costs are at least lower semicontinuous, a behaviorally meaningful notion of user-equilibrium can still be defined which reduces to Wardrop equilibrium in the continuous case. In addition, it is shown that such equilibria are guaranted to exist under fairly general conditions (Bernstein and Smith 1994).

In order to apply the model it is only necessary to choose a fare policy (fixing the coefficient of the parking curves), the budget allocation and the available location for all parking lots. In fact, in most cities it is not possible to build parking lots of any dimension everywhere. The parking plan design (Fig.3) can be set initially in terms of available areas and maximum dimensions allowed in the areas.

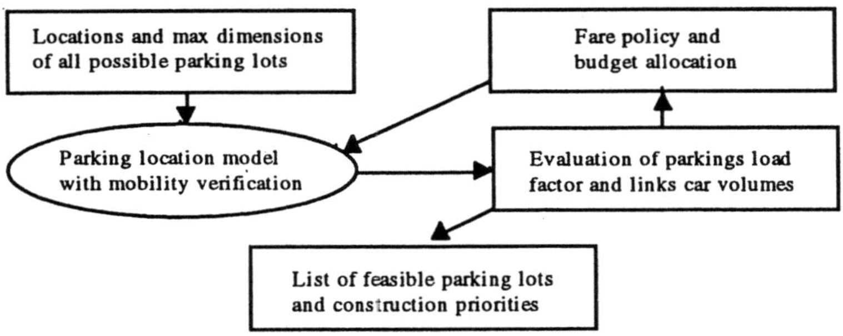

Figure 3. Parking plan design using the proposed model

The outputs of the model are the occupation status of all parking lots available and car volumes on all links of the road network; from these it is easy to find travel times, costs and all the usual performance indicators. By changing the fare policy (modifying the coefficients of the parking curves), occupation status and car volumes on network links can be substantially altered. There are then two choice problems concerning:

- which parking lot has to be built;
- the optimal fare policy to introduce.

In fact, it is not always possible to build all the feasible parking lots, but some priority problems must be faced. However, it has to be considered how the proposed model can simulate the development of all the parking and traffic system in correct terms and how it can provide for outputs (occupation status, rotation, parking accumulation, congested links) that enable effective choices.

Last, it is possible to simulate the distribution of parking volumes consequent to different fare policy, which establish different fares for each parking lot.

5. Case Study of the Model Application

In order to assess the performance of the proposed model, some applications to real cases of different dimensions are foreseen:

1) small size town;
2) medium size town;
3) metropolitan urban area

In the following, the results of a study carried out in the town of Viterbo (an example of case 1), the first one for which a usable data base was available, are presented. However, a realistic demand function was not available and so a fixed demand assignment was the only one possible.

The actual public transportation system presents headways greater than 30 minutes, and this condition is incompatible with a Park & Ride (P&R) hypothesis. Given the size of town, a realistic P&R hypothesis exists only if it is possible to set up a specific high frequency system (e.g. minibus), connecting the P&R lots with the main attractive poles.

Given the structure of the different time-volume functions used, including the proposed parking time-volume ones, all of the model's activities can be seen in the flow chart of Figure 4.

Viterbo has about 60,000 inhabitants. The study area, including the historical center (CBD) inside the ancient walls and its immediate surroundings, is divided into 23 zones.

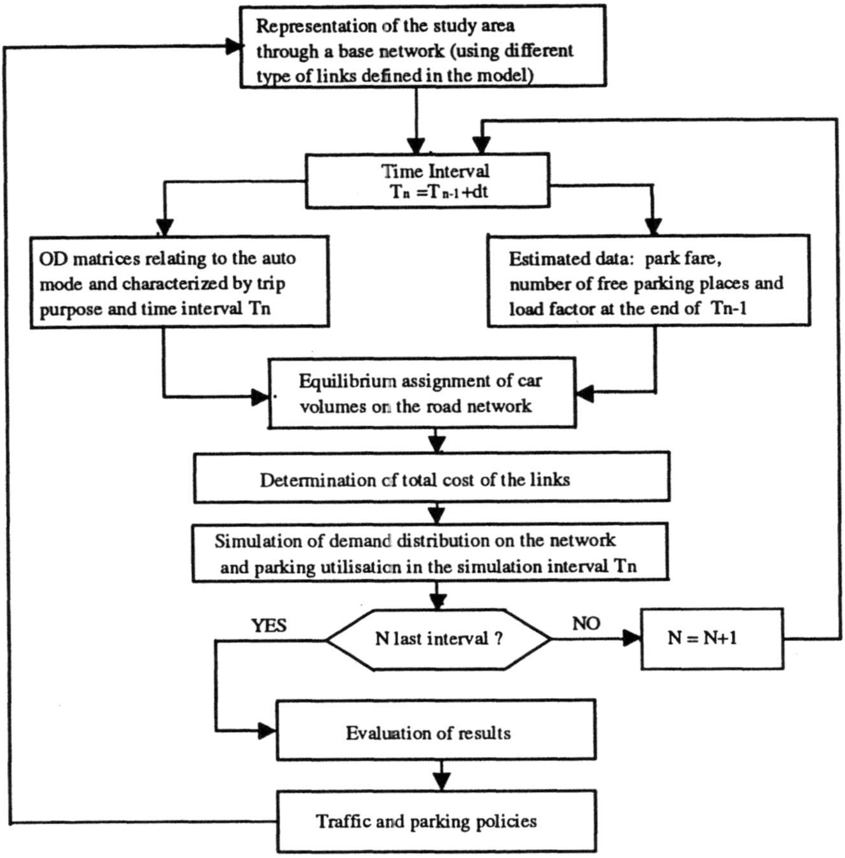

Figure 4. Flow chart of activities of the parking location model

The base network consists of 64 nodes and 266 links. First, a main road network is characterized by 86 links and a secondary road network is formed by 180 links. This set also contains 42 "parking links", and their capacity represents the number of actual parking places in the single zones. These parking links (there can be more for each zone) also represent the access link to the parking lot from the network; further pedestrian links make it possible to reach destinations even from the neighboring zones.

A window of the transportation network is represented in figure 5.

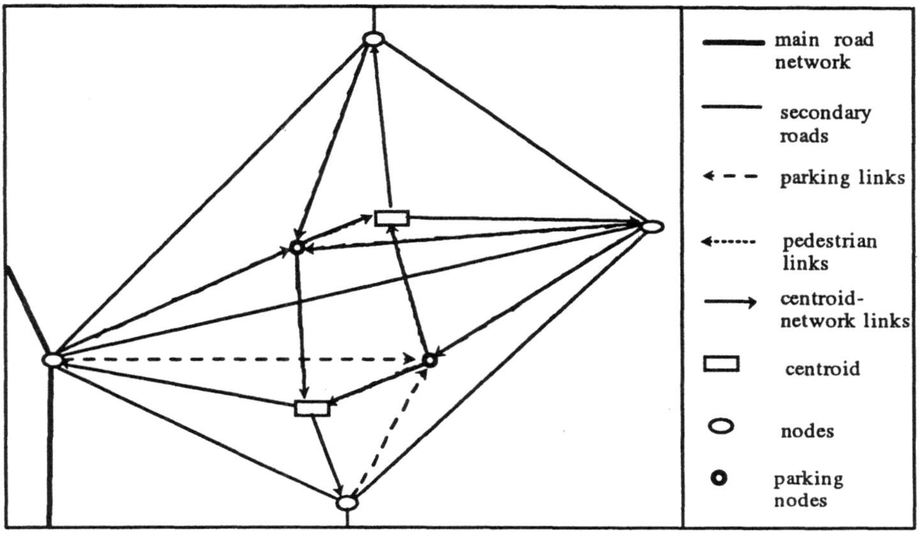

Figure 5. Network representation: car, parking and pedestrian links

In the first simulation the actual situation has been represented, through the peak hour matrix which, due to access limitation to the center, goes from 10.20 to 11.20. This simulation regards parking users-behavior and the supply function represents the situation of parking availability at the beginning of the peak hour. The number of free spaces derives from the calculation of the accumulation in every zone, due to the previous hours O/D flows.

In figure 6 the capacity of each parking link is shown; the total sum of all parking places available is 3,492 places (including 420 toll places). Then the equilibrium assignment is performed. In figure 7 the flow-capacity ratio on each parking link is shown. Results show that 15 parking links reach the capacity, while 27 links exceed it. It means that 873 users park illegally. In figure 8 the pedestrian flows of users that park in a zone which is near to their destination are shown; the total sum of users that park in a zone which is not their destination is 825persons/h.

In the second simulation 5 new parking lots have been added inside the ancient walls; they have been extracted from the last parking plan of the town. It can be easily observed that there is a volume redistribution of parking flows in many links.

In figure 9 the project capacity of 5 parking links is shown: the total sum of all parking places available is now 4,942 places (including 1,350 new places). In figure 10 the project flow-capacity ratio on each parking link is shown. Results of the project simulation show that only 13 links are in oversaturation and 261 users park illegally. Figure 11 shows that pedestrian flows reach the value of 1,332 persons/h.

Evaluation of the results can be conducted on the basis of congested parking links, parking and pedestrian volumes. In fact due to the availability of only accumulation and rotation parking data and due to the lack of a parking duration cluster analysis, a comparison between times would not be interesting without the simulation of a realistic fare policy.

The comparison between the two simulations shows the following results: as it is shown in table 1, the effects of increasing the number of free parking places (+39%) are: decrease of the number of oversaturation links (-52%)and illegal parking users (-70%) and increase of pedestrian flows (+61).

	PARKING CAPACITY	*OVER SATURATION LINKS*	*ILLEGAL PARKING USERS*	*PEDESTRIAN*
I SIMULATION NO PROJECT	*3492*	*27*	*873*	*825*
II SIMULATION PROJECT	*4842*	*13*	*261*	*1332*
DIFFERENCE S (II - I) / I	*+39%*	*-52%*	*-70%*	*+61%*

Table 1. Comparison of results

Finally, the model has provided low occupation status for the toll parking lot with a decrease in the second simulation due to the increase of free of toll capacity. Anyway, the model is able to show satisfying differences in results according to the different fare policies which can be simulated.

5. Conclusions

In this study the analysis of the remarkable effects is limited only to those concerning the system performances of the transportation supply and the mobility demand. A general agreement exists on this point among the analysts (Bonsall et al. 82, Polak et al. 90): the models traditionally used are often inadequate whereas the characteristics of an "ideal" model should be:

- mathematical consistency in the representation of route and location choice;
- use of easily available data;
- explicit modeling of the demand / supply interaction;
- explicit temporal approach;
- multi-user approach;
- multi-modal approach and modeling of the demand-elasticity at least regarding the modal choice level .

The "key" variables considered in the parking planning are:

- capacity, typology and location of the parking lots;
- fare structure;
- parking restriction and regulation;
- modeling of illegal staying and overstaying.

It must be stated that the proposed model cannot determine directly which parking lots have to be built, but it can only describe the occupation status of each one; in addition, it computes the car volumes of all the integrated transportation network components. It is interesting to observe that the model contains the ideal "key variables" and also presents most of the major characteristics of the "ideal" model. Only the explicit modeling skill of demand-elasticity is not included. A limit of the model is the need for further clustering in relation to human behavior (trip purpose) when stop duration is the same.

Application of the model to further real cases is foreseen.

Acknowledgments

The application of the model to the town of Viterbo would not have been possible, without the data base relating to demand modeling, which has been provided by ENEA. The authors hope to obtain more demand and parking data in order to develop their model.

References

[1] Austin T. W., (1973). "Allocation of parking demand in a CBD", *Transportation Research Record*, n. 444, pp. 1-8.

[2] Austin T. W., (1973). "Estimation of potential use of peripheral parking for Los Angeles CBD", *Transportation Research Record*, n. 444, pp. 21-26.

[3] Avarello L. e Carrese S. (1992) "Integrazione tra pacchetti esistenti e modelli di simulazione per nuove necessità: il caso del Park & Ride". *13° Corso in: Comunicazioni, Trasporti e Governo della Mobilità*. Capri, giugno 1992.

[4] Bates J., (1972). "A gravity allocation model for parking demand". *Transportation Research Record*, n. 395, pp. 1-5.

[5] Beinborn E. A. e Gur Y. J., (1984). "Analysis of parking in urban centers: equilibrium assignment approach", *Transportation Research Record*, n. 957, pp. 55-62.

[6] Bernstein D., Smith T. E., (1994). "Equilibria for networks with lower semicontinuous costs: with an application to congestion pricing", *Transportation Science* Vol. 28, No. 3, August 1994, pp.221-235

[7] Bifulco G.N., (1993). “A stochastic user equilibrium assignment model for the evaluation of parking policies”, *EJOR*, vol.71-2

[8] Bonsall et al., (1982). “A model of the combined parking location mode split/journey timing decision”, *paper of I.T.S. Leeds University* (UK).

[9] Branston D., (1976). "Link Capacity Functions: a review", *Transp. Res.*, Vol 10.

[10] Carrese S. e Gori S. (1993) “Elementi per un piano parcheggi compatibile con i problemi di congestione urbana”, *Seminario su rilievi e modellizzazione del traffico veicolare*. Università di Napoli, Giugno 1993

[11] Daganzo C.F., (1993). "Stochastic Network Equilibrium with multiple vehicle types and asymmetric, indefinite link cost iacobians", *Transp. Science*, pp. 282-300

[12] Ellis R. H. e Rassam P. R., (1970). "Structuring a system analysis of parking", *Transportation Research Record*, n. 317, pp. 1-13.

[13] Ellis R. H. e Rassam P. R. e Bennet J. C., (1972). "Development and implementation of a parking allocation model", *Transportation Research Record*, n. 395, pp. 5-20.

[14] Florian M., (1986). "Non linear cost network models in transportation analysis". *Mathematical Programming Study* 26, pp.167-196.

[15] Frank M. and Wolfe P., (1956). "An algorithm for quadratic programming". *Naval Res. Logist.* Q., 3, pp.95-110

[16] Gattuso D., (1992). “Un modello di assegnazione dinamica della domanda di sosta. Applicazione per la valutazione del Programma Urbano Parcheggi della città di Reggio Calabria”, *Modelli e Metodi per l'analisi ed il Controllo delle Reti di Trasporto*, a cura di *Cascetta E.* e *Salerno G.* S.I.D.T. 1992 Angeli F. Collana Trasporti.

[17] Gray V.O., (1972). “Parking space allocation by computer model”, *Transportation Research Record*, n. 395, pp. 21-30.

[18] Hunt J.D. e Teply S., (1993). “A nested logit model of parking location choice”, *Transportation Research*, Vol. 27B,n. 4, pp. 253-265.

[19] Matsoukis E.C. (1980). "Speed/flow relationships in the central Glasgows area", *Transp. Planning and Tech.*, Vol. 6, pp. 21-26.

[20] Odoni A.R. e Larson R.C., (1981). "Urban Operation Research", *Prentice-Hall, Inc., Englewood Cliffs, N.J.07632,* pp. 383-480.

[21] Papola N., (1993). "Trasporti urbani" *Atti del dipartimento I.T.S. Trasporti n° 10, ESAGRAFICA*, Roma.

[22] Perl J. e Ho P.K. (1990). "Public facilities location under elastic demand", *Transportation Science*, Vol. 24, n. 2, pp. 117-136.

[23] Polak J. et al., (1990). "The application of CLAMP to the analysis of parking policy in Birmingham city centre", *Proc. of the PTRC Summer Annual Meetings*

[24] Spiess H. and Florian M. (1989). "Optimal strategies: a new assignment model for transit networks", *Transp. Res.* B, Vol. 23B, pp. 83-102.

[25] Whitlock M., (1973). "Use of linear programming to evaluate alternative parking sites", *Transportation Research Record*, n. 444, pp. 9-19.

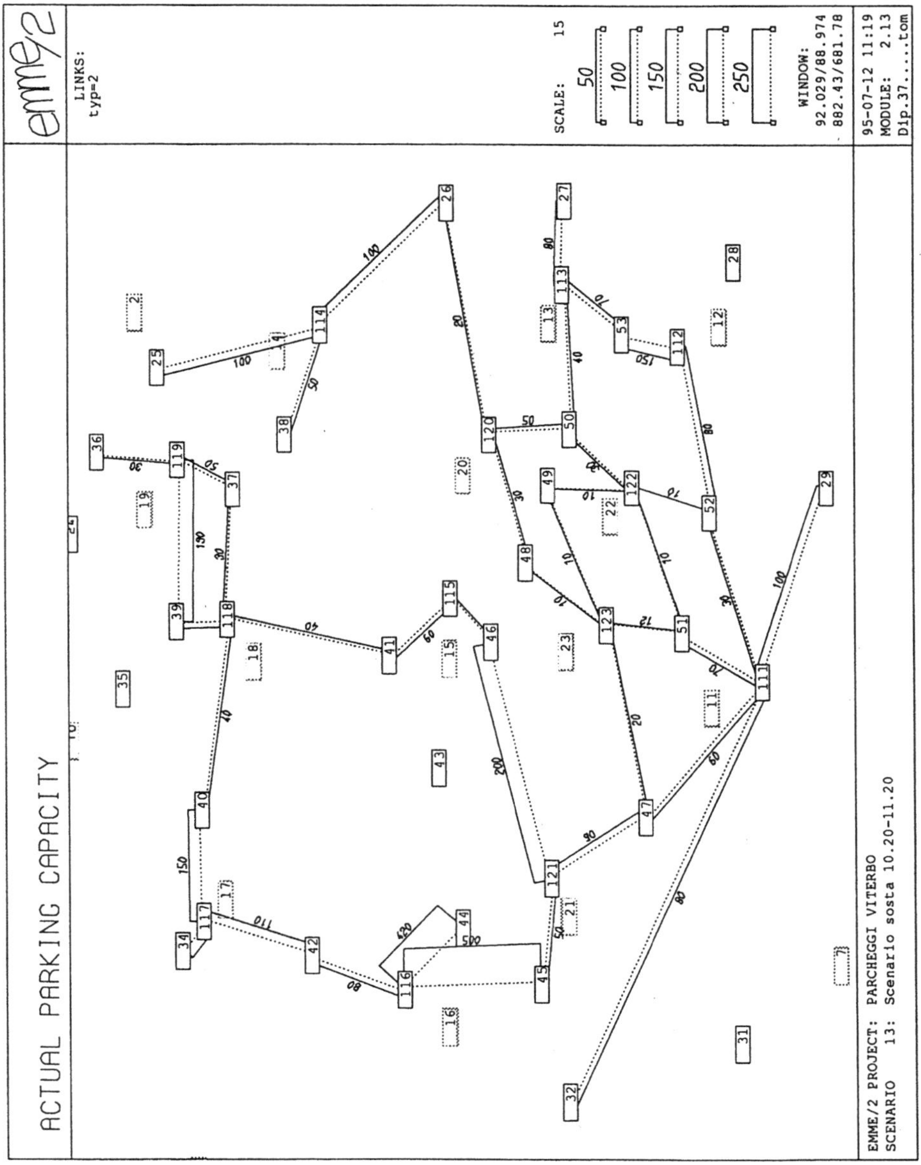

Figure 6. First simulation: actual parking capacity

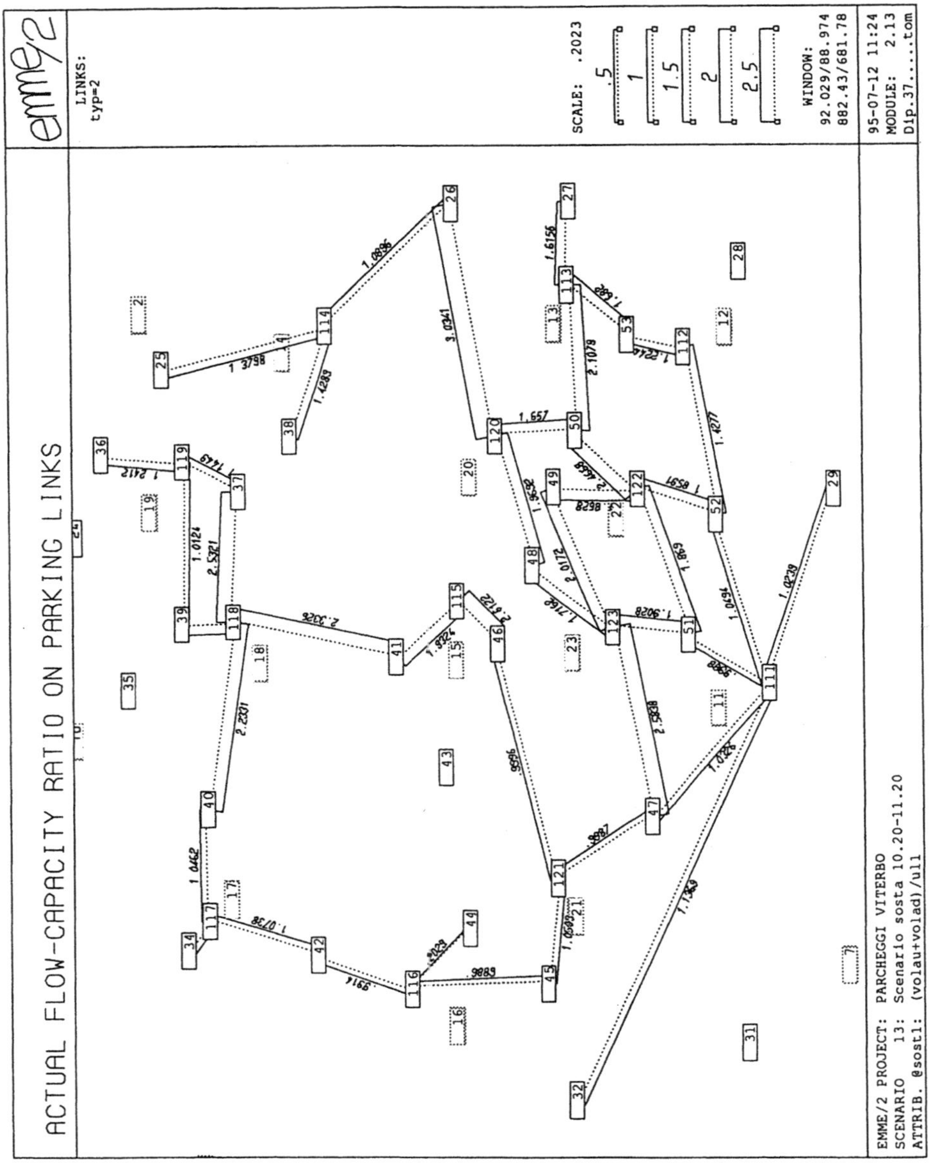

Figure 7. First simulation: actual flow-capacity ratio on parking links

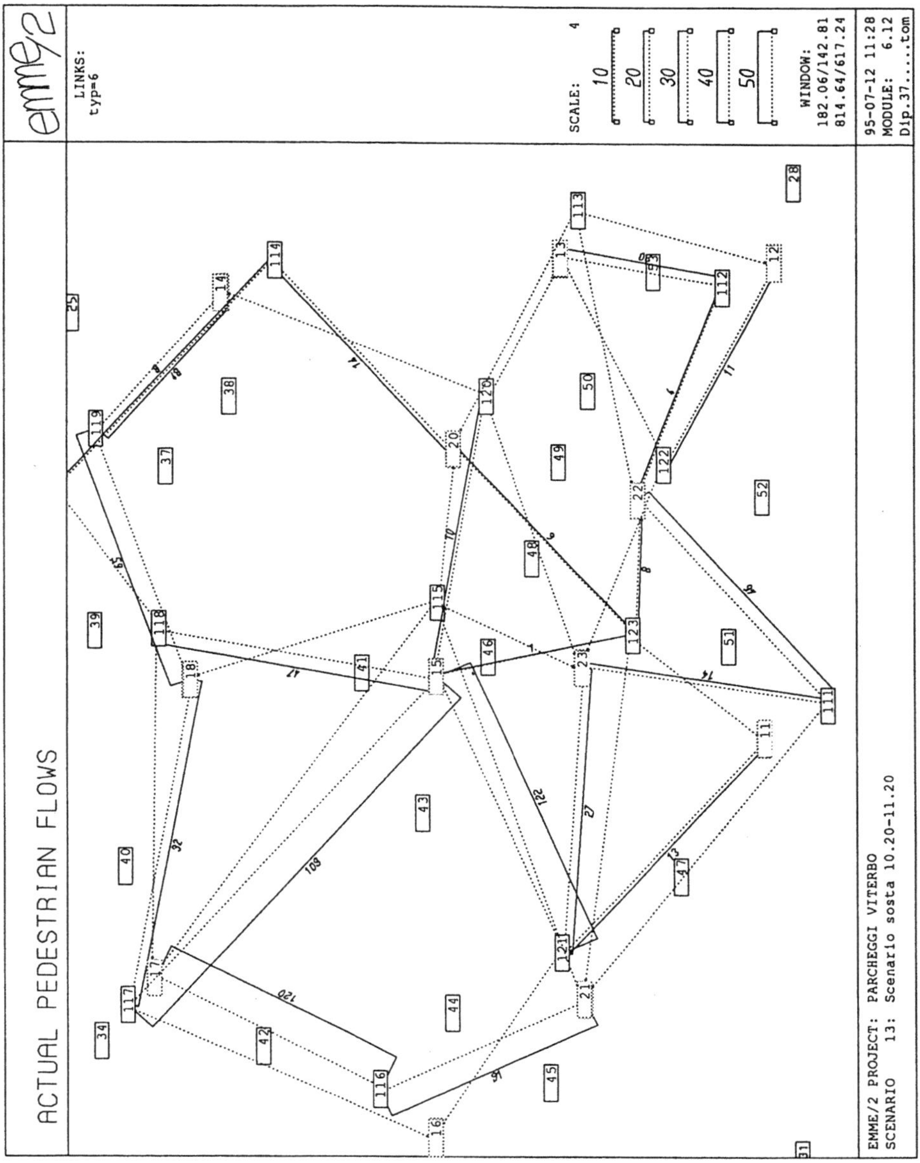

Figure 8. First simulation: actual pedestrian volumes distribution

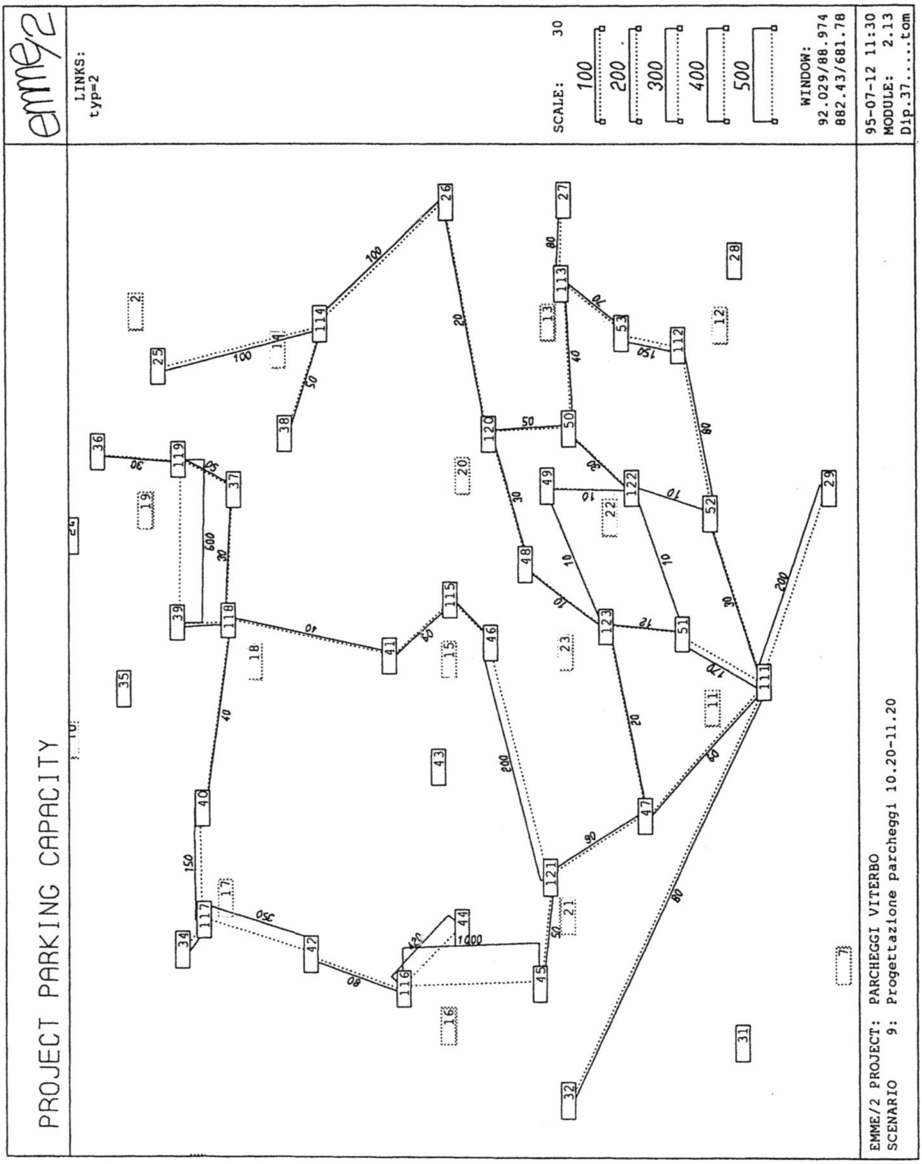

Figure 9. Second simulation: project parking capacity

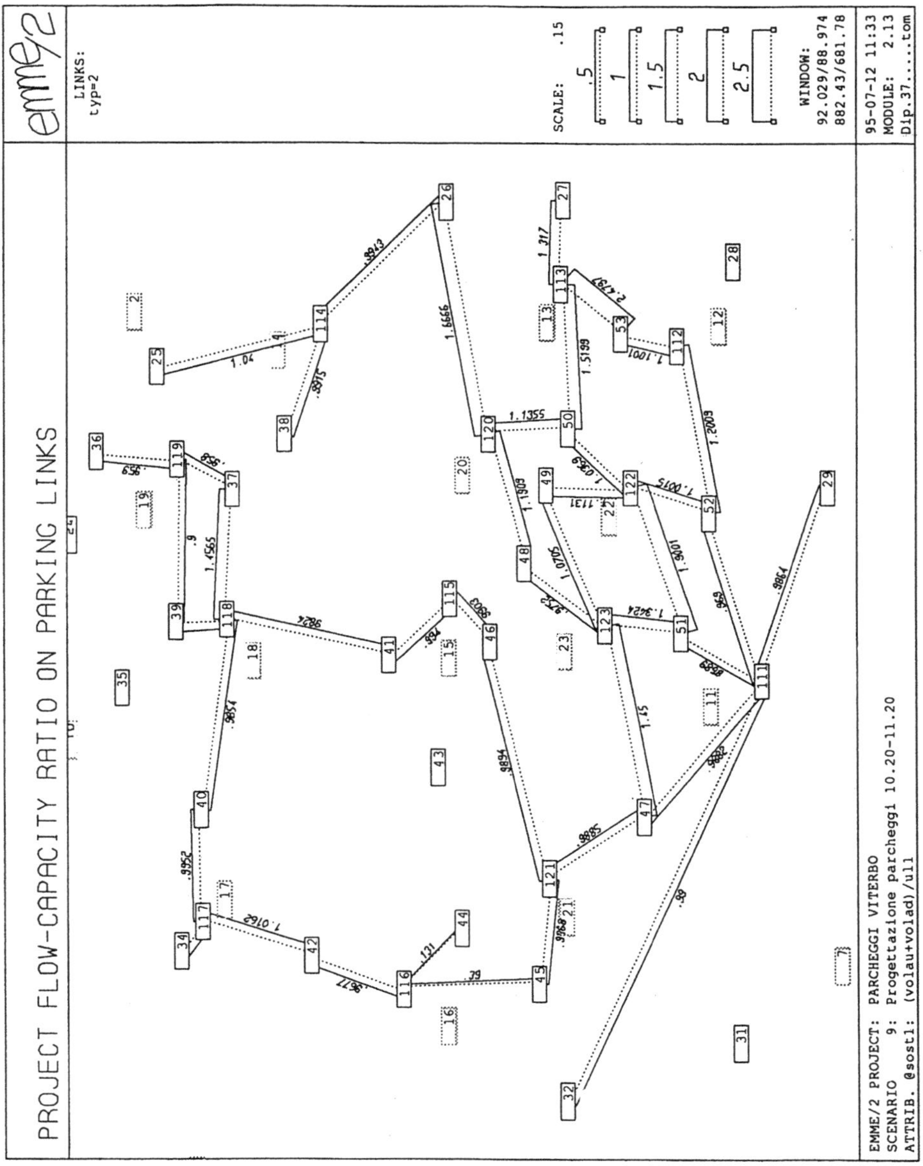

Figure 10 Second simulation: project flow-capacity ratio on parking links

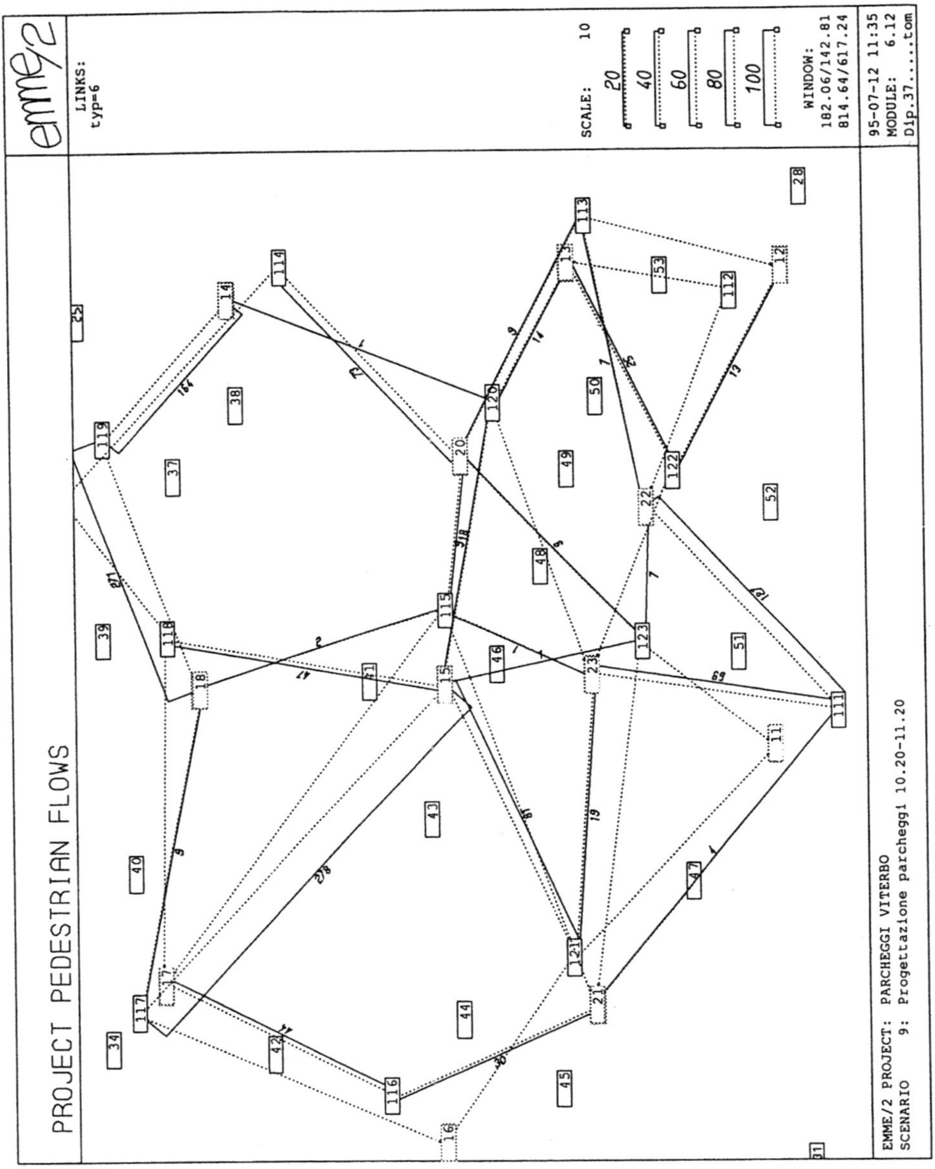

Figure 11. Second simulation: project pedestrian volumes distribution

A DSS PROTOTYPE FOR URBAN INTERMODAL PATH PLANNING WITH PARKING MANAGEMENT[1]

Anna Sciomachen

Istituto di Matematica Finanziaria, Universita' di Genova, Via Bertani 1, 16125 Genova, Italy

In this paper the problem of guiding car drivers to parking facilities in an urban intermodal transportation network is considered. Origin-Destination (O-D) shortest paths are determined in different situations, corresponding to possible constraints in the network, while looking for available parking spaces. Presently, time, transit, and parking capacity constraints are considered. The proposed approach, which is based on a Constraint Satisfaction Problem (CSP) formalism, allows the integration of the parking management and the multimodal path planning. In particular, if no parking space is available within a "walking neighbourhood" of the required destination, the shortest O-D path is computed both on the private and on the public transportation networks.
A software prototype of a Decision Support System (DSS), that can be considered as a first resolution step to a more complete and rigorous research aimed at implementing an on-board parking reservation service, is described.
An experimental evaluation of the performances of this approach is reported together with a test-bed related to traffic informations of a rush-hour in the city of Milan.

1. Introduction and Problem Definition

In this paper the problem of guiding a car driver to a parking facility is considered [2]. In particular, a car driver in an urban area has to reach the required destination and to choose a parking space with the aim of minimizing the total travelling time.

---

[1] This work has been partially supported by the Italian National Research Council (CNR) "Progetto Finalizzato Trasporti 2".

Parking facilities close to the destination may not be available due to saturation or traffic restrictions. In order to make the correct choice, the car driver has to know the actual availability of a parking space within a "walking neighbourhood" of D. If no parking space is available at D, he/she has to know the availability of other parking spaces that make possible to reach the required destination using the public transportation system. In this case the required O-D path is hence computed integrating the private and public transportation networks, while guaranteing to the car driver a parking space in a given transit node, where the changing between public and private transportation takes place. The idea is to suggest a suitable O-D path while looking for an available parking space, in such a way that a set of possible constraints (e.g. time, transit or user visiting requirements) is satisfied.

The proposed approach allows the integration of the parking management and the multimodal path planning: the benefit of such an integration is the possibility of supplying shortest time paths, thus resulting in a decrease of the traffic congestion and air/noise pollution.

The main goal of the paper is to present the basic idea of a prototype of a DSS for implementing an on-board "parking reservation" tool that allows the car driver to make a reservation for the selected parking facility based on its actual availability; in this way, the car driver will be able to find the previously reserved parking space independently on a possible delay on the O-D path, thus avoiding that the parking becomes full when he/she gets to it. A more general case, where the probability of finding available parkings spaces is considered, is presented in [3].

The transportation network is modelled on a digraph $G = (V, E)$, where node set V consists of properly selected relevant locations in the urban area, and arc set E consists of possible road connections between different locations. (For an accurate introduction to urban transportation the reader is referred to [13, 16, 17]).

In this paper a Constraint Satisfaction Problem (CSP) methodology [10, 14] is proposed for properly handle all the different situations arising in modelling this problem. A wide range of practical problems can be solved using the CSP and Constraint Logic

Progamming (CLP) approaches. In particular, the Applause project [12] has developed applications that use a parallel implementation of a CLP language for monitoring planning, tourist advice and environment monitoring and control. Moreover, CSP applications to, among others, car sequencing [7] and floor planning [11] have been presented. However, up to now, at least at the author's knowledge, only very few real-size problems have been solved using this technique in the area of traffic and transportation (see e.g. [5] for air traffic control problems).

The proposed CSP approach is thought as a possible basis of a DSS for the urban path planning [17].

The organization of the paper is as follows. The representation of an urban intermodal transportation network together with some basic definitions and notations are given in Section 2. Section 3 briefly describes the CSP methodology and presents the proposed modelling approach. Section 4 concerns with the computation of the optimal O-D path. The basic structure of the DSS under development is described in Section 5 along with a case study related to a rush-hour in the city of Milano. Finally, some concluding remarks and comments on future research are given in Section 6.

2. The Transportation Network

The first step in modelling a multimodal path planning problem is to represent the urban area under consideration by a transportation network in which both the private and the public transportation systems are represented [16].

Starting from G, let $G' = (V', E')$ denote the digraph representing the *private* transportation network and $G'' = (V'', E'')$ denote the digraph representing the *public* transportation network, such that G thus results superimposing G' on G'', i.e. $G = (V' \cup V'', E' \cup E'')$, provided that $I = V' \cap V'' \neq \emptyset$.

Each node $i \in I$ is labeled with its current parking availability $Q(i)$. The weight associated with each arc $(i,j) \in E$ represents the travelling time between nodes i and j. In particular, $t'(i,j)$ gives the weight associated with arc (i,j) $\forall$ $(i,j) \in E'$, corresponding to the car travelling time between i and j, and $t''(i,j)$ denotes the weight associated with arc (i,j) $\forall$ $(i,j) \in E''$, corresponding to the public transportation travelling time between i and j. As it will be explained in more details in Section 5, the travelling time associated with each arc $(i,j) \in E$ are computed with respect to different rush hours and properly fit the classical downflow functions derived from the literature (see e.g. [4]).

In this paper only the travelling time is considered as arc cost, however more general cost functions may be considered as well.

Finally, let $T(i,j)$ be the time related to the shortest path between each pair i,j of connected nodes in V, where the shortest path is computed on G using both the private and the public transportation networks without any additional constraint. For this purpose, any of the shortest path algorithms presented in literature can be used: see Gallo and Pallottino [8] for a detailed survey of such methods considered from a computational point of view.

3. The CSP Model

In the present problem the decision variables are all the possible paths on G; they are modeled as vectors

$$X = (x_1, y_1, x_2, y_2, \ldots, y_{n-1}, x_n) \tag{3.1}$$

i.e. alternate sequences of nodes $x_i \in V$ and arcs $y_j \in E$, connecting the origin node O (x_1) to the destination node D (x_n). Nodes and arcs in G represent candidate *values* for the elements of the path vector X.

It is worth mentioning that a given origin node x_1 is necessarily a node from which the driver leaves by car, while a destination node x_n is a node where he/she must go to, possibly by car. Let a transit node x_i , i = 2, ..., n-1, be defined as any node between x_1 and x_n.

Within path X there must be a parking node, say x_k, such that k = 1 means that the whole path is made by using the public transportation network G", while k = n implies that only the private transportation network G' is used; for 1 < k < n a multimodal path is selected.

Different situations, representing additional constraints in the network, are modeled as relations among the components of the path vector X. In the present problem, the following six different types of constraints are considered:

C_1. Node-avoiding constraint: external accidents, congestions or road interruptions can result in a modification of the usual path between the O-D pair of nodes. In particular, if a given node $a \in V$ is not reachable, the transit through it must be forbidden, i.e.

$$x_i \neq a \;\forall\; x_i \in X \tag{3.2}$$

C_2. Arc-avoiding constraint: as in C_1, if a given arc $(i,j) \in E$ is not practicable, then

$$y_j \neq (i,j) \;\forall\; y_j \in X \tag{3.3}$$

C_3. Node-forcing constraint: the car driver can explicitly require to visit a given node $a \in V$ in between x_1 and x_n, i.e.

$$\$ \; x_i \; \hat{I} \; X \; \mathrm{I} \; x_i = a \tag{3.4}$$

C_4. Arc-forcing constraint: as in C_3, the car driver can explicitly require to pass through a given arc $(i,j) \in E$, i.e.

$$\$ \; y_j \; \hat{I} \; X \; \mathrm{I} \; y_j = (i,j) \tag{3.5}$$

<u>C_5. Parking-capacity constraint:</u> if $x_k \neq x_1$, i.e. if the car driver leaves the origin by car, at least one parking space must be available at the candidate parking node x_k, $k = 2, \ldots, n$:

$$\$\ x_k\, \hat{\mathrm{I}}\ X \mid Q\,(x_k\,) > 0 \tag{3.6}$$

<u>C_6. Total-travelling-time constraint:</u> the car driver may specify the maximum available time T_{max} for going from x_1 to x_n. In this case, for all the n-1 arcs connecting the pair of nodes O-D, relation (3.7) must hold:

$$\sum_{j=1}^{k-1} t'(y_j) + \sum_{j=k}^{n-1} t''(y_j)\ \pounds\ T_{max} \tag{3.7}$$

Note that constraint C_6 does not affect the shortest-path search strategy, as it is described in the next section. In fact, once the shortest path has been found, condition (3.7) is tested whenever a T_{max} value is specified, in order to either accept or reject the selected path.

The car driver asks for the desired path by specifying the origin and the destination nodes, the maximum time T_{max} available for the trip, and possibly some place / road passing requirements together with the preferred transportation modality (for instance, the use of the only public transportation network could be required). The following four different *travelling modes* are hence foreseen:

<u>M_1. Private modality</u>: the optimal (x_1-x_n) path is computed only on the private transportation network G'. In this case the parking node x_k is the destination node and it must satisfy condition (3.6) for $k = n$.

M_1 is hence applicable only if x_1, x_n Î V' and Q(x_n) > 0.

<u>M_2. Private and walking modality:</u> if it is not possible to reach the required destination by car because either condition (3.6) is not satisfied or node x_n is not reachable (due to traffic restrictions or congestions), the goal is to look for an available parking space in an

intermediate node x_k that is located within a "walking distance" neighbourhood of the destination and such that $y_k = (x_k, x_n)$. i.e. x_k is directly connected to x_n in G. Note that in this case the maximum allowable walking time T_w is supposed to be given (it may be reasonably about 10 minutes).

M_2 is hence applicable only if x_1 Î V', x_k Î V', $Q(x_k) > 0$ and $t^w(y_k) < T_w$, where $t^w(y_k)$ is the on-foot travelling time between the parking node and the destination.

M_3. Public modality: the optimal $(x_1\text{-}x_n)$ path is computed only on the public transportation network G''. In this case, the parking node x_k is the origin node and it must satisfy condition (3.6) for $k = 1$.

M_3 is hence applicable when x_1, x_n Î V'' and $Q(x_1) > 0$. In particular, if the car driver explicitly asks for this travelling mode, it is assumed that an available parking space at x_1 has been already found.

M_4. Intermodality: the optimal $(x_1\text{-}x_n)$ path is computed both on the private and on the public transportation networks, i.e. on G. Indeed, starting from x_1 on the private network, it is to look for an intermediate node x_k belonging to $V' \cap V''$ such that a parking space is available at x_k and there is the possibility of going from x_k to x_n by using the public transportation.

M_4 is hence applicable when x_1 Î V', x_k Î I, $Q(x_k) > 0$ and x_n Î V'' .

If the car driver requires to use necessarily the car, either travelling mode M_1 or, if it is not the case, M_2 is considered, while if the car driver does not ask for any particular travelling mode, then the shortest O-D path among the four allowed travelling modes is computed.

4. The Resolution Approach

The proposed approach for finding the desired O-D path has been implemented using the Prolog language, that is particular suitable for easily enlarging the set of constraints expressed in a CSP therminology. Prolog is a non-procedural language [6], therefore it is not necessary to specify the solution approach but it is enough to describe facts and rules relevant for the problem for letting the Prolog internal search algorithm provide the solution.

Dealing with a transportation network, facts (predicates) are the elements of G used to describe the urban area together with arc and node attributes (e.g. travelling times and parking capabilities), while rules (clauses) are the restrictions imposed on the path so that all the involved contraints are satisfied.

The available knowledge about the urban area under consideration is summarized in a small set of predicates describing both the private and the public networks in terms of their nodes and arcs. In particular, each node i Î V is described by the predicate "node" having the following notation

node (Name, Available_parking_space, File_name)

where "Name" is the node name (e.g. a street name), "Available_parking_space" is the value of $Q(i)$, i.e. the total amount of available parking space in the zone identified by node i, and "File_name" is the file where the information about the effective location of the parking facilities are stored.

Note that the predicate "node" provides a single notation both for parking nodes and transit ones where no parking is available: only the "Available_parking_space" parameter enables to characterize them. Moreover, the predicate "node" does not specify whether a node belongs to the private or to the public network: in fact, only the definition of the predicates related to E' and E'' takes into a proper account the difference between the private and the public networks. In particular, an arc belonging to E' is described by the "private_arc" predicate

private_arc (Origin_node, Arc_name, Destination_node, Time) while an arc belonging to E'' is described by the "public_arc" predicate

public_arc (Origin_node, Path, Destination_node, Time)

where "Origin_node" and "Destination_node" are in both predicates the two nodes defining the arc, "Arc_name" is the description of the private arc y_j Î E' (in terms of the sequence of corresponding streets, squares, and so on), "Path" is the description of the public arc y_j Î E'' (in terms of line numbers and changing stops), and "Time" is the value of $t'(y_j)$ in case of a "private_arc" and of $t''(y_j)$ otherwise.

The resolution algorithm consists of a set of clauses. The main clause is the following:

program $(x_1, x_n, F, A, T_{max}, M_h)$

where x_1 and x_n are as before, F is the list of possible node/arc visiting requirements given by (3.4) and (3.5), A is the list of possible node/arc avoiding constraints given by (3.2) and (3.3), T_{max} is the maximum available time for the path as it is given in (3.7) and M_h, h = 1,...,4, specifies whether restriction on G are given as for the travelling mode is concerned.

For a better understanding of how the "generate and test" approach works, the optimal path search strategy is here below reported in a Pascal-like notation, in the case of travelling mode M_4 (similar steps are used in the case of any other travelling modes).

In the procedure "program", the Dijkstra's shortest-path algorithm is used and defined by the "SP" procedure

SP $(x_i, x_j, G, T_{max}, T_{new})$

where x_i and x_j are the origin and destination nodes of the desired path, G is the urban transportation network model, T_{max} and T_{new} are the maximum available time and the shortest path time, respectively.

Procedure PROGRAM (x_1, x_n, F, A, T_{max}, M_4);
begin
comment: "delete" all the elements of the node/arc avoiding set A
for all $x_j \in A$ delete_node(x_j, V');
for all $y_j \in A$ delete_arc(x_j, E');
if T_{max} **is** not specified **then set** T_{max} **to** ∞ ;
for all $x_k \in G'$ **begin**
comment: search a candidate parking node x_k
if $Q(x_k) > 0$ **begin**
comment: search the shortest path on G'' from x_k to x_n and store the time in T_{public}
SP(x_k, x_n, G'', T_{max}, T_{public});
comment: compare T_{public} and T_{max}
if $T_{public} < T_{max}$ **then begin**
comment: store the path from x_k to x_n in X
update_path(x_k, x_n, X);
set T_{ref} **to** $T_{max} - T_{public}$;
comment: compute the time available for the path from x_1 to x_k and store it in T_{ref}
if $R = \varnothing$ **then SP**(x_1, x_k, G', T_{ref}, $T_{private}$) **else...**
comment: if the transit requirement set R is not empty, split the x_1, x_k path...
if $T_{private} < T_{ref}$ **then begin**
comment: store the second part of the path in X
update_path(x_1, x_k, X);
set T_{max} **to** $T_{public} + T_{private}$;
end;
end;
end;
end;
show_path(X);
end.

It is easy to note that the main computational effort arises from the first external loop "for all x_i..." at line 4 of the above procedure. In order to improve its computational performances, a sort of "short-cut" procedure has been implemented such that the maximum available time T_{max} is given as a parameter with the aim of avoiding unuseful computations that will lead to paths that exceed the maximum available time. This is done breaking the loop whenever the "find min..." statement detects as minimum time T(O,D) a value that is greater than T_{max}..

Moreover, in order to efficiently compute the desired O-D path, the whole problem has been split into different subproblems, in such a way that to each subproblem corresponds a "feasible" subgraph $S_i \subseteq G$, i = 1,..., 6, with respect to a given constraint C_i. Note that C_1, C_2 and C_5 are related only to the private transportation network G', and that C_6 does not affect the search space.

Once all the predicates have been specified, only the corresponding digraph is considered as the parameter of the procedure "SP". In particular, if any of constraints C_1 or C_2 is active, either all the nodes, together with their corresponding incident arcs, specified by the node avoiding constraint C_1 are deleted from G, or all those arcs specified by the arc avoiding constraint C_2 are deleted from G. Moreover, if the node forcing constraint C_3 is active and the transit through a given node $a \in V$ is required, the search for the optimal O-D path is split into the search for two different paths, namely from O to a and from a to D. Analogous considerations can be done for the satisfiability of constraint C_4.

Finally, the parking capacity constraint C_5 is represented by considering in G' as candidate parking nodes x_k only those nodes having at least one parking space available. i.e. such that $Q(x_k) > 0$.

5. The DSS Prototype: An Application to the City of Milan

A DSS prototype has been developed aimed at integrating the CSP modelling approach, the procedures for computing the shortest paths and the input/ output information related to both the private and public transportation networks.

To get an idea of how the DSS sofware prototype works, all the main steps that have been implemented are described in details together with an application to the search for O-D shortest paths with the parking selection in the city of Milan.

As a first step, a multimodal transportation network has been derived starting from the logistic and traffic data related to the city of Milan [16]. The area of Milan is about 180 km^2: for a better representation, it has been divided into three main traffic areas, namely center, first ring and suburb. Within each traffic area, a certain number of relevant locations, or centroids, has been selected (depending mainly on the parking capacity and the traffic density) in such a way that each location corresponds to a node $v \in V = V' \cup V''$. Thus the representation of the urban area of Milan results in a graph G having 102 nodes (see Figure 1). Note that the center is characterized by a large number of nodes, while in the suburb only the main approaching routes have been considered.

Once the node set V is defined, the next step is the specification of the two arc sets E' and E'' with their corresponding weights. Without entering in too many details, the process that has been required for associating a weight with each arc of the digraph G can be synthesized in the following way.

The so-called downflow functions, giving the travelling time of each arc depending on different car flows at different rush hours have been defined on the basis of the static features of the arcs, such as length and width and traffic area, and the dynamic one, that is the flow. Note that the weights of the arcs have been computed considering different rush hours and traffic conditions but without considering a priori any congestions. Other factors affecting the travelling time in shortest path problems can be found in [1, 15].

The transportation demand has been taken into account by considering the well known Origin/Destination (O/D) flow matrix, that is a $|V|$ dimensional square matrix, where the (i,j) component gives the hourly flow f_{ij} of drivers going from node i to node j. Different O/D matrices have been considered with respect to different daily time periods.

Matching the downflow functions and the flow matrices for each time period, the speed and the travelling time for each arc can be obtained. Figure 2 shows how the travelling time on G' depends upon the flow for each traffic area.

The specification of the public arc set E'' is quite different. As a first step, the most significant public transportation lines have to be selected: the selection criterion is usually made on the basis of the line length. In the present case, since the main goal is an integrated path planning and parking management across the whole city, public lines related only to few nodes in the suburb area of Milano have been disregarded. In particular, 26 public lines within the Milan urban area have been selected. Data about the travelling times and the average waiting time for each selected line have been collected. Each line has been represented as a node sequence and the travelling time $t''(i,j)$ required to go from node $i \in V'$ to its adjacent node $i \in V'$ $\forall\ (i,j) \in E''$ has been obtained through the real line-scheduling.

Finally, the last preprocessing operation is the computation of the shortest path time $T(i,j)$ for each pair of nodes in V on both the private and the public transportation systems without any active constraint C_1 - C_6. These values are usefully used in order to reduce the feasible region and the computational time (see Secton 4).

As we have already seen, the O/D flow matrix is static within each time period and describes a steady-state traffic condition. However, unusual traffic conditions may be considered as well. For instance, if an accidents occurs, or if a road interruption modifies the "usual" shortest path, the unavailable nodes and/or arcs can be given as in (3.2) and (3.3).

It is worth mentioning that the proposed approach has been designed for being included in a DSS that is able to act as a system supervisor for updating traffic conditions and properly taking into account reliable times when roads are congested. Note in fact that the possible path modifications represented in the model by C_1 and C_2 are not considered here as random events, therefore the information given at the driver when he/she leaves the origin could disappear during the journey, whereas others could take place along the path. In order to overcome to this situation, a stochastic approach could be considered (see, for instance, [9]). However, note that the proposed DSS software prototype is thought for enabling the reservation of a parking space in the preferred parking facility in advance, so that it will be available for the car driver also in the case of possible delays along the path.

Presently, the DSS is designed to supply the required path by processing the O-D nodes pair together with other additional requirements like the available time, and some real-time informations, like the current parking capacity at each node and the nodes/arcs practicability. The real-time information management requires a centralized system supervision both for the collecting and the preprocessing of the data.

A software prototype integrating the modelling and methodological approach has been implemented using the LPA Windows Prolog within the Microsoft "Windows" environment on a 386 IBM PC compatible. The traffic information related to the morning-peak period data of the city of Milan have been considered as test bed.

The DSS software prototype has an user-friendly interface that allows the user to select all the query elements, that is the pair (x_1,x_n), the maximum available time T_{max}, the selected modality and the elements of the F and A lists. For instance, the first multiple-choice-box allows to select the origin node among the whole node list V, as it is reported in Figure 3.

Then, the optimal O-D path is computed and the best path to follow in both the private and public network is given as an alternate sequence of nodes and arcs. Once the optimal path is computed, a more detailed information about the actual location of the parking facility is provided. In particular, the selected parking node x_k is then exploded in order to choose the parking place among the available ones within the zone closest to the predecessor node x_{k-1} (see Figure 4). Remember that during the computation of the optimal path all the parking spaces available in a given zone are grouped together in a single node representing the centroid of the zone under consideration: the centroid x_k is labelled with its total parking capacity $Q(x_k)$. This feature is designed to allow the real-time parking management: once the parking node has been selected, the car driver is able to book a parking space in it and the corresponding parking capacity is automatically updated accordingly. Obviously, the software prototype is merely a demonstrator of the methodological approach and the real-time parking information management and parking booking are simulated: these features have been designed for a future use of the prototype as a DSS.

As an information about the running time, the average CPU time required for finding O-D paths with different requirements on a graph G representing about 40 km^2 in the urban area of Milan is reported in Table 1.

O-D shortest path on G (1)	1.5
(1) + C_1, C_2	5.0
(1) + C_3, C_4	3.2
(1) + C_1, C_2, C_3, C_4	5.4
(1) + C_5	2.2
(1) + C_5 + M_1/M_2	2.3
(1) +M_1/M_2	6.2
(1) + C_1, C_2 + M_1/M_2	4.8

Table 1. CPU time (in seconds) on a PC 386

6. Concluding Remarks

In this paper the basic structure of a DSS prototype for urban path planning able to guide a car driver to a parking facility is presented. The underlying methodology of the proposed DSS is based on a CSP formalism together with the classical shortest-path algorithms. Even if the implementation of the DSS is still in progress and needs some more testing, the computational experimentation aimed at validating the methodological approach has shown that it seems to be very promising for an effective use in urban areas where the parking problem is a crucial one, and particularly suitable to be included in a more advance system supervisor for handling both the topological and traffic information and the user transit requirements.

References

[1] Abkowitz M.D., Engelstein I., 1983, "Factors Affecting Running Time on Transit Routes", *Transportation Research-A*, V.17,N.2, pp.107-113.

[2] Archetti F., Bertocchi S., Sciomachen A., 1994, "Path planning to a parking facility: a Constraint Satisfaction Approach", Proc. *Second Meeting of the EURO Working Group on Urban Traffic and Transportation*, Actes INRETS N. 38, pp.441-454.

[3] Baricelli P., Sciomachen A., 1995, "Parking Classification in Urban Intermodal Transportation Networks", *Research Report 10.95*, I.M.F, University of Genova.

[4] Branston D., 1976, "Link Capacity Function: a review", *Transportation Research-A*, V.10, pp.223-236.

[5] Codognet P., Fages F., Jourdan J., Lissajoux R., Sola T., 1992, "On the Design of Meta(F) and its Applications in Air Traffic Control", *Proc. JICSLP Workshop on Constraint Logic Programming*, pp.28-35.

[6] Cohen, J., 1990, "Constraint logic programming languages", *Communications of the ACM*, V.33, N.7, pp. 52-67.

[7] Dincbas M., Simonis H., Van Hentenryck P., 1988, "Solving the car sequencing in Constraint Logic Programming", *European Conference on Artificial Intelligence*, p. 290-294, August.

[8] Gallo G., Pallottino S., 1988, "Shortest Path Algorithms", *Annals of Operations Research*, 13, pp.3-79.

[9] Gendreau M., Laporte G., Seguin R., 1995, "An Exact Algorithm for the Vehicle Routing Problem with Stochastic Demands and Customers", *Transportation Sciences*, V.29, N.3, pp.143-1 54.

[10] van Hentenryck P., 1989, "Constraint Satisfaction in Logic Programming", MIT Press.

[11] Kanchanasut K., Sumetphong C., 1992, "Floor Planning Applications in CLP($\Re$)", *Proc. JICSLP Workshop on Constraint Logic Programming*, pp.36-44.

[12] Liang-Liang et al., 1993, "APPLAUSE: Applications Using the ElypSys Parallel CLP System", *Proc. 10th International Conference on Logic Programming, pp.847-848.*

[13] Meyer M. D., Miller E. J., 1984, *Urban transportation planning*, Mc Graw Hill, New York.

[14] Nadel B. A., 1989, "Constraint satisfaction algorithms", *Computational Intelligence: an international journal*, V.5, N.4, pp. 188-224.

[15] Orda A., Rom R., 1990, "Shortest-path and minimum-delay algorithms in networks with time-dependent edge-length", *Journal of the Association for Computing Machinery*, V.37, N. 3.

[16] Sheffi Y., 1985, *Urban transportation networks*, Prentice-Hall Inc., Englewood Cliffs.

[17] Wentworth J., 1994, "Developing expert systems in transport", in "*Artificial Intelligence Applicatios to Traffic Engineering*", Bielli, Ambrosino and Boero eds., pp. 29-42, VSP International Publisher.

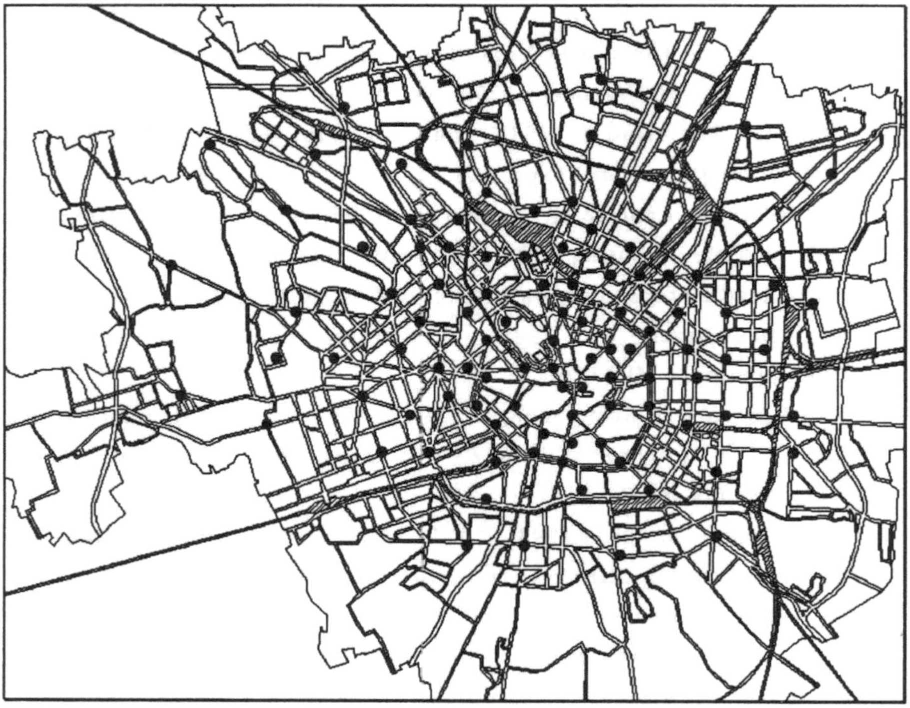

Figure1. Nodes/centroids location in Milan

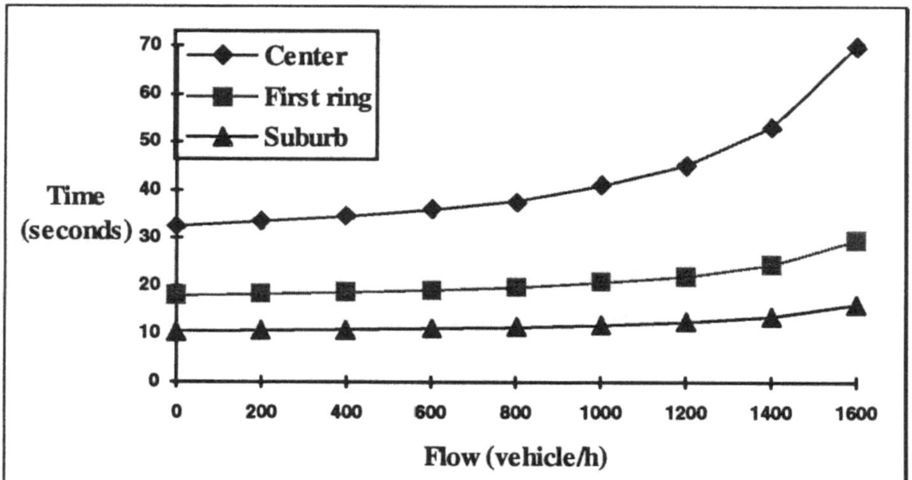

Figure 2. Flow/travelling-time diagram of the downflow functions for each traffic area

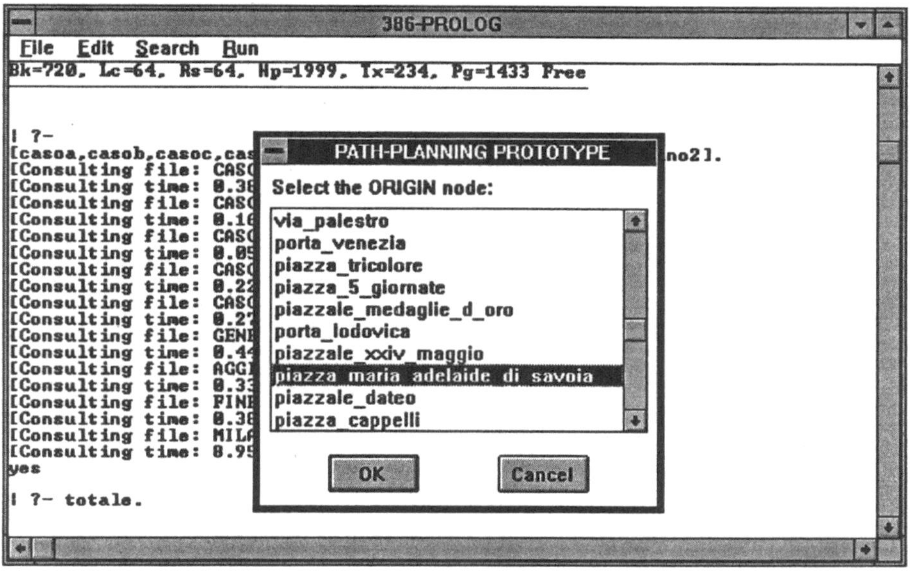

Figure 3. Origin-selection window

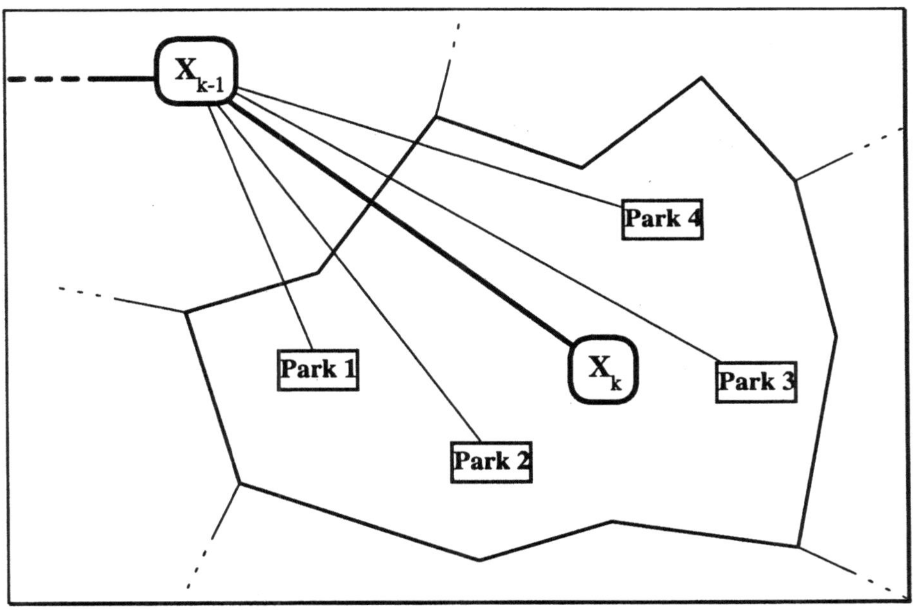

Figure 4. Actual location of the parking facilities within the zone

STRUCTURE OF A DYNAMIC NETWORK LOADING MODEL FOR THE EVALUATION OF CONTROL STRATEGIES

Massimo Di Gangi and Vittorio Astarita

Dipartimento di Ingegneria dei Trasporti
Università degli Studi di Napoli "Federico II"
Via Claudio 21, 80125 Napoli, Italy

The efforts spent by researchers in the last few years in traffic modelling have been focused on the modelization of dynamic behaviour of the several components making up a transportation system. In the field of traffic assignment a large amount of models and procedures have been proposed in order to perform Dynamic Network Loading (DNL) that is the reproduction of within-day variable link performances once given a corresponding O/D demand and users' choice model. These models can be used to simulate the effects of regulation strategies on users' behaviour. In this paper a continuous dynamic network loading model is described; it removes some of the drawbacks of other packet approach models proposed in literature and explicitly allows the en-route modification of followed path. An algorithmic development of the model and a set of examples on link modelization, while-trip re-routing modelization and evaluation of control strategies are also proposed.

1. Introduction

The increasing congestion in all urban area has forced to the development of new information based control strategies. Those strategies relay on the explicit modelization of time-dependent flows and oversaturation queues, with time-dependent demand (due for instance to rush hour) and/or changes in supply (due for instance to accidents or weather conditions), which can be only simulated through *within-day or intra-periodal dynamic* models.

The extension of within-day static models to take into account within-day dynamics is by no means straightforward, since within-day dynamic supply modelling requires completely new definitions and formulation of the problem.

Several approaches have been proposed for the solution of the DNL, and are briefly reviewed in the following after the description of some requirements that a fully satisfactory DNL method should meet:

[a] link and path flows should be explicitly and consistently defined, which means that the argument of the link performance functions should be consistent with the relationship between link and path flow (**internal consistency**);

[b] user journey should be consistently simulated assuring that no overtake can occur between two users having entered a link in different times while they are moving on the link (**no overtake**);

[c] queuing phenomena due to oversaturation at limited capacity links, possibly caused by incidents or spill-back, should be explicitly simulated (**bottlenecks**);

[d] the DNL method should allow for en-route re-definition of followed paths according to the adopted en-route users' behaviour model (**while-trip re-routing**).

In addition the model should be able to manage different classes of users with different types of available information and/or choice rules (**multi-class network loading**).

Most of the existing methods do not clearly define link flows and/or can hardly be extended to include en-route diversions from the initially chosen path, because of the computational burden of keeping the identity of diverted path flows. Moreover some of them rely on assumptions which do not necessarily rule out overtaking between users following the same path with different departure times (heuristic generalisation of within-day static methods: Hammerslag [9], Janson [10]; exit-function methods: Merchant and Nemhauser [12], Carey [4] [5], Friesz et al. [8], Wie et al. [15], Papageorgiou [13], Boyce et al. [3]; discretization of differential equations: Ben Akiva et al. [1][2], Vythoulkas [14]).

Other methods, more suitable for within-day dynamic applications, are based on a packet-approach, in which users following the same path and entering the network during the same interval are grouped and moved as a single unit experiencing the same trip of the leader. Under these assumptions these methods may satisfy all the described requirements. (CORQ proposed by Yagar [16]; CONTRAM proposed by Leonard and Gower [11]; the method proposed by Cascetta and Cantarella [6]).

In this paper a model dealing with DNL by adopting a continuous packet approach where users are assumed uniformly spread between the first point and the last point of the packet is described. In particular, flow evaluation has been generalised in order to evaluate both the flow crossing a specified section of a link and the characteristic flow on the link. An explicit modelization, that allows both to take into account users' diversions from the initially chosen path due to traffic and/or network conditions and to manage different classes of users, with different types of available information and/or choice rules, has been developed.

In the following section 2 is dedicated to the description of DNL problem, section 3 presents a simple example of the link modelization and how an incoming flow pattern changes into a different outflow pattern depending on the link speed function and capacity. Section 4 is then dedicated to an example on while-trip re-routing and on the capability of the model to be an evaluation tool for control strategies. Two different control strategies for a very simple temporary oversaturated network are presented together with the complete results of the simulation: packet positions, link flows and queues. A short appendix is also included in order to summarise definitions and notations adopted within the text.

2. The Proposed DNL Model

A DNL model essentially reduces to a set of relationships able to evaluate some link quantities (i.e. flow, density, queue) once known the path-flows crossing the considered links. This implies the formulation of both "*behavioural rules*" expressing users' movement on the network and "*computing rules*" concerning the definition of the evaluated link quantities.

Starting from the "behavioural" point of view, let $K_r = \{1,..,n_r\}$ the set of the n_r paths related to O/D pair r; the set of all the N^{jk} users leaving in the same interval j and following the same path $k \in K_r$ will be here called **packet** (j,k), the set of all the d^{jr} users leaving during interval j and belonging to the OD pair r will be called **cluster** (j,r).

The proposed one can be defined as *continuous-packet approach*, since departures of the users of a packet are assumed uniformly and continuously distributed along the time interval duration T yielding to a path flow $F^{jk} = N^{jk}/T$. In other words, following this approach, a packet can be considered as formed by an infinite number of point users.

In particular, users belonging to the same packet are assumed **uniformly distributed over time** between two particular point forming the edges of the packet itself: the first one, named **head-point**, and the last one, named **tail-point**. Starting from the origin, a packet divides itself in other packets at each diversion node as shown in figure 1.1.

The proposed model allows the evaluation of the link quantities, that will be further on defined, related at time interval h, following the evolution of each packet along its followed path during the interval given by the movements of the above defined two points. So, link flows, densities and queues during an interval can be evaluated from the positions of these two points at the beginning and at the end of the considered interval.

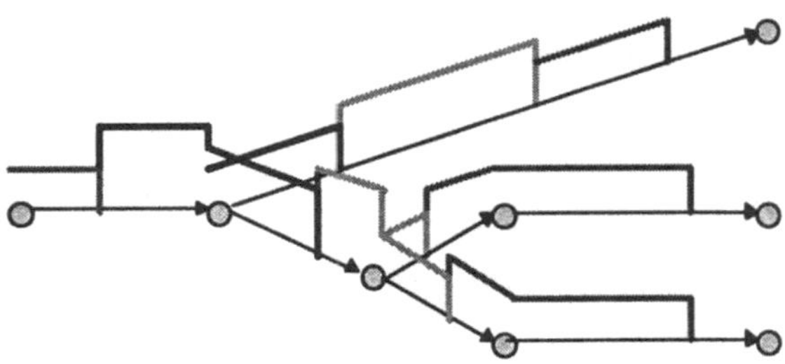

Figure 1.1. Packet distribution in the proposed approach

2.1. Network Modelling

The network is composed of nodes and links. The considered physical characteristics of each link are the length and the capacity. Points on the links move during each interval with a speed that depends on characteristics flow that will be defined in the following. A vertical queue is located in the ending node of each link. When a point has moved past the running section of the link it faces a queue which depends on the outflow of the running section and on the capacity of the link (see figure 2.1).

Given an O/D pair r and the set of paths K_r related to it, let $\Omega_r(L_r, N_r)$ be the sub-graph formed by the sets of links L_r and nodes N_r belonging to the paths connecting o/d pair r. The set K_r can also be represented by mean of a rooted tree whose root is the common origin. Each vertex and each arc of the tree univocally correspond to a node and a link respectively of the sub-graph Ω_r. Let R_v the set of arcs departing from a generic vertex v of the tree.

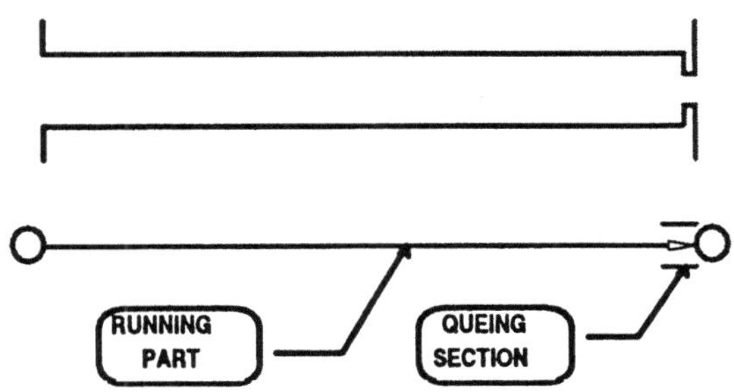

Figure 2.1. Link modelization adopted in the proposed approach

It can be worth noting that since each vertex has only one entering arc, as soon as two paths diverge, they are represented, starting from the *diversion vertex*, by two completely different succession of arcs no matter if the paths have, in the following, some links in common.

The obtained tree has a number of leaves equal to the number of paths related to the considered O/D pair. It is important to say that this representation does not require a significative increase in storage effort than other more "classical" representations of paths as link sequence enumeration.

In the following arc characteristic will be evaluated.

2.2. Flow Definitions

Let the definition of time-distance be here introduced. The time distance between two generic points p1 and p2 in a fixed instant (say τ), is defined as the time needed to point p1 starting from its location at time τ to reach the location where point p2 was at time τ considering all the link characteristics (speed, flow, queues) fixed at the values evaluated at time τ.

In order to evaluate the flow crossing section s of arc a during current interval h due to a generic packet (j,k), let τ be a generic point of the time interval h $\tau \in [0,T]$, so that $\tau=T$ for interval h coincides with $\tau=0$ for interval h+1. Assuming as origin the position of tail point on the followed path at time τ, let $\lambda_s(\tau)$ be the time-distance between the section s of the considered arc a and the position of tail point at time τ (this means that, if the considered tail point at time τ has just passed section s, the value of $\lambda_s(\tau)$ is negative) and let $\Lambda(\tau)$ be the time-distance between the head point and the tail point at time τ, that is the time that tail point needs to reach the location of the head point, starting from its own position, considering the speeds at interval h. Figure 2.2 depicts positions of head and tail point for $\tau=0$ and $\tau=T$ for a generic packet and graphically indicates the values of $\lambda_s(\tau)$ and $\Lambda(\tau)$.

Flow crossing section **s** of arc **a** during period $[0,\tau]$ of interval **h** due to packet (j,k) can be evaluated as a fraction $\beta_{ah}^{jk}(s,\tau)$ of the packet flow F^{jk}. The value of $\beta_{ah}^{jk}(s,\tau)$ can be expressed as [7]:

$$\beta_{ah}^{jk}(s,\tau)=\frac{T}{\tau}\left\{\min\left(\frac{\lambda_s(0)}{\Lambda(0)},\ 1\right)-\max\left(\frac{\lambda_s(\tau)}{\Lambda(\tau)},\ 0\right)\right\} \tag{2.1}$$

therefore the flow crossing section at abscissa **s** of link **l** belonging to path **k** during interval **h** due to the packet left during interval **j** can be expressed as:

$$f_{ah}^{jk}(s)=\beta_{ah}^{jk}(s,T)\cdot F^{jk} \tag{2.2}$$

with $\beta_{ah}^{jk}(s,T)$ given by (2.1).

A ***characteristic flow*** is defined for each arc as the average value of flow crossing each running section of arc a during interval h:

$$f_{ah}^{jk}=\frac{1}{L_a}\int_0^{L_1}\beta_{ah}^{jk}(s,T)\cdot F^{jk}ds \tag{2.3}$$

equation (2.3) can also be written as:

$$f_{ah}^{jk}=b_{ah}^{jk}\cdot F^{jk} \tag{2.4}$$

where:

$$b_{ah}^{jk}=\frac{1}{L_a}\int_0^{L_a}\beta_{ah}^{jk}(s,T)ds \tag{2.5}$$

is the *crossing fraction* of the packet (j,k) on arc a evaluated for the interval h.

It must be stressed that, in all the above formulations involving flows, consistency between path-flow and link flow must be assured in terms of measure units.

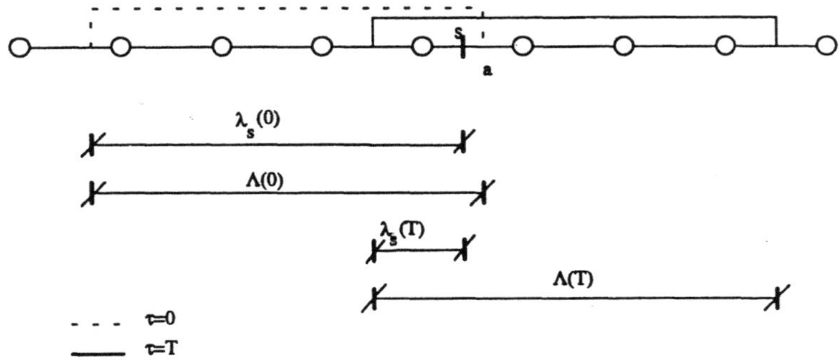

Figure 2.2. Time distances related to a packet at the beginning and at the end of an interval

2.3. Packet Movement

The motion of each packet is tracked, as said before, by following only the head and the tail point. The structure of the point tracing procedure can be described as follows depending if the points are moving on a running or queuing section:

- <u>*running section*</u>

Let v_{ah} be the *running speed,* representing the operative speed on link corresponding to arc a during the interval h and L_a be the *length* of link corresponding to arc a. Point p located at abscissa s of arc a must cover an arc segment extended of $s^p_{ah} = L_a - s$ and the time t^s needed to cover this space is given by $t^s = s^p_{ah} / v_{ah}$.

If point residual time is enough to entirely cover the arc, that is $t^* \geq t^s$, point p leaves the arc and its residual time is updated subtracting t^s to its previous value elsewhere point p cover on arc a the space Δs given by $\Delta s = t^* \cdot v_{ah}$ and a length $s^p_{ah+1} = s^p_{ah} - \Delta s$ must be covered by the point in next intervals.

The value of s^p_{ah} that is the space that point p must cover on the arc, is set, when point p enters the arc, at the value of L_a.

- *queuing section*

Let Q_{ah} be the arc capacity, that is the maximum exiting flow from the arc a, q_{ah} the number of users making up the queue that point p has in front of it at the beginning of interval h. The time needed to exit the arc is given by $t^s = q_{ah}/Q_{ah}$.

Even in this case if point residual time is enough to serve the whole queue present in the arc in front of point p itself, that is $t^* \geq t^s$, point p leaves the arc and its residual time is updated subtracting t^s to its previous value; otherwise point p remains on the arc and a number of users $\Delta q = t^* \cdot Q_{ah}$ before point p leave the queuing arc. The value of the queue in front of point p is then updated as $q_{ah+1} = q_{ah} - \Delta q$.

The value of the queue in front of point p is set, when point p enters the arc, at the value $q_{ah}(\psi) = \mathrm{MAX}\{0, q_{ah-1} + (\varphi_{ah} - Q_{ah}) \cdot \psi^p_{ah}\}$ where φ_{ah} is the in-flow and ψ^p_{ah} is the entering time of point p on the arc.

2.4. While-Trip Re-Routing

While-trip re-routing represents the possibility of simulating the modifications made by users to the initially chosen path during the course of their trip, considering either exogenously given suggestions or casual factors. Such a behaviour can be taken into account in the proposed model introducing, for each arc of the tree, an *absolute* and a *relative cluster rate*. The *absolute cluster rate* $\pi^{jr}_a \in [0,1]$ represents the fraction of users belonging to cluster (j,r) that will cross arc a. This value can be expressed as:

$$\pi^{jr}_a = \frac{1}{d^{jr}} \sum_{k \in K_r} N^{jk} \cdot \delta^k_a \qquad (2.6)$$

with $\delta_a^k = 1$ if arc a belongs to path k, =0 otherwise. It can be noted that N^{jk}/d^{jr} can represent the probability that users belonging to cluster (j,r) choose path k.

The *relative cluster rate* $\tilde{\pi}_a^{jr}$ is introduced referring to all the arcs exiting a vertex v; it is defined as:

$$\tilde{\pi}_a^{jr} = \pi_a^{jr} \Big/ \sum_{a' \in R_v} \pi_{a'}^{jr} \qquad \forall a \in R_v \tag{2.7}$$

where R_v is the set of all the arcs departing from vertex v; thus, for each vertex, it follows that:

$$\sum_{a \in R_v} \tilde{\pi}_a^{jr} = 1 \tag{2.8}$$

The mechanism of re-routing can be simulated by means of the cluster rates of packets coming out from a diversion point. As a matter of fact, given new values of relative cluster rates (supplied by a path choice model), flow evaluation is conducted taking into account these new values. Congruence in flow evaluation is guaranteed making the hypothesis that, if the head-point of a packet has crossed or simply occupied an arc, the cluster rates of all the arcs previously crossed by head-point cannot be changed (the behaviour of the whole packet is the same of its head-point). This means that, for cluster (j,r) whose head point is located at time t on arc a, re-routing affects, starting from time t+dt, only those arcs belonging to the *sub-tree* generated by arc a.

To allow the model to manage different classes of users, with different types of available information and/or choice rules, each packet is supposed being composed by different classes of users. An example of users' classes can be:

- *non informed* users whose probability of choosing path k is given by the absolute cluster rates at interval j of departure and does not change during the trip;

- *informed* users that are aware of the situation of the system and so can react changing their choice probability during the trip.

Each cluster is composed of $d^{jr} = \sum_{k \in K_r} N^{jk}$ users divided in c classes. π^{jr}_{ca} is the absolute rate of class c users of cluster(j,r) that will cross arc a. At each diversion vertex V the cluster separate itself in smaller packets. The absolute rate π^{jr}_{ca} have to be recalculated each times in the following way:

$$\pi^{jr}_{ca} = \tilde{\pi}^{jr}_{ca} \cdot \pi^{jr}_{cF(a)} \tag{2.9}$$

where: F(a) is the father of arc a; and $\tilde{\pi}^{jr}_{ca}$ is the relative cluster rate for users of class c that will cross the generic arc a departing from vertex Vo.

The value for π^{jr}_{a} is:

$$\pi^{jr}_{a} = \sum_{\forall c} \pi^{jr}_{ca} . \tag{2.10}$$

In this way it is possible to cope with every kind of behaviour that users may have during their trip, depending on their knowledge of the system.

3. Link Modelling

Each link is composed of two separate sections: a running section followed by a queuing section: Points move on the running section according to link speed and on the queuing section according to link capacity. It is possible do define a resistance term Z for each section to be crossed: in running sections Z is the space left to be covered by a point and in queuing sections Z is the queue in front of the point that has still to be dissipated. A moving rate W can be defined for running sections as the link speed and for queuing sections as the link capacity. When points enter a new section their resistance are initialised to the total length of link to be crossed for running section, or to the queue in front of the point at the moment of its arrival. The total time needed to completely traverse the section is then t=Z/W. As better shown in table 3.1.

Characteristic	*Running section*	*Queuing section*
Resistance z^p_{ah}	Space to be covered s^p_{ah}	Queue in front of the point q^p_{ah}
Moving rate w_{ah}	Arc speed v_{ah}	Arc capacity Q_{ah}
Initialisation	$s^p_{ah} = L_a$	$q_{ah} = MAX\{0, q_{ah-1} + (\varphi_{ah} - Q_{ah}) \cdot \psi^p_{ah}\}$

Table 3.1. Characteristics of point movement

An example that explains the adopted modelization for the arcs is shown in the following. They differs for the adopted speed-flow relationship: in the first one speed is constant, in the second one speed depends on the characteristic flow.

Simulations on those single links have been carried out using a 15 minutes time step and the adopted in-flow profile has been the same for both of them; the results of these simulations are sketched in figures 3.1 and 3.2. In the link with constant speed the outflow from the running section (that is also the inflow of the queuing section) has the same pattern of incoming flow, the points are just translated in time due to the constant time delay needed by the packets to cross the running section.

The value of characteristic flow, as expected, is always between the input flow and the outflow except than in the time period between 105 and 120 sec where it is higher than the two other flows; this is due to the fact that some packets, travelling in the middle of the link, give a contribution only to the value of characteristic flow.

In the queuing section the outflow is the same as the inflow until capacity value is reached; then a queue start to grow and the outflow gets to capacity level where remains until the queue disappears.

In the link with variable speed the results are almost the same, except that the outflow profile of the running section has a different pattern because packets have been delayed in their travel and the outflow shape appears shifted to the right. This yields to a larger queue and its later disappearance.

This link modelization does not suffer the same problems of exit link function formulations. As a matter of fact, in models based on exit function, users experience the same value of flow for all the different locations of the same link, since link flows are supposed to be characteristic of the whole link. This means that users entering a link are immediately spread and added to the number of users already on the link, and that some user may experience an infinite propagation speed, since they reach the end of the link at the same time they enter. Furthermore exit functions do not seem able to guarantee the respect of FIFO rule (Fernandez [17]). In our model points and packets spend always a finite and realistic time (depending on link speed) travelling on running sections. So the modelization is closer to a true representation of the physical phenomena.

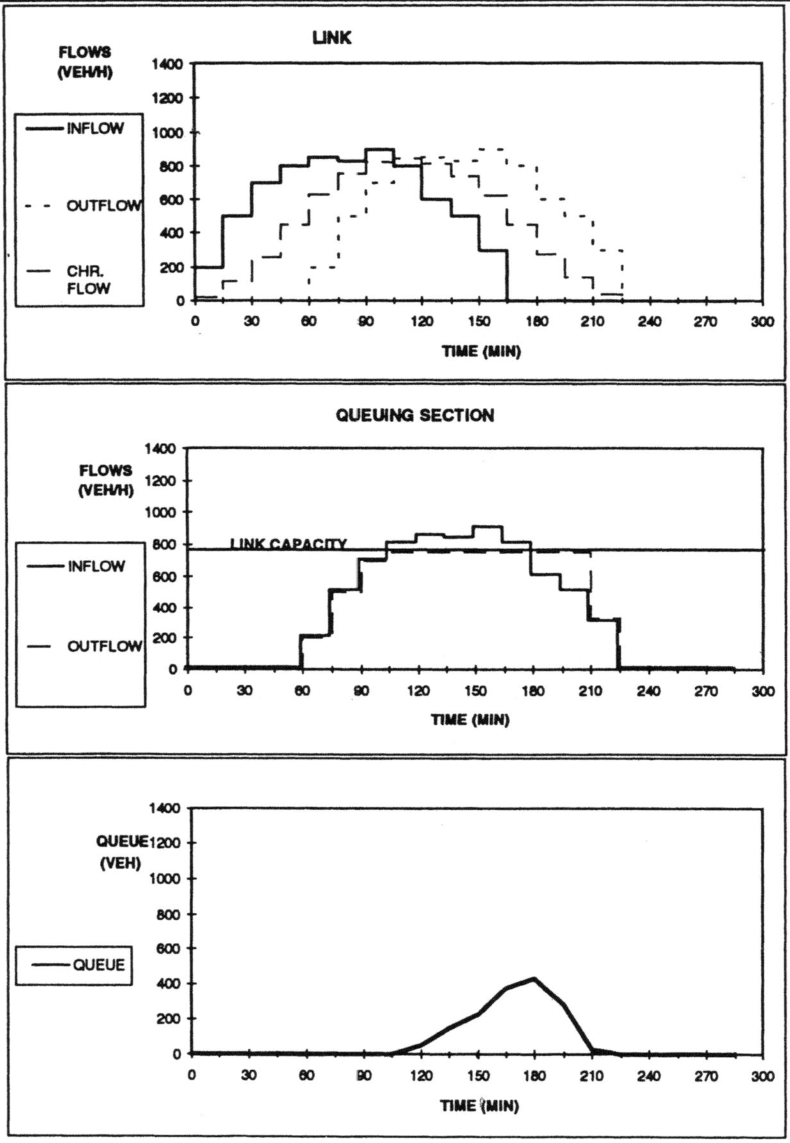

Figure 3.1. Link modelization with constant speed: flows and queue generated by an in-flow dynamic pattern

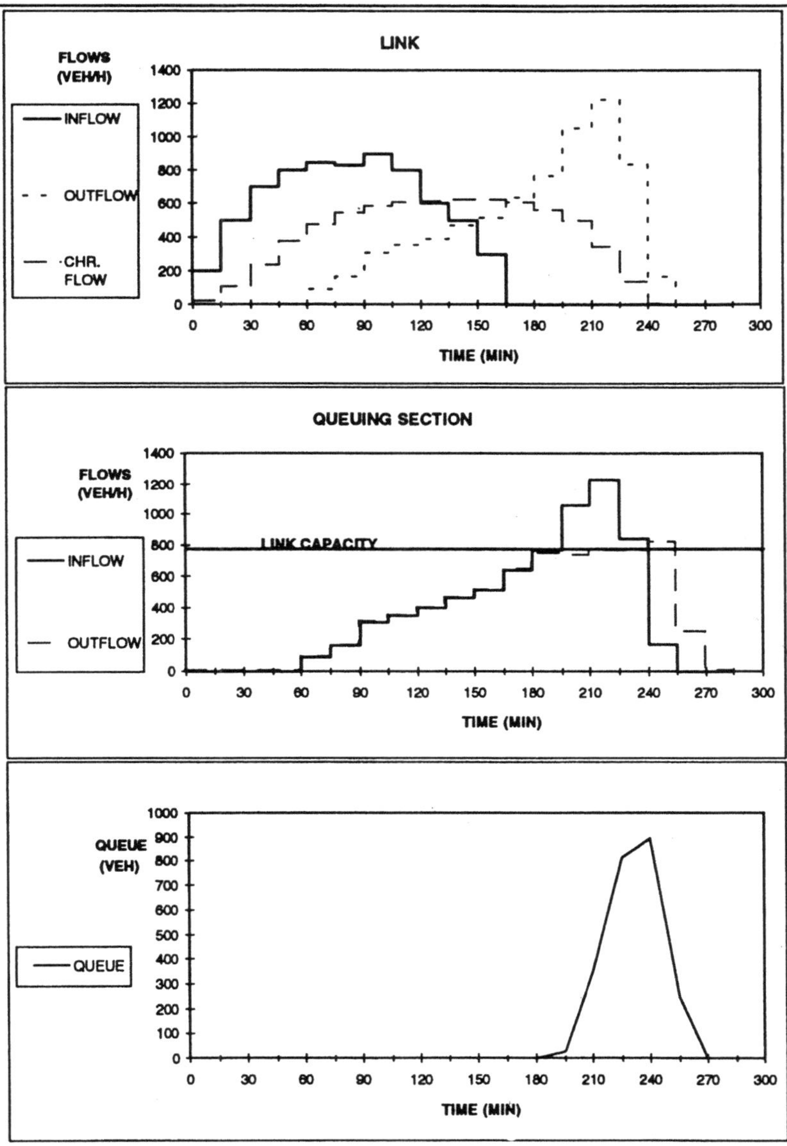

Figure 3.2. Link modelization with speed depending on characteristic flow: flows and queue generated by an in-flow dynamic pattern

4. Illustrative Applications

The following simple applications show both how the procedure can model while-trip re-routing and the model suitability to the evaluation of control strategies. A small network is supposed to be oversaturated for a while. Two exogenous different strategies of control are supposed and are evaluated together with the situation without control. The speed is fixed in such a way that the motion of packets can be easily followed: each point moves exactly from the beginning to the end of each link during one time interval. The network and the three possible paths are shown in the figure 4.1.

The model works as an assignment model and do not take into consideration how information are supplied to the drivers, how the drivers process those information and how they react depending on their personal knowledge and behaviour. All the user of the system are supposed to be informed in real time and in the right way and all of them will follow the suggestions.

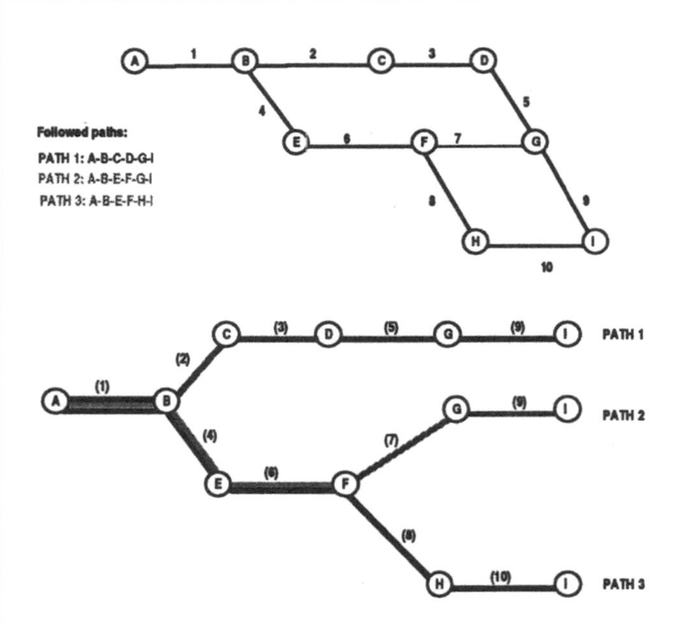

Figure 4.1. The test network and the considered three paths

A failure in link capacity (i.e. an accident) is supposed to be localised at link n. 9 which belongs both to path one and two. Without any control and before of this failure each path is supposed to absorb one third of the total flow; when capacity of link n. 9 decreases suddenly, it is supposed that some strategies are adopted moving the users from path 1 and path 2 to path 3 as described in tab. 4.1. It is worth noting that the aim of this example is only to show how the proposed model can simulate the effects of control strategies, whatever they are.

The three different situations are outlined in figures 4.2, 4.3 and 4.4 where the positions of the packet at the and of each period and the queue at link 9 are shown.

Without any control the queue on link 9 starts to grow at the beginning of second interval and, when the capacity reduces, it continues to grow until the capacity recover its initial value at the beginning of interval 7; at this point the inflow on the link is equal to the capacity and so the queue remains unchanged because dissipation is possible (figure 4.5 for details on link 9).

Path	No Control	Strategy #1	Strategy #2
1	0.33	0.33	0.00
2	0.33	0.00	0.33
3	0.33	0.66	0.66

Table 4.1. Use percentage of paths for each simulated strategy

With control strategy #1 at the beginning of third interval the flow is diverted after link 6 only on link 8 and nothing changes at the diversion point located at the end of link 1 (33% of the flow still goes to link 2 and 66% to link 4); the effects of control on link 9 appear at the beginning of interval 5, during which the queue does not grow.

When the capacity on link 9 returns to the original value, the queue begins to be dissipated until the end of interval 9 when it disappears (figure 4.6). Control strategy #1 is unable of avoiding the queue formation, because the effects of control suffer from two delays: the first one is due to the fact that the control begins after one time interval and this is supposed to reproduce real situation control strategies that will always suffer from a delay due to the time needed to the control system to react to non predicted situation; the second one is due to the control set of actions which are obviously finite and affect the traffic from a point located before the capacity failure so there is a "transport" delay of the effects of control. An example of this problem is shown with control strategy #2 whose effect take place at a point more distant to the point of congestion than the one of strategy #1; the outcome of the control is less effective due to this kind of delay.

With control strategy #2 the effects of control are localised in two diversion points: at the end point of link 6 66% of the flow goes to link 8 and the remaining to link 7 ,at the end point of link 1 100% of the flow goes to link 4. The delay makes the queue grow for a longer period of time and so it reaches a value that is higher than the one obtained with the previous examined control strategy and it is dissipated only at the end of interval 11 (figure 4.7).

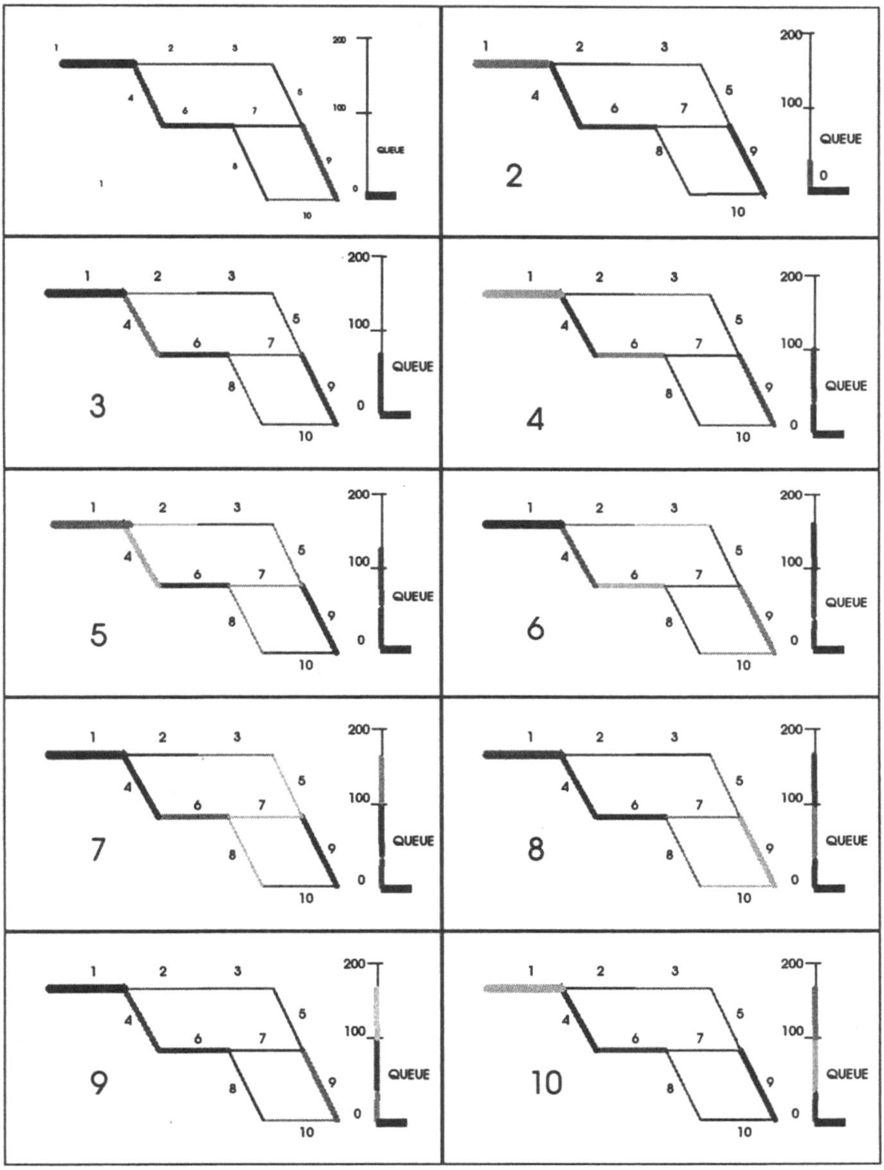

Figure 4.2. Positions of packets at the end of each interval
Without control

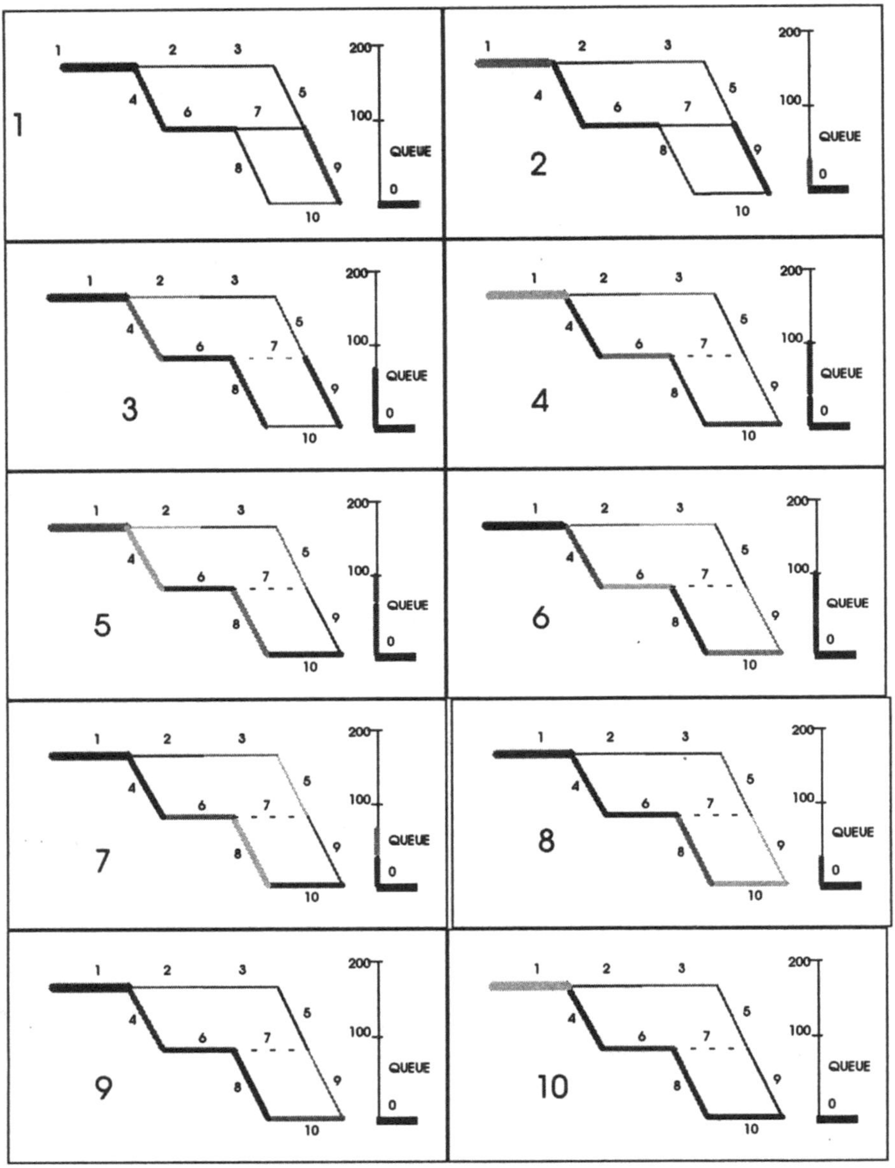

Figure 4.3. Positions of packets at the end of each interval
Control strategy #1

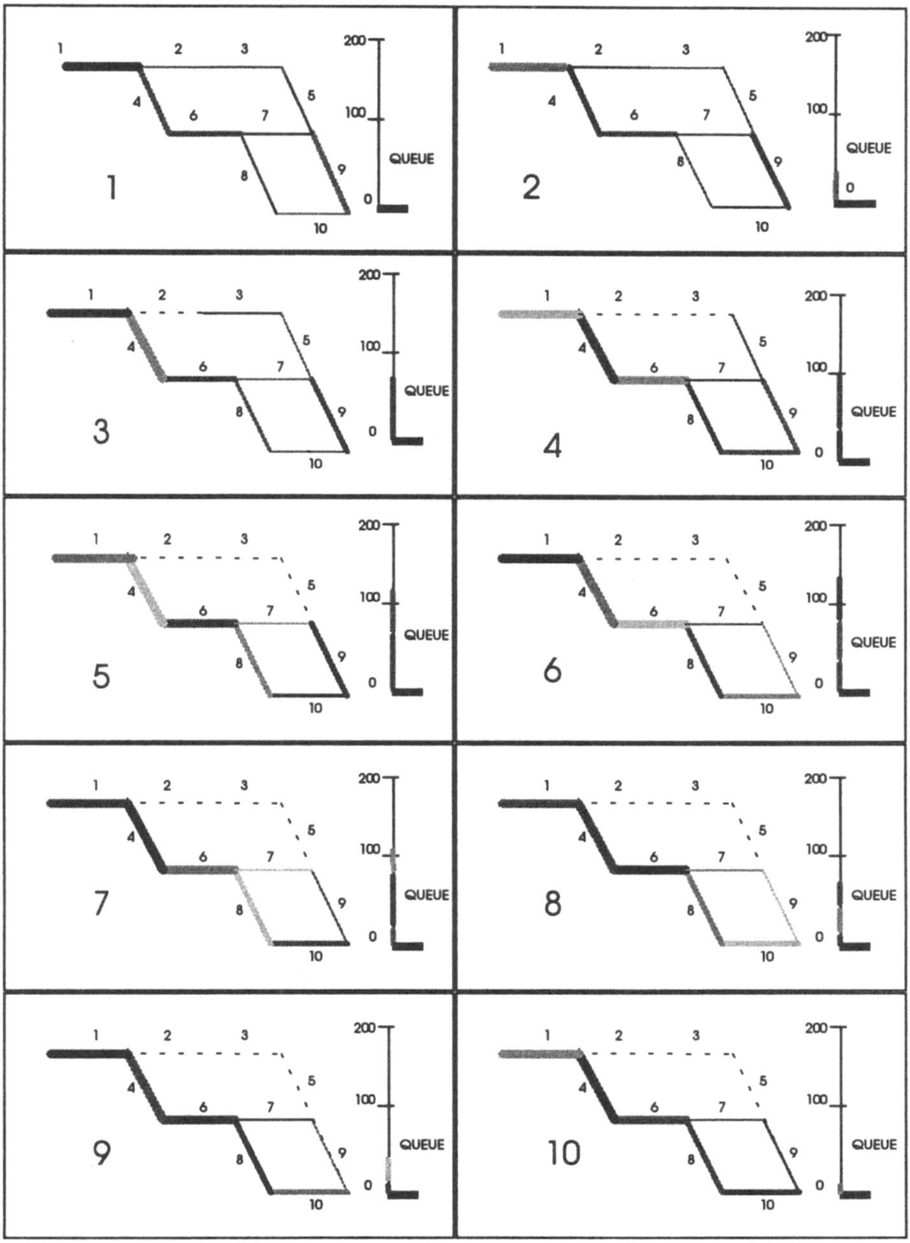

Figure 4.4. Positions of packets at the end of each interval
Control strategy #2

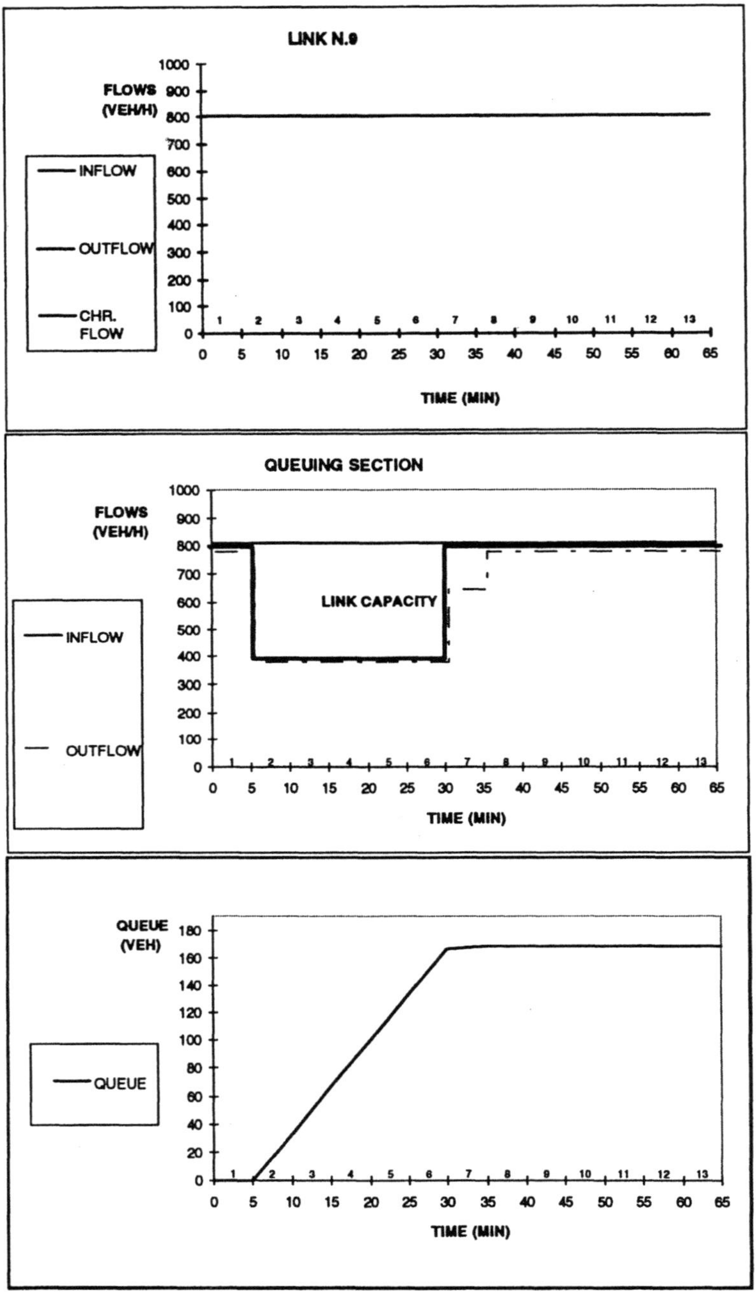

Figure 4.5. Flows and queue on link n. 9 - Without control

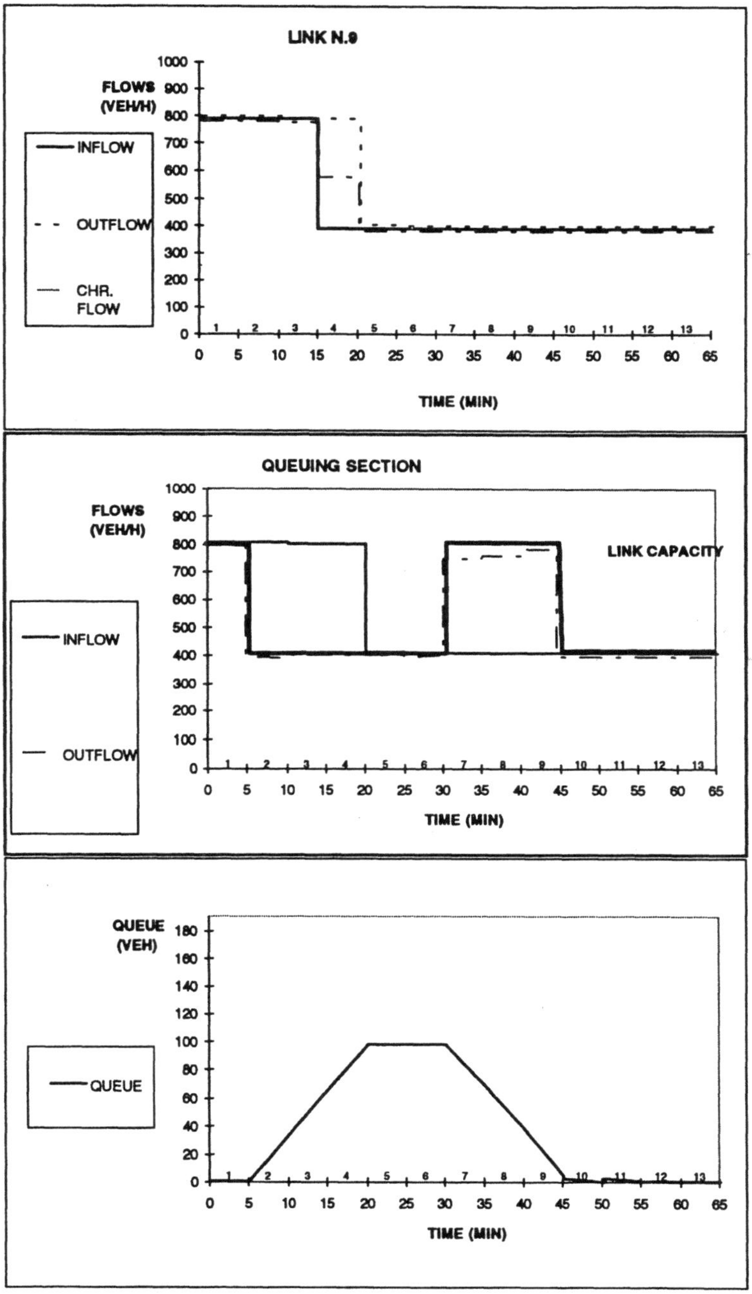

Figure 4.6. Flows and queue on link n. 9 - With control strategy #1

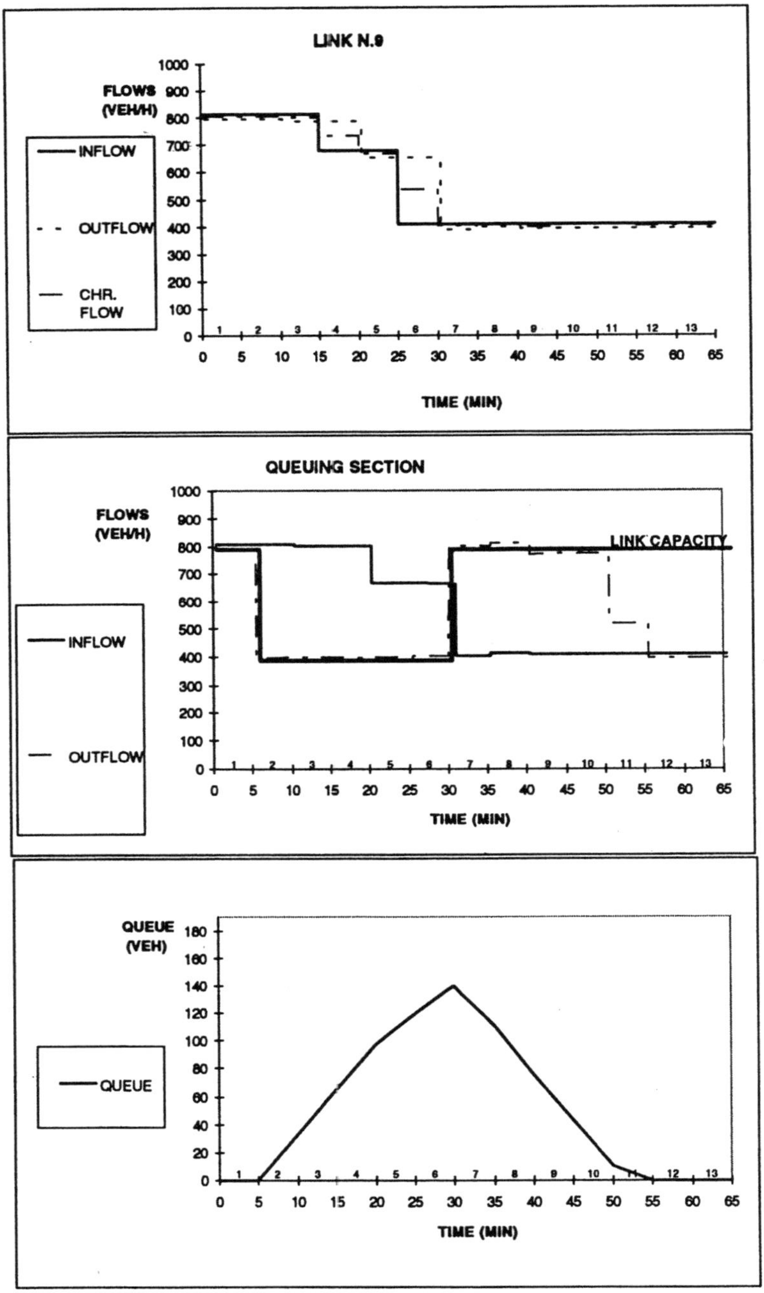

Figure 4.7. Flows and queue on link n. 9 - With control strategy #2

5. Conclusions

In this paper a model that performs Dynamic Network Loading by using a continuous packet approach and that allows en route modifications of chosen paths is described. It seems that the proposed model is suited both for planning operation, in order to check the feasibility and effectiveness of control strategies, and for real time applications such as short-term forecasts to be used by traffic control centres. Examples have been given to show how this model may be used in order to reproduce the effects due to regulation strategies on users' behaviour.

Acknowledgements

This work has been developed in the context of Progetto Finalizzato Trasporti 2 of the National Research Council of Italy (CNR) under grant no. 93.01801.PF74.

References

[1] Ben Akiva M., Cyna M., and De Palma A. (1984), "Dynamic Model of Peak Period Congestion", *Transportation Research,* 18B (4/5), 339-355.

[2] Ben Akiva M., De Palma A., and Kanaraglou P. (1986), "Dynamic Model of Peak Period Traffic Congestion with Elastic Arrival Rates", *Transportation Science,* 20 (2), 164-181.

[3] Boyce D.E., Bin Ran,LeBlanc L.J. (1991), "Dynamic User-Optimal traffic Assignment model: A new Model and Solution Technique", submitted to *Transportation Science.*

[4] Carey, M. (1986), "A constraint qualification for a dynamic traffic assignment model", *Operations Research,* 35 No. 5, 55-58.

[5] Carey, M. (1987), "Optimal time-varying flows on congested networks", *Operations Research,* 35 No. 5, 58-69.

[6] Cascetta E. and Cantarella G.E. (1991), "A Day-to-day and Within-day Dynamic Stochastic Assignment Model", *Transportation Research,* 25a (5).

[7] Di Gangi M. (1993), "A model for DNL allowing while-trip re-routing", submitted to *Mathematical modelling of Systems.*

[8] Friesz T.L., Luque J., Tobin R.L., Wie B.W., (1989), "Dynamic network traffic assignment considered as continuous time optimal control problem", *Operations Research,* 37, 893-901.

[9] Hammerslag R. (1988), "A three-dimensional assignment in the time-space", *UTSG Annual Conference 1988*, London.

[10] Janson B.N. (1989), "Dynamic traffic assignment for urban road networks", *Transportation Research,* 25B (2/3), 143-161.

[11] Leonard D.R., Tough J.B., Baguley P.C., (1978), "CONTRAM: A traffic assignment model for predicting flows and queues during peak periods", *TRRL Report 841*, Crowtorne.

[12] Merchant D.K., Nemhauser G.L. (1978), "A model and an algorithm for the dynamic traffic assignment problems", *Transportation Science*, 12 (3), 183-207.

[13] Papageorgiou M. (1990), "Dynamic modelling, assignment, and route guidance in traffic networks", *Transportation Research* 24B (6), 471-495.

[14] Vythoulkas P.K. (1990), "A Dynamic Stochastic Assignment Model for the Analysis of General Networks", *Transportation Research*, 24B (6), 453-469.

[15] Wie B.-W., Friesz T.L., and Tobin R.L. (1990), "Dynamic user optimal traffic assignment on congested multi-destination networks" *Transportation Research,* 24B(6), 431-442.

[16] Yagar S. (1975), "CORQ - A Model for Predicting Flows and Queues in a Road Corridor", *Transportation Research Record 533*, pp. 77-87, TRB, National Research Council, Washington D.C.

[17] Fernandez J.E., De Cea J. (1994), "Flow propagation description in dynamic network assignment models", *Proceedings of TRISTAN II TRiennal International Symposium on Transportation ANalysis* pp. 517-531, Capri, June 1994.

Appendix A Definitions and Notations

The standard nomenclature for both equations and relationships described within this paper is shown in fig. A.1. In this paper vectors and matrices are indicated by means of bold characters. Subscripts are usually adopted to specify the link of the network or the arc of the tree and the period of time to which the parameter refers whilst superscripts are generally used to denote followed path and the period of time at which users begin their trip.

name	meaning
Ω_r	sub-graph formed by the paths connecting o/d pair r
$\beta_{ah}^{jk}(s,\tau)$	fraction of F^{jk} crossing section s of arc a during time $[0, \tau]$ of interval h
b_{ah}^{jk}	crossing fraction of the packet (j,k) on arc a evaluated for the interval h
d^{jr}	number of users belonging to the cluster (j, r)
f_{ah}^{jk}	flow on arc a at interval h due to the packet (j, k)
F^{jk}	path flow of users of the packet (j, k)
h	current interval of the day
(j, k)	packet leaving at interval j and following path k
K_r	set of paths related to o/d pair r
$\Lambda(\tau)$	time distance between the head and the tail points of a packet at time τ
L_a	length of arc a
$\lambda_s(\tau)$	time distance between section s and the position of tail point at time τ
N^{jk}	number of users belonging to the packet (j, k)
π_a^{jr}	absolute cluster rate of arc a for cluster (j, r)
$\tilde{\pi}_a^{jr}$	relative cluster rate of arc a for cluster (j, r)
r	generic o/d pair
R_v	set of arcs departing from vertex v of the tree
s	generic section of arc a $[0 \leq s \leq L_a]$
τ	generic instant of time interval $[0 \leq \tau \leq T]$
T	time length of an interval
v_{ah}	speed on arc a during interval h

Fig. A.1 List of the main variables adopted within the paper

DYNAMIC TRAFFIC ASSIGNMENT IN CONGESTED NETWORKS

Giuseppe Bellei

Dipartimento di Idraulica, Trasporti e Strade
Università degli Studi di Roma "La Sapienza"
Via Eudossiana 18 - 00184 Roma, Italy

Maurizio Bielli

Istituto di Analisi dei Sistemi ed Informatica
Consiglio Nazionale delle Ricerche
Viale Manzoni 30 - 00185 Roma, Italy

The evaluation of potential benefits of new information systems to road users and of demand management measures, like limited traffic zones and road pricing, in congested areas, requires representing time of the day dynamic of traffic flows on the road network and users' behaviour. This can be accomplished by taking into account several aspects such as users' information level and travel demand distribution over time, with the objective to get a forecast of the expected smoothing effects of adopted measures and information systems. Modelling approaches presented to perform this assessment are briefly reviewed in this paper, with the aim of improving traffic congestion representation in a dynamic framework. In particular, queuing phenomena are studied and explicitly introduced in a simple link performance model embedded in a dynamic assignment procedure. Moreover, properties of the proposed model are investigated, the computation algorithm is defined and some preliminary results of its application to small size test networks are presented and discussed.

1. Introduction

The model and assignment technique presented carry out previous work on mixed (discrete/continuous) heuristic approach to within-day dynamic route choice, where a discrete representation is adopted for departure time and a continuous one for arrival time at nodes and destinations. This feature of the model suggest reviewing the scientific literature on dynamic assignment focusing on the way time is represented.

1.1 State of the Art

Several continuous dynamic traffic assignment problem formulations have been proposed by, among others, Friesz et al. (1989), Ran and Shimazaki (1989), Wie et al. (1990), Boyce et al (1992). They derived continuous equilibrium models as optimal control problems based on the principle of selfish path time minimization. In these works path times were evaluated as the sum of current link times at the moment path choice is performed or revised, so defining an instantaneous user optimum, or reactive equilibrium in the terminology of Papageorgiu (1990), who first acknowledged the distinction between the two forms of equilibrium conditions arising in a dynamic framework, naming predictive equilibrium the one based on link times to be encountered when travelling.

Vythoulkas (1990) presented, generalizing previous results obtained by Ben-Akiva et al. (1984) on a parallel links network, another continuous route and departure time choice model, not making any explicit reference to optimal control theory and defining, in analogy with the static case, stochastic predictive equilibrium conditions as a steady state for an implicit Markovian model of day to day demand adjustment.

More recently Ran et al. (1992a,1992b) defined and studied, still in an optimal control theory framework, also the properties of continuous models for an "ideal dynamic user optimum", that is for predictive equilibrium, extending these models to the choice of time when to travel and to the case of stochastic route travel times. Nevertheless, they didn't present for this last models any solution algorithm, while it should be noted that also for reactive equilibrium the solution algorithms and numerical result available are obtained on small "toy" networks.

Models based on continuous time representation, although allowing easier derivation of useful mathematical properties for the corresponding problems, are difficult to solve, up to the point that numerical results presented for all the quoted references are obtained from a discretized version of the problem, rising some doubt about the chances of finding a trade-off point between accuracy of travel time representation and feasibility of a close time-of-the-day discretization. Moreover, the most complete theoretical results have been obtained on reactive equilibrium, an unsound representation of tripmakers' behaviour.

On the other side, some kind of discretization was adopted in the earliest attempts to define a dynamic traffic assignment model, from the pioneering simulation models by Yagar (1976), to the first mathematical programming models by Merchant and Nemhauser (1978a,1978b), dealing with single destination, system optimal route choice. This kind of models was studied and improved in more recent years by several authors, like Carey (1986,1987) and Ho (1980,1990); they were also extended to time dependent departure patterns and tested on a practical application to a corridor network by Mahmassani and Chang (1986,1988), always keeping a discrete time representation. Carey (1992) also pointed out the relevance of no overtaking conditions in dynamic traffic assignment and Smith (1993) proposed a discrete approach ensuring the respect of such conditions.

The comment already made about solution by discretization of problems originally defined as continuous obviously applies also to wholly discretized approaches. It should be added only that no overtaking condition, although more easily amenable in this framework, requires however cumbersome and heavily computer time demanding procedures to be dealt with.

A significantly different approach was taken by Cascetta and Cantarella (1990), who developed the Markovian model implicitly assumed by Vythoulkas and obtained a comprehensive non-equilibrium description of traffic phenomena. Such a description, although more general, in principle, with respect to equilibrium based ones, supplies transient and steady states whose utilization in a planning process is less direct and easily acknowledged than the definition of an equilibrium state.

In another group of models, time varying traffic flows are determined by directly applying an assignment technique defined in analogy with the static case, like in Hamerslag (1988), or by heuristically solving a dynamic assignment problem in its turn defined by analogy, like in Bellei and Bielli (1990,1992) and Janson (1991). It is however only from heuristics and simulation approaches that came, up to now, numerical results on full size networks. These approaches lack however of a sound theoretical basis and in general rely upon utilization of volume delay functions; a choice that, while appealing because volume delay functions may be available from previous studies in many situations, hiddens to some degree the dynamic features of traffic phenomena, where queuing plays an important role.

Even if approaches focusing on queues representation to deal with link performance in a dynamic assignment framework can be found in Drissi-Kaïtouni and Hameda-Benchekroun (1992) and in Bernstein et al. (1993), they either rely upon massive network expansion which doesn't guarantee, in our opinion, no overtaking condition, or build up a comprehensive variational inequalities based theoretical model, but ignore spill-back queuing phenomena and need however an heuristic approach for problem solving.

Following up previous, wholly heuristic, approach, the goal of this work is to begin framing already defined and newly developed dynamic assignment techniques into a fixed point equilibrium formulation, where queuing phenomena representation implies no overtaking condition, while testing the degree of accuracy requested when solving the dynamic network loading (DNL) subproblem. Future extension to a more complete representation of spill-back queues, at present only approximately represented, is also allowed in this framework. Moreover, applicability to full size networks is kept as a constraint for algorithmic developments, even if numerical results presented refer to small, but not trivial, networks.

The ever growing scientific literature on the subject makes it difficult to definitely assert anything about the point reached by other researchers in the achievement of such a goal, but the part which has been examined is relevant enough to state that the goal by itself is worth a research effort and its partial achievement innovates the state of the art with regard to such issues like fixed point formulation of dynamic assignment, interdependencies between queue representation and platoons' overtaking and potential gains in algorithmic efficiency from approximate solution of DNL.

1.2 Structure of Presentation

The model and dynamic assignment technique which are developed in the following are based on a model for link performance which takes into account queuing and checks (at present without exhaustive representation) oversaturation phenomena. At first, this model is developed independently by the network framework where within-day dynamic assignment of travel demand, representing daily route choices in an aggregate way, is carried on. To keep as simple as possible the exposition the link indexes are suppressed, the inflow is assumed as given and piecewise constant, while link exiting flow is limited, like the number of vehicles queuing on it. The link is thus represented as a bottleneck with given output and storage capacity.

Notwithstanding this limitations, which are however, at least in part, easy to overcome, this link performance model is able to give a picture of traffic flowing on congested link, supplying a quantitative description of what happens to any entering vehicle, be it smoothly travelling from initial to final node, or held for some time in a queue, even if vehicles are not individually represented. The description of this model will follow, while the network model and assignment technique developed, some preliminary numerical results and guidelines for further development are the object of the subsequent chapters.

2. Congested Link Performance Model

Let's assume that, on any link of a road network, vehicles may be only in one of the following two conditions: travelling at a constant speed V, eventually a suitable fraction of free flow speed, or queuing along the link. If inflow entering the link is lower than its outflow capacity OC (eventually a suitable fraction of actual link capacity as determined by physical capacity and green splits), all vehicles travel at speed V, no queuing take place and vehicles exit the link at the same rate they enter it, otherwise vehicles exit the link at a constant outflow C and the queue occupies a section of the link, going backward from final node for a variable in time length (queue occupancy) OQ, determined by average spacing between queuing vehicles D, number of lanes M and variable-in-time entering flows F^i, assumed as constant within each time interval from time T^{i-1} to time T^i, belonging to a time interval set I defining the time horizon when link performance is represented . Queue occupancy OQ has a maximum, corresponding to link length L, proportional to overall storage capacity SC (maximum number of vehicles queuing on a link), given by $L \cdot M/D$.

2.1 Non Zero and Limited Queue Length Model

At first, the limiting assumption that queue occupancy is non zero and shorter than link length is made and a linear arrival time function is derived for each interval. In this way spill-back queues don't need to be considered, link performances are independent from each other and both queue occupancy and arrival times at the end of the link can be determined as functions of the vector of link inflows **F** and of link entering time t. These functions OQ(**F**, t), TA(**F**, t) result to be, with the assumptions made, linear within each interval i and piecewise linear with respect to t on the whole time horizon considered. Moreover, they can be recursively determined, with the only requirement that an initial state of the link (in practice an uncongested state) is known.

In fact, if for some interval i the time TQ^{i-1} at which the vehicle entering the link at time T^{i-1} joins the queue and queue length at that time, OQ^{i-1}, are known, queue evolution can be traced as long as vehicles entering the link during interval i join it, that is up to time TQ^i when the vehicle entering the link at time T^i joins in its turn the queue. During any time unit following TQ^{i-1} and up to TQ^i it happens that FQ^i vehicles are added and OC vehicles are subtracted to the queue, each one determining a queue lengthening or shortening of D/M length units.

The effect of entering flows can thus be interpreted as a lengthening speed of the queue $VQ^i = FQ^i \cdot D/M$ and the effect of exiting flows as a shortening speed $VS = OC \cdot D/M$, while their difference gives the relative speed of queue's tail. Flow FQ^i, influencing queue lengthening speed, is different from inflow F^i because the vehicles entering the link within interval i $(T^{i-1}, T^i]$ join the queue within interval $(TQ^{i-1}, TQ^i]$, whose length is generally different from the length of interval i.

To identify time TQ^i the corresponding queue occupancy OQ^i may be derived in two different ways. First, it is observed that the length of the part of the link that is not occupied by the queue at time TQ^i is equal to the distance travelled, at a speed V, by the vehicle joining the queue at that time and entering the link at time T^i:

$$L - OQ^i = (TQ^i - T^i)V \tag{1}$$

Then the length OQ^i of the queue at time TQ^i is expressed as the sum of its length OQ^{i-1} at time TQ^{i-1} and of (relative) lengthening:

$$OQ^i = OQ^{i-1} + (VQ^i - VS)(TQ^i - TQ^{i-1}) \tag{2}$$

Denoting by LQ^i the length of the part of the link that would be occupied by vehicles entered during interval i, if queuing:

$$LQ^i = \frac{D}{M} F^i (T^i - T^{i-1}) = \frac{D}{M} FQ^i (TQ^i - TQ^{i-1}) = VQ^i (TQ^i - TQ^{i-1}) \tag{3}$$

time TQ^i and queue length OQ^i are given, independently by each other, combining formulas(1) and (2) with some further manipulation, by:

$$\text{a)}\, TQ^i = TQ^{i-1} - \frac{LQ^i - (T^i - T^{i-1})V}{V - VS}; \text{b)}\, OQ^i = OQ^{i-1} + \frac{LQ^i - (T^i - T^{i-1})VS}{1 - VS/V} \tag{4}$$

Correspondingly, if only vehicles entering after time T^{i-1} and until time $t \leq T^i$ are considered, the length:

$$LQ(t) = \frac{D}{M} F^i (t - T^{i-1}) \tag{5}$$

of the part of the link that would be occupied by these vehicles, if queuing, may be substituted to LQ^i and t may be substituted to T^i in the left -hand sides of (4) to get TQ(t) and OQ(t) within interval i:

$$\text{a)}\; TQ(t) = TQ^{i-1} - \frac{LQ(t) - (t - T^{i-1})V}{V - VS}; \quad \text{b)}\; OQ(t) = OQ^{i-1} + \frac{LQ(t) - (t - T^{i-1})VS}{1 - VS / V} \tag{6}$$

Also arrival times at final node of the vehicle entering the link at time T^i and of a generic vehicle entering at time t within interval i, TA^i and TA(t), are easily found, considering that after joining the queue the vehicles move at a speed VS:

$$\text{a)} \quad TA^i = TQ^i + \frac{OQ^i}{VS} \qquad \text{b)} \quad TA(t) = TQ(t) + \frac{OQ(t)}{VS} \tag{7}$$

Alternatively, if times when vehicles join the queue, or even queue occupancies, are not needed, also arrival times can be independently determined by a recursion. Such a recursion may be derived combining (4) and (6) with (7), or directly by considering that an observer at final node would see a section of queue of length LQ^i, moving at speed VS, while vehicles entering during the whole interval i exit the link, and a section of length LQ(t) while vehicles entering during sub-interval (T^{i-1}, t] are exiting. Hence, if arrival time of vehicle entering the link at time T^{i-1} is TA^{i-1}, it is:

$$\text{a)}\; TA^i = TA^{i-1} + \frac{LQ^i}{VS} = \frac{F^i (T^i - T^{i-1})}{OC}$$

$$\text{b)}\; TA(t) = TA^{i-1} + \frac{LQ(t)}{VS} = \frac{F^i (t - T^{i-1})}{OC} \tag{8}$$

and all time functions that may play a role in describing, under the assumptions made, link performance as a function of inflow, are defined.

2.2 Queue Vanishing or Exceeding Storage Capacity

The event of queue's vanishing is easily checked through expressions (4b) and (6b), and a zero length may be attributed to the queue whenever negative values would be supplied by these expressions:

$$OQ^i = \max\left\{0; OQ^{i-1} + \frac{LQ^i - (T^i - T^{i-1})VS}{1 - VS/V}\right\}$$

$$OQ(t) = \max\left\{0; OQ^{i-1} + \frac{LQ(t) - (t - T^{i-1})VS}{1 - VS/V}\right\} \qquad (9)$$

but such an event can be recognized also in (8) by remarking that no vehicle can travel along the whole length of the link at a speed greater than V, so that minimum travel time on the link is L/V:

$$TA^i = \max\left\{T^i + \frac{L}{V}; TA^{i-1} + \frac{F^i(T^i - T^{i-1})}{OC}\right\}$$

$$TA(t) = \max\left\{t + \frac{L}{V}; TA^{i-1} + \frac{F^i(t - T^{i-1})}{OC}\right\} \qquad (10)$$

If queue vanishes during some interval i, however, even if (9a) and (10a) still hold, the functions OQ(t) and TA(t) are non linear within the interval and time TV^i when vanishing takes place can be determined by setting OQ(t) equal to 0 in (6b) and determining $TV^i = t$:

$$TV^i = +T^{i-1} + \frac{OQ^{i-1}(1/VS - 1/V)}{(1 - F^i/OC)} \qquad (11)$$

The vehicle entering the link at time TV^i takes a time L/V to arrive at final node, thus either (TV^i, TV^i + L/V) is added to the sequence of (T^i, TA^i) pairs to define the piecewise linear arrival time function or the function (10b) has to be used to compute arrival times whenever storage capacity constraint is satisfied, instead of simply utilizing linear functions between a sequence of points in the departure/arrival times space.

In case the queue should reach maximum link length L, it spills back into upstream links, so that interaction between link performances take place; an approximate representation of link congestion phenomena could avoid taking into account such an interaction by considering vehicles which can't be accommodated along the link (on the "horizontal" queue) as waiting into a fictitious tank where a FIFO rule is respected (in other words, a "vertical" queue) located at initial node. This approximate representation makes expression (10a) hold whether there is a vertical queue or not, and arrival time function keeps linear even within intervals when storage capacity is exceeded. Nevertheless, the time TE^i when storage capacity is exceeded, may be needed for a capacity adjustment procedure to be introduced in the following. In such a case, it is obtained by setting OQ(t) equal to 0 in (6b) and determining Te^i=t:

$$TE^i = +T^{i-1} + \frac{(L - OQ^{i-1})(1/VS - 1/V)}{(F^i / OC - 1)} \tag{12}$$

In conclusion, if the origin of departure times axis is arbitrarily fixed at the beginning of time horizon ($T^0 = 0$) and an uncongested state is assumed at that time, arrival times and queue occupancies may be recursively determined by setting initial values $TA^0 = L/V$, $OQ^0 = 0$ and utilising:

- (10a) to get the sequence of arrival times TA^i for vehicles departing at T^i;
- (10b) to get arrival times TA(t) for vehicles departing at any time $t \in (T^{i-1}, T^i]$;
- (9a) to get the queue occupancies OQ^i encountered by vehicles departing at T^i;
- (9b) to get queue occupancies OQ(t) encountered by vehicles departing at any time $t \in (T^{i-1}, T^i]$;

- (11) and (12), if needed, to determine departure times TV^i and TE^i of vehicles joining the queue when vanishing and storage exceeding takes place, if negative and greater than L OQ^i values, respectively, are supplied by (9a).

2.3 Monotonicity Properties of Arrival Time Function

Arrival times at final link nodes given by (10) reduce to the sum of departure time and of the L/V constant whenever there is no queue so that its derivative is equal to 1. Otherwise it is easy to derive that, if $t \in (T^{i-1}, T^i]$:

$$\frac{\partial TA(\mathbf{F},t)}{\partial t} = 1 + \frac{F^i - OC}{OC} = \frac{F^i}{OC} \tag{13}$$

which is clearly non negative, being the ratio of two non negative constants.

It is easy to show that arrival time function is also non decreasing with respect to each average inflow F^i. In fact, in the case there is no queue on the link at time t, arrival time is trivially not depending by any inflow, otherwise it depends directly only from inflow F^i, with i such that $t \in (T^{i-1}, T^i]$ and it is:

$$\frac{\partial TA(\mathbf{F},t)}{\partial F^i} = \frac{t - T^{i-1}}{OC} > 0 \tag{14}$$

while indirect dependence from inflows other than F^i take place through the term TA^{i-1} in (10b), in its turn directly depending from F^{i-1} and indirectly from other inflows through TA^{i-2}, and so on. Dependence from any inflow $F^{i'}$, with i' < i, thus, leads by induction to verify again (14), with i' instead of i, or to exclude further dependencies.

3. Network Model

Link performance model plays here the role that performance functions play in static equilibrium models and, in some earlier work of the authors (Bellei and Bielli, 1990,1992), also in within day dynamic assignment. This model represents how an element of transportation system behave when it has to carry a given traffic load, so that an overall model has however to be built, to determine the traffic load for each element. Such a model must be consistent with link performance model and able to represent the time, space and flow relationships among elements, in such a way that transportation offer of the whole network, together with relationships between transportation offer and demand, are also represented. In short, this overall model is a network model. The general formulation of this model adopted here is not by itself completely new, but there are several of its features (fixed point formulation, explicit modelling of queues, mixed discrete-continuous representation of time variables, definition of relationships between demand and network flows, aggregate and approximate solution technique) which in the authors' belief give or suggest, both separately and as a whole, a different and simpler solution to some relevant problems identified by the scientific literature on dynamic assignment, with respect to the previously proposed ones.

3.1 Time of the Day Transportation Demand and Offer

Space relationships among elements of transportation system are represented as usual as a directed graph G(N, L) where N is a set of nodes, representing locations in space, and L a set of links directly connecting some ordered node pairs mn, supplying communication between adjacent locations. It is assumed that any node pair is indirectly connected by at least one path p, defined here as an ordered sequence of K_p nodes $\{n^1{}_p, n^2{}_p, \ldots, n^{K_p}{}_p\}$, corresponding to K_p - 1 links, such that no node appear more than once in the sequence and a link exists, for any pair of nodes $(n^{k-1}{}_p, n^k{}_p)$, from $n^{k-1}{}_p$ to $n^k{}_p$. Each path represent a route which can be chosen to travel between nodes and belongs to a path set P. Since only some locations in space are usually assumed to be origin or destination of a trip, identified by subsets R and S of N, the only relevant paths either belong to some subset P_{rs} of P or are part of paths belonging to one of these subsets.

Time is represented in different ways for different purposes, both as a non negative real variable and as intervals, whose definition is introduced to allow two important simplifying assumptions:

a) departures taking place during any interval j $(TD^{j-1}, TD^{j}]$ are grouped, to represent vehicles travelling on the network as platoons departing together at some fixed time (i. e. time TD^{j} for interval j);

b) average flow determined by platoons arriving at a node m during any interval i $(TN^{i-1}, TN^{i}]$ and travelling on link mn is dealt with as a constant inflow, entering the link during the interval (times TN^{i} are the same as times T^{i} of link model).

Intervals are defined by assuming also $TD^{0} = TN^{0} = 0$ and leaving to times TD^{j} and TN^{i} the task of identifying the intervals belonging to the interval set I (J). Even if different sequences may be defined for each origin and node, only intervals with respect to departure from origins and to arrival at nodes are distinguished up to this point. The time of the day dimension of a trip on path p, that is the "journey" corresponding to that trip (Addison and Heydecker, 1993), is described by non negative real arrival times TA^{j}_{mp} at node m, when departing at TD^{j} and travelling along p so that also travel time $TA^{j}_{mp} - TD^{j}$ is a non negative real number.

Link flow variables are the inflows introduced at previous chapter as a vector **F**, conceptually defined by assumption b) above. Obviously, when considering a link mn inflow vector **F** becomes $\mathbf{F_{mn}}$ and its ith element (average inflow entering link mn during interval i) is denoted by F^{i}_{mn}, and also arrival time functions $TA(\mathbf{F}, t)$ become $TA_{mn}(\mathbf{F_{mn}}, t_{m})$ while the vector $\mathbf{\Phi}$ of inflows for each interval and link may be considered to express dependencies in a more compact form. Path flows FD^{j}_{p} for any departure time TD^{j} are also defined to represent route choice by travellers departing from the origin of p during interval j, while inflows are the only link flows which are explicitly defined, since space and time continuity of flows is guaranteed if path flows and link inflows are consistent, so that outflows and conservation equations involving inflows and outflows are not needed (in the following, however, consistency is only approximated to avoid path enumeration).

Transportation demand is simply defined as a three dimensional OD matrix given in terms of number of vehicles ND^j_{rs}, departing from origin r to destination s at time TD^j, having grouped, consistently with assumption a), flows GD^j_{rs} departing during interval j in platoons departing at time TD^j.

3.2 Route Choice and Network Flows

Differentiating between path flows (defined with respect to departure time) expressing route choice and consistency with travel demand, and link flows (defined with respect to arrival time at nodes) loaded to the network and determining its performance, is a distinctive feature of time of the day dynamic assignment model presented. This difference arises because the number of vehicles departing from an origin r to travel along a path p during a time interval do not pass through nodes within the same time interval, so that direct relationships between mobility demand and network flows, usually adopted in the static case, are no longer exhaustive in representing mobility phenomena. Vehicles departing from their origin at time TD^j and travelling along p, whose number is denoted as ND^j_p, correspond to an average departure time flow FD^j_p during interval j:

$$FD^j_p = \frac{ND^j_p}{TD^j - TD^{j-1}} \tag{15}$$

and consistency with travel demand may be given for both absolute values and flows:

$$\text{a) } ND^{j}_{rs} = \sum_{r\in R}\sum_{s\in S}\sum_{p\in P_{rs}} ND^{j}_{p}; \quad \text{b) } GD^{j}_{rs} = \sum_{r\in R}\sum_{s\in S}\sum_{p\in P_{rs}} FD^{j}_{p} \tag{16}$$

The departure time flow defined by (15) give to the flow passing through node m a contribution which is, on the average, equal to $ND^{j}_{p}/(TA^{j}_{mp} - TA^{j-1}_{mp})$. This average contribution during the interval $(TA^{j-1}_{mp}, TA^{j}_{mp}]$, in general different from any fixed $(TN^{i-1}, TN^{i}]$, would not be constant along the interval, even if departure time flow were assumed as constant along interval j.

This happens because the contribution given depends by the "history" of the ND^{j}_{p} vehicles, whose initially constant flow would change several times when travelling on any link, being influenced by any change in flows sharing that link during the time needed to go from initial to final node. It is easily acknowledged that, due to additive and multiplicative effects when passing from link to link, an explicit representation for the history of departure time flows on any non trivial network is quite complex. Thus, even if an ingenuous, but cumbersome, method to deal with platoon history, oriented to ensure no overtaking, has been proposed by Smith (1993), our approach is oriented to find some kind of approximation, avoiding such a representation, but allowing to distinguish between flows defined with respect to departure time and node arrival time.

Such an approximation is based on assumptions a) and b) above. To accept approximation a), the difference in traffic conditions encountered by the vehicles loosing their identity to become part of a compact platoon must be negligible, as it happens when departure time interval j and/or platoon size ND^{j}_{p} are small enough. In principle this condition is easily satisfied, because, if platoons are defined as ND^{j}_{p}, their size is negligible for all practical purposes. To accept approximation b), no a priori assumption being possible about platoons' distribution within node arrival time intervals, interval length has to be assumed small enough to make negligible the effect of averaging. Formally the average inflow into link mn during interval i can be expressed dividing the sum of the number of vehicles departing at time TD^{j}, choosing to travel along any path p including link mn, which are on the whole ND^j_p, arriving at node m during interval i, by interval length. Utilising, to express the fraction of the ND^j_p vehicles assumed as arriving at m during i, the notation α^{ij}_{mp}, the average node arrival time inflow into link mn during interval i is obtained, denoting by δ_{mnp}, as usual, the element mn,p of link-path incidence matrix, as:

$$F^i_{mn} = \sum_{j\in J}\sum_{p\in P}\alpha^{ij}_{mp}\delta_{mnp}ND^j_p \Big/ (TN^i - TN^{i-1}) \tag{17}$$

If the length of departure and node arrival time intervals is assumed as equal and everyone of the vehicles is assumed to get the same node arrival times, so that δ^{ij}_{mp} defined as:

$$d^{ij}_{mp} = \begin{cases} 1 & if \quad TA^j_{mp} \in \left(TN^{i-1}, TN^i\right] \\ 0 & otherwise \end{cases} \tag{18}$$

is substituted to α^{ij}_{mp}, the relationship between interval-link and departure time-path flows may be written in such a way that it closely resembles correspondent static formulation:

$$F^i_{mn} = \sum_{j\in J}\sum_{p\in P}\delta^{ij}_{mp}\delta_{mnp}FD^j_p \tag{19}$$

The symbols δ^{ij}_{mp} and α^{ij}_{mp}, however, are conceptually different from δ_{mnp} because they are endogenous variables, depending by node-path arrival times TA^j_{mp}, and could be also denoted as $\delta^{ij}_{mp}(\mathbf{T}^j_\mathbf{p})$ and $\alpha^{ij}_{mp}(\mathbf{T}^j_\mathbf{p})$, making explicit the dependence from the vector $\mathbf{T}^j_\mathbf{p}$ of arrival times at nodes when travelling along path p and departing at time TD^j. On the whole hence (19) defines a vectorial function $\mathbf{\Phi} = \Delta(\Pi, \mathbf{T})$ expressing the dependence of link-interval inflow vector $\mathbf{\Phi}$ from node-path arrival times vector $\mathbf{T}$ and departure time-path flow vector Π. This function, which will be utilised in the following, is consistent with a point-packet assumption, where temporal and spatial extension of platoons is ignored.

3.3 The Dynamic Network Loading (DNL) Problem

It has been already noted that the vectors $\mathbf{F}_{mn}$ of inflows defined by (18), or (19), for $i \in I$ are, together with departure time t_m from initial node m of a link mn, the arguments of arrival time functions $TA_{mn}(\mathbf{F}_{mn}, t_m)$ supplying arrival time t_n at final node n. These functions allow establishing a relationship between average interval link flows $\mathbf{\Phi}$ and node arrival times $\mathbf{T}$; assuming, without loss of generality, path p as represented by the sequence $\{1, 2, \ldots, k, \ldots, K_p\}$, arrival time TA^j_{kp} at node k departing at time TD^j from node 1 and travelling along p is given by the recursive formula:

$$TA^i_{np} = TD^j \quad for \quad n = 1$$

$$TA^i_{np} = TA_{n-1,n}(\boldsymbol{FA}_{n-1,n}, TA^j_{n-1,p}) \quad for \quad n = 2, \ldots, k \tag{20}$$

Each node arrival time TA^j_{mp} can thus be considered as a function of the vector $\mathbf{F}_p$ of inflows on the links of path p, defined by (20), while, on the whole, node arrival times $\mathbf{T}$ are a function $T(\mathbf{\Phi})$ of link inflows (dependence from intermediate departure times vanishes when compounding link functions along the path, while departure time from origin is assumed as fixed).

It is evident how a circular dependence (similar to the one existing between flows and times in static assignment at equilibrium) takes place, in dynamic assignment, also in defining the relationship between users choices and network flows, even if for these choices no equilibrium concept has been introduced at this stage. The problem of finding mutually consistent average inflows and node arrival times can thus be written, for any path choice vector Π, as the fixed point problem:

$$\mathbf{\Phi} = \Delta[\Pi, T(\mathbf{\Phi})] \tag{21}$$

where Φ is defined on real n-dimensional space $\mathbf{R}^n$ (n = n° of links x n° of arrival intervals), since all constraints on inflows are implicitly satisfied by function $\Delta(\cdot)$.
Problem (21) is the form specific to this model of what is commonly known as the dynamic network loading problem. Function $\Delta(\Pi, \mathbf{T})$, defined by (19) and utilised in defining problem (21), however, even if easy to implement and allowing relationship between path and link flows to be expressed in a familiar form, is not continuous with respect to $\mathbf{T}$, as it is function (20) with respect to Φ. If point-packet assumption is relaxed, however, and any platoon is split between intervals utilising any continuous function of the difference $TA^j{}_{mp}$ - TN^i to get platoon arrival fraction $\alpha^{ij}{}_{mp}$, also function (20) can be made continuous. In such a way, existence of a solution to problem (21) could be proofed, but a conceptual consistency problem has to be solved since a unique node arrival time would be attributed to a platoon which, for the purpose of computing link inflows, is splitted between different time intervals. Taking into account this difficulty, the practical advantages of a point-packet approach already pointed out, and the fact that solution of problem (21) is however approximated to avoid path enumeration, function (20) is retained. The approximate solution of DNL problem (21) is dealt with together with the flow aggregation problem in the following.

3.4 Arrival Time Function and No Overtaking Constraint

The respect of a no overtaking constraint is ensured by arrival time functions, which results to be piecewise linear and monotone non decreasing with respect to time and to each of the $\mathbf{FA}_{mn}$ elements. Representing homogeneous vehicles on a network and taking into account time of the day, in fact, requires that vehicles arrive to any node, destination included, in the same order they depart from any other node, origin included, or, in other words, that no overtaking occurs (Carey, 1992). Among other things, if, when travelling between any two nodes, arriving earlier by starting later would be possible, stopping in the first one would be the most rational tripmakers' behaviour.

The no overtaking constraint is easily satisfied for each path if it is satisfied for each link, because arrival times at nodes are compound functions of link arrival time functions. Taking as example a simple (r-m-n-s) path p, where r and s are not necessarily origin and destination nodes, arrival time at s travelling along p, t_{sp}, for a departure time t_r from r is:

$$t_{sp} = TA_{ns}\left\{\boldsymbol{FA}_{ns}, TA_{mn}\left[\boldsymbol{FA}_{mn}, TA_{rm}(\boldsymbol{FA}_{rm}, t_r)\right]\right\} \tag{22}$$

where $t_r = TD^j$ and $t_{sp} = TA^j_{sp}$ if r is an origin node and platoon departing at TD^j and travelling along p has to be considered.

If, as it has been shown, derivatives with respect to time and to each element FA^i_{mn} of the vector $\mathbf{FA_{mn}}$ of the functions $TA_{mn}(\cdot)$ are non negative, this is true also for the derivatives of function (22) with respect to t_r and FA^i_{mn}:

$$\frac{\delta t_{sp}}{\delta t_r} = \frac{\delta TA_{rm}}{\delta t_r} \cdot \frac{\delta TA_{mn}}{\delta t_m} \cdot \frac{\delta TA_{ns}}{\delta T_n} \geq 0 \tag{23}$$

$$\frac{\delta t_{sp}}{\delta FA^i_{mn}} = \frac{\delta TA_{rm}}{\delta t_r} \cdot \frac{\delta TA_{mn}}{\delta FA^i_{mn}} \geq 0 \tag{24}$$

utilising the symbols t_m and t_n, in analogy with t_r, to express the fact that derivatives are taken, for each link, with respect to initial node entering time. Conditions (23) and (24) are easily generalised to any path p, not necessarily from an origin to a destination, whose nodes are assumed, as above, to be numbered $\{1, 2, \ldots, k, \ldots, K_p\}$:

$$\frac{\delta t_{k_p p}}{\delta t_1} = \prod_{k=2}^{K_p} \frac{\delta TA_{k-1,k}}{\delta t_k} \geq 0 \tag{25}$$

$$\frac{\delta t_{k_p p}}{\delta FA^i_{k-1,k}} = \prod_{k=2}^{K} \frac{\delta TA_{n-1,n}}{\delta t_{n-1}} \cdot \frac{\delta TA_{k-1,k}}{\delta FA^i_{k-1,k}} \geq 0 \tag{26}$$

where $t_1 = t_r = TD^j$ and $t_{K_p p} = TA^j_{sp}$ if 1 is an origin r, K_p a destination s and departure time TD^j is considered. The no overtaking constraint, by consequence, doesn't need to be explicitly formulated and arrival times at nodes are non decreasing functions of each average interval link flow.

3.5 Route Choice Flow Aggregate Representation

To avoid path enumeration, compact platoons must be defined at a more aggregate level, with respect to paths, taking care that relevant platoons' size keeps negligible. If aggregation is performed with respect to path and destination the number of vehicles departing from origin r at time TD^j and choosing to travel along any path including link mn is defined as:

$$ND^i_{rmn} = \sum_{s \in S} \sum_{p \in P_{rs}} \delta_{mnp} ND^j_p \tag{27}$$

Inflow FA^i_{mn} can't be expressed, in general, in terms of ND^j_{rmn} since arrival times at nodes defining δ^{ij}_{mp} may differ among paths, even if these paths share the origin r. Departure time flow from origin r at time TD^j on link mn is:

$$FD^j_{rmn} = ND^j_{rmn} / (TD^j - TD^{j-1}) \tag{28}$$

and also for departure time flow FD^j_{rmn}, as for ND^j_{rmn}, no relationship with inflow, holding in any case, can be established.

If however the minimum among arrival times at node m travelling along paths from origin r and departing at time TD^j is determined:

$$TM^j_{rm} = \min_{s \in S, p \in P_{rs}} \left\{ TA^j_{mp} \right\} \tag{29}$$

defining, in analogy with (18):

$$d^{ij}_{rm} = \begin{cases} 1 & if \quad TM^{j}_{rm} \in (TN^{i-1}, TN^{i}) \\ 0 & otherwise \end{cases} \tag{30}$$

the average node arrival time inflow into link mn during interval i, based on minimum arrival times may be derived from ND^{j}_{rmn}, like FA^{i}_{mn} is derived from ND^{j}_{p}in (17), case a), or from FD^{j}_{rmn}, if intervals lengths are assumed as equal, as in (19), case b):

$$\text{a)}\ FM^{i}_{mn} = \frac{\sum_{j \in I} \sum_{r \in R} \delta^{ij}_{rm} ND^{j}_{rmn}}{TN^{i} - TN^{i-1}} \qquad \text{b)}\ FM^{i}_{mn} = \sum_{j \in I} \sum_{r \in R} \delta^{ij}_{rm} FD^{j}_{rmn} \tag{31}$$

It should be noted that, while average inflows FA^{i}_{mn} would be actually loaded to links mn in correspondence to intervals i if node arrival times were given by TA^{j}_{mp}, inflows FM^{i}_{mn} wouldn't, unless only minimum time paths are chosen. As it will be stated more formally in the following, however, this condition is satisfied if an equilibrium state is attained, so that the difference between the two above defined inflows would vanish at equilibrium. Hence, if an equilibrium is reached through aggregate (link-origin level) representation of route choice it will be such also for a disaggregate (path level) one.

Also inflows FM^{i}_{mn}, however, are in general not consistent with node arrival times and may be taken as a initial value for an iterative network loading procedure having approximated solution of DNL problem (21) as a purpose. Two versions of this procedure have been developed, both initialised by inflows FM^{i}_{mn} and aimed at platoons' assignment to node arrival intervals consistent with node arrival times on paths belonging to some fixed or iteratively updated tree.

In the first case the tree which results to be a shortest path tree (SPT) when arrival times are computed from initial inflows is kept as fixed, iteratively performing, along the whole algorithm, network loading on the basis of arrival times computed on that tree, while in the second SPT is updated at each network loading, determining shortest paths on the basis of current inflows. If resulting inflows are denoted by FC^{i}_{mn}, their derivation from ND^{j}_{rmn} is described by the following algorithm, where vector notation is utilised throughout to suppress time interval indexes:

Step 0 Initialisation - Set IT = 0; $\mathbf{FC_{mn}}^{(0)} = \mathbf{FM_{mn}}$; $\forall mn \in L$; **(A1)**

Step 1 Arrival time function definition - Determine $TA_{mn}[\mathbf{FC_{mn}}^{(IT)}, t_m]$ $\forall mn \in L$ utilising (10a);

Step 2a Node arrival times calculation - Determine node arrival times from origins $\mathbf{TC_{rm}}^{(IT)}$ applying functions $TA_{mn}[\mathbf{FC_{mn}}^{(IT)}, t_m]$ to the paths belonging to SPTs consistent with initial inflows $\mathbf{FM_{mn}}$; (1st case-no SPT updating)

OR

Step 2b Node arrival times calculation - Determine node arrival times from origins $\mathbf{TC_{rm}}^{(IT)}$ finding minimum node arrival times through a SPT search with functions $TA_{mn}[\mathbf{FC_{mn}}^{(IT)}, t_m]$; (2nd case-SPT updating)

Step 3 Network loading - Compute interval link inflows $\mathbf{FC_{mn}}^{(IT+1)}$ loading flows $\mathbf{FD_{rmn}}$ in accordance with node arrival times $\mathbf{TC_{rm}}^{(IT)}$

$$d^{ij\,(IT)}_{rm} = \begin{cases} 1 & if \quad TC^{j\,(IT)}_{rm} \in (TN^{i-1}, TN^{i}] \\ 0 & otherwise \end{cases}$$

$$FC^{i\,(IT+1)}_{mn} = \sum_{j \in I} \sum_{r \in R} \delta^{ij\,(IT)}_{rm} FD^{j}_{rmn}$$

Step 4 Stopping criterion - If IT = MAXIT set $\mathbf{FC_{mn}} = \mathbf{FC_{mn}}^{(IT)}$ and STOP;

Else go to Step 1.

The two alternative options which have been left open for the node arrival times calculation, define the above mentioned versions for this procedure. These options, and the simple iteration number based stopping criterion chosen, are in accordance with the purpose of testing the effectiveness of different degrees of accuracy in solving DNL problem.

It may be useful to observe that, since the coherence between travel, or arrival time and inflows throughout the paths may be efficiently sought only with reference to paths included in a tree, which is in origin a SPT, a choice has to be made if to keep the tree fixed throughout the algorithm (so that this coherence may be more easily approximated on it, even if changes in inflows will generally lead to a different SPT) or to determine a new SPT at each iteration (so that reaching arrival time and inflows coherence may be more difficult to get, but arrival times are always defined on SPTs).

There is not, however, so a large difference between the two options with regard to accuracy, since both may ensure DNL problem solution only at equilibrium, when only shortest paths are utilised and node arrival times become independent from the chosen path, while also computational burden is less different than at first sight because visiting a tree, as it is required in the first option, is not so much easier than finding a SPT in the second one.

3.6 Model Definition

The arrival times at nodes **T** recursively determined by (20) may be considered as argument of an assignment mapping supplying a vector of path flows Π = A(**T**). Even if, generally speaking, such a mapping could be assumed to be a biunivocal continuous function of path travel times (arrival times at destinations minus travel times) as it happens, for instance, when applying Dial's algorithm (Sheffi, 1985) to probabilistic fixed time assignment, no dynamic version of this assignment method was available. Moreover, deterministic shortest path assignment mapping results to be more consistent with link performance model and a dynamic version had already been developed by the authors modifying a method originally proposed by Ford and Fulkerson (1962). Such a mapping, however, when there is more than one shortest path for the same origin destination pair, is neither univocal (origin destination flow may be assigned in an infinite number of ways among shortest paths) nor continuous (increasing by an infinitely small amount travel time of a shortest path reduces to zero the flow previously assigned to that path). The use of shortest path assignment mapping to define a deterministic dynamic user equilibrium problem as a fixed point problem thus, doesn't allow, at this stage, existence of solutions to be proofed. This means that, although some interesting properties of the proposed approach were found out and set into an analytical framework, the assignment method proposed should still be considered as an heuristic.

Nevertheless, if DNL problem is assumed as solved, the resulting flow vector $\mathbf{\Phi}$ can be considered as a function of path flows Π appearing as argument of the $\Delta(\cdot)$ function. Denoting by $\Lambda(\Pi)$ such a function, dynamic assignment problem in terms of path flows may be expressed (Daganzo, 1983) as a generalized fixed point problem:

$$\Pi \subseteq A\{T[\Lambda(\Pi)]\} \tag{32}$$

where Π belongs to the convex polyhedron on $\mathbf{R}^n$ (n = n° of paths x n° of departure intervals) defined by (16b) and non negativity conditions.

The arrival times TA^j_{sp} allow to define equilibrium conditions, equivalent to fixed point formulation (32), as a straightforward extension to a time of the day dynamic framework of Wardrop's first principle requiring equal minimum travel times on utilised paths. Expressing travel times TT^j_{rs} from origin r to destination s when departing at TD^j as the difference between departure time from origin r and arrival time at destination s of a generic path $p \in P_{rs}$, these conditions are:

$$\begin{aligned} &(TA^j_{sp} - TD^j - TT^j_{rs})FD^j_p = 0 \\ &TA^j_{sp} - TD^j \geq TT^j_{rs} \end{aligned} \tag{33}$$

where TT^j_{rs} is the equilibrium (minimum) travel time between r and s departing at TD^j.

3.7 Network Assignment Method

The assignment method applied to problem (32) is derived from the method of successive averages for static traffic assignment, the main differences arising from the direct representation of link performance in time through link performance model and from the distinction made between choice flows, defined with respect to departure time, and network inflows, defined with respect to arrival time at nodes.

The first difference is mainly a conceptual one, because a link travel time value is computed by (8a) for each link and arrival interval, so that the only practical difference with respect to calculation of travel times in a static framework is, as obvious, the fact that different values must be computed for each interval. From a conceptual point of view, however, it is worth noting again that these values are not separately computed and attributed to arrival intervals, but result, instead, in the definition of arrival time functions which give a global representation of link state and performance in time.

The second difference is relevant both from a conceptual and a from a practical point of view, since it derives from a specific feature of dynamic assignment and leads to apply successive averaging to choice flows, while network inflows are computed again at each iteration by approximate solution of DNL problem through algorithm (A1).

The resulting algorithm is defined as follows, utilising vector notation to suppress time indexes as in (A1):

<u>Step 0</u> <u>Initialisation</u> **(A2)**

<u>0-1</u> Define arrival time functions $TA_{mn}^{(0)}(\cdot)$ as $t_n = t_m + L_{mn}/V_{mn}$;

0-2 Determine shortest paths and node arrival times $\mathbf{TM_{rm}}^{(0)}$ as in step 2;

0-3 Compute choice flows $\mathbf{FD_{rmn}}^{(0)}$ assigning travel demand to shortest paths identified at previous substep;

0-4 Determine average inflows $\mathbf{F_{mn}}^{(0)}$ as in step 4, set IT = 0;

Step 1 Arrival time function definition-Determine arrival time functions $TA_{mn}^{(IT)}(\cdot)$ as $TA_{mn}[\mathbf{F_{mn}}^{(IT)}, t_m]$ utilising (10a);

Step 2 Shortest path in time search-Identify SPTs with variable in time travel times supplied by arrival time functions $Ta_{mn}{}^{(IT)}(\cdot)$ and correspondent node arrival times $\mathbf{TM_{rm}}^{(IT)}$;

Step 3 Choice flows updating-Compute tentative choice flows $\mathbf{FT_{rmn}}^{(IT)}$ assigning travel demand to shortest path identified at previous step and set:
$\mathbf{FD_{rmn}}^{(IT+1)} = (1-\alpha)\mathbf{FD_{rmn}}^{(IT)} + \alpha\mathbf{FT_{rmn}}^{(IT)}$ with $\alpha = 1/(IT + 1)$;

Step 4 Dynamic network loading-Apply (31b) to compute link inflows $\mathbf{FM_{mn}}^{(IT)}$ consistent with node arrival times $\mathbf{TM_{rm}}^{(IT)}$; and algorithm (A1) to get link inflows $\mathbf{FC_{mn}}^{(IT)}$ approximately solving DNL problem, set $\mathbf{F_{mn}}^{(IT+1)} = \mathbf{FC_{mn}}^{(IT)}$;

Step 5 Stopping criterion-Check if IT = ITMAX or $\mathbf{F_{mn}}^{(IT+1)} \cong \mathbf{F_{mn}}^{(IT)}$ or any criterion capable to verify approximate respect of equilibrium conditions is satisfied, if it is STOP, otherwise set IT = IT + 1, go back to Step 1

Also in algorithm (A2), like in (A1), there are options which are left open for testing; one, not explicitly stated in its definition, is not utilising at all algorithm (A1) at step 4, setting directly $\mathbf{F_{mn}}^{(IT+1)} = \mathbf{FM_{mn}}^{(IT)}$. Such an option leaves solution of DNL problem to arrival time and flow adjustments among upper level iterations of algorithm (A2), without dealing specifically with it. Another open option is about stopping criterion, where maximum iteration number is given as an alternative to a measure of closeness between inflows at successive iterations, which could be specified, for instance, as the variation coefficient $VC^{(IT)}$:

$$VC(IT) = \frac{\sqrt{\sum_{i \in I} \sum_{mn \in L} \left[F_{mn}^{i\ (IT+1)} - F_{mn}^{i\ (IT)} \right]^2}}{\sum_{i \in I} \sum_{mn \in L} F_{mn}^{i\ (IT)}} \tag{34}$$

and to a measure of the difference between total travel times computed as average interval link inflows × average interval link travel times minus the same quantity expressed as OD flows × travel times on shortest paths, which correspond to what in static assignment is the gap function of equivalent nonlinear programming problem utilised to solve equilibrium problem. In static assignment, the value of this function gives an upper bound for the difference between objective function value at current iteration and at optimum.

4. Preliminary Numerical Results

Algorithms (A1) and (A2) have been tested, up to now, only on test networks, even if the results obtained indicate that, with adequate hardware and some software improvement, they could be applied also to real size networks. The chance of distinguishing between node arrival and departure time intervals has been exploited by taking equal intervals (more precisely, 8 intervals of 15 minutes each, plus other 8 intervals to represent inflows on farthest links) as a default and obtaining departure time intervals by dividing node arrival time ones by a factor r > 1 to reduce platoon dimension. At first, testing on an elementary network with only one OD pair and two symmetric alternative paths has been carried on to check convergence to the right solution, which in such a simple network, may be identified exploiting symmetry to determine path choice and solving DNL by simple calculations.

This test was not entirely satisfactory, because the symmetry of the network determined an high level of instability on flow and time patterns successively generated and also stopping criteria tested were unable to check that very good patterns were found at every of the first even iteration, starting right from the second one. Moreover, to get a fair approximation of known solution (it must be remarked that such a solution is not unique since first departures may choose, in a symmetrical uncongested network, both paths without affecting travel times, but giving rise to different interval link inflows) for actually constant departure flows, a very high number of departure intervals (r = 100) was needed. Nevertheless, after unstable patterns at odd iterations had slowly been overcome, the algorithm showed a tendency to converge toward a good approximation of "right" solution with about 100 iterations, while ensuring the respect of implicit constraints like outflow capacity.

Test on a slightly larger network gave, as expected, better results, since platoon dimension resulted less critical when they are assumed to merge and divide on a more complex network structure. This network, represented in Figure 1, included 10 nodes (all origin and destinations) and 25 links. Tests on it were carried on with a factor r=10. OD pairs where systematic choice (and consequent loading) of different paths took place at various iterations were sought, to verify if, for these paths and another unchosen path, equilibrium conditions (33) were satisfied. Only a maximum number of 1 and 2 iterations was considered (because no such improvement of results as to justify the computational burden increase was found for larger iteration numbers) as a stopping criterion for algorithm (A1), both for the option of updating shortest paths and for the opposite one, in addition to test skipping algorithm (A1) at all. The results of these tests are synthesised in columns of table 1, where the ranges of values within the last six iterations are shown in the first three rows for the following items:

- values of the different stopping criteria (multiplied by 10^4 and 10^2 for better readability);
- travel time at the end of each outer iteration (min:sec) on minimum time path for above mentioned OD pairs;
- travel time difference between minimum time path and the alternative utilised path (seconds).

In addition, in the fourth row of table 1, running time (min:sec) on a IBM ThinkPad 386 PC with mathematics coprocessor of the program implementing algorithms (A1) (A2) is reported.

It is also worth noting that while flow variation coefficient is not monotonically decreasing, even if varying on a very tight range, relative difference in total time decreases very smoothly. The results of one of these tests, i. e. the application of 1 iteration of algorithm (A1) without shortest path updating, is reported in some more detail by Figures 2÷5, where values of stopping criteria (Fig. 2) and of travel times on selected paths at each iteration are shown. The graphics are thus consistent with figures in second column of Table 1, excluding running time, but evidentiating also an OD pair (5-8 in Fig.5) where convergence to equilibrium looks quite slow.

5. Conclusions

The preliminary tests have shown that the proposed approach is consistent and allows dealing with real size networks, even if application on such networks has not actually been carried on. Also quite crude approximation in the solution of DNL problem with flow aggregation at origin-departure time level looks, at this stage, a promising strategy for dynamic assignment, even if coherence between arrival times and inflows, which appears to be ensured also trough such an approximation, is for sure worth wider and more specific testing. The potential for further developments is given mainly by the fact that the assumption of equal node arrival intervals for all links is by no mean necessary to the approach and its dropping out may take place quite easily in this framework. Node specific arrival intervals (equivalent to link specific ones if the network is expanded to represent each turning by a link) allow mitigating crudeness of inflow averaging and imbedding the overall procedure in an adjustment process where arrival intervals and outflow capacities become endogenous, in such a way to eliminate vertical queues and to respect storage capacity constraints by explicit representation of spill-back phenomena.

References

[1] Addison J.D. ,Heydecker B.G., "A Mathematical Model For Dynamic Traffic Assignment Transportation and Traffic Theory", C. F. Daganzo ed. 171-183 *Elsevier* 1993.

[2] Bellei G., Bielli M., "Dynamic Assignment for Trip Planning Systems Assessment", *Atti del Convegno Nazionale ANIPLA L'Automazione nei Sistemi di Trasporto* 1990.

[3] Bellei G., Bielli M., "Sensitivity Analysis of a Dynamic Equilibrium Model for Route and Arrival Time Choice", *presented at the 2nd Capri Seminar on Urban Traffic Networks* 1992.

[4] Ben-Akiva M., Cyna M., De Palma A., "Dynamic Model of Peak-period Traffic Congestion", *Transp. Res.* 18B 339-355 1984.

[5] Bernstein D., Friesz T.L., Tobin R. L., Wie B. W., "A Variational Control Formulation of the Simultaneous Route and Departure-Time Choice Equilibrium Problem", *Transportation and Traffic Theory* C. F. Daganzo ed. 107-126 *Elsevier* 1993.

[6] Boyce D.E., Ran B., LeBlanc L.J., "Solving Dynamic User Optimal Traffic Assignment Model", *Urban Transportation Center Advance Working Papers Series N.11 (submitted to Transpn. Sci.)* 1992.

[7] Carey M., "A Constraint Qualification for a Dynamic Traffic Assignment Model", *Transpn. Sci.* 20 55-58 1986.

[8] Carey M., "Optimal Time-varying Flows on Congested Networks", *Oper. Res.* 35 pp.58-69 1987

[9] Carey M., "Nonconvexity of the Dynamic Traffic Assignment Problem", *Transpn. Res.* 26B 127-133 1992

[10] Cascetta E., Cantarella G.E., "A Day-to-Day and Within-Day Dynamic Stochastic Assignment Model", *Transpn. Res.* 25A .277-291 1991.

[11] Daganzo C. F., "Stochastic Network Equilibrium with Multiple Vehicle Types and Asymmetric, Indefinite Links Cost Jacobians", *Transpn. Sci.* 17 282-300 1983.

[12] Di Gangi M., "Continous Flow Approach in Dynamic Network Loading", *presented at the 2nd Capri Seminar on Urban Traffic Networks 1992.*

[13] Drissi-Kaïtouni O., Hameda-Benchekroun A., "A Dynamic Traffic Assignment Model and a Solution Algorithm", *Transpn. Sci.* 26 119-128 1992.

[14] Ford L.R., Fulkerson D.R., "Flows in Networks", *Princeton University Press.* Princeton N. J. 1962.

[15] Friesz T.L., Luque F.J., Tobin R.L., Wie B.W. "Dynamic User Optimal Traffic Assignment Model Based on Optimal Control Approach", *Op. Res.* 37 pp.893-901 1989.

[16] Hamerslag R., "Dynamic Assignment in Three-Dimensional Time Space", *Transpn Res.* Rec. 1220 28-32 1988.

[17] Ho J.K., "A Successive Linear Optimization Approach to the Dynamic Traffic Assignment Problem", *Transpn. Sci.* 14 pp.295-305 1980.

[18] Ho J.K., "Solving the Dynamic Traffic Assignment Problem on a Hypercube Multicomputer", *Transp. Res.* 24B pp.443-451 1990.

[19] Janson B.N., "Dynamic Traffic Assignment for Urban Road Networks", *Transpn. Res.* 25B pp.143-161 1991.

[20] Mahmassani H.S., Chang G.L., "Experiments with Departure Time Choice Dynamics of Urban Commuters", *Transpn. Res* 20B pp.297-320 1986.

[21] Mahmassani H.S., Chang G.L., "Travel Time Prediction and Departure Time Adjustment Behaviour Dynamics in a Congested Traffic System", *Transpn. Res* 22B pp.217-232 1988.

[22] Merchant D.K., Nemhauser G.L., "A Model and Algorithm for the Dynamic Traffic Assignment Problem", *Transpn. Sci.* 12 183-199 1978a.

[23] Merchant D.K., Nemhauser G.L., "Optimality Conditions for a Dynamic Traffic Assignment Model", *Transp. Sci.* 12 200-207 1978b.

[24] Papageorgiou M., "Dynamic Modelling, Assignment and Route Guidance in Traffic Networks", *Transp. Res.* 24B pp.471-495 1990.

[25] Ran B., Shimazaki T., "A General Model and Algorithm for the Dynamic Traffic Assignment Problem", *Proceedings of the 5th World Conference on Transportation Resarch* - Yokohama, Japan 1989.

[26] Ran B., Boyce D.E., LeBlanc L.J., "Dynamic User-Optimal Departure Time and Route Choice Model: A Bilevel, Optimal Control Formulation", *submitted to Ann. of Op. Res., special issue on Advances in Equilibrium Modelling, Analysis and Computation* 1992a.

[27] Ran B., Boyce D.E., LeBlanc L.J., "Dynamic User-Optimal Route Choice Models Based on Stochastic Route Travel Times", *presented at the 2nd Capri Seminar on Urban Traffic Networks* 1992b.

[28] Smith M.J., "A New Dynamic Traffic Model and the Existence and Calculation of Dynamic User Equilibria on Congested Capacity-constrained Road Networks", *Transp. Res.* 27B pp 49-64 1993.

[29] Vythoulkas P.K., "A Dynamic Stochastic Assignment Model for the Analysis of General Networks", *Transp. Res.* 24B pp.453-469 1990.

[30] Wie B.W., Friesz T.L., Tobin R.L., "Dynamic User Optimal Traffic Assignment on Congested Multidestination Networks", *Transp. Res.* 24B pp.431-442 1990.

[31] Yagar S., "Emulation of Dynamic Equilibrium in Traffic Networks", *Traffic Equilibrium Methods* 240-264 (Florian M. Editor) Springer-Verlag 1976.

Table 1 Summary of Convergence Test Results					
	No Dyn.Net. Load. (DNL)	DNL w/o SPU 1 Iteration	DNL w/o SPU 2 Iterations	DNL w SPU 1 Iteration	DNL w SPU 2 Iterations
Flow Var. Coeff. E4	3.06 - 5.95	1.96 - 4.21	1.74 - 4.90	4.29 - 5.92	2.83 - 6.81
Rel. Diff. Tot. time E2	3.00 - 3.27	2.50 - 2.72	2.60 - 2.77	2.54 - 2.73	2.54 - 2.76
Min. Time OD 3 - 6	20:08 - 20:10	20:28 - 20:39	20:23 - 20:37	20:15 - 20:21	20:16 - 20:24
Altern. Path Time Diff.	34 - 40	27 - 38	22 - 39	37 - 45	33 - 44
Min. Time OD 4 - 7	43:49 - 43:59	44:07 - 44:22	43:59 - 44:22	44:01 - 44:13	44:03 - 4:13
Altern. Path Time Diff.	7 - 22	8 -19	2 - 26	18 - 30	12 - 23
Running Time	4:48	7:04	9:23	8:05	11:21

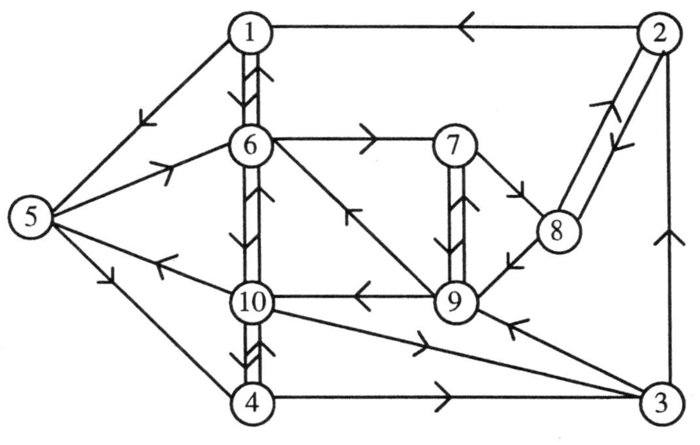

Figure 1 Test Network Graph

Figure 2 Convergence Indexes

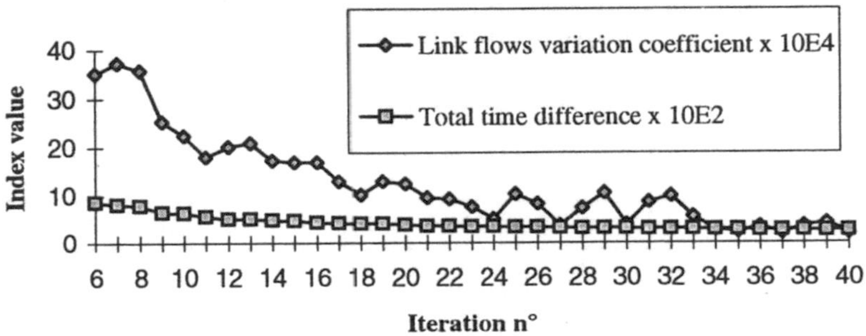

Figure 3 Travel times convergence among selected paths (OD 3-6)

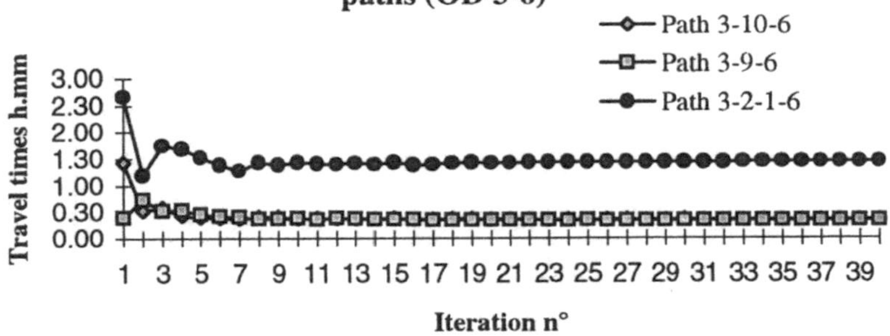

Figure 4 Travel times convergence among selected paths (OD 4-7)

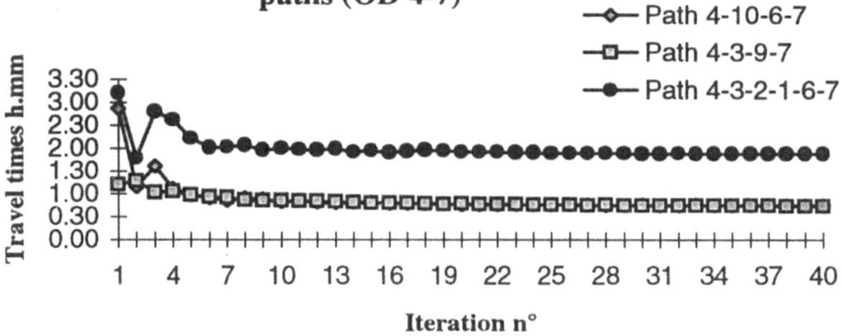

Figure 5 Travel time convergence among selected paths (OD 5-8)

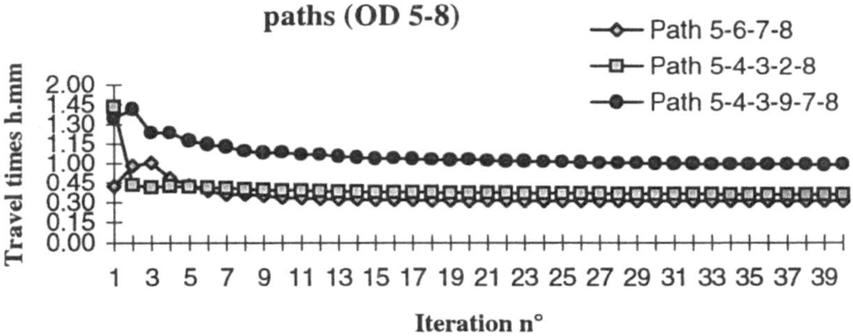

A SYSTEM OPTIMAL TRAFFIC ASSIGNMENT MODEL WITH DISTRIBUTED PARAMETERS

E. Codina and J. Barceló

Departament d' Estadística i Investigació Operativa
Facultat d' Informàtica, Universitat Politècnica de Catalunya
Pau Gargallo, 5 - 08028 Barcelona, Spain

In this paper a multidestination system optimal dynamic traffic assignment model with distributed parameters is examined. This model can be considered an extension of previous and well known models such as those of Merchant and Nemhauser (1978) and Friesz (1990). Flow dynamics of the model is based on an extension of the simple continuum model for flows composed by several commodities with equal propagation characteristics. An important property of flows following this model is that no overtaking can occur between flows of different commodities. In the case of flow propagating at constant speed and vertical queues at the end of links it is possible the *stable* approximation of the proposed dynamic system optimal model by means of optimal control problems. For one of the possible stable approximations it is shown how a strengthened Courant-Friedrichs-Levy condition ensures no overtaking and FIFO observance at vertical queues of the approximating optimal control problem. Finally, and for the case of a single destination in the network, the application of an extremals calculation method for optimal control problems developed by the authors in previous papers is also shown and favorable conditions for its application are discussed.

1. Introduction

In the last years dynamic behaviour of urban traffic flows under route guidance systems has been modeled by some authors using a deterministic optimal control approach formulation. This formulation has emerged as a dynamic extension of wardropian equilibrium principles widely used in urban transportation planning. Deterministic optimal control based formulation was stated initially by J.F. Luque and T.L. Friesz [8] (1980) coming from a Merchant and Nemhauser model [9], [10] for networks with a single destination. M. Carey [2], [3] analyzed the constraints qualification of Merchant and Nemhauser's model. Also models with multiple destinations have been proposed by W. Wie, Y.L. Friesz and R.L. Tobin [15]. All these models had the common characteristic of using the so called *exit link functions* in order to model traffic dynamics. Models without exit link functions also based on deterministic optimal control have been proposed by B. Ran, D. Boyce and Leblanc L.J. [13]. Although models with and without exit link functions provided a compact formulation they presented some non adequate properties for the evolution of traffic flows. Thus, models without exit link functions presented unproper characteristics when applied to traffic flows and in [7] E. Codina shows paradoxes of the solutions given by dynamic traffic assignment models without exit link functions that lead to an extension of the Dafermos-Sparrow theorem for static traffic assignment models. Instantaneous propagation characteristics of flow dynamics in models with exit link functions are analyzed in [6] and it is shown how this unproper characteristic vanishes as a simple continuum model is approximated. Another unproper characteristic for multidestination models already shown by M. Carey in [3] consists of the conflict arising when flows with different destinations exit from links and the observation of FIFO queueing disciplines by these flows. However, as suggested by Papageorgiou in [12], models using exit link functions can be viewed as an extension of a model with distributed parameters (E. Codina [7]).

This paper presents a general continuous System Optimal Dynamic Traffic Assignment model for multidestination networks. Flows of this model verify the inviscid PDE for stationary and irrotational flows also known as the simple continuum model and this ensures that neither instantaneous propagation nor overtaking may occur between flows of different commodities at links. Also, because of its basic traffic behaviour characteristics, i.e. finite propagation speed, wave dispersion and effects of congestion (i.e. speed reduction and spillback) this model can be taken as a reference. In fact, it is shown how Merchant and Nemhauser's model can be considered a discrete approximation to the continuous system optimal model presented here for the case of density dependent propagation speed and how Friesz's and Wie's models are also approximations by means of continuous time optimal control in the case of constant propagation speed on links. Second section of this paper gives a brief description of Merchant and Nemhauser's model and Carey's work and the problem of overtaking between flows of different commodities at links. Section 3 starts with a basic description of the simple continuum model and discusses *unstability* characteristics of their first order regressive difference approximations, i.e. Euler explicit method and implicit method and the need of vertical queues at the exit of links when these approximations are used. Next, an extension to the multicommodity case of the simple continuum model on a link is presented. The no overtaking of trajectories property of this extended model is remarked and it is shown in lemma 2 that, when constant propagation speed and vertical queues at the end of links hold, a first order implicit regressive finite difference approximation gives solution flows with no overtaking at links and FIFO observance at vertical queues, provided that a strengthened Courant-Friedrichs-Levy condition is observed by the discretization mesh. A remarkable fact is that lemma 2 needs *convergence* of the approximation used to the simple continuum model to prevent overtaking. Previous authors have suggested non convex formulations to overcome the problem of overtaking and homogeneous exit from queues at the end of links (see, for instance [4]). In this paper a set of *linear* inequalities are developed to prevent the problem of overtaking and preservation of homogeneity of exits from a queue. These set of linear inequalities are used subsequently in numerical approximations to the system optimal dynamic traffic assignment model proposed in section 3. Because of unstabilities due to nonlinearities a continuous time optimal control approximation is presented, only for the case of constant propagation speed and vertical queues at the end of links. In order to guarantee no overtaking and no destination selective output form vertical queues at links, results obtained in section 3 regarding elimination of nonconvexities are used and a nonlinear optimization problem with *linear* constraints is formulated to solve the optimal control approximation. Finally and for the case of a single destination network, conditions that permit the application of an extremals calculation method for optimal control problems developed in [7] and that decompose the previous nonlinear programming formulation are briefly discussed.

2. Merchant and Nemhauser's Model. Preliminaries

In 1978 Deepak K. Merchant and George L. Nemhauser (refs. [9] and [10]) published two papers describing a system optimal dynamic traffic assignment model formulated as a nonlinear optimization problem for networks with a single destination. Because of the nonconvexity of their model Merchant and Nemhauser could not analyze constraints qualification and uniqueness aspects. This was done by M.Carey ([2] , [3]) by means of a modified model. This model is also a precursory one of "continuous time" models presented by Friesz *et al.* ([8]), B.W.Wie *et al.* ([15]). Also an optimal control distributed parameters model presented in this paper and suggested in ([7]) can be considered as an extension of Merchant and Nemhauser's model. Because of their historical relevance we shall give a summarized description of Merchant and Nemhauser and M.Carey's models as well as of their properties.

Merchant and Nemhauser's System Optimal Model

Let a network be represented by a directed graph $G = (N, A)$ consisting of a set of nodes N and a set of links A with traffic flows moving towards a single destination $q \in N$ in the graph.

Let us denote by $I\ (i)$ and by $E\ (i)$ the set of links entering and emerging from node $i \in N$ respectively. Let's supose a finite horizon of time [0, T] divided in N time subintervals of equal length δ = T/N and let us consider the following quantities:

$x_{a,j}$	Number of vehicles in link $a \in A$ (or the average density of vehicles) and time subinterval j.
$u_{a,j}$	Number of incoming vehicles at link $a \in A$ and time j.
$v_{a,j}$	Number of outgoing vehicles from link $a \in A$ and time j.
$p_{i,j}$	Number of vehicles entering the network at node $i \in N$ in the j-th time subinterval with destination q.
$X_{a,\ j}(\bullet)$	Cost functions for link $a \in A$ and the j time subinterval.

Exit link flows are assumed to verify the following hypotheses:

1. $v_a = w_a(x_a)$, i.e. exit flow from a link depends on the number of vehicles at the j-th time subinterval on the link $a \in A$. Merchant and Nemhauser refer to $w_a(\bullet)$ as "exit link functions" (ELF).
2. Functions $w_a(x_a)$ are non decreasing, continuous and concave and verify $0 \leq w_a(x_a) \leq x_a$.
3. Functions $w_a(\bullet)$ verify a *saturation* property: $\lim_{\omega \to \infty} d\, w_a(\omega)/d\omega = 0$.

With all these hypotheses, Merchant and Nemhauser model can be reformulated as the following optimization problem:

M-N 1 (Refomulation of Merchant-Nemhauser´s model **M-N 1** using exit link functions as side constraints.)

$$
\begin{aligned}
& Min_{x_j,\, u_j,\, v_j, s_j} \quad \sum_{j=1}^{N} \sum_{a \in A} X_{a,j}(x_{a,j}) \\
& x_{j-1} - x_j - v_j + u_j = 0 \\
& B_+ u_j - B_- v_j + e_q s_j = p_j \\
& v_{a,j} = w_a(x_{a,j-1}) \\
& u_j \geq 0, \quad v_j \geq 0, \quad x_j \geq 0, \qquad s_j \geq 0 \quad i = 1, \cdots, N \\
& (x_0 \quad known)
\end{aligned}
$$

The first group of constraints is a vector notation for $x_{a,\, j-1} - x_{a,\, j} - w_a(x_{a,\, j-1}) + u_{a,\, j} = 0,\ a \in A$, and can be viewed as state equations. The second group of constraints, $B_+\, u_j - B_-\, w(x_j) + e_q\, s_j = p_j$, are balance equations for flows at nodes being B_+ and B_- matrices derived from the incidence matrix B of graph G with $(B_+)_{i,j} = 1$ if $B_{i,j} = 1$ and zero otherwise $(B_-)_{i,j} = 1$ if $(B)_{i,j} = -1$ and zero otherwise and s_j arrival flow at destination q in the j-th time subtinterval. It must be pointed out that because of $0 \leq w_a(x_a) \leq x_a$ and $u_{a,j} \geq 0$ nonnegativity constraints on variables x_j are *redundant*. In this formulation exit flow variables $v_{a,j}$ are used explicitly and side constraints are derived from the hypotheses of exit flow at links thus suggesting a network structure.

We shall call "time-expanded networks" the networks originated by the linear constraints in problem **M-N 1** and we shall refer to G as the "original network".

Because of analytical difficulties of Merchant and Nemhauser's model, M. Carey considered in [2] and in [3] a similar model that can be formulated as **M-N 1** "almost equal". The only difference is that exit flows on links are *limited* by the exit link functions: $0 \leq v_{a,j} \leq w_a(x_{a,j})$ ($\leq x_{a,j}$). It is worth noting that redundancy of nonnegativity constraints for state variables $x_{a,j}$ remains. Carey denominates the slack $\sigma_{a,j} = w(x_{a,j}) - v_{a,j}$ as "control" for the link a and provides in [3] two sets of sufficient conditions on the objective costs (**CFC** conditions) and the exit link functions (**EFC** conditions) in order to solutions of the modified Merchant Nemhauser's model **M-N 1** provide *null* controls $\sigma_{a,j}$. For this modified model M. Carey proved that the modified model satisfies Kuhn Tucker conditions. It must be noted that Model **M-N 1** and also the modified model due to Carey is restricted to a single destination in the network and also that **CFC** conditions on cost functions may not hold in many practical cases.

For the extension to multiple destinations these models present a nondesirable property when their solutions are compared to traffic flows behavior. This nondesirable property is commonly known as "FIFO discipline violation". On a road network, traffic of different types entering the same link at approximately the same time t will pass each other but their speeds will not differ too much and therefore, approximately equal travel times to traverse links for flows with different destinations in the network must be expected. Consequently it will appear as "anomalous" any solution for which flows entering the link at time $t' > t$, (after cars entering at time t) leave the link at time $\tau' < \tau$, i.e., *before* cars that entered at time t, thus reflecting that an overtaking occurs between cars of different commodities.The practical implications of the "FIFO discipline violation" in a multidestination situation are of great relevance. If solutions of the model are to be applied on *real* traffic networks (i.e. for route guidance purposes), then, as imposed by "controls" $\sigma_{a,j}^q$, $q \in D$, a *selective* exit of cars from a link would be required accordingly to the different car destinations (some cars with destination q would have to wait or travel at a lower speed in order to allow cars with destination q' exit the link) and this strategy seems impossible to implement in practice or, at least, adds huge difficulties in real traffic networks, where the natural quantities that can be controlled are the aggregated exit flow from link a, $v_a = \Sigma_{q \in D} v_a^q$ or input flow, $u_a = \Sigma_{q \in D} u_a^q$. As stated in [3], constraints of the type $w_a(\Sigma_{q \in D} x_{a,j}^q) \geq \Sigma_{q \in D} v_{a,j}^q$, $1 \leq j \leq N$, allow any order in exiting a link for the commodities and only ensure that exit flow throuputs, $w_a(\bullet)$, at links are not violated. In [4] Carey formulates four classes of constraints that ensure the FIFO discipline of multicommodity flows or are consequence of the observation of the FIFO discipline. They all require that problems be formulated with flow variables $h_{a,k,\underline{k}}$ entering a link at time subinterval k and leaving the link at time subinterval $\underline{k} \geq k$. The first of these classes of constraints needs additionally the use of binary decision variables. Each of the four classes of constraints results in a nonconvex constraint set and results in nonlinear-integer programming formulations. Empirically, Carey suggests in [4] to solve dynamic assignment problems without constraints for FIFO observance, analyze the degree of overtaking or FIFO violations, and introduce FIFO constraints if necessary. Also, through empirical work, Carey intuits that a certain degree of correlation exists between sharp flow fluctuations and FIFO discipline violation and thus, the technique previously described is specially suggested for the case of traffic flows, because their *smooth fluctuations in time* result in the observation of FIFO discipline or in its approximate verification.

Thus far, we have summarized with some detail a set of drawbacks of Merchant and Nemhauser's model. In section 3 a system optimal model is presented overcoming them and an optimization problem approximating the model is developed. However we need first to discuss properties of the simple continuum model as well as of some numerical methods used to solve it.

3. Considerations on State Equations in Dynamic Models

State equations in Merchant and Nemhauser's model can be considered an explicit first order approximation of the well known "simple continuum model" based on inviscid flow equation (see, for instance [1]) with the role of exit link functions being parallel to the role of flow-density relationship in the simple continuum model. In this section, after an exposition of the fundamentals on the simple continuum model, we analyze properties and inconvenients of its approximation by explicit and implicit first order regressive-differences.

We show in this section that implicit approximations, when propagation speed is constant, present the desirable property that solution flows present only a reduced degree of overtaking that vanishes as the discretization mesh increases and a strengthened Courant-Friedrichs-Levy (CFL) condition is observed by the discretization mesh (see, for instance reference [11]) and that this is a consequence of convergence properties of the approximation method to the exact solutions of the model. On the other hand we outline the well known handicap of the unstability of the first order approximations in presence of nonlinearities (non constant propagation speed) and congestion. Implicit approximations, when propagation speed is constant, are *numerically stable*. As the approximation method is *consistent* then, by Lax's equivalence theorem (see, for instance [1]) then, there exists *convergence* of the solutions of the implicit approximation to the exact solutions of the PDE as the grid is refined.

Let us consider a link as an unidimensional continuum. Let $x(z,t)$ be the vehicle density of the link and $j(z,t)$ be the flow on the link at time t in position z, respectively, and let speed ω be assumed to depend directly on density following the flow-density relationship $j = x\omega(x)$. The simple continuum model can be stated as:

$$x_t + \varphi(x)\, x_z = 0, \; (\varphi(x) = \omega(x) + x\, d\omega/dx) \qquad (1)$$

We shall assume a maximum capacity on this link $\hat{x}$ and we shall consider two classes of speed functions $\omega(x)$. The first class are simply positive constant functions. The second class of functions verify the following conditions: 1) $\omega(x)$ is at least once differentiable on $[0, \hat{x}]$ and is decreasing on $[0, \hat{x}]$ with $\omega(0) > 0$ and $\omega(\hat{x}) = 0$. 2) Flow function $\underline{j}(x) = x \bullet \omega(x)$ has a single stationary point (maximum) at density $\underline{x} \in]0, \hat{x}[$. We shall refer to $\underline{x}$ as *critical density.* We shall use the term "congestion" when densities are greater than the critical density $\underline{x}$.

Boundary Conditions for PDE (1).

The Boundary conditions that we will consider in order to determine a unique solution of the previous PDE on $[0, \hat{z}] \times [0, T]$ are: 1) Initial density distribution on $[0, \hat{z}]$, $x(z, 0) = h(z)$, 2) Input flows to the link at each time t: $\underline{j}(0, t) = u(t)$, 3) *Either* exit flows from the link at each time t: $\underline{j}(\hat{z}, t) = v(t)$ *or* 4) Use an additional variable κ, the total number of accumulated vehicles at the end of the link, ruled by the differential equation $d\kappa / dt = x(\hat{z}, t)\omega(x(\hat{z}, t)) - v(t)$.

Together with conditions 1) and 2), condition 3) can not determine a solution in the case of a positive constant speed-density relationship. However it will be possible to determine a solution if boundary condition 3) is substituted by 4). Conditions 1), 2) and 4) are equivalent to model a *vertical queue* at the end of the link. Boundary conditions at $z = \hat{z}$ are clearly related with the effects of a throuput capacity reduction at the link exit (i.e. because of reduction in the number of lanes in the next link) and it must be outlined that when it is possible to impose boundary conditions of the type 3) it is then possible to model spillback. It is worth noting at this point that modifications made by Carey to Merchant and Nemhauser model are able to reproduce horizontal queues as exit flows on links are subject to inequalities $\underline{j}^i_k \leq x^i_k\ \omega(x^i_k)$ instead of equalities. However controls or "slacks" $\sigma^i_k = x^i_k\ \omega(x^i_k) - \underline{j}^i_k$ are guaranteed to be nonnull only under very restrictive assumptions on the cost functions.

Having in mind potential theory for bidimensional stationary and irrotational flows in fluid mechanics (see, for instance[14]), density $x(z, t)$, solution of $x_t + \varphi(x)\ x_z = 0$, and its associated flow $\underline{j}(z, t) =$

$x(z,t) \bullet \omega(x(z,t))$,are the components of a vector field $\vec{E} = (-x, j)$ orthogonal to flow trajectories, and $\vec{E}$ comes from an at least twice differentiable potential function $U(z,t)$, ($-x = \partial U \backslash \partial z$, $j = \partial U \backslash \partial t$), at regions where functions x , j are differentiable. On the other hand conservation of flow ensures continuity of potential function $U(z,t)$ at points where continuity of density function $x(z,t)$ fails. Thus, even if functions $u(t)$ or $v(t)$ determining boundary conditions for $x_t + \varphi(x)\, x_z = 0$ have a finite set of discontinuity points, flow trajectories shall *never intersect* and a strict observation of the FIFO principle must hold.

Finite difference methods to solve PDE (1) rely on the approximation made to the term $\partial j \backslash \partial z$. If rectangle $[0,L] \times [0,T]$ is discretized in a grid mesh of size (δ_z , δ_t) , then approximations to the solution of (1) are made at points (z_k, t_i) , k=0, 1, 2, ... , $M = L/\delta_z$ and i=0, 1, 2, ... , $N=T/\delta_t$. A first order regressive approximation to the term $\partial j \backslash \partial z$ leads to explicit or implicit methods:

$$x_k^{i+1} = x_k^i - \nu \bullet (j_k^i - j_{k-1}^i) = \nu\, j_{k-1}^i + x_k^i(1 - \nu\omega(x_k^i)) \qquad \text{(Euler explicit)}$$

$$x_k^{i+1} = x_k^i - \nu \bullet (j_k^{i+1} - j_{k-1}^{i+1}) \Rightarrow$$
$$\Rightarrow x_k^{i+1}(1 + \nu\omega(x_k^{i+1})) = x_k^i + \nu j_{k-1}^{i+1} \qquad \text{(implicit)}$$

$$(\text{ as } j_k^i = x_k^i\, \omega(x_k^i), (\nu = \delta_t / \delta_z))$$

It must be noted that methods based on regressive approximations of the term $\partial j \backslash \partial z$ *do not* allow the inclusion of boundary conditions at $z = \hat{z}$ of the type 3, i.e. exit flow $v(t)$ can not be imposed at $z = \hat{z}$ and can only reproduce boundary conditions of the type 4). Thus, the use of regressive approximations to the term $\partial j \backslash \partial z$ allows only to model vertical queues at the end of the link by considering an extra relationship expressing total number of vehicles κ^i at the i-th period as, for instance: $\kappa^{i+1} = \kappa^i + \delta_t\, (x^i_M\, \omega\, (x^i_M) - v^i)$.

Explicit methods for PDE (1) require the observance of CFL condition (see, for instance [11]). This condition requires that discretization mesh (δ_z, δ_t) observes the inequality $\hat{\omega}\ \delta_t/\delta_z \leq 1$, with

$\hat{\omega}$ = $\omega(0)$ the maximum speed. This condition ensures two desirable properties for Euler explicit method: 1) Flows do not propagate at speeds higher than the maximum speed $\hat{\omega}$. 2) If densities at $t = 0$ and input flows at $z = 0$ are nonnegative then solutions x^i_k are also nonnegative.

Unfortunately it is known that Euler explicit method is *unstable even* if CFL condition is observed (see [1]) no matter the class of function for the speed-density relationship. We can consider a link subdivided into M sublinks and identify $\nu\ x\ \omega(x)$ with an exit link function and the term $\nu\ j^i_k$ with inputs u during time period δ_t in Merchant and Nemhauser's model. Thus, as Merchant and Nemhauser's model uses Euler explicit method to approximate PDE (1) it is subject to *numerical unstabilities* even for time discretization length $\delta_t \rightarrow 0$.

Previous results regarding FIFO properties can be shown to hold in a countinuous model when flow on a link is composed by a set of commodities $q \in D$, being thus possible to enunciate lemma 1 below. For brevity we enunciate it here without proof.

LEMMA 1 *(FIFO observation for trajectories of the system* $x^q_t + (x^q \bullet \omega(x))_z = 0,\ q \in D$*)*

Let $x^{*q}(z,t)$, $q \in D$, *be solutions of the PDE system:*

$$\frac{\partial\ x^q}{\partial\ t}+\frac{\partial\,(x^q\omega(x))}{\partial\ z}=0, \qquad q\in D, \qquad (x=\textstyle\sum_{q\in D}x^q) \tag{2}$$

fixed by initial conditions $x^q(0,t) = u^q(t)$, $x^q(t).\ v^q(t)$, $t \in [0,T]$ *and* $x^q(z,0) = h^q(z)$, $\forall\ z\ [0,L]$, $\forall\ q \in D$, *with* $u^q(t)$, $u^q(t)$ *and* $h^q(z)$ *functions with discontinuities in a finite set of points of their respective intervals. Then:*

1. *Trajectories in* $[0,L] \times [0,T]$ *are common to all commodity flows and are given by* $dz/dt = \omega(x^*(z,t))$ *and no intersection occurs between two trajectories, even at points where densities* $x^{*q}(z,t)$ *may be discontinuous.*

2. *If a point (z,t) is a discontinuity point for density $x^{*q}(z,t)$ then it is a discontinuity point for all other densities $x^{*q'}(z,t)$, $q' \in D$. Even more, discontinuity points for densities, if they exist, verify differential equation:*

$$\frac{d\rho}{d\tau}=\frac{j_+^q-j_-^q}{x_+^q-x_-^q}\Bigg|_{(\rho,\tau)} \qquad \left(=\frac{j_+^{q'}-j_-^{q'}}{x_+^{q'}-x_-^{q'}}\Bigg|_{(\rho,\tau)}, \quad \forall q'\in D\right) \tag{3}$$

with j^q_+, x^q_+, j^q_-, x^q_- flows and densities at both sides of the curve $\rho(\tau)$. ($(\rho(\tau), \tau)$ can be considered a "collision" point of two flow waves).

3. *Composition ratios $r^q(z,t)$ given by:*

$$r^q(z,t)=\frac{j^q(z,t)}{j(z,t)}=\frac{x^q(z,t)}{x(z,t)} \qquad \textit{if} \qquad x(z,t)>0 \tag{4}$$

($r^q(z,t) = 0$ if $x(z,t) = 0$) *are continuous functions along flow trajectories.*

As a consequence, two vehicles entering a link at times t and t', respectively, ($t' > t$) will leave the link at times τ and τ' and always $\tau' > \tau$, independently of their destination $q \in D$. The first result in previous lemma appears directly when approximating system (2) in the case of *constant propagation speed* ω, with an implicit regressive differences method of the type $x^{q,i+1}_j - x^{q,i}_j + \nu\,\omega \bullet (x^{q,i+1}_j - x^{q,i+1}_{j-1}) = 0$. Solutions of this implicit approximation can be represented by means of a grid time expanded network. Next lemma shows that amongst the set of paths in the grid time expanded network of a first order regressive implicit approximation it is always possible to find a subset of paths for which no crossing can occur between flows either of different or the same commodity, entering the link at different time subintervals and that therefore there will exist FIFO observance.

LEMMA 2 *(FIFO observance for solutions of implicit first order approximation to multicommodity system (2)). Let system (2) in the case of constant propagation speed* ω *be approximated by an implicit method of the type:*

$$x^{q,i+1}_{j} - x^{q,i}_{j} + \nu\omega\bullet(x^{q,i+1}_{j} - x^{q,i+1}_{j-1}) = 0, \ \nu=\delta_t / \delta_z \qquad \textbf{(5)}$$

then two sufficient conditions in order to solution flows of (5) with different commodities do not intersect inside a link are:

1. CFL parameter $\nu\omega < 1/2$.

2. An enough dense grid (δ_z, δ_t) *is used for the approximation.*

Proof. When approximating (2) by means of a grid (δ_z, δ_t), link $a \in A$ of the original network splits as shown in figure 1. Solution flows of (5) for different commodities $q, q' \in D$ can intersect each other *inside* link a either at an L-node or at an x-link. Intersecting flows a, α at L-nodes are linked by the following relationship $\alpha^q - a^q = x^q_0 - v^q$ (R1) whereas flows a, α at x-links are linked by $\alpha^q - a^q = x^q_0 - v^q_1 - v^q_2$ (R2). Thus, for fixed x^q_0, v^q_1, v^q_2 there exists a non unique decomposition of this quantities in a^q , α^q at both L-nodes and x-links. Then the no intersecting condition for flows of the same commodity can be stated as: 1) There exist α^q, a^q verifying (R1) such that $\alpha^q a^q = 0$ and the no intersecting condition at x-links or L-nodes for flows of different commodities can be stated as: 2) there exist α^q, a^q verifying (R2) such that $\alpha^q = 0$, $a^q \geq 0$, $\forall q \in D$ or, alternatively, 2') $\alpha^q \geq 0$, $a^q = 0$, $\forall q \in D$. Because of the *convergence* of method (5) to the exact solution of PDE system (2), as the grid is more dense $\alpha^q - a^q = x^q_0 - v^q = x^q_0 - \nu\,\omega\, x^q \rightarrow x^q_0\,(1-\nu\,\omega) > 0$ if $\nu\,\widehat{\omega} < 1$ and $\alpha^q - a^q = x^q_0 - v^q_1 - v^q_2 = x^q_0 - \nu\,\omega \bullet (x^q_1 + x^q_2) \rightarrow x^q_0\,(1 - 2\,\nu\,\omega) > 0$, if $\nu\,\widehat{\omega} < 1/2$ implying thus that there exists $\alpha^q \geq 0,\ a^q = 0$.

The final aspect that has to be considered is composition of exit flows at links in the multicommodity case. Previous lemma 1 states the fact that composition ratios are continuous functions on trajectories. This *homogeneity* is implicitely assumed for exit flows, i.e: $v^q(t) = j^q(\hat{z}+,t) = r^q(\hat{z}-,t) \bullet j(\hat{z}+,t)$. In the case of constant speed ω when system (2) is approximated by means of a finite difference method *homogeneity* must be *imposed* on exit flows. As the case of constant speed can not reproduce boundary conditions 4) unless by means of vertical queues, the following additional relationships imposing homogeneity are needed:

$$\kappa^{q,i+1} = \kappa^{q,i} + \delta_t(x_M{}^{q,i}\ \omega(x_M{}^{q,i}) - v^{q,i}) \qquad \text{(Vertical queues)}$$

$$v^{q,i} = v^i\ \kappa^{q,i} / \kappa^i, \text{ if } \kappa^i > 0,\ v^{q,i} = 0, \text{ if } \kappa^i = 0 \qquad \text{(Homogeneity)} \tag{6}$$

However homogeneity condition at vertical queues, when expressed as in (6) leads to nonconvex formulations of the feasible set of flows of a dynamic traffic model, with the corresponding analytic and algorithmic difficulties in the resulting optimization problems. It is possible to reformulate homogeneity condition (6) by means of the flow decomposition at L-nodes and ϰ-links used in the proof of lemma 1 which ensures the existence of a subset of paths without crossing in the case of $x_0{}^q - v_1{}^q - v_2{}^q \geq 0$, $v_1{}^q, v_2{}^q \geq 0$ so that no crossing or selective output from the queue may exist:

$$\kappa^{q,i-1} \geq \delta_t\ (v_i{}^q + v_{i+1}{}^q),\ i = 1, \ldots N-1,$$

$$\kappa^{q,N-1} \geq \delta_t\ v_N{}^q,\ \forall q \in D \tag{7}$$

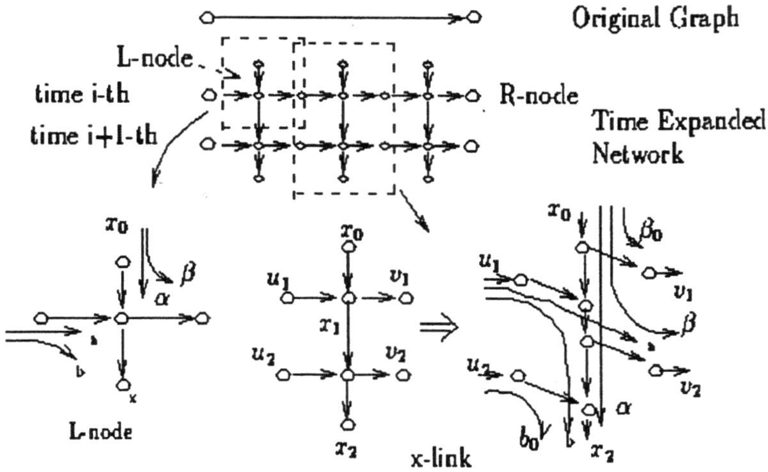

Figure 1. Time expanded network for link a of the original graph when an implicit first order regressive approximation to PDE system (2) is used and the resulting flow decomposition at x-links and L-nodes.

4. A System Optimal Model with Distributed Parameters

State equations of Merchant and Nemhauser's Model can be considered an approximation to the simple continuum model. This model reproduces basic traffic behaviour characteristics: finite propagation speed of flows, dispersion, backward propagation or spillback and no intersection between trajectories even when flows are composed by a set of commodities. For an initial distribution of densities, input and exit flows on a link determine univoquely flow propagation and can be considered as controls. Therefore a generic System Optimal Dynamic Traffic Assignment model with separability of the costs consists of the minimization of a given functional of the nonnegative solutions of systems (2). Solutions of these systems are linked each other by means of input and exit flows on links verifying a set of balance equations at nodes of the original graph. This model can be formulated as:

DOCP1 (*Extended multidestination* **M-N 1** *model*)

$$Min_{x_a \cdot j_a} \quad \sum_{\forall a \in A} \int_0^T \int_0^Z F_a(x_a(z_a,t), j_a(z_a,t), z_a, t)\, dz_a\, dt$$

$$\frac{\partial \; x_a^q}{\partial \; t} + \frac{\partial \; j_a^q}{\partial \; t} = 0, \qquad (\forall a \in A, \quad \forall q \in D)$$

$$B_+ j^q(0,t) - B_- j^q(\hat{z},t) + e_q s^q(t) = p^q(t)$$

$$j_a^q(z_a,t) = x_a^q(z_a,t) \cdot \omega_a(x_a(z_a,t))$$

$$j_a^q(z_a,t) \geq 0$$

$$(x_a(z_a,t) = \sum_{\forall q \in D} x_a^q(z_a,t))$$

Initial state $x_a^q(z_a, 0) \geq 0$ is assumed to be known. Nonnegativity constraints on the densities $x^q(z_a, t) \geq 0$ are redundant provided that speed functions $\omega_a(\bullet)$ verify $\omega_a(0) > 0$ and $\omega_a(\hat{x}_a) = 0$ and decreassing in $[0, \hat{x}_a]$. In the case of constant propagation speed ω and vertical queues at the end of links, it is possible the approximation of model **DOCP 1** by means of an optimal control problem as numerical stability of the resulting state equations is ensured. Approximation of systems (2) can be made by means of discretizing each ``z-continuous'' link a in M_a sublinks and approximating systems (2) by means of systems of ODE's as described in the previous section. For instance, in the case of a first order regressive approximation of the term $\partial j / \partial z\}$ the approximating ODE system would be:

$$\begin{pmatrix} \dot{x}_1^q \\ \dot{x}_2^q \\ \vdots \\ \dot{x}_{M_a}^q \\ \dot{k}_a^q \end{pmatrix} = -\frac{\omega_a}{\delta_{z_a}} \begin{pmatrix} -1 & & & & 0 \\ 1 & -1 & & & 0 \\ & 1 & \ddots & & \vdots \\ & & \ddots & -1 & 0 \\ & & & 1 & 0 \end{pmatrix} \begin{pmatrix} x_1^q \\ x_2^q \\ \vdots \\ x_{M_a}^q \\ k_a^q \end{pmatrix} + \begin{pmatrix} \bar{u}_a^q(t) \\ 0 \\ \vdots \\ 0 \\ -\bar{v}_a^q(t) \end{pmatrix}$$

$\bar{u}^q{}_a = 1/\delta_{za}\, u^q{}_a$, $\bar{v}^q{}_a = 1/\delta_{za}\, v^q{}_a$ or in more compact form: $dy^q{}_a / dt = - \rho_a \; C\, y^q{}_a + b^q{}_a(t)$, $(\rho_a = \omega_a / \delta_{za})$, verifying input and exit flows at links the set of balance equations: $B_+ \, u^q(t) - B_- \, v^q(t) + e_q \, s^q(t) = p^q(t)$

Having into account a limited capacity at vertical queues and imposing homogeneity on exit flows from the vertical queue, problem **DOCP 1** can be approximated by the following OCP (optimal control problem):

OCP 1

$$Min_{y,u,v} \sum_{a\in A}\left\{\delta_{z_a} \sum_{j=1}^{M_a}\int_0^T F_a\left(x_{j,a},\ w_a x_{j,a}\right)\ dt\right\}$$

s.t.

$$\begin{gathered}
\dot{y}_a^q = \rho_a y_a^q + b_a^q(t)\\
B_+ u^q - B_- v^q + e_q s^q = p^q(t)\\
\kappa_a v_a^q = \kappa_a^q v_a\\
0 \le \kappa_a \equiv \sum\nolimits_{q\in D} \kappa_a^q \le \hat{\kappa}_a\\
v_a^q \ge 0,\ u_a^q \ge 0,\ s^q \ge 0\\
\textit{Known initial state } x_{a,j}^q(0)
\end{gathered}$$

It can be easily checked that nonnegativity constraints on the state variables are redundant: for a link $a \in A$ and the first sublink, $dx^q{}_1/dt = -\rho_a x^q{}_1 + u^q{}_a(t)$ holds. As $u^q{}_a(t) \ge 0$ and $x^q{}_1(0) \ge 0$ then $x^q{}_1(t) \ge 0$, for $t \ge 0$. For the second sublink $dx^q{}_2/dt = -\rho_a x^q{}_2 + \rho_a x^q{}_1(t)$, as $x_1{}^q(t) \ge 0$ and $x^q{}_2(0) \ge 0$ then $x^q{}_2(t) \ge 0$, for $t \ge 0$ ... For the vertical queue length $d\kappa^q{}_a/dt = \rho_a x^q{}_M - v^q{}_a(t)/\delta_z$ and because of the homogeneity conditions at the exit of the link this equation can be rewritten *as* $d\kappa^q{}_a/dt + v_a(t)\kappa^q{}_a(t)/\kappa_a(t) = \rho_a x^q{}_M \ge 0$. As $v_a(t) \ge 0$ and $\kappa_a(t) \ge 0$ then $\kappa^q{}_a(t) \ge 0$.

If this OCP is discretized in t with step δ_t, observing the strengthened CFL condition at each link, i.e. $\omega_a \delta_t/\delta_{za} < 1/2$, $\forall a\in A$, as required by lemma 1, homogeneity conditions at vertical queues are expressed as in (7), and first order regressive approximation to $\partial j/\partial z$ are used, then the following optimization problem arises:

ΔOCP 1

$$Min_{y,u,v}\ \delta_t \sum_{a\in A} \delta_{z_a} \sum_{j=1}^{M_a} \sum_{i=1}^{N} F_a\left(x_{j,a},\ w_a x_{j,a}\right)\ (LM)$$

$$y_a^{q,i} - y_a^{q,i}-1 + v_a \omega_a\, y_a^{q,i} - b_a^{q,i} = 0,\quad \lambda_a^{q,i},\ v_a = \frac{\delta_t}{\delta_{z_a}}$$

$$B_+ u^{q,i} - B_- v^{q,i} + e_q s^{q,i} = p^{q,i}, \qquad i = 1,\ldots,N$$

$$\kappa^{q,i-1} \geq \delta_t \left(v^{q,i} + v^{q,i+1}\right), \qquad \varsigma_i^q,\ \ i = 1,\ldots,N-1$$

$$\kappa^{q,N} \geq \delta_t v^{q,N} \qquad \varsigma_N^q$$

$$0 \leq \kappa_a \equiv \sum\nolimits_{q\in D} \kappa_a^q \leq \hat{\kappa}_a \qquad \mu_a^i,\ \overline{\mu}_a^i$$

$$v_a^{q,i} \geq 0,\ u_a^{q,i} \geq 0,\ s^{q,i} \geq 0$$

$$(Known\ initial\ state\ x_{a,j}^{q,0} \geq 0)$$

$$\left(\delta_t < Min_{a\in A} \left\{\frac{\delta_{z_a}}{2\omega_a}\right\}\right)$$

Difference state equations of this optimization problem are now a *convergent* approximations to solutions of systems (2). It must be emphasized that all constraints are *linear*. We shall consider now the case of *strict* convex costs functions $F_a(x_a,\ j_a)$. It is then possible then the partial dualization of homogeneity constraints and limited capacity at vertical queue constraints $0 \leq \Sigma_{q\in D}\ \kappa^q{}_a \leq \hat{\kappa}_a$. resulting the dualized objective:

$$\Psi(x,\omega\cdot x,\varsigma,\mu,\overline{\mu}) = \delta_t \sum_{a\in A} \delta_{z_a} \sum_{j=1}^{M_a} \sum_{i=1}^{N} F_a\left(x_{j,a}^i, \omega_a \cdot x_{j,a}^i\right) + \sum_{i=1}^{N} \overline{\mu}_a^{i^T}\left(\hat{\kappa}_a^i - \kappa_a^i\right) -$$

$$- \sum_{i=1}^{N-1} \varsigma_i^{q^T}\left(\kappa^{q,i-1} - \delta_t\left(v^{q,i} + v^{q,i+1}\right)\right) + \varsigma_N^{q^T}\left(\kappa^{q,i-1} - \delta_t v^{q,i}\right) - \sum_{i=1}^{N} \mu_a^{i^T} \kappa_a^i \tag{9}$$

The Case of a Single Destination

In the case of a single destination in the network it is possible the application of a decomposition method developed in [7] for nonlinear programming problems approximating optimal control problems, which is equivalent to an implicit integration method for the extremals of such problems. Given a discrete-time optimal control problem of the type:

ΔOCP2

$$Min_{x_i \ u_i} \quad \delta \sum_{i=1}^{N} f_0(x_i, u_i) \qquad (L.M.)$$

$$s.t. \quad \begin{array}{ll} \delta\, Au_i + x_{i-1} - x_i = 0 & \lambda_i \\ Cu_i = D(t_i) & \alpha_i \\ u_i \geq 0 & \varsigma_i \\ (x_0 \ known),\ (i = 1, \cdots, N) & \end{array}$$

The decomposition method results in the solution of successive nonlinear optimization subproblems of the type **ΔOCP2 1** shown below, each of them for a time subinterval *[(i-1)δ, iδ]* . In [7] it is proved how first order conditions of these subproblems approximate the extremals of an optimal control problem provided that subproblems present *sensitivity* of their solutions for certain parameters λ' which are proved to be an approximation to the adjoint variables of the optimal control problem at time $(i+1)\delta$.

ΔOCP2 1

$$Min_{x,\ u} \quad \delta\ f_0(x\ ,u\ ) - \lambda'^{T} x \qquad (L.M.)$$

$$s.t. \quad \begin{array}{ll} \delta\, Au + x_{k-1} - x\ = 0 & \lambda\ (\lambda') \\ Cu = D(t_i) & \alpha(\lambda') \\ u \geq 0 & \varsigma_i(\lambda') \end{array}$$

We write below the structure of subproblems Δ **OCP2 1** when the decomposition method is applied to problem Δ **OCP 1**:

ΔOCP2 2

$$
\begin{aligned}
& Min_{y,u,v} \quad \psi_{i}(x,\omega x,\varsigma,\mu,\overline{\mu}) - \lambda^{\prime T} y^i \\
& \quad y_a^i - y_a^{i-1} + v_a \omega_a y_a^i - b_a^i = 0 \\
s.t. & \quad B_+ u^i - B_- v^i + e_q s^i = p^i \\
& \quad u^i \geq 0, \quad v^i \geq 0, \quad s^i \geq 0, \quad i = 1,\cdots,N
\end{aligned}
\qquad (8)
$$

with the objective function Ψ_k containing only terms of the right of (9) corresponding to a given time superscript k.

In [7] and [5] it is shown the network structure of the linear constraints of Δ **OCP2 2**. Additionally, for the case of networks with a single destination, if cost functions $F_a(x, \underline{j})$, depending on the total flow $\underline{j}$ and density x, are *strictly* convex on $\underline{j}$ and x it is possible to guarantee sensitivity of solutions of problem Δ **OCP2 2** respective parameter λ'. Unfortunately, the use of this decomposition method requires that subproblems Δ **OCP2 2** be solved with accuracy enough because, as shown in [7], errors propagate as the extremal calculation progresses. Submanifold optimization algorithm (see, for instance [16]) seems a suitable candidate in the case of quadratic positive definite costs to overcome accuracy requirements of this decomposition method.

5. Remarks

This paper presents a generic System Optimal Dynamic Traffic Assignment model for multidestination networks based on an extension of the simple continuum model. This model can be considered as a reference because of basic characteristics of its flow dynamics, no overtaking between flows of different destinations, finite speed propagation, dispersion and effects of congestion such as spillback. For the case of constant propagation speed and vertical queues at the end of links numerically stable approximations to the system optimal model by means of optimal control are presented. For the flow dynamics based on regressive difference approximations of the term $\partial \underline{j}/\partial z\}$ it is possible the formulation of a nonlinear programming problem with *linear* constraints, thus avoiding nonconvexities of constraints for FIFO observance in previous formulations.

References

[1] Anderson D.A., Tannehill J.C., Pletcher R.H (1984) "Computational Fluid Mechanics and Heat, Transfer", *Hemisphere Publishing Corporation* 1984.

[2] Carey M. (1986), "A Constraint Qualification for a Dynamic Traffic Assignment Model", *Transportation Science*, 20, pp 55-58.

[3] Carey M. (1987), "Optimal Time-Varying Flows on Congested Networks", *Operations Research* Vol.35, No 1, January-February 1987.

[4] Carey M. (1991), "Nonconvexity of the Dynamic Assignment Problem", *Transpn Res.B* Vol 26B, No 2, pp. 127-133.

[5] Codina E., Barceló J. (1992), "Extremals Calculation for the Dynamic Traffic Assignment Problem", *Presented at 39th North American Meeting of the RSAI*, Chicago USA 1992.

[6] Codina E., Barceló J. (1992), "Dynamic Traffic Assignment: An Annotated Review of Modelling and Algorithmic Approaches", Universitat Politècnica de Catalunya. Departament d'Estadística i Investigació Operativa, *Document de Recerca* DR 94-01.

[7] Codina E. (1993), "Un Algoritmo para Problemas de Control Optimo y su Aplicación a la Asignación Dinámica de Tráfico", *Tesis Doctoral*. Universitat Politècnica de Catalunya, Departament d'Estadística i Investigació Operativa.

[8] T.L.Friesz, F.J.Luque, R.L.Tobin, B.W.Wie (1989), "Dynamic Network Traffic Assignment Considered as a Continuous Time Optimal Control Problem", *Op. Research* Vol 37-6 pags 58-69 USA.

[9] Merchant D.K., Nemhauser G.L. (1978), "A Model and an Algorithm for the Dynamic Traffic Assignment Problems", *Transportation Science* Vol.12 No 3 August 1978.

[10] Merchant D.K., Nemhauser G.L. (1978), "Optimal Conditions for Dynamic Traffic Assignment Model", *Transportation Science* Vol.12 No 3 August 1978.

[11] Mitchel A.R., Griffiths D.F. (1980), "The Finite Difference Method in Partial Differential Equations", *John Wiley and Sons*, New York (1990).

[12] Papageorgiou M. (1989), "Dynamic Modeling, Assignment, and Route Guidance in Traffic Networks", *Transpn Res.* Vol 24B, No 6, pp. 471-495 (1990), From US Italy Joint Seminar Preprints at Capri June (1989).

[13] Ran B., David E.B., LeBlanc L.J. (1990), "A New Class of Instantaneous Dynamic User-Optimal Traffic Assignment Models", *Operations Research* Vol. 41, No. 1993.

[14] Shames I.H. (1962), "Fluid Mechanics", *Mac Graw Hill* (1962).

[15] B.W.Wie, T.L.Friesz, R.L.Tobin (1990), "Dynamic User Optimal Traffic Assignment on Congested, Multidestination Networks", *Transportation Research* 24B 6 December 1990.

[16] Zangwill W.I. (1969), "Nonlinear Programming: A Unified Approach" *Prentice-Hall, Englewood Cliffs*, NJ, 1969.

STOCHASTIC ASSIGNMENT MODELS FOR TRANSIT LOW FREQUENCY SERVICES: SOME THEORETICAL AND OPERATIVE ASPECTS

Agostino Nuzzolo

Professor of Transportation Planning, Department of Civil Engineering, Tor Vergata University of Rome, Italy

Francesco Russo

Assistant of Transportation System Design
Department of Systems Engineering and Mathematics
University of Reggio Calabria, Italy

In this paper the problem of assignment for low frequency transit services is dealt with in the framework of the stochastic approach, with random utility models of path choice and with "time-space" (diachronic) network models which explicitly consider timetables. In the first sections of the paper, user run choice models are considered and the problem of the demand-supply interaction is examined; subsequent sections present time-space network models, and in the final section some applications of this approach are described.
This work was partially supported by the National Research Council of Italy, grant n. 94.01447.PF74.

1. Introduction

From the assignment model point of view, transit systems can be classified on the basis of some characteristics concerning service, chiefly the frequency of the offered service and the punctuality of vehicle arrivals compared to the scheduled timetable.

In general, high frequency and low punctuality, typical in urban areas, lead to a random arrival of the user at stops with a broad margin of adaptability in path choice. The models of supply representation, path choice and assignment which are more suited to this type of service and which have been developed in the literature originated in the concept of hyperpath are in [16,17,7,25,22].

For services with medium-low frequency and high punctuality, typical of national, regional and, in general, extraurban systems, the hypothesis of adaptable choice proves incongruous and is replaced with the hypothesis of a completely pre-trip choice. Because of the medium-low frequency of the service, which may entail considerable anticipation or delay times relative to departure and arrival target times, a specific run choice model that takes into account such user target times has to be used. Furthermore, if capacity constraints occur, further anticipations or delays connected with them must be allowed for.

The assignment problem for low frequency services can be tackled by using a deterministic approach [14,19] or a stochastic approach as the one considered in this paper, with random utility models of run choice and "time-space" network models, which explicitly consider timetables.

In section 2 the user's run choice models are considered; section 3 covers the problem of the demand-supply interaction and section 4 deals with the time-space (diachronic) network models, which allow path alternatives and their relative attributes to be more easily determined. In the final section some applications of the approach are described.

2. Run Choice Models

For the definition of run choice models, which belong to the class of models known in the literature as joint choice of departure time and path, two main problems arise: the definition of the choice set and the simulation of the choice among the alternatives present in the set.

For the generation of the choice set, one or more criteria can be used, such as minimum travel time, minimum cost, minimum schedule delay and minimum transfer number.

The most widely used choice models in the literature are of random utility type and the hypotheses on the user's behaviour are:

a) that the choice concerns simultaneously path (boarding, alighting and transfer stops) and departure time;

b) that the user is a rational decision-maker who considers all the paths which connect the O-D pair, and chooses the alternative which involves the perceived maximum utility; this utility is a random variable. In low frequency transit systems the random utility model maintain their validity in spite of the punctuality of the trips and the user's hypothesised knowledge of the timetables, for the following reasons:

1) different users can evaluate in a different way the same attributes (for example, users are more or less informed of access and egress mode attributes or of the transfer time);
2) some attributes evaluated by users may not have been included in those used by analyst and defined in point C (vehicle characteristics, etc);
3) some attributes can be evaluated by the analyst in an approximate way (transfer times, access and egress times, etc.);

c) in order to choose between the numerous alternatives, the user considers as utility components: 1) access time from origin to first stop; 2) waiting time at stops; 3) on-board time; 4) time and inconvenience for any eventual transfer; 5) egress time from the last stop to the destination; 6) fare; 7) comfort related to the degree of crowding; 8) disutility for the anticipation or delay compared to the origin or destination target time, usually termed schedule delay.

The differences between various existing random utility models derive from the hypothesis concerning the distribution of the random term involved in the user's utility. However, the models employed in practice mainly refer to two distributions: Weibull-Gumble (WG) for logit and nested logit models and MultiVariate Normal (MVN) for probit models.

A possible classification concerning existing models, generally using logit approach is relative to the number of considered modes.

The following consider two or more modes:

- Abkowitz [1] includes in the choice set 12 departure time alternatives each using car and bus, and specific linear and non-linear terms for anticipation and delay.
- Hendrikson and Plank [13] include up to 28 alternatives obtained with 4 modes and 7 different departure times and allow for non-linear terms of anticipation and delay.

The following models consider one mode, generally car:
- Ben Akiva, De Palma and Kanaraglou [3] present 3 possibilities: leaving and arriving early to avoid the traffic, leaving and arriving late, taking an alternative longer path; the authors specify linear terms but introduce a discontinuity by means of an indifference time band on arrival;
- Alfa [2] explicitly introduces the variable DTT (destination target time), and verifies the existence of a correlation between the flexibility of arrival at work and the uniformity of demand in the peak-hour; he considers linear terms without discontinuity;
- Mannering [15] hypothesises a Poissonian distribution of path changes, including the non-change option; he considers flexibility concerning the time at which one starts to work;
- Biggiero, Cascetta and Nuzzolo [5] consider the WG distribution, and the width of the indifference band is estimated compared to the DTT, thereby using a discontinuous functions.

The authors (1994) calibrated probit models for home to work and home to school travel by transit, and Cascetta, Nuzzolo and Biggiero [9] used a tree-logit model for the choice of service, run and class on a railway service network. A synthesis of these two models is reported in section 5.

3. General Formulation of Stochastic Assignment Models

3.1 Problem Definition

The approach to the problem of demand-supply interaction, in the case of stochastic assignment for low frequency transit systems, may be traced back to the consolidated approach for urban road networks [23,6].

The relations between vectors of path flows $\underline{F}$ and link flows $\underline{f}$, and the demand vector $\underline{d}$ through the path utility vector $\underline{U}$, are expressed by:

$$\underline{F} = \underline{P}(\underline{U})\ \underline{d} \tag{3.1}$$

$$\underline{f} = \underline{A}\ \underline{P}\ (\underline{U})\ d \tag{3.2}$$

in which $\underline{P}$ is the path choice probability matrix, function of path utility, and $\underline{A}$ is the link-path incidence matrix.

In general, path utilities $\underline{U}$ may be expressed through link utility $\underline{u}$. Moreover, in the presence of congestion, link utilities are function of the link flow vector. Hence:

$$\underline{U} = \underline{U}\ [\ \underline{u}(\underline{f})]$$

Consideration will be given below first to the case of non-congested systems, and then to that of congested systems.

3.2 The Case of Non-Congested Systems

In non-congested systems, link utilities are not dependent on link flows. Thus, to calculate matrix $\underline{P}$ of (3.1) it is sufficient to define the hypotheses on the generation of the path choice set and the hypotheses on the distribution of perceived utilities.

In the case where the choice set is generated with a monocriterion approach, and the utility function for the choice model is the same as that used for choice set generation, and where it is also possible to express linearly the relation between path utility $\underline{U}$ and link utility $\underline{u}$ according to:

$$\underline{U} = \underline{A}^T\ \underline{u} \tag{3.3}$$

flows may be directly calculated without explicit path enumeration, using the usual stochastic network loading algorithms (Logit-Dial, Probit-Montecarlo) set up for road networks.

In the case where choice sets is generated with multicriterion or monocriterion alternatives with a different utility function for choice set generation and for path choice model of alternatives, or where path attributes are used which cannot be expressed linearly as a function of link attributes, the various feasible paths for each O-D pair must be explicitly enumerated.

3.3 The Case of Congested Systems

Congestion produces two main effects: with an increase in crowding, user comfort decreases and, once capacity is reached, the "run" may become inaccessible to new users.

The flows which can be achieved on the congested network must be subject to conditions imposed by demand and supply.

It is possible to determine the flows using cost functions with a steep slope after maximum real capacity, with implicit treatment of capacity restraints.

As is well known for the car network, the feasibility set of $\underline{F}$ and of $\underline{f}$ with regard to the demand vector $\underline{d}$, using an appropriate cost functions is:

$$S_F = \{\underline{F}: \underline{F} \geq \underline{0};\ \underline{BF} = \underline{d}\}$$

and in terms of $\underline{f}$:

$$S = \{\underline{f}: \underline{f} = \underline{AF}, \underline{F} \in S_F\}$$

where $\underline{A}$ is the incidence link-path matrix previously defined and $\underline{B}$ is the incidence matrix of paths between O-D pairs.

As it is well known [23,6] in equilibrium stochastic assignment, for the equilibrium solution $\underline{f}^*$, the condition of existence of the solution is satisfied by the continuity of cost functions $\underline{c}(\underline{f})$ and the condition that $\underline{J}[\underline{c}(\underline{f})]$ the Jacobian matrix of the cost function vector is positive definite, is sufficient for $\underline{f}^*$ to be unique.

Cost functions are not the subject of this paper. Nevertheless, it is appropriate to examine the specifications which may be obtained in relation to "waiting", "standing", and "run" links (Fig. 1) that represent the generic run at the "run-stop", since they are relevant to the treatment of congestion.

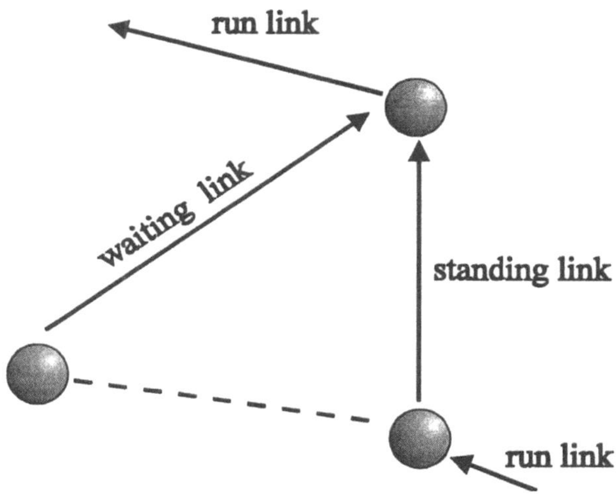

Fig. 1 Network specification at run-stop

In networks where it may be assumed that the service still remains available to "waiting" users, the congestion effect is translated merely into a decrease in on-board comfort. Thus the separable cost function below may be used for the "standing" and "run" links:

$$c_i(f_i) = k_1 + k_2(f_i / Q_r)^{k_3} \tag{3.4}$$

c_i and f_i being the cost and flow, and Q_r the capacity of run r on link i, with $k_1 \geq 0$, $k_2 \geq 0$ and $k_3 \geq 1$ parameters which depend on the characteristics of the system.

In networks where the implicit capacity constraint is active, with f_j and f_i respectively the flows on the waiting and the incident standing links, and Q_r the vehicle capacity of run r, the following non-separable and asymmetric cost function may be used for the waiting links :

$$c_j(f_i, f_j) = k_4 + k_5 f_j + k_6 \left(\frac{f_i + f_j}{Q_r} \right)^{k_7} \tag{3.5}$$

where $k_4 \geq 0$, k_5 and $k_6 > 0$ and $k_7 \geq 1$.

For the other link types in the network, in general "separable" cost functions may be used.

In networks where, for the waiting-boarding, standing and run links, only (3.4) type cost functions are brought into effect, and therefore all the cost functions are of a separable continuous type, the sufficient condition for the existence of $\underline{f}^*$ is immediately verified.

The Jacobian matrix of the cost functions is in this case a diagonal matrix, with positive elements, producing a positive definite Jacobian matrix, which therefore verifies the condition for uniqueness. Thus, in this case, the equilibrium vector f^* is unique.

With the symmetrical Jacobian matrix, $\underline{f}^*$ can be defined as an optimum problem solution in different ways [21,11].

It must be noted that the main limitation of this procedure is that the users of run links where $f_i > Q_i$, are equally penalised irrespective of boarding order, and therefore the ensuing path flows cannot be the true ones. In order to calculate the equilibrium flows, if we use an adequate network, the usual algorithms used in road assignment model can be transferred, as is the case with the MSA algorithm (Method of Successive Average, [23]).

In networks in which type (3.5) cost functions are used for the waiting-boarding links and type (3.4) functions for the standing and run links, the sufficient condition for existence, which has already been mentioned, is verified, functions type (3.4) and type (3.5) being continuous.

In the case of a stochastic asymmetric equilibrium, it emerges that sufficient conditions for uniqueness of $\underline{f}^*$ have not been formulated.

However, a vector $\underline{f}^*$ can be calculated by means of algorithms which are analogous to those of diagonalization.

4. Spatio-Temporal Network Models

From the above analysis, there emerges the utility of using a space-time network model which allows:

a) application of Logit-Dial or Probit-Montecarlo algorithms without explicit path enumeration;
b) in the case of algorithms with explicit path enumeration, straightforward execution of the following operations, taking timetables into consideration:
   - path identification;
   - calculation of path attributes relative to point C in section 2, and in particular
   - evaluation of transfer time;
c) more straightforward treatment of congestion.

Certain aspects of cost graphs and functions relative to such models are covered below.

4.1 The Graph of the Network

The spatio-temporal network model (Fig. 2) represents each run of each line; this network is called the diachronic network [19]. The diachronic graph consists of two subgraphs: one relates to run representation and the other is dependent on the demand structure.

For every stop j there is a temporal axis on which, for every run r, a node n_j^r is located which is relative to the stopping time of run r. This node is connected by means of a waiting-boarding link to node n_{jb}^r, which is relative to the departure time of run r. It is also connected by means of an alighting link from node n_{ja}^r, relative to the arrival time of run r. The link (n_{ja}^r, n_{jb}^r) represents the standing time at stop j of run r;

the link $\left(n_{jb}^{r}, n_{ka}^{r}\right)$ represents the connection between stop j and stop k by means of run r.

The second subgraph consists of various subgraphs, with at least one for a zone centroid, in relation to the structure of the O/D matrix, which is supposed known for trip purpose and destination target time (DTT) or origin target time (OTT).

In general, the distribution of the DTT or OTT can be discretized with reference to a certain number of finite time intervals, where it may be accepted that the demand distribution is concentrated at the final point of the interval. A temporal axis is then considered for every zone, and a temporal centroid, positioned in the axis on the final point of each interval, will correspond to each time interval. Every temporal centroid can therefore be the origin and destination of user trips, and every trip can be with the DTT and/or OTT.

Given an O/D matrix of dimensions $\left[n_O * (i_{OTT}+1), n_D * (i_{DTT}+1)\right]$, where n_O is the number of spatial origins, n_D the number of destinations, i_{OTT} and i_{DTT} the number of intervals in which the time bands referring to origin and destination for the target departure and arrival times are divided, in each zone it is possible to associate as many temporal centroids concerning origin n_{OTT} and destination n_{DTT} as there are specified time intervals with an additional one for the trips without target.

In order to connect the temporal centroid axis and the spatial centroids with the run subgraph, access and egress links need to be introduced. Such links have one extremity on the centroid temporal axis or on the spatial centroid and the other extremity on the temporal axis of the stop, each of which guarantees connection with an accessible run in the zone. The first extremity of the generic access link $\left(n_O^r\right)$ represents the time at which it is necessary to leave the point of origin in order to reach stop j (second extremity) in time to take the choice run $r\left(n_j^r\right)$.

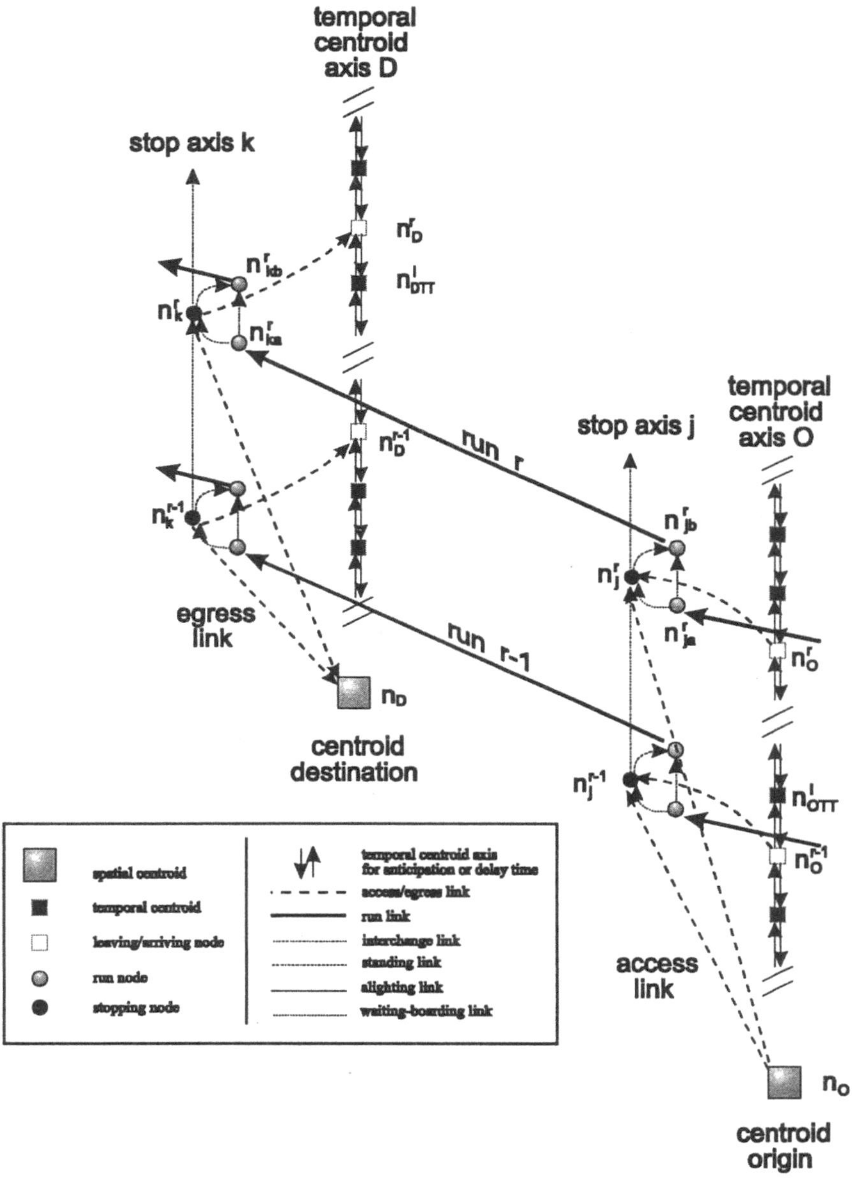

Fig. 2 Example of diachronic network

The time difference between the first extremity n_O^r of the access link and the temporal centroid of origin n_{OTT}^i represents exactly the disutility of anticipation or delay connected to the utilisation of the generic run r taken into account. A similar consideration is also true for the egress link.

4.2 The Cost Functions

A qualifying characteristic of the diachronic network derives from the possibility of introducing penalty functions of anticipation and/or delay compared to the user DTT and/or OTT, directly on the network. Furthermore, it is possible to consider the different weights of delay and anticipation for the different user categories and different purposes, using a temporal axis for each purpose or category.

A second qualifying characteristic is given by the possibility of considering the flow on every run, thereby treating congestion explicitly.

In the literature, penalty functions are expressed both in a linear and in a non- linear form; and in addition they can be further divided into continuous or discontinuous form, according to whether the introduction of an indifference time band within which the user does not suffer any penalty, has been allowed for or not.

The general forms of the three cost function types for the DTT case (the OTT case is similar) are:

- linear

$$t_l = \beta_1 \, t_{ear} + \beta_2 \, t_{lat} \tag{4.1}$$

- non-linear

$$t_n = \beta_3 \, t_{ear}^{\beta 4} + \beta_5 \, t_{lat}^{\beta 6} \tag{4.2}$$

- discontinuous

$$t = \begin{cases} 0 & t < \Delta \\ t_l \text{ or } t_n & t > \Delta \end{cases} \tag{4.3}$$

where:

t_{ear} = time difference between the destination target time and the time of arrival at destination;

t_{lat} = time difference between the time of arrival at destination and the target time;

Δ = width of indifference band.

The β values can be obtained from a departure time and path choice model like those recalled in section 2 and in section 5.

The other cost function related to "waiting", "standing" and "run" links defined in section 3.2 are directly useable in the diachronic network.

5. Some Applications

The stochastic approach with a random utility model of run choice and with a diachronic service network model was used by the authors in several operative planning activities, with the simplifying hypothesis of non congested systems. Brief descriptions are given below of two of such applications: the first is at a regional scale, the second at a national scale.

5.1 A Regional Scale Assignment Model

The application at regional scale [12] was effected with regard to home-to-school and home-to-work trips on the bus service network in the province of Potenza, southern Italy, with a global population of 310,000 in 70 zones (one for each town). Seven time bands are considered, obtaining a total of 490 time centroids, while the bus service network consists of 806 runs. The complete space-time graph consists of a total of 20,739 links and 8,546 nodes.

The run choice model used [18] was of Probit random utility type.

The data used for calibration were obtained by means of a destination survey with more than a hundred interviews with workers and students. A maximum of four alternatives were considered: two that gave destination or origin delay time and two that gave anticipation time, inside the band of an hour and a half around the target time. Various models were calibrated in connection with the typical potentiality of the Probit to introduce different variances among the alternatives and non-zero covariances: the best models obtained are reported in Tab. 1.

The assignment was carried out with a Stochastic Network Loading (SNL) model, without explicit path enumeration, using a Montecarlo simulation algorithm which at each iteration determines the perceived utility value of each link, by sampling from a normal random variable with mean u_{ij} (link utility) and with variance θ u_{ij}.

attributes (unit)	*home-work*	*home-school*
On board time (h)	-2.81	-0.41
Early time (h)	-1.41	-4.39
Delay time(h)	-0.64	-4.21
Cost (1000 Lit.)	-0.18	-1.49
Access and Egress (h)	-7.39	-5.37
σ^2	1.64	1.64
σ_{ij} (runs with delay time)	0.50	0.50
σ_{ij} (runs with early time)	0.20	0.20
RHO^2	0.21	0.46

Tab. 1. Probit models for work and school [18]

The path utility is thus still distributed according to a normal one with the mean given by the sum of u_{ij} belonging to the path and with variance proportional to path utility according to the parameter θ. The path choice set is generated with a monocriterion approach (minimum disutility), with a linear relation between type (3.3) link costs and path costs.

The utility function considered for the anticipation and delay links belongs to type (4.1); besides, the use of more complicated functional forms than the linear ones of the anticipation and delay time does not offer great advantages in terms of the model's performance [18]. The parameters and attributes of penalty functions are those of the path choice model summarised in Tab. 1.

In formal terms, the path choice model used in the assignment is dissimilar to the calibrated one which instead assumes a constant variance and a choice set defined only by two early runs and two late runs with respect to the target time. As regards the constant variance, in practical terms the differences are minimal because other calibrations were carried out modifying σ^2 values and obtaining similar parameter vector to that reported in Tab. 1. With regard to the choice set, further paths than the four considered in the calibration of path choice model could be generated by the SNL-Montecarlo algorithm, but only when very high values are used for the variances of perceived times.

For the characteristics described so far and with a CPU Pentium/90 Mhz/16 Mb RAM computer and programs in VB3, approximately 10" CPU are required to generate the diachronic network and roughly 6" CPU for each all-or-nothing assignment within the SNL assignment.

5.2 A Decision Support System for Railway Service Planning

In order to analyse market requirements, improve traffic returns and reduce operating costs of intercity services, the Italian Railways (FS) are defining and testing a decision support system called SASM (System of Analysis and Simulation for Operative Marketing) [24,8,20].

For a correct simulation of user behaviour in response to the variations in service characteristics and thus for a better evaluation of returns and flows on the network, it is necessary to represent explicitly the services with run departure times and to use demand models which simulate user choices in each market segment. In the DSS, market segmentation takes into account the origin or destination target time, the car availability, the household income and the number of users travelling together.

So the DSS uses a diachronic network model of the services, and the assignment model to the rail service network uses a Stochastic Network Loading assignment model and a Tree-logit [4] path choice model with explicit path enumeration. For a given O-D pair and a given market segment the run choice set includes (Fig. 3) for each service (low speed and high speed) and for each class (I and II class), two run alternatives: the last scheduled run before the target time (run 1) and the first scheduled run after the target time (run 2).

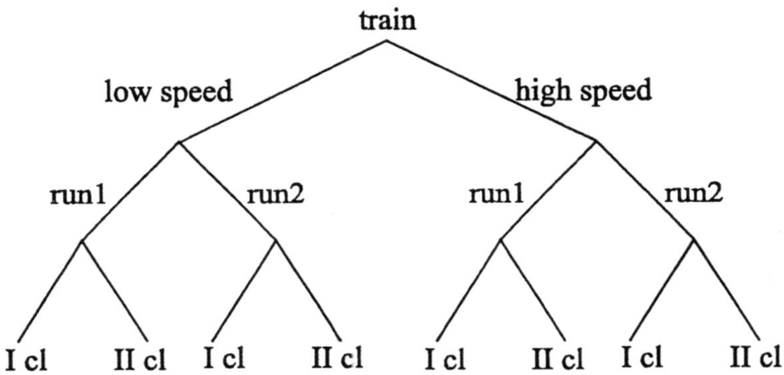

Fig. 3 The path choice model representation

An application test of the DSS is under way on the Turin-Milan-Venice railway corridor. In this application, Italy was divided into 359 traffic zones, 178 of which were in the area most directly affected by the sample corridor and 181 in the rest of the country. The used diachronic network models of the services are of a considerable dimension: for example, the inter-regional and regional rail service network consists of 11,000 nodes and 41,000 links.

Calibration of the path choice models is being carried out, using the data of an on-board train survey with about 10,000 interviews on the corridor. For example, for the business travel purpose the first results of the calibration [9] give the following utility function:

$$V = \beta_1 Tb + \beta_2 Cmh + \beta_3 Cl + \beta_4 EA + \beta_5 LA + \beta_6 Ht + \beta_7 IcIL + \beta_8 IcIH$$

where:

Tb = on-board travel time (h)

CMH = on-board travel cost for medium-high income users (1000Lit)

Cl = on-board travel cost for low income users (1000Lit)

EA = early arrival time of the run with respect to the target time (h)

EL = late arrival time of the run with respect to the target time (h)

Ht = specific preference for high speed trains

IcIL = specific for the Ist fare class in low speed trains

IcIH = specific for the Ist fare class in high speed trains

The calibration results are reported in table 2.

coef	*Tb*	*Cmh*	*Cl*	*EA*	*EL*	*Ht*	*IcIL*	*IcIH*	*θrun*	*θclas*
β	-4.11	-0.21	-0.51	-2.48	-8.75	1.76	0.23	0.48	0.99	0.37
t-St	-1.95	-4.01	-3.06	-2.06	-2.72	2.65	0.83	2.04	0.01	4.21

rho squared = 0.30 rho bar squared = 0.27

Tab. 2. The path choice model for business travel [9]

References

[1] M.D. Abkowitz, "An analysis of the commuter departure time decision", *Transportation*. pp. 283-297, 1981.

[2] A.S. Alfa, "A review of models for the temporal distribution of peak traffic demand", *Transportation Research* 20B, pp. 491-499, 1986.

[3] M. Ben-Akiva, A. De Palma, P. Kanaraglou, "Dynamic Model of Peak Period Traffic Congestion with Elastic Arrival Rates", *Transportation Science* 20, 1986.

[4] M. Ben-Akiva, S.R. Lerman, "*Discrete choice analysis*", Mit Press, Cambridge Mass 1985.

[5] L. Biggiero, E. Cascetta, A. Nuzzolo, "Analysis and modelling of commuters departure time and route choices in urban networks", *Sixth World Conference on Transportation Research*, Lyon 1992.

[6] E. Cascetta, "Metodi quantitativi per la pianificazione dei sistemi di trasporto", *Cedam*, Padova 1990.

[7] E. Cascetta, A. Nuzzolo, "Uno schema comportamentale per la modellizzazione delle scelte di percorso nelle reti di trasporto pubblico urbano", *Internal Paper of the Department of Transportation Engineering. University of Naples "Federico II"*, 1988.

[8] E. Cascetta, A. Nuzzolo, G. Beltrami, "A system of demand and supply models for the marketing of intercity passenger services", *Seventh World Conference on Transportation Research*, Sydney, 1995.

[9] E. Cascetta, A. Nuzzolo, L. Biggiero, "A within-day dynamic tree logit model for mode-service choice with explicit representation of time-table", *Internal paper of the Department of Transportation Engineering*, University "Federico II", Naples 1995.

[10] C.F. Daganzo, "Multinomial probit: the theory and its application to demand forecasting", *Academic Press*, New York, 1979.

[11] C. Fisk, "Some developments in equilibrium traffic assignment" *Transportation Research* vol. 14B, pp. 243-255, 1980.

[12] R. Giordano, A. Riso, F. Russo, "Una procedura di assegnazione al trasporto extraurbano: Applicazione ad un bacino di media dimensione" *Atti del II Convegno Nazionale PFT2-CNR*.. Genova, 1995.

[13] C. Hendrikson, E. Plank, The flexibility of departure times for work trips. *Transportation Research* 18A, pp. 25-36. L23, 1984.

[14] F. Malucelli, S. Pallottino, A. Schettino, "Passenger assignment in a regional area transit system by active paths enumeration". *Preprint euro working group on Transportation*. Barcellona 1995.

[15] M. Mannering, "Poisson analysis of commuter flexibility in changing routes and departure times", *Transportation Research* 23B, pp. 53-60, 1989.

[16] S. Nguyen, S. Pallottino, "Assegnamento dei passeggeri ad un sistema di linee urbane: determinazione degli ipercammini minimi", *Ricerca Operativa* n°38, 1986.

[17] S. Nguyen, S. Pallottino, "Equilibration Traffic Assignment for Large Transit Networks", *Eur. J. of Oper. Res.* 37, pp. 176-186, 1988.

[18] A. Nuzzolo, F. Russo, "Departure time and path choice models for intercity transit assignment" *Preprint IATBR-94*. Valle Nevado, Santiago, Cile, 1994.

[19] A. Nuzzolo, F. Russo, "An equilibrium assignment model for Intercity Transit Networks", *Preprint TRISTAN Conference*. Capri 1994.

[20] A. Nuzzolo, F. Russo, U. Crisalli, "A decision support system for railway service planning" *Preprint euro working group on Transportation*. Barcellona 1995.

[21] W.B. Powell, Y. Sheffi, "The convergence of equilibrium algorithms with predetermined step" *Transportation Science* Vol. 16, pp. 45-55, 1982.

[22] F. Russo, "Un modello di assegnazione alle reti di trasporto collettivo. Aspetti teorici ed applicazione a casi reali" In *Metodi e modelli per la pianificazione e la gestione dei sistemi di trasporto collettivo*, A. Nuzzolo F. Russo. eds. Franco Angeli. Milano. 1995.

[23] Y. Sheffi, "Urban Transportation Networks", *Englewood Cliffs Prentice-Hall*, 1985.

[24] Sintra, *SASM*. "Sistema di analisi e simulazione per il marketing operativo", *Progetto esecutivo. Ferrovie dello Stato. 1994.*

[25] H. Spiess, M. Florian, "Optimal strategies: A new assignment model for transit networks", *Transportation Research* 23B, pp. 83-102, 1989.

A PARALLEL APPROACH TO LARGE-SCALE NONLINEAR NETWORK OPTIMIZATION

Lucio Grandinetti, Francesca Guerriero, Roberto Musmanno

Dipartimento di Elettronica, Informatica e Sistemistica, Università della Calabria, Rende (CS) - Italy

In this paper we present an innovative method which usefully explores modern information technologies like high performance parallel computers to solve large-scale nonlinear network flow problems arising in several applications, including transportation. The classical iterative methods are implemented on conventional computers and, therefore, they do not seem to be particularly well-suited to solve problems whose computational overhead is mainly due to their high dimension.

The idea of using parallel computation in large-scale network flow optimization is remarked in this paper. In particular, we discuss on some numerical aspects of the dual relaxation algorithm on shared-memory parallel architectures, to focus on the effectiveness of the parallel approach especially when applied to large-scale transportation problems, based on a dynamic allocation of the nodes to processors, on the basis of which each node is scheduled to a different processor at the beginning of each iteration.

We present and analyze computational results which demonstrate that asynchronous implementations always outperform the synchronous and excellent speedup factors may be achieved for the majority of test problems.

1. Introduction

In this paper we consider the problem of solving a particular class of nonlinear network flow problems which are involved in several kind of applications, including transportation. Recent developments in network optimization take into account the idea of parallel computing in the design, analysis and implementation of new strategies for solving large-scale problems. As we will show in the sequel, the structure of the network problems plays a crucial role in a parallel computation context, and it may be amenable for both synchronous and asynchronous implementations.

We choose to focus on pure network problem with a special case of the objective function, assumed here separable. Given a directed network $G = \{N,A\}$, where $N = \{i \mid i = 1, 2, \ldots, n\}$ is the set of nodes and $A = \{(i,j) \mid i , j \in N\}$ is the set of arcs, the problem (P) can be formally defined as follows:

$$\text{(P)} \qquad \textit{minimize} \sum_{(i,j) \in A} c_{ij}(x_{ij}) \tag{1}$$

s.t.

$$\sum_{\{j:(i,j)\in A\}} x_{ij} - \sum_{\{j:(j,i)\in A\}} x_{ji} = b_i \quad , \forall\, i \in N \tag{2}$$

$$l_{ij} \leq x_{ij} \leq u_{ij} \qquad , \forall\, (i,j) \in A \quad , \tag{3}$$

where x_{ij} is the flow on the arc (i,j), c_{ij}: $A \rightarrow \Re$ is a strictly convex cost function of one variable (assumed to be differentiable), b_i is the supply at node i, l_{ij}, u_{ij} are respectively the lower and upper bound for x_{ij} on the arc (i,j); therefore, (2) e (3) specify conservation of flow constraints and capacity constraints.

An iterative method to solve (P), based on the solution of the dual problem, is considered. In section 2 we present the methodological ideas for the design of the relaxation method. In section 3 we explain how the relaxation method can be usefully implemented on parallel computers. Finally, numerical results, collected using an Alliant with 8 processors on significant test problems, are reported and discussed in section 4.

2. The Relaxation Method

A dual problem of (P) can be formulated as follows. We assume primarily that (P) is feasible, that is:

$$\sum_{i \in N} b_i = 0 \ . \tag{4}$$

This means that there exists a unique optimal solution, given the hypothesis on the cost functions c_{ij} and the fact that the region defined by the constraints (2)-(3) forms a compact set.

Attaching a vector of lagrangian multipliers $\pi \in \Re^n$ to the equality constraints (2), we obtain the following lagrangian function:

$$L(x,\pi) = \sum_{(i,j)\in A} c_{ij}(x_{ij}) + \sum_{i\in N}\pi_i\left(\sum_{\{j:(j,i)\in A\}} x_{ji} - \sum_{\{j:(i,j)\in A\}} x_{ij} + b_i\right)$$

$$= \sum_{(i,j)\in A} (c_{ij}(x_{ij}) + (\pi_j - \pi_i)x_{ij}) + \sum_{i\in N} \pi_i b_i \ . \qquad (5)$$

Assigned $\pi \in \Re^n$, the value of the dual function $q(\pi)$ can be obtained by minimizing $L(x,\pi)$ over all flow distributions x which satisfy the lower and upper bounds. Therefore, the dual (D) of (P) can be formulated as follows:

$$\text{(D)} \qquad \textit{maximize}\ q(\pi) \qquad (6)$$

$$\pi \in \Re^n \ ,$$

where:

$$q(\pi) = \min_{l_{ij}\le x_{ij}\le u_{ij}} L(x,\pi) = \sum_{(i,j)\in A} q_{ij}(\pi_i - \pi_j) + \sum_{i\in N}\pi_i b_i, \qquad (7)$$

and

$$q_{ij}(\pi_i - \pi_j) = \min_{l_{ij}\le x_{ij}\le u_{ij}} c_{ij}(x_{ij}) - (\pi_i - \pi_j)x_{ij} \qquad (8)$$

A pair (x,π) is optimal for the primal and dual problems if and only if:

a) x is feasible for (P), that is:

$$g_i = \sum_{\{j:(j,i)\in A\}} x_{ji} - \sum_{\{j:(i,j)\in A\}} x_{ij} + b_i = 0, \quad \forall\, i \in N,$$

where g_i indicates the *surplus* of node $i \in N$;

b) π is feasible for (D);

c) the following *complementary slackness conditions* are satisfied:

$$t_{ij} \geq \frac{\partial c_{ij}(x_{ij})}{\partial x_{ij}}\bigg|x_{ij} = u_{ij} \Rightarrow x_{ij} = u_{ij} \; ; \tag{9a}$$

$$t_{ij} \leq \frac{\partial c_{ij}(x_{ij})}{\partial x_{ij}}\bigg|x_{ij} = l_{ij} \Rightarrow x_{ij} = l_{ij} \; ; \tag{9b}$$

$$t_{ij} = \frac{\partial c_{ij}(x_{ij})}{\partial x_{ij}}\bigg|x_{ij} = \bar{x}_{ij} \; , \; l_{ij} \leq \bar{x}_{ij} \leq u_{ij} \Rightarrow x_{ij} = \bar{x}_{ij} \; , \tag{9c}$$

where we assume that $t_{ij} = \pi_i - \pi_j$.

Furthermore, if we consider the partial derivative $\frac{\partial q(\pi)}{\partial \pi_i}$ of the dual function q with respect to the i-th price π_i, we obtain:

$$\frac{\partial q(\pi)}{\partial \pi_i} = \sum_{(m,n)\in A} \frac{\partial q_{mn}(\pi_m - \pi_n)}{} + b_i =$$

$$= - \sum_{\{j:(j,i)\in A\}} \nabla q_{ji}(\pi_j - \pi_i) + \sum_{\{j:(i,j)\in A\}} \nabla q_{ij}(\pi_i - \pi_j) + b_i =$$

$$= \sum_{\{j:(j,i)\in A\}} x_{ji} - \sum_{\{j:(i,j)\in A\}} x_{ij} + b_i \; , \tag{10}$$

that is, $\frac{\partial q(\pi)}{\partial \pi_i}$ corresponds to the surplus g_i of the node i associated with the unique flow distribution x for which the pair (x,π) satisfies the complementary slackness conditions.

Differentiability and convexity of q play a crucial role for the definition of iterative methods for solving (D), since, starting from a generic estimate of the dual solution we can improve the dual function along a direction defined by a single component π_i of π. The greatest improvement along the coordinate direction π_i corresponds to $\frac{\partial q(\pi)}{\partial \pi_i} = 0$, that is, when the surplus $g_i = 0$.

The sketch of k-th iteration of the relaxation algorithm for maximizing $q(\pi)$ is depicted in Fig. 1.

The asymptotic convergence of the method can be proven ([1]), starting from any estimate π of the solution π^*.

1. Assigned $\pi^{(k)}$ and computed $x^{(k)}$ by means of complementary slackness (9), *choose* a node $i \in N$ such that $\left|g_i^{(k)}\right| > \varepsilon$, where ε is a tolerance defined by the user; if such node does not exist, STOP, $x^{(k)}$ is optimal.

2. *Compute* $\pi^{(k+1)}$ such that $\pi_j^{(k+1)} = \pi_j^{(k)}$, $\forall\, j \neq i$, with $\pi_i^{(k+1)}$ chosen in such a way $q(\pi)$ could be improved along the i-th coordinate.

3. Determine $x^{(k+1)}$, that is $x_{ij}^{(k+1)} = x_{ij}^{(k)}$ e $x_{ji}^{(k+1)} = x_{ji}^{(k)}$, if the node j is not adjacent to i, whereas for adjacent nodes j, $x_{ij}^{(k+1)}$ is chosen in such a way the complementary slackness conditions are satisfied:

$$\frac{\partial c_{ij}(x_{ij})}{\partial x_{ij}}\bigg|_{x_{ij} = x_{ij}^{(k+1)}} = \pi_i^{(k+1)} - \pi_j^{(k+1)}.$$

4. Compute the surplus $g_j^{(k+1)}$ for $j = i$ and for all nodes j adjacent to i according to the flow distribution $x^{(k+1)}$.

Figure 1. k-th iteration of the relaxation method.

3. Parallel Implementations of the Relaxation Method

The relaxation method is particularly suited for both synchronous and asynchronous parallel implementation, since the iterations could be performed for many nodes concurrently; in the synchronous case, the computation proceeds in rounds of parallel iterations.

During each round, at most *p* nodes are considered (where *p* is the number of processors), from a list of candidate nodes for which the optimality conditions are not satisfied. The processors then execute in parallel the steps 2, 3, and 4 of the method described in Fig. 1 and a new round begins once all processors have finished. Such scheme works without altering the relaxation method only if we never iterate simultaneously on the dual variables of adjacent nodes. This involves the idea of a coloring of the network graph $G = \{N,A\}$ for partitioning the node set N into subsets such that any two nodes of the same color are not adjacent in the network. Since the computation of the dual variable π_i is totally independent on the values of any variables associated to nodes having the same color of i, it turns out that the synchronous scheme does not affect the course of the algorithm. Moreover, the synchronous approach works better on transportation problems, for which only two colors are required.

The parallel synchronous method is depicted in the following figure, which shows the synchronization point at the end of the surplus updating procedure. We observe that, even though at each iteration we never select adjacent nodes, we may contemporarily update the surplus of a node j, as shown in Fig. 3. In this case we need a lock to the shared memory location of the surplus to ensure that no other processor can change simultaneously the value.

One way to accomplish this is that processor i could locally determine the variation $\Delta x_{ij} = x_{ij}^{(k+1)} - x_{ij}^{(k)}$ of the flow along the arc (i,j) and update the surplus $g_j = g_j + \Delta x_{ij}$, by using a lock. When all processors complete this procedure, we will obtain $g_j^{(k+1)} = g_j$.

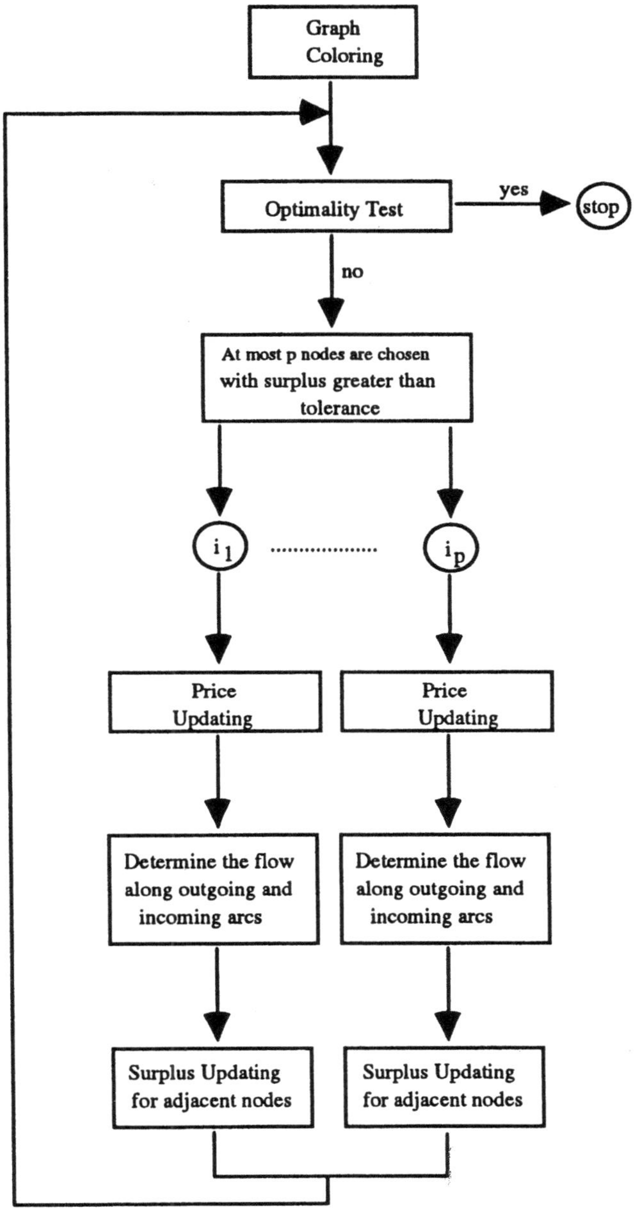

Figure 2. Sketch of the parallel synchronous relaxation method.

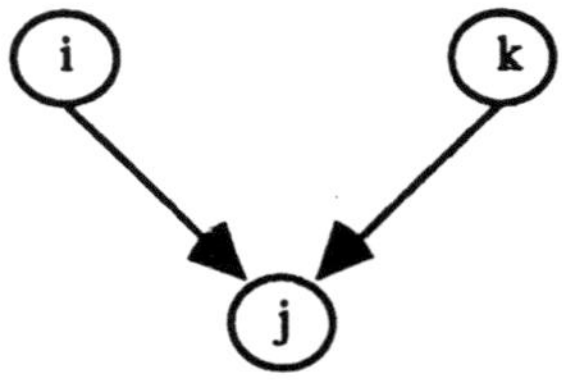

Figure 3. The surplus g_j could be updated simultaneously starting from nodes i and k.

Several variations of the procedure for selecting nodes at each iteration could be considered. In this paper two different extreme strategies are investigated:

a) cyclical node selection; the nodes are chosen in a cyclical order, by starting from the node selected at the previous iteration;
b) largest surplus selection; at each iteration, for each color ξ, the candidate nodes are the *p* nodes which correspond to the largest surplus, the sum of which is S_ξ; the selected node are chosen among the candidates by determining the color $\xi^* = \arg\max_\xi \{S_\xi\}$.

We refer to these two strategies as dynamic allocations of nodes to processors; there exists also an alternative procedure to assign a priori each node to a processor, using a procedure executed at the beginning of the first iteration; we have, in this case, a static allocation of nodes to processors. We found that, at least for transportation problems, the dynamic allocation outperforms the static allocation because of the reduced number of colors. In [2] the authors have shown that the static allocation is more preferable than dynamic for the particular class of test problems used in their experiments.

In an asynchronous algorithm there is no notion of rounds, and a new node can be considered from the candidate list by some processor while other processors are still executing on different nodes.

The convergence of the parallel asynchronous methods can be proven under weak assumptions. For more details see [1]. From the computational point of view the asynchronous methods seem to be more attractive, because they are not delayed by synchronizations requirements. In fact, we have found empirically that the asynchronous versions always outperform the synchronous counterpart. The main drawback is due to the need of a more sophisticated procedure at the end of the algorithm, through which we compute again the flow vector and the surplus for each node even if the optimality test is passed. This is due to the fact that during the steps of the asynchronous algorithms the information used by each processor in the computational phases could be outdated.

Another crucial computational aspect in the parallel relaxation method is, for each selected node i, the procedure for calculating the price π_i at k-th iteration. It could be implemented according to one of the following ideas:

a) maximization of q with respect to π_i ;
b) updating of π_i, by the relation: $\pi_i^{(k+1)} = \pi_i^{(k)} + \delta$.

According to the first possibility, we have implemented two different procedures, on the basis of which the price $\pi_i^{(k+1)}$ is determined in such a way $g_i^{(k+1)} = 0$ (in the sequel we refer to these procedures as *Newton-like method* and *exact line search*; for a more detailed description of the two methods, see [2], [3], [4], and also [5]).

It is worth while to characterize a generic arc (i,j)∈ A in the following way:

- (i,j) is called *active* if: $t_{ij} > \frac{\partial c_{ij}(x_{ij})}{\partial x_{ij}}\Big|_{x_{ij} = u_{ij}}$;
- (i,j) is called *balanced* if: $t_{ij} = \frac{\partial c_{ij}(x_{ij})}{\partial x_{ij}}\Big|_{x_{ij} = \bar{x}_{ij}}$, $l_{ij} \leq \bar{x}_{ij} \leq u_{ij}$;
- (i,j) is called *inactive* if: $t_{ij} < \frac{\partial c_{ij}(x_{ij})}{\partial x_{ij}}\Big|_{x_{ij} = l_{ij}}$.

The idea of the exact line search is that the surplus of node i is a non-differentiable, non-increasing function of the price π_i, as pictorially represented in Fig. 4. The linear segments parallel to the π-axis correspond to the range of π_i for which all arcs are either active or inactive. The nonlinear segments correspond to the range of value of π_i for which at least one arc is balanced.

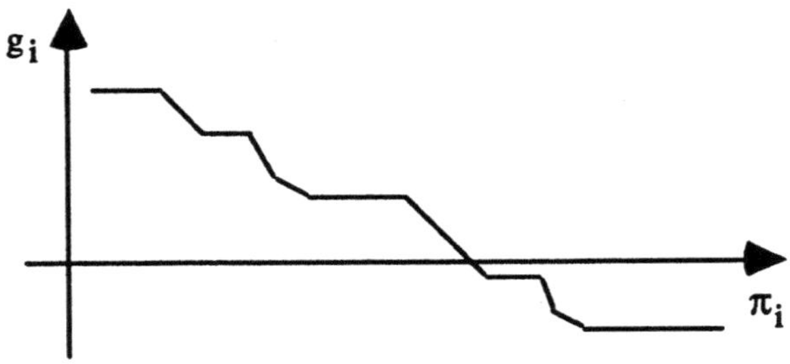

Figure 4. The surplus g_i as function of the price π_i.

We can easily implement a procedure to find the largest range of prices for which the surplus will change sign, by sorting the vector of the dual break points and using a logarithmic search to avoid the computation of the surplus in all break points. Thus, we can use a more efficient line-searching procedure in a tighter interval (the advantage of this approach is clear especially when we consider quadratic cost function for each arc; in this case it is trivial to find the price for which the surplus will be zero using a simple interpolation step between the extremes of the segment).

According to the second possibility, we have used similar approaches proposed in [2], and [3], by implementing the approximate line search of Tseng, the approximate Newton-like method, and the Bregman-step method (we generally refer to these methods as *approximate line searches*). For a more detailed description of these approaches, we remind to [2], [6], and [7] .

4. Computational Experiments

Computational experiments have been carried out with the aim to conduct an accurate comparative analysis among versions of the algorithm which use different computational strategies for updating the dual prices and for scheduling the nodes to be processed. In particular, a set of 10 transportation problems are considered. The dimension of the problems is reported in the next table. The cost function c_{ij}, $\forall\ (i,j) \in A$, is a quadratic-type. The problems are divided in two groups, each of them characterized by a fixed number of source and sink nodes. The number of arcs is selected so that the density of the networks is approximately 0.05, 0.1, 0.15, 0.20, and 0.25. All the problems have been generated by using a modified version of the public domain software named Netgen ([8] and [9]).

Test	Nodes	Sources	Sinks	Arcs
1	1000	300	700	10500
2	1000	300	700	21000
3	1000	300	700	31500
4	1000	300	700	42000
5	1000	300	700	52500
6	10000	7000	3000	105000
7	10000	7000	3000	210000
8	10000	7000	3000	315000
9	10000	7000	3000	420000
10	10000	7000	3000	525000

Figure 5. List of test problems.

We have used an Alliant FX/80, a shared memory multiprocessor system with 8 processors, each of them characterized by a peak performance of 23 Mflops and a core memory of 64 Mbytes. The compiler used is the FX/Fortran 4.2.40.

The computational results are summarized in the next two tables and figures, in which the results of the best sequential, synchronous, and asynchronous implementations of the relaxation method (in terms of execution time) are reported.

The tolerance ε is set to $\varepsilon = 10^{-3}\ \eta$, where $\eta = \frac{\sum_{i \in N} b_i}{2m}$.

The parameter η is a scaling which takes into account the difference between the demand and supply, which may substantially vary for each node;

First of all, we remark that all versions converged to a solution which satisfies the tolerance criterion. For the sequential versions, we observe that the cyclical node selection, for the first class of problems, is the most efficient, even if the number of iterations is remarkably increased. Such behavior may be explained observing that, since the cost per iteration is strictly dependent on the degree for each node, the computational workload is determined by the procedure for selecting the nodes, at least for low values of density. This means that it is preferable to implement a more sophisticated node selection strategy only when the computation of the prices will be more expensive, that is, for networks with higher values of density or for large-scale problems.

As a matter of fact, when we increase the number of nodes, the range of the density value for which the cyclical node selection strategy is more efficient, is reduced (with n=1000 the largest surplus node selection is cheaper starting with density of 0.01).

In the majority of cases, the most efficient line search procedure to compute the dual prices is the Newton-like method, a procedure which resembles the very-well-known method for nonlinear unconstrained optimization problems. This method guarantees a good balance between the need of a substantial reduction of the number of iterations (obtained with an exact line search) and the need to avoid to increase so much the computational cost per iteration.

However, the efficiency of the Newton-like method is mainly due to the particular type of the arc cost function considered in our computational experiments, assumed to be quadratic. In more general cases, we found that the exact line search procedure illustrated above is much more useful.

In the parallel case, we do not have a substantial reduction in the execution time for the synchronous versions with respect to the sequential counterpart (at least if we use as a term of comparison the best sequential implementation). This behavior can be seen in any experiment, that is, also when we increase the density of the networks, and there is not a big difference in terms of speedup using 8 processors instead of 4. However, the gap in the average cost per iteration between the two node selection strategies is relatively small, whereas in the sequential case the same difference is much more consistent. For several test problems it seems to be cheaper, in terms of execution time, using more sophisticated node selection strategies. This is essentially due to the availability of additional processors, which guarantee the possibility to use parallelism in an effective way, when the largest surplus must be computed at each iteration (this represents a remarkable computational workload especially for the sequential versions).

In our experiments we have considered also a numerical comparison between the static and dynamic allocation of the nodes. We have obtained that only with low density values the maximum speedup was achieved with the static allocation. For more significant test problems, with higher dimensions, the dynamic allocation procedure always outperforms the static counterpart. For this reason the results with the static allocation are not reported in this paper.

Also in the parallel case, the most efficient line search procedure is the Newton-like method. However, when the number of nodes and the density is increased, it seems to be more useful using the exact line search (this may be explained by the fact that the number of dual break points is increased and, as a consequence, the fact that we avoid to compute the surplus in all dual break points will be useful compared with the time spent for sorting the vector of the break points; for more details see [5])

The most important result obtained in our experiments is that the parallel asynchronous versions of the relaxation method are considerable faster that their synchronous counterparts. As a matter of fact, the values of speedup achieved are excellent for the majority of the test problems used.

A final remark is that, when the density of the networks is increasing, the speedup is decreasing. This is due to the fact that, when the density of the problems overcomes some limit, the probability that we simultaneously process adjacent nodes becomes relatively high (if we do not fix any sort in the node selection strategy, as in our parallel asynchronous codes). In this case, since it is possible that more processors attempt to update the flow on the same arc, we have to consider the extra overhead due to the synchronization mechanism introduced for locking the shared memory location during the writing procedure.

However, numerical experiments carried out on sparse graphs have shown that the availability of 8 processors is extremely useful in terms of speedup achieved.

Test Problem	Sequential	Parallel			
		Synchronous		Asynchronous	
		4 Processors	8 Processors	4 Processors	8 Processors
1	31.75 / 18205	18.67 / 4594	13.25 / 2329	10.44 / 4901	6.21 / 2880
2	52.82 / 13638	28.54 / 3439	19.35 / 1742	16.51 / 3447	9.35 / 1987
3	112.74 / 22319	59.52 / 5743	38.47 / 2870	39.05 / 5884	20.65 / 3043
4	76.47 / 5423	27.37 / 21	19.18 / 21	19.24 / 1485	11.82 / 809
5	94.91 / 4946	32.45 / 23	25.8 / 23	23.17 / 1307	14.3 / 1073

Figure 6. Time in secs / number of iterations required to solve the test problems 1, 2, 3, 4, and 5. The number of iterations in the parallel implementations is computed on the processor with the maximum workload.

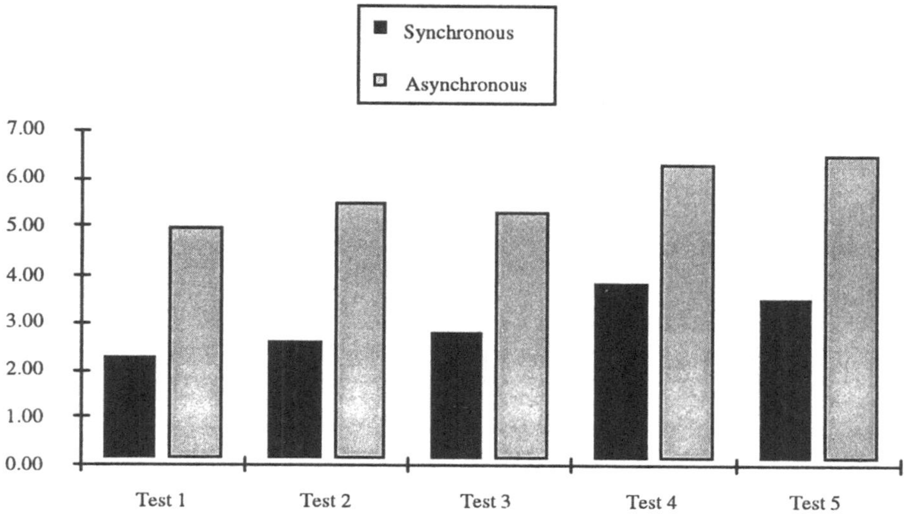

Figure 7. Speedup values for the synchronous and asynchronous parallel codes (8 processors).

Test Problem	Sequential	Parallel			
		Synchronous		Asynchronous	
		4 Processors	8 Processors	4 Processors	8 Processors
6	600.77 / 137903	318.32 / 120	196.04 / 120	165.49 / 34487	94.62 / 17252
7	1158.2 / 91611	413.34 / 109	266.02 / 109	337.16 / 26075	232.12 / 14549
8	1596.5 / 79418	537.41 / 110	355.47 / 110	558.62 / 22630	322.38 / 11256
9	1906.8 / 68333	601.52 / 53	357.25 / 53	742.65 / 19628	432.79 / 10009
10	2298.5 / 64028	709.13 / 72	401.7 / 49	645.93 / 16517	457.9 / 9348

Figure 8. Time in secs / number of iterations required to solve the test problems 6, 7, 8, 9, and 10. The number of iterations in the parallel implementations is computed on the processor with the maximum workload.

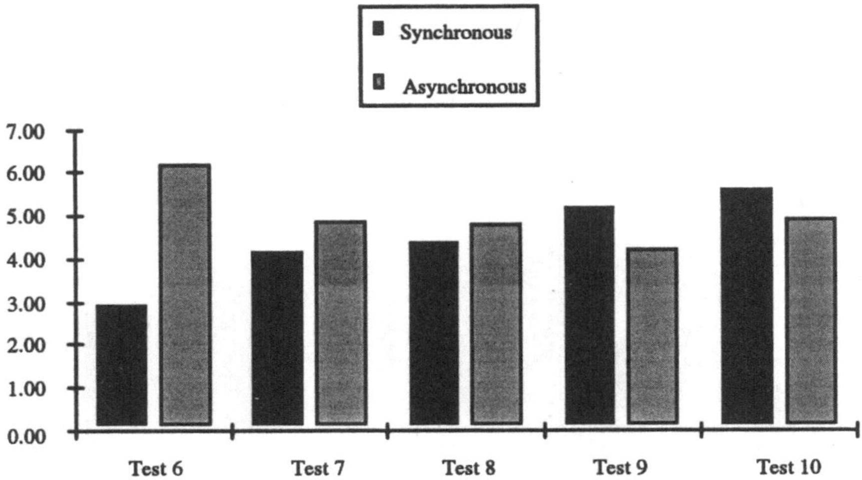

Figure 9. Speedup values for the synchronous and asynchronous parallel codes (8 processors).

Acknowledgment

We would like to acknowledge the director and the staff of CERFACS, Toulouse, France, for the possibility to use the Alliant FX/80 for obtaining our computational results and Enrico Vena, for his fruitful collaboration given to us in collecting all the results.

References

[1] Bertsekas D.P. and Tsitsiklis J.N., "Parallel and Distributed Computation (Numerical Methods)", *Prentice-Hall, Englewood Cliffs*, New Jersey, 1989.

[2] Chajakis E.D. and Zenios A.S., "Synchronous and Asynchronous Impementations of Relaxation Algorithms for Nonlinear Network Optimization", *Parallel Computing*, (17), 873-894, 1991.

[3] Zenios S.A. and Nielsen S.S., "Massively Parallel Algorithms for Singly Constrained Nonlinear Programs", *Report no. 90-03-01, Decision Sciences Department*, The Wharton School, University of Pennsylvania, Philadelphia, May 1990.

[4] Zenios S.A. and Mulvey J.M., A Distributed Algorithm for Convex Network Optimization Problems, *Parallel Computing* (6), 43-56, 1988.

[5] Zenios S.A. and Mulvey J.M., "Relaxation Techniques for Strictly Convex Network Problems", *Annals of Operations Research*, Vol. 5, pp. 517-538, 1985/6.

[6] Guerriero F., and Musmanno R., "Un algoritmo parallelo iterativo per il problema di flusso a costo minimo non lineare", *Technical Report C.N.R. no. 1/164* - Progetto Finalizzato "Sistemi Informatici e Calcolo Parallelo", Sottoprogetto 1: "Calcolo Scientifico per Grandi Sistemi", 1993.

[7] Grandinetti L., Guerriero F., and Musmanno R., "Parallel Techniques for Large-Scale Nonlinear Network Optimization", in: *Lecture Notes in Computer Science* (796), High Performance Computing and Networking, Wolfgang Gentzsch, Uwe Harms (Eds.), Springer Verlag, 1994.

[8] Klingman D., Napier A., and Stutz J., "NETGEN - A Program for Generating Large Scale (Un)Capacitede Assignment, Transportation and Minimum Cost Flow Network Problems", *Management Science* (20), 814-822, 1974.

[9] Musmanno R., and Vena E., "GRICORN: un programma per la generazione di reti di flusso basato sul software di pubblico dominio Netgen", *Technical Report C.N.R. no. 1/183* - Progetto Finalizzato "Sistemi Informatici e Calcolo Parallelo", Sottoprogetto 1: "Calcolo Scientifico per Grandi Sistemi", April 1994.

DATA MANAGEMENT OF LARGE-SCALE TRANSPORTATION NETWORKS

N. Christofides H.O. Badra Y.M. Sharaiha G. Salkin

Operational Research and Systems - The Management School
Imperial College of Science, Technology and Medicine
53 Prince's Gate, London SW7 2PG, United Kingdom

There has been growing interest in computer graphics and Geographical Information Systems, which has made it necessary to develop efficient data structures and algorithmic procedures for handling large-scale graphical information in real-time. A number of tree data structures have been successfully applied to manage vector images and graphics data. Among them, the k-d tree (Bentley [1975]), 2^N-division tree (Samet [1980] and Meagher [1982]), and BD-tree (Ohsawa and Sakauchi [1983]) are noteworthy from the view point of coping with a large amount of data in a dynamic environment. In particular, such representations are computationally efficient in terms of retrieval, processing and manipulation. It is often the case, particularly in the representation of large-scale networks, that the entire tree cannot be maintained in main memory, and thus paging from disk becomes necessary. A variety of schemes have been proposed (see *eg,* Shaffer *et al* [1993] and Henrich *et al* [1990]) to overcome this problem by paging from disk. In this paper, we first present a new dynamic structure for the representation of large-scale networks (F-tree) and then present a scheme for its memory management.

1. Data Structures for Transportation Networks

In Christofides, Badra & Sharaiha [1993], we developed two complementary data structures for the representation of networks: a *topological* and a *topographical* representation. Together, they support efficient graph-theoretic and geometric operations. A topological representation facilitates the implementation of graph-theoretic optimization algorithms under dynamic on-line editing operations. A new graph structure, termed the *Dynamic Forward Star (DFS),* was developed and based on the well-known *(Static) Forward Star (SFS)* so the algorithms in the literature can be readily modified. A topographical representation is necessary to support efficient spatially-based geometric operations that require a different labelling scheme from that of the graph structure. We classify geometric operations into two categories: (i) Standard geometric operations of network *editing* which involve the identification of *"nearest"* arc (and/or vertex), and their *insertion (or deletion)*; and (ii) advanced geometric operations involving high-speed *range retrieval*, and *splitting* and *merging* of sub-networks.

The topographical data structure, termed the *Folded tree (F-tree)*, presented in Christofides *et al* [1993] is an extension of the *Binary Division tree (BD-tree)*, first introduced by Ohsawa & Sakauchi [1983]. The F-tree is used for the representation of networks by recursive partitioning of the space into rectangular regions. Figure 1 presents an example of an F-tree. Figure 1(a) shows the spatial decomposition of points (vertices or arc centroids of the network), and Figure 1(b) shows the tree structure representing the decomposition process. Each region is constructed dynamically by adhering to some predetermined rules regarding decomposition. The resulting *binary tree* structure is composed of *internal* and *leaf* nodes. The tree structure continues to grow as additional data is represented. A *zone identifier* is used to illustrate how the regions have been partitioned. This is a binary string generated by assigning (recursively) 0 if the region lies in the left half of the bisector parallel to the y axis or lower half of the bisector parallel to the x axis, and assigning 1 if it corresponds to the right or upper half. The binary sequence also represents the path from the root node to any other node, with 0 for the path to the left node and 1 for the path to the right node. Moreover, each internal node also has a *circumscribed quadrilateral* which includes all data located under the node (see, for example, nodes 5 and 11 in Figure 1(b)). In that paper, algorithms for the application of the above geometric operations on the F-tree were presented. The data structure was evaluated on the following functions: (i) The properties of the tree structure; (ii) the efficiency of the data structure in supporting standard geometric operations of insertion, deletion and nearest arc; and (iii) the efficiency of the data structure in performing advanced geometric operations of range retrieval. The data structure was tested on a case study of the road map of Italy. Figure 4 illustrates the data-file content of the largest network size (case I), with an image size of $(2^{15} x 2^{15})$, and a graph size of *n_12603* vertices and *m_20693* arcs. The three network sizes (II, III, and IV) represent sections of the complete road network (I) reflecting smaller regions chosen in the northern part of Italy (36, 16 and 8% of the original network size). Table 1 includes the main characteristics of the constructed vertex and arc F-trees for all network sizes. Computational results for range retrieval on the complete road network (I) are presented in Section 4.

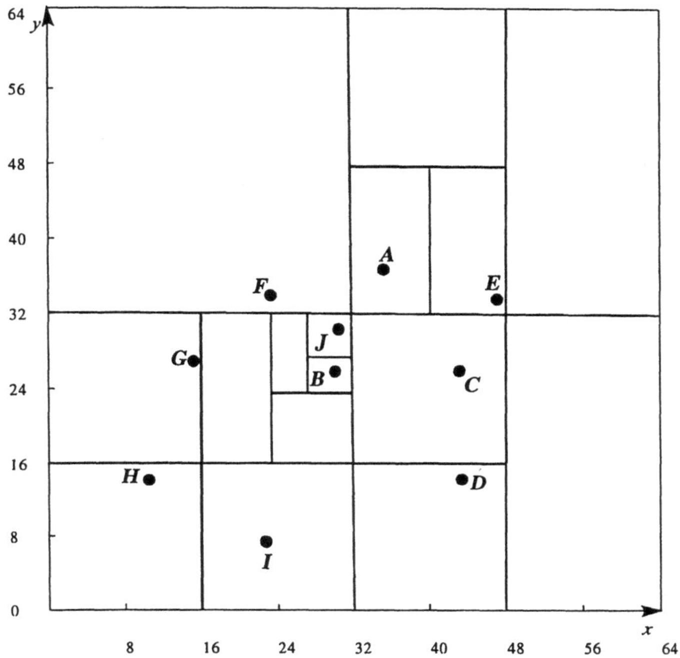

(a) Space decomposition

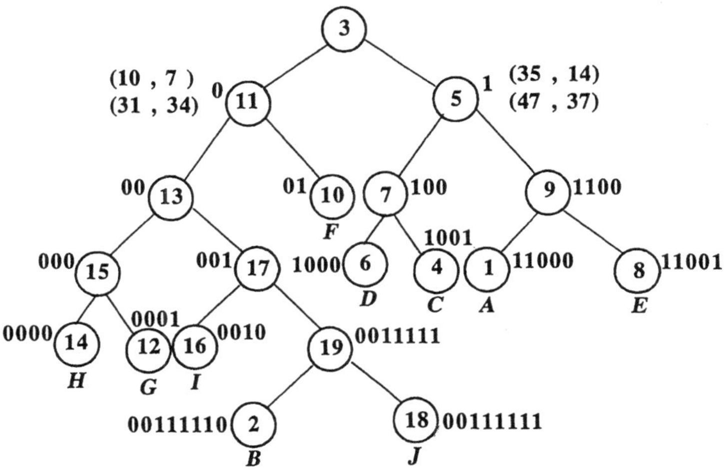

(b) Tree structure

Figure 1. Example of the F-tree structure using binary division of the space

2. Management of the Data Structure

For large-scale networks (*eg*, size I), it is often the case that the entire F-tree cannot be stored in *main memory* and *paging* from *disk* becomes necessary. In this section, we propose a procedure for efficient memory management based on a systematic partitioning of the tree during its construction into smaller *subtrees*, each subtree representing a corresponding *subnetwork*. A *tree of subroots* is created dynamically which has the property of preserving the *neighbourhood* relationships of the original F-tree. This scheme becomes particulary suitable for search algorithms in which a minimum number of external pages must be traversed since paging from disk is computationally expensive. Figure 2 represents the tree of subroots for the F-tree of Figure 1(b). The tree of subroots is constructed as follows. A fixed *capacity* Q is chosen to be the maximum number of nodes in each subtree. A subtree is defined by the tree rooted at its subroot upto, but not including, any other subroot. A *weight* $w(i)$ is associated with each node i, where $w(i)$ is the number of nodes in the subtree rooted at i, including i itself. Clearly, the weight of any leaf node is equal to one. Next, we describe the management of the tree of subroots.

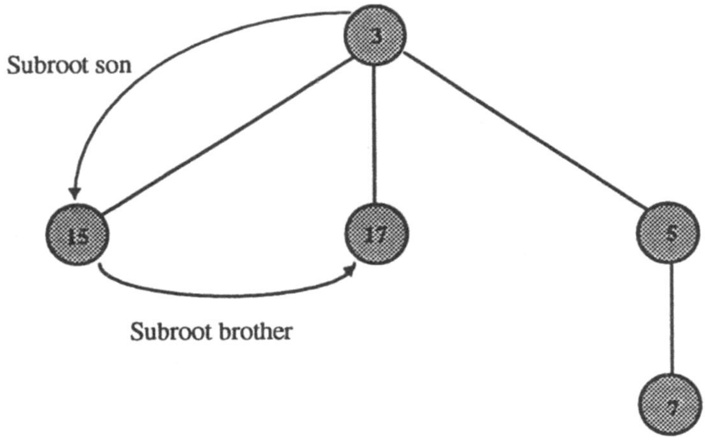

Figure 2. Tree of subroots for the F-tree in figure 1

2.1 Subtree Management

Let r be the subroot of a subtree $T_r \subset T$, where T is the entire F-tree. For any subroot r, $w(r) \leq Q$, where Q is user-defined such that it utilizes the maximum storage space occupied by a given subtree to be paged to and from disk. Furthermore, we aim to maintain the average weight of each subroot r to be approximately $\frac{Q}{2}$. We also assign a lower bound L on the size of each subtree. We choose $L\_\frac{Q}{4}$ in order that we limit the number of subtrees that contain a small number of nodes. This is consistent with the requirement of maintaining a balance in the amount of data stored in each subtree and thus reduce the number of major adjustments on *overflow* and *underflow* of pages under dynamic updates to the tree.

Splitting of two subtrees

Let r be the subroot of the subtree T_r. If at any time, for a subroot r , $w(r) > Q$, then the subtree rooted at r is *split* into two subtrees, an *upper level* subtree $T_{r'}$ and a *lower level* subtree T_{r^*}. This may occur after an addition of a node, or after a deletion of a node leaving a subtree with very few nodes which may then be a candidate for merging with the tree above which, in turn, would create the upper bound bottleneck. The choice of r^*, the subroot of the lower level subtree, is dependent on the node weights. As noted above, this is chosen to be the node in the subtree to be split which has a weight closest to $\frac{Q}{2}$. This subroot is identified by starting with r and traversing down the upper level subtree along the branch containing the heavier node, until the split node (or new subroot) is met. Figure 3(a) illustrates how the two subtrees, $T_{3'}$ and $T_{1'7}$, are split from T_3. The weight of all nodes on the path (highlighted in Figure 3(a)), between the two subroots (nodes 3 and 17) is updated.

Merging of two subtrees

Let r^* and r be two subroots such that r is the subroot of the subtree just above r^*. If at any time, $w(r^*) \leq L$ then we check if $w(r^*)+w(r) \leq Q$. If this is the case, then the subtrees T_{r^*} and T_r are *merged* into $T_{r'}$ only. This may occur after deletion of a node from the subtree rooted at r^* which has very few nodes. However, if r^* is the main root of T then r does not exist and it is excluded from merging; thus no action takes place. On the other hand, the root may be involved in merging if it is r. The two subtrees, T_{r^*} and T_r , are also merged if $w(r^*)\_1$, regardless of whether T_r has reached its capacity.If the resulting tree has $w(r) > Q$, it is split using the splitting rules described before. Figure 3(b) illustrates how two subtrees, T_3 and T_{17}, are merged.The weight of all nodes on the path (highlighted in Figure 3(b)) between the two subroots (nodes 3 and 17) is updated by adding the weight of the subroot in the lower level subtree (node 17) to all node weights on that path.

Tree of subroots

As illustrated in Figure 2, the tree of subroots constructed using the rules above (splitting and merging) is generally not a binary tree. However, the *pointer system* of the tree of subroots preserves the original concept of the F-tree. Two pointers are required for each subroot to store the tree of subroots.

Subroot Son: The subroot son pointer of an internal node points to its *left son* node in the tree of subroots. In the case of a leaf node, the subroot son pointer points to itself, flagged by a negative sign for distinction from the internal nodes.

Subroot Brother: The subroot brother pointer of a node points to its immediate *right brother* node. If the node has no right brother, then the subroot brother pointer points to its *father* node, and is flagged by a negative sign for distinction. The subroot brother pointer of the root node is set to zero.

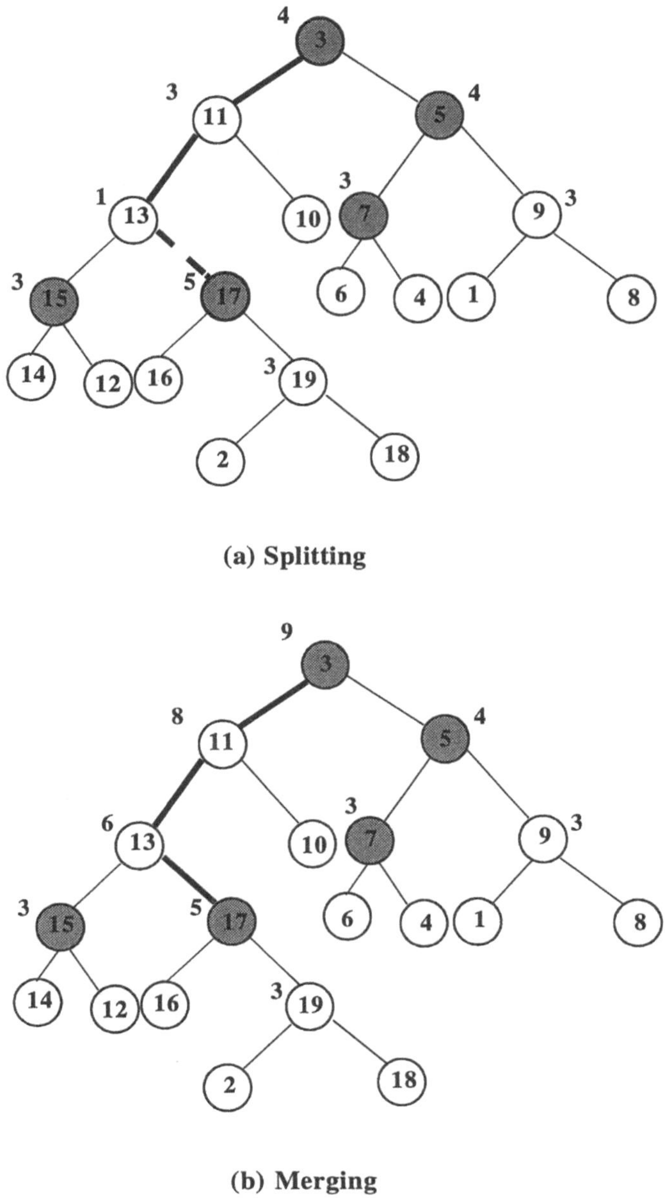

Figure 3. Updating of weights after (a) splitting; (b) merging

2.2 Cache Management

In this section, we describe a caching system for maintaining data in main memory and addressing data on disk as required.

2.2.1 Data on Disk

The entire F-tree is stored on disk in a *subtree-file* which is divided into a collection of *pages*. Each page contains a complete self-contained subtree indexed by its subroot node number in the tree of subroots, and contains all information on each of its nodes.

2.2.2 Data in Main Memory

The tree of subroots (*subroots-table*) and a subset of the total number of subtrees and their associated information (*nodes-table*) are both kept in main memory.

Nodes-table

The nodes-table is broken into a collection of *blocks*. Each block holds the latest information on a single subtree. The number of subtrees kept in main memory at any one time and therefore the size of the nodes-table can be chosen according to the type of machine used. In order to perform all editing operations in main memory prior to splitting of subtrees, we choose the capacity of each block to be $(Q+2)$. The nodes of a subtree are maintained in sorted order in each block so that a *binary search* can be used to locate a particular node within a block once it is addressed

Subroots-table

The tree of subroots is always kept in main memory in the subroots-table. For each subroot, the table consists of the following arrays:

(i) The *subroot* node number in the F-tree.

(ii) The *page* number in the subtree file, indicating the storage position of that subtree on disk.

(iii) The *block* number in the nodes-table, indicating the storage position of that subtree in main memory. If the subtree is not in main memory, this is set to zero.

(iv) The pointers *Subroot Son* and *Subroot Brother* required to define the structure of the tree of subroots as described above.

(v) The *ranking* of the subtree according to the "*least recently used*" rule. In addition, a negative sign identifies whether the subtree has been *modified* since it was last read in to main memory.

2.2.3 Mode of Caching

Once a node is addressed and is not in main memory, the subtree which contains it is accessed from disk. The mode of *caching* we use is first based on the identification of whether the node in question is in main memory or on disk. We represent the node number using a 32-bit binary string. The first 19-bits of this binary string contain the actual node number and the remaining 13-bits contain the subroot node number to which the node belongs. Thus, given this representation, we can identify from the subroots-table whether the node is in main memory or on disk using the Block*(.)* array. If the node is not in main memory, the subtree on disk which contains it is cached. It then replaces a subtree in main memory based on the "least recently used". If the node is in main memory, a binary search is carried out on its node-table block to locate the node and its details.

3. Computational Results

In this section, we examine the properties of the subtrees and the tree of subroots created by applying the partitioning scheme described above. We also investigate the efficiency of this scheme in implementing range retrieval (or zooming) algorithms when caching from disk is necessary. This is the case for large zoom areas of the road map of Italy (Figure 4), which was used as a case study. The retrieval algorithm implemented was run on a PC 80486 co-processor.

In Table 2, the main properties of the resulting vertex and arc subtrees for varying values of Q are presented. The resulting average size (column (2) and (6)) is consistent with the aim of creating subtrees of average size approximately equal to $\frac{Q}{2}$. It is also guaranteed that each subtree has more than $\frac{Q}{4}$ nodes. This is an important factor since *ill-balanced* subtrees may cause inefficient memory utilization for node splitting and merging. Table 2 also provides an indication of the balance of each of the subtrees in terms of maximum (column (3) and (7)) and average (column (4) and (8)) *height*, where the height is defined by the branch cardinality of the path from the subroot to a leaf node. The experimental results support the claim of good balance using this partitioning scheme. The main properties of the resulting vertex and arc tree of subroots are also presented in Table 3 for varying values of Q. In all experimentation, we use a constant number of blocks (i.e., subtrees in main memory) to be fifty for all values of Q. The tree of subroots constructed is then used for testing the retrieval algorithm.

In testing the retrieval algorithm, we specify a rectangular *window* varying in size from 5 to 80% for zooming. For each window size, the number of arcs which overlap wholly or partly with that window is identified. We note that this wide window size variation allows for a balanced testing of this algorithm since the output number of arcs, given as a percentage of the total number of arcs, varies between less than 1% (for window size of 5%) upto 80% (for window size of 80%). Table 4 shows the real-time efficiency of the procedure. The table also shows how the CPU time decreases rapidly as the maximum size of each subtree increases.

4. Conclusion

The primary motivation for our work was to effectively manage the topolographical storage of large-scale transportation networks requiring the use of paging to and from disk. This paper presented an overview of a new scheme for efficient memory management of the F-tree data structure by partitioning the entire tree in to smaller subtrees. We have described this partitioning scheme in relation to the F-tree and presented a memory-managed data structure for its implementation. However, with minor modifications, the scheme can be adapted to manage a variety of tree-based data structures. We showed how each subtree can be handled as a separate entity of the network each indexed by a subroot. Moreover, the tree of subroots preserves the topology and neighbourhood relationships of the original tree. The scheme was tested against road map data from Italy. Computational results presented include the characteristics of the tree-based structures and confirm the real-time efficiency of the retrieval algorithm.

References

[1] Bentley, J. (1975), "Multidimensional binary search trees used for associative searching", *Communications of the ACM,* Vol. 18, No. 9, 509-517.

[2] Christofides, N., Badra, H.O. and Sharaiha, Y.M. (1993), "Data structures for topological and geometric operations on large-scale networks", *Imperial College Technical Paper No SWP9326/OR.*

[3] Henrich, A., Six, H. and Widmayer, P. (1990) "Paging binary-trees with external balancing", *Lecture Notes in Computer Science,* Vol. 411, 260-276.

[4] Meagher, D. (1982), "Geometric modelling using octree encoding", *Computer Graphics and Image Processing,* Vol. 19, No. 2, 129-147.

[5] Ohsawa, Y. and Sakauchi, M. (1983) "The BD-tree - A new N-dimensional data structure with highly efficient dynamic characteristics" *The Proceedings of the IFIP 9th World Computer Congress,* 539-544.

[6] Samet, H. (1980) "Region representation: quadtrees from boundary codes", *Communications of the ACM,* Vol. 23, No. 3, 163-170.

[7] Shaffer, C. and Brown, P. (1993) "A paging scheme for pointer-based quadtrees", *Lecture Notes in Computer Science,* Vol. 692, 89-104.

Figure 4. Road map of Italy: Network size I

Table 1. Dimensions of the constructed vertex and arc F-trees

Network Size	Vertex F-tree			Arc F-tree		
	(1)	(2)	(3)	(4)	(5)	(6)
	# Nodes	Max height	Avg height	# Nodes	Max height	Avg height
I (100%)	25205	28	16.52	41385	29	17.42

Legend:

(1) and (4) : Number of nodes in the vertex and arc F-trees.

(2) and (5) : Maximum height (in number of branches in the path from the root to the leaf nodes) of the vertex and arc F-trees.

(3) and (6) : Average height (in number of branches in the path from the root to the leaf nodes) of the vertex and arc F-trees.

Table 2. Dimensions of the constructed vertex and arc subtrees

Max Capacity Q	Vertex Subtrees				Arc Subtrees			
	(1)	(2)	(3)	(4)	(5)	(6)	(7)	(8)
	Max size	Avg size	Max height	Avg height	Max size	Avg size	Max height	Avg height
20	19	11.77	14	5.89	19	11.57	13	5.78
60	59	35.06	31	17.53	59	34.49	33	17.24
100	99	58.34	51	29.17	99	57.80	56	28.90
400	389	223.05	197	111.52	399	237.84	200	118.92
600	583	315.06	292	157.53	599	339.22	300	169.61

Legend:

(1) and (5) : Maximum size of the vertex and arc subtrees.

(2) and (6) : Average size of the vertex and subtrees.

(3) and (7) : Maximum height of the vertex and arc subtrees.

(4) and (8) : Average height of the vertex and arc subtrees.

Table 3. Dimensions of the constructed vertex and arc tree of subroots

Max Capacity *Q*	Vertex Tree of Subroots				Arc Tree of Subroots			
	(1)	(2)	(3)	(4)	(5)	(6)	(7)	(8)
	Max height	Avg height	Max width	Avg width	Max height	Avg height	Max width	Avg width
20	13	5.75	471	164.62	11	6.20	896	325.09
60	9	4.63	158	79.78	12	5.14	277	99.92
100	8	4.30	122	48.11	9	4.11	208	79.44
400	6	3.57	34	18.67	7	3.33	58	24.71
600	8	3.40	26	9.88	7	3.28	41	17.29

Legend:

(1) and (5) : Maximum height of vertex and arc tree of subroots.

(2) and (6) : Average height of vertex and arc tree of subroots.

(3) and (7) : Maximum width of vertex and arc tree of subroots.

(4) and (8) : Average width of vertex and arc tree of subroots.

Table 4. Computational time (CPU sec) for range retrieval

Max Capacity Q	Window Size				
	5%	10%	20%	40%	80%
20	0.22	0.33	0.66	3.46	24.78
60	0.22	0.22	0.33	1.65	11.32
100	0.22	0.22	0.39	1.37	8.68
400	0.22	1.65	0.22	1.15	5.28

WHY REGULATE PRICES IN FREIGHT TRANSPORTATION MARKETS?

W.J. Hurley

Department of Business Administration,
The Royal Military College of Canada and School of Business,
Queen's University, Kingston, Ontario, Canada

E.R. Petersen

School of Business, Queen's University,
Kingston, Ontario, Canada

In bulk commodity rail freight markets (grain, coal, potash, etc.), specific origins and destinations are served by one, or perhaps two, carriers. Economic theory suggests that, when supply is limited, prices are not likely to be competitive. Applying this wisdom, railroads are likely to charge high prices, and hence shippers (producers) will move quantities less than the socially optimal quantities. This is the argument for the two standard economic solutions: Marginal Cost Pricing and Ramsey Pricing. We examine the efficiency of Marginal Cost Pricing and Ramsey Pricing relative to Deregulated Prices, those which obtain when carriers and shippers (producers) are free to contract without regulatory impediment. Our model suggests that:

- Marginal Cost Prices will not enhance economic efficiency relative to Deregulated Prices; and
- Ramsey Prices will degrade economic efficiency relative to Deregulated Prices.

Hence the standard regulatory pricing solutions do not enhance efficiency relative to a regime of commercial freedom.

1. Introduction

In the future, it is unlikely that transportation firms, particularly railroads, will escape the well-meaning scrutiny of economists and legislators bent on ensuring that these firms do not take advantage of market power, especially in markets where producers are captive.

To suggest otherwise would be to ignore the lessons of economic history and the cyclic nature of regulatory sentiment.

In bulk commodity rail freight markets (grain, coal, potash, etc.), specific origins and destinations are served by one, or perhaps two, carriers. Economic theory suggests that, when supply is limited, prices are not likely to be competitive. Applying this wisdom, railroads are likely to charge high prices, and hence shippers (producers) will move quantities less than the socially optimal quantities. This is the argument for the two standard pricing solutions: Marginal Cost Pricing and Ramsey Pricing.

In this paper we examine the efficiency of Marginal Cost Pricing (MCP) and Ramsey Pricing (RP) relative to Deregulated Prices (DP), those which obtain when carriers and shippers (producers) are free to contract without regulatory impediment. Our model suggests that:

- Marginal Cost Prices will not enhance economic efficiency relative to Deregulated Prices; and
- Ramsey Prices will degrade economic efficiency relative to Deregulated Prices.

These strong statements fly in the face of the conventional wisdom about regulation in freight transportation markets.

The key to these results is the way we model shipper-carrier interaction. Most, if not all, welfare analyses of freight rates derive from two assumptions. First, "consumer" behavior is summarized by a demand curve for freight transport. And second, the supply side is characterized by one of a variety of industrial structures, e.g. monopoly, duopoly, perfect competition, etc. The results of this modeling approach are well known. For instance, if a carrier has a monopoly in a particular OD market, shippers will pay the monopoly price, and therefore ship less than the social optimum. Accordingly, there is room for a regulatory authority to force the carrier to charge a lower price.

Our approach does not begin with the usual demand curve for transportation. Typically such a demand curve summarizes the requirements of a group of shippers in a particular OD market. Rather, we focus on one shipper. The profit function of this shipper is modeled explicitly, so that his demand for transportation is implicit in his first-order condition for profit maximization. Hence we focus on the way a single shipper contracts with one or more carriers. This permits us to consider the welfare problem in the context of the vertical exchange literature. Tirole (1988) provides an excellent introduction to this literature.

We study a bilevel game in which carriers choose tariffs, and based on these tariffs, each shipper decides how much output to move to market over each carrier. Not surprisingly, the optimal form of carrier tariff is nonuniform. That is, a carrier can always make a higher profit by charging a nonuniform unit price rather than a constant unit price. In relatively simple vertical structures, a characteristic of this optimal nonuniform price is that *the shipper will choose the same output he would have had the carrier priced a marginal cost.* Thus, relative to Deregulated Prices, economic efficiency is not enhanced by moving to a regime of Marginal Cost Pricing.

A similar argument applies to Ramsey Pricing. In the case where the carrier is forced to use a Ramsey Price which exceeds marginal cost, the shipper will choose a *lower* output than he would have had the carrier been free to choose a profit-maximizing price. Thus the imposition of Ramsey pricing degrades economic efficiency.

2. A Digression on Welfare and Pricing

The argument for Marginal Cost Pricing is well known. In a perfectly competitive economy, prices will be equated with marginal cost and efficiency will be maximized. However, if a market is imperfectly competitive, there is a role for market intervention. If a single imperfect market is viewed in isolation, the usual solution is to force the offending firm(s) to price at marginal cost, and thereby enhance economic efficiency. Of course, if there is more than one imperfect market in an economy, then it may not be optimal to impose marginal cost pricing in these markets. This argument is outlined in the seminal work of Lipsey and Lancaster (1956).

The rationale for Ramsey Pricing in freight transport markets, particularly rail markets, is based on the nature of a carrier's costs. Suppose a carrier has a production technology exhibiting increasing returns to density or scale. If the carrier is forced to price at marginal cost, it will suffer a financial loss because average cost will be below marginal cost. Thus, if a regulator wished to maximize efficiency by forcing the carrier to price at marginal cost, the carrier would soon be bankrupt. Ramsey prices are intended to solve this problem: they maximize economic efficiency *subject to the constraint that the carrier must breakeven*. In general, Ramsey prices are uniform prices set above marginal cost. They have the property that the carrier's margin as a percentage of price, (P-MC)/P, at the Ramsey Price, varies inversely with the elasticity of transport demand.

Ramsey pricing has its origin in Frank Ramsey's (1927) seminal paper. Among others, Baumol and Bradford (1970) refined Ramsey's ideas. ŀThe suggestion that Ramsey pricing be applied to transportation markets is more recent. For instance, Braeutigam (1979) shows how to construct Ramsey prices where two transportation firms offer differentiated products (service) in the same market. Roberts (1983) provides a good discussion of Ramsey prices in the context of railroad maximum rate control. In the early eighties a number of prominent economists (Baumol (1981), Willig (1981), Goldfeld (1981), Arrow (1981), Moses (1981), Braeutigam (1981) and Wecker (1981)) argued that the application of Ramsey prices to railroad ratemaking, the so-called Inverse Elasticity Rule, offered a number of advantages. Faulhaber and Baumol (1988) comment on the importance of Ramsey pricing:

> Ramsey pricing is a clear example of a principle that derives from the literature and has (recently) achieved a good deal of attention among government agencies. Indeed it casts its shadow on

> virtually every hearing on price regulation by a federal agency in the United States, and it has apparently arisen in similar circumstances in the United Kingdom, Canada, France, Australia, and New Zealand. (page 594.)

They go on to assess the economic profession's acceptance of Ramsey prices:

> Eventually, however, frequent reiteration by economists in regulatory proceedings and the profession's general (but not perfectly complete) acceptance of Ramsey pricing as the theoretically correct rule for regulation of the prices of a multiproduct monopoly left an impression on regulators.(page 595)

One of the notable exceptions to this support of Ramsey pricing is the paper by Tye and Leonard (1983). They argue that, since freight transportation demand elasticities are hard to measure, freight rates based on Ramsey prices would be subject to substantial error, so that Ramsey price regulation may lead to an efficiency loss as a result of measurement error.

3. A Theory of Deregulated Prices

The theory is motivated by a typical market for bulk commodity transportation. Canadian potash is moved from a small number of Saskatchewan mineheads to a collection of some 5,000 fertilizer plants in the U.S. Midwest by Canada's two major railroads, CN and CP. Here, we refer to a potash mine as a *shipper*, CN and CP as *carriers*, and fertilizer plants in the Midwest as *consumers*. Sometimes we refer to this fertilizer market as a *delivered market*.

Consider two carriers competing to move a shipper's freight to consumers at a delivered market. We do not restrict the carriers to have the same costs. Assume that the cost function of the lower cost carrier, termed the LC-carrier, is given by $m_L x_L$, where x_L is the output the shipper moves with the LC-carrier. This cost function is an incremental cost function. It gives the addition to the LC-carrier's total costs as a result of the shipper using the carrier, holding all other traffic on the LC-carrier's network fixed. The cost function of the higher cost carrier, termed the HC-carrier, is $m_H x_H$, where x_H is the output moved with the HC-carrier, and $m_H \geq m_L$. In all, the shipper moves $x = x_L + x_H$ to the delivered market.

Consumers are prepared to pay the shipper a price $p_0(x)$ if output x is moved to this market. Hence, shipper revenue is $xp_0(x)$. We assume that $p_0(x)$ is a strictly monotone decreasing function.

The shipper has two costs. $C_0(x_L + x_H)$ is the shipper's cost of manufacturing, and represents all costs to get the product to the plant door. The other cost is what the shipper must pay the carriers to move his product to the delivered market. Let $T_L(x_L)$ be the total amount the LC-carrier charges the shipper to move x_L units of output to the delivered market. We refer to $T_L(x_L)$ as a *tariff* or *contract*. In a similar fashion, we define $T_H(x_H)$.

Given these assumptions, the shipper's profit is given by

$$\pi_0(x_L, x_H | T_L, T_H) = (x_L + x_H) p_0(x_L + x_H) - C_0(x_L + x_H) - T_L(x_L) - T_H(x_H). \tag{1}$$

The LC-carrier's profit is

$$\pi_L(T_L | x_L, x_H) = T_L(x_L) - m_L x_L. \tag{2}$$

and the HC-carrier's profit is

$$\pi_H(T_H | x_L, x_H) = T_H(x_H) - m_H x_H. \tag{3}$$

The bilevel structure of the game is as follows. The two carriers form the upper level, each selecting a tariff. The shipper forms the lower level, and given the carrier tariffs, selects an output to move over each carrier. Suppose $(\hat{x}_L, \hat{x}_H)$ solves the shipper's problem.

At the higher level, the LC-shipper solves

$$\max_{T_L} \pi_L(T_L | \hat{x}_L, \hat{x}_H), \tag{4}$$

and the HC-carrier solves

$$\max_{T_H} \pi_H(T_H | \hat{x}_L, \hat{x}_H), \tag{5}$$

and these problems are solved simultaneously. We label this game $G(2C,1S)$ and suppose an equilibrium for it is $(x_L^B, x_H^B, T_L^B, T_H^B)$.

At this point we discuss some of the implicit assumptions of the model. First, we assume common knowledge - all players (carriers and shipper) know the parameters of each other's profit function, and, furthermore, all players know that the other players know this, and so on. A second assumption concerns the nature of carrier costs. To this point, we have assumed that carrier cost functions are linear. However, at a later point in the paper, we will extend our arguments to general cost functions, $C_L(x_L)$ and $C_H(x_H)$. These cost functions are assumed to be incremental cost functions. That is, $C_i(x_i)$ represents the incremental cost to carrier i of moving x_i units across carrier i's network. Note that this cost function does not depend on other traffic on carrier i's network. Hence we are assuming that this bulk commodity traffic does not impose costs on other traffic the carrier moves. In other words, there are no network externalities. Finally, we remark that the model covers the case of intermodal competition. For instance, the LC-carrier might be barge service and the HC-carrier, unit train service.

Before identifying the equilibrium, we need to define several surplus functions. Let $\Gamma(x_L, x_H)$ be the vertically integrated profit function of all three players:

$$\Gamma(x_L, x_H) = (x_L + x_H) p_0(x_L + x_H) - C_0(x_L + x_H) - m_L x_L - m_H x_H. \tag{6}$$

We assume that $\Gamma(x_L, x_H)$ is maximized at (x_L^*, x_H^*) where $\Gamma(x_L^*, x_H^*) > 0$.

The vertically integrated profit of the LC-carrier and shipper, assuming the HC-carrier moves 0 output, is $\Gamma(x_L,0)$. We assume that $\Gamma(x_L,0)$ is maximized by $\bar{x}_L^* > 0$ where $\Gamma_L^* = \Gamma(\bar{x}_L^*,0)$. In a similar way we define the vertically integrated profit of the HC-carrier, assuming the LC-carrier moves 0 output. We assume that $\Gamma(0,x_H)$ is maximized by $\bar{x}_H^* > 0$ where $\Gamma_H^* = \Gamma(0,\bar{x}_H^*)$. We are now in a position to give the equilibrium.

Proposition 1. *Suppose $m_L < m_H$. Then an equilibrium for $G(2S,1C)$ is*

$$
\begin{aligned}
&x_L^B = x_L^* \\
&x_H^B = 0 \\
&T_L^B(x_L^B) = \begin{cases} \Gamma^* - \Gamma_H^* - m_L x_L & \text{if } x_L > 0 \\ 0 & \text{if } x_L = 0 \end{cases} \qquad (7) \\
&T_H^B(x_H) = m_H x_H .
\end{aligned}
$$

Proof Consider the case where the carriers have generalized cost functions, $C_L(x_L)$ and $C_H(x_H)$, and suppose that an optimal form of carrier tariff is $T_i(x_i) = b_i + C_i(x_i)$ for $i=$ L,H. Given these tariffs, the shipper has 3 alternatives: he can use both carrier tariffs, or only the LC-carrier's tariff, or only the HC-carrier's tariff.

Suppose the shipper chooses to use both tariffs. Then we can write the shipper's profit function as

$$\pi_0(x_L,x_H) = \Gamma(x_L,x_H) - b_L - b_H, \qquad (8)$$

and, if b_L and b_H are sufficiently small, the shipper will choose (x_L^*,x_H^*) since, by definition, this point maximizes $\Gamma(x_L,x_H)$. At this output profits are

$$
\begin{aligned}
\pi_0(x_L^*,x_H^*) &= \Gamma^* - b_L - b_H \\
\pi_L(x_L^*) &= b_L \qquad (9) \\
\pi_H(x_H^*) &= b_H
\end{aligned}
$$

Now suppose b_L and b_H are such that it is optimal for the shipper to choose only the LC-carrier. Then shipper profit is

$$\pi_0(x_L, x_H) = \Gamma(x_L, 0) - b_L, \tag{10}$$

so the shipper chooses $\bar{x}_L^*$, and profits are

$$\begin{aligned} \pi_0(\bar{x}_L^*, 0) &= \Gamma_L^* - b_L \\ \pi_L(\bar{x}_L^*) &= b_L \end{aligned} \tag{11}$$

The final possibility is that the shipper chooses to use only the HC-carrier. In this case, output is $\bar{x}_H^*$ and profits are

$$\begin{aligned} \pi_0(0, \bar{x}_H^*) &= \Gamma_H^* - b_H \\ \pi_H(\bar{x}_H^*) &= b_H \end{aligned} \tag{12}$$

We now examine the conditions under which the shipper will choose to use both carriers. This will happen if

$$\pi_0(x_L^*, x_H^*) > \pi_0(\bar{x}_L^*, 0) \quad and \quad \pi_0(x_L^*, x_H^*) > \pi_0(0, \bar{x}_H^*) \tag{13}$$

or if

$$\begin{aligned} b_L &< \Gamma^* - \Gamma_H^* \\ b_H &< \Gamma^* - \Gamma_L^* \end{aligned} \tag{14}$$

(There is an indifference point when one or both these conditions hold with equality.) The shipper will choose only the LC-carrier if

$$\begin{aligned} b_H &> \Gamma^* - \Gamma_L^* \\ b_L - b_H &< \Gamma_L^* - \Gamma_H^* \end{aligned} \tag{15}$$

Finally, the shipper will choose only the HC-carrier if

$$b_L > \Gamma^* - \Gamma_H^*$$
$$b_L - b_H < \Gamma_L^* - \Gamma_H^* \tag{16}$$

We now claim that the carriers will choose

$$b_L = b_L^B = \Gamma^* - \Gamma_H^*$$
$$b_H = b_H^B = \Gamma^* - \Gamma_L^* \tag{17}$$

To see this, suppose the LC-carrier chooses b_L^B and the HC-carrier chooses $b_H < b_H^B$. Then the HC-carrier can increase profit by raising b_H to b_H^B since the shipper will continue to choose both carriers. A similar argument shows that the LC-carrier will never choose $b_L < b_L^B$. Now suppose the HC-carrier chooses $b_H > b_H^B$. Then the LC-carrier can choose $b_L = b_H + \Gamma_H^* - \Gamma_L^* - \varepsilon$, where $\varepsilon > 0$, and at b_L, the shipper will choose to move 0 output over the HC-carrier. Thus HC-carrier profit will go to 0. Hence the HC-carrier will not choose $b_H > b_H^B$. A similar argument can be made for the LC-carrier.

Now we argue the optimality of the assumed form of carrier price. The question is whether the LC-carrier can find another from of price scheme which increases profit, holding the form of the HC-carrier's fixed at $b_H + C_H(x)$. Suppose there is such a scheme and that this scheme gives the LC-carrier profit of

$$b_L > \Gamma^* - \Gamma_H^* \tag{18}$$

Whatever this price scheme, the shipper must make profits at least those he would make if he used only the HC-carrier, or

$$\pi_0 \geq \Gamma_L^* + \Gamma_H^* - \Gamma^*. \tag{19}$$

The implication of (18) and (19) is that

$$\pi_0 + \pi_L > \Gamma_L^* \tag{20}$$

which is impossible since $\pi_0 + \pi_L$ can be no greater than Γ_L^* by definition.

Thus there is no price scheme which will make the LC-carrier's profit higher, and $b_L^B + C_L(x)$ is an optimal form for the LC-carrier's price. Along similar lines, we argue the optimality of the HC-carrier's price.

Having completed the argument for the general case, we now argue the proof for the cost functions $C_L(x_L) = m_L x_L$ and $C_H(x_H) = m_H x_H$. In the case where $m_L < m_H$, $\Gamma^* = \Gamma_L^*$. Therefore $b_H^B = 0$. Examining the shipper's first-order condition for profit maximization gives $x_L^B = x_L^*$ and $x_H^B = 0$. Consequently $b_L^B = \Gamma^* - \Gamma_H^*$.

Thus, in a regime of commercial freedom, carriers offer nonuniform prices which cause the shipper to choose outputs, x_L^* and x_H^*, which maximize the joint profit of all three agents (shipper and both carriers). In the case where carrier cost functions are linear and different, the shipper will contract only with the lower cost carrier.

In the proof of the proposition we showed that an optimal form of price is a nonuniform cost-plus price. For linear carrier costs this optimal price is a two-part tariff. However in the rail freight markets we are familiar with, the nonuniform price of choice is a *quantity-forcing* price which takes the form

$$T_Q(x) = \begin{cases} (p+\Delta)x & \text{if } x < \bar{x} \\ px & \text{if } x \geq \bar{x} \end{cases} \tag{21}$$

where $\Delta > 0$. The following proposition, offered without proof, demonstrates that a quantity-forcing tariff can be an equilibrium tariff for $G(2C,1S)$.

Proposition 2. *Suppose* $m_L < m_H$. *Then an equilibrium for* $G(2C,1S)$ *is*

$$\begin{aligned} &x_L^B = x_L^* \\ &x_H^B = 0 \\ &T_L^B(x_L^B) = \begin{cases} (m_L + u^* + \Delta)x_L & \text{if } x_L < x_L^* \\ (m_L + u^*)x_L & \text{if } x_L \geq x_L^* \end{cases} \\ &T_H^B(x_H) = m_H x_H. \end{aligned} \tag{22}$$

where Δ *is a large positive number and* $u^* = (\Gamma^* - \Gamma_H^*)/x_L^*$.

Now we consider efficiency. For the game defined, the standard measure of *economic efficiency* is

$$W(x_L, x_H) = \int_0^{x_L + x_H} p_0(z)dz - C_0(x_L + x_H) - m_L x_L - m_H x_H \tag{23}$$

where the first term measures consumer surplus in the delivered market. The variable of interest is *output*, which, in this case, is the vector (x_L, x_H). However, when $m_L < m_H$, $W(x_L, x_H)$ is maximized at $x_H = 0$. Therefore, given the specific carrier cost functions we have assumed, $W(x_L, x_H)$ increases as x_L increases. Hence, to compare the options under consideration -- Marginal Cost Pricing, Ramsey Pricing, and Deregulated Prices -- we need only compare on the basis of which gives the highest value for x_L.

First, we compare MCP and DP. We assume that each carrier is forced to offer a constant unit price of transportation equal to marginal cost. Thus the LC-carrier offers m_L, and the HC-carrier, m_H.

Lemma *If each carrier prices at marginal cost, the shipper will choose outputs* $x_L = x_L^*$ *and* $x_H = 0$.

Proof Given marginal cost pricing, the shippers profit function is

$$\pi_0(x_L,x_H) = (x_L + x_H)p_0(x_L + x_H z) - C_0(x_L + x_H) - m_L x_L - m_H x_H \\ = \Gamma(x_L,x_H). \qquad (24)$$

Hence, the shipper chooses the output which maximizes $\Gamma(x_L,x_H)$, which is $(x_L^*,0)$.

So with Marginal Cost Pricing, the shipper chooses the same output he does under Deregulated Pricing, and therefore, each regime provides the same welfare. Hence, the imposition of Marginal Cost Pricing will not enhance economic efficiency.

Now consider Ramsey pricing. Suppose the regulatory authority, for whatever reason, fixes a unit price of transportation, p^R, which exceeds the marginal cost of the LC-carrier, m_L. (In theory, our choice of linear carrier cost functions does not require Ramsey pricing. However we could simply add a fixed component to each carrier cost function and the "conditions" for Ramsey pricing would be met.) Then, letting (x_L^R,x_H^R) be the resulting output, it is easy to show that

$$x^R = x_L^R + x_H^R < x_L^*. \qquad (25)$$

That is, under Ramsey Pricing the shipper will choose a lower output. Hence, Ramsey Pricing will lower economic efficiency relative to Deregulated Pricing.

4. Conclusions

We have argued that Marginal Cost Pricing and Ramsey Pricing in freight transport markets cannot enhance efficiency. The rationale is that, relative to a regime of Deregulated Pricing, neither Marginal Cost Pricing nor Deregulated Pricing influence a shipper to choose a higher output. A carrier's optimal nonuniform price has the characteristic that a shipper will pick the same output had the carrier priced at marginal cost. Hence, the critical assumption is that a carrier will try to maximize profit, which is not difficult for most economists and regulators to swallow.

While our argument has been theoretical, there is empirical support (albeit limited). First, our model suggests that carriers will use quantity-discount tariff structures. In fact, every bulk commodity tariff (coal, potash, sulfur) that we have studied, offers a quantity discount. Hence, carriers are at least using the profit-maximizing form of tariff. A second implication of the model is that shipper profit will be higher, the closer the HC-carrier's cost is to the LC-carrier's. In fact the existing empirical evidence on freight rates suggests that a shipper's rates will be lower, the more competition there is in the shipper's origin-destination market. Hence our model appears to be consistent with the available evidence on freight rates.

We conclude that the economic rationale for regulating freight rates can only appeal to fairness.

References

[1] K. Arrow, "Verified Statement On Behalf of the Eastern Railroads", in Ex Parte No. 347 (Sub No. 1), *Coal Rate Guidelines - Nationwide*, May, 1981.

[2] R.R. Braeutigam, "Optimal Pricing with Intermodal Competition", *American Economic Review*, 69:38-49, 1979.

[3] R.R Braeutigam, "Verified Statement On Behalf of the Eastern Railroads" in Ex Parte No. 347 (Sub No. 1), *Coal Rate Guidelines - Nationwide*, May 11, 1981.

[4] W.J. Baumol, "Verified Statement On Behalf of the Eastern Railroads", in Ex Parte No. 347 (Sub No. 1), *Coal Rate Guidelines - Nationwide*, May 11, 1981.

[5] W.J. Baumol and D.F. Bradford, "Optimal Departures from Marginal Cost Pricing", *American Economic Review*, 60:265-283, 1970.

[6] G.F. Faulhaber and W.J. Baumol, "Economists as Innovators", *The Journal of Economic Literature*, 26:577-600, 1988.

[7] S. Goldfeld, "Verified Statement On Behalf of the Eastern Railroads", in Ex Parte No. 347 (Sub No. 1). *Coal Rate Guidelines - Nationwide*, May 11, 1981.

[8] W.J. Hurley and E.R. Petersen, "A Noncooperative Equilibrium for the Freight Network Equilibrium Problem", *School of Business, Queen's University*, 1993.

[9] R.G. Lipsey and K. Lancaster, "The General Theory of the Second Best", *Review of Economic Studies*, 24:11-32, 1956.

[10] L.N. Moses, "Verified Statement On Behalf of the Eastern Railroads", in Ex Parte No. 347 (Sub No. 1). *Coal Rate Guidelines - Nationwide*, May 11, 1981.

[11] F.L. Ramsey, "A Contribution to the Theory of Taxation", *Economic Journal*, 37:47-61, 1927.

[12] M.J. Roberts, "Railroad Maximum Rate and Discrimination Control", *Transprt J*, 22:23-33, 1983.

[13] J. Tirole, "The Theory of Industrial Organization", *MIT Press*, Boston, 1988.

[14] W.B. Tye and H.B. Leonard, "On the Problems of Applying Ramsey Pricing to the Railroad Industry with Uncertain Demand Elasticities", *Transportation Research*, 7A:439-450, 1983.

[15] W.E. Wecker, "Verified Statement On Behalf of the Eastern Railroads", in Ex Parte No. 347 (Sub No. 1) *Coal Rate Guidelines - Nationwide*, May 11, 1981.

[16] R.D. Willig, "Pareto-Superior Nonlinear Outlay Schedules", *Bell Journal of Economics*, 9:56-69, 1978.

[17] R.D. Willig, "Verified Statement On Behalf of the Eastern Railroads", in Ex Parte No. 347 (Sub No. 1) *Coal Rate Guidelines - Nationwide*, May 11, 1981.

OPTIMAL FREIGHT TRANSPORT PRICING AND THE FREIGHT NETWORK EQUILIBRIUM PROBLEM

W. J. Hurley

Associate Professor, Department of Business Administration, Royal Military College of Canada, Kingston, Canada

E. R. Petersen

Professor, School of Business, Queen's University, Kingston, Canada

The freight network equilibrium problem consisting of K shippers and a network of M carriers is solved for both the case when carriers act noncooperatively and when they cooperate. In the noncooperative game, the carriers select a cost-plus fixed fee tariff, where the fixed fee is charged if the shipper uses any part of the carrier's network. If carriers cooperate, then the vertically efficient coalition can offer a single tariff that can be of several forms: two-part, cost-plus, quantity-discount, or minimum-charge. Both formulations are equivalent, in the sense that an equilibrium exists for each that has identical shipper output, network flows, and shipper and carrier profits. Equilibrium network flows are obtained by solving a single level traffic assignment problem that maximizes the sum of all agents profit. The division of benefits is obtained by solving a linear programming problem. Examples are presented to illustrate the results.

1. Introduction

Consider the freight network equilibrium problem (FNEP) consisting of multiple carriers and shippers, each acting as a noncooperative profit maximizing agent. Carriers set freight rates (tariff schedules) to maximize profit. Each shipper chooses the output level and routing that maximizes its profit, given the freight rates charged and the inverse delivered demand function for each origin-to-destination shipment.

The literature on freight networks has borrowed heavily from the urban traffic problem. Early work by Petersen and Fullerton[11] assumed fixed O-D demand and routed the traffic to minimize total transportation cost using the system equilibrium criterion. Taborga and Petersen[12] extended this to variable demands and obtained a solution that had supply-demand equilibrium with traffic routed to minimize total cost. These approaches gave the carriers no active role, nor did they address the problem of what tariffs carriers should charge and the profitability of each agent. In a world of regulated transportation this seemed like a reasonable solution.

With the emphasis on deregulation, it was necessary to consider an active role for carriers. Friesz[3], Friesz and Harker[4], and Harker[6] formulated the FNEP using game theory, where the equilibrium flows and prices are the result of each agent maximizing profit. These approaches have all resulted in models that are difficult to solve. For example, Friesz, Gottfried and Morlok[5] use a sequential shipper-carrier model, while Harker[7] uses a mixed behavior model to capture the effects of varying shipper market power. Fisk[1] minimizes the carrier(s) cost subject to constraints that describe user behavior, however, the constraints are in general nonlinear, which makes the problem difficult to solve. While the need to solve for the equilibrium flows simultaneously has been widely recognized, no computationally tractable procedures have been available. Furthermore, an assumption common to all the literature is that the carrier sets a linear tariff.

Our approach to the FNEP differs from that in the literature in a number of important ways. The principal difference is our assumption about the form of carrier pricing. Generally, the transportation literature has assumed that carriers use linear tariffs (that is rates independent of traffic volume). However, we allow carriers to choose tariffs which depend on volume. There are two reasons for doing so. First, the relationship between shipper and carrier is one of vertical exchange. This exchange is between two independent agents and leads to vertical relationships that are often richer and more complex than those usually studied between a firm and a group of consumers. Vertical exchange transactions are not restricted to use linear pricing as required when there are many customers, but can be contracts that may take many different forms. If carriers are allowed to choose the form of tariff, then they will never choose a constant tariff rate since, in general, constant tariff rates are suboptimal (see Hurley[8], Rey and Tirole[13], or Tirole[14]). Second, the freight transportation industry has consistently used some form of nonlinear tariff, be it two-part (tapered), quantity-discounts or quantity-forcing.

We examine two game structures which differ primarily in the way carriers behave. The first, termed the C-game, has many elements of a cooperative game among carriers, (Hurley and Petersen[9]). Carriers, individually or in coalition, set freight rates (tariff schedules), and each shipper then selects an output and a carrier or coalition of carriers, given these freight rates. The key element is that carriers are able to form coalitions. The second game structure, termed the N-game, is the full noncooperative game between shippers and carriers (Hurley and Petersen[10]). Each carrier chooses tariffs, and given these tariffs, shippers choose output and route the shipment through the freight network.

Given the more general carrier pricing, both game structures have a remarkable equivalence to a problem we term the Integrated System Optimization Problem (ISP). The ISP assumes that the shippers and carriers are controlled by a single agent, and this agent chooses a network flow vector, x^*, termed the vertically efficient flow vector, to maximize the joint profit of all carriers and shippers. Denoting the equilibrium flow vector for the C-game by x^C, and for the N-game by x^N, our basic result is that

$$x^C = x^N = x^*,$$

This result is important because it allows us to calculate the FNEP flows by solving the ISP, a concave optimization problem which retains the structure of the traffic assignment problem, for which there are good algorithms.

While the ISP is important for calculating FNEP flows, it says nothing about the profit distribution among shippers and carriers.

For each game structure, profit distributions are determined endogenously by solving a linear programming problem.

The solution of the ISP identifies a subset of carriers over which shipment w moves. We term these carriers the vertically efficient carriers for shipment w. In the C-game, the coalition of vertically efficient carriers offers the shipper a two-part tariffs (see Hurley and Petersen[10] for a discussion of quantity-discount, cost-plus or minimum-charge tariff structures). A two-part tariff has the general form $R(q)= b+aq$, where $R(q)$ is carrier revenue if quantity q is moved. The parameter a influences the output the shipper chooses and is set to the carrier's marginal cost for the shipment. Thus the two-part tariff has the same efficiency properties as marginal cost pricing. The parameter b effects a profit distribution between shipper and carrier and will depend primarily on whether there is competition among carriers to move the shipment, and secondly, the relative cost of these competing carriers. In general, if carrier costs are close, b will be small.

The solution of the N-game has a number of interesting properties besides the flow property in (1). Each vertically efficient carrier will offer a nonlinear cost-plus tariff of the form:

$$R_m^w(q) = \pi_m^w + c_m(q),$$

where $c_m(q)$ is the network cost incurred by carrier m to move a quantity q of shipment w, and π_m^w is the profit earned by carrier m. (In general, the two-part tariffs, quantity discounts, and other forms of nonlinear tariffs are not optimal for the N-game formulation.) However, the profit realized by each carrier is the same in the N-game as in the C-game.

In both formulations, equilibrium tariffs must be vertically efficient and distributes the profit from each shipment so that the individual carrier's profit is maximized. The approach may be summarized as: make the pie as large as possible by using a vertically efficient tariff, and then maximize the size of each carrier's slice. Thus the original approach of assigning freight traffic to minimize total cost, or with variable demands, to maximize total system profit was correct and resulted in the proper shipper outputs and network flows.

The paper is organized as follows. First the FNEP is formulated followed by a description of the ISP and a calculation of the profitability of each shipment. Equilibrium solutions for the different forms of carrier behavior are then obtained. Finally a number of examples are presented to illustrate the theory.

2. The Freight Network Equilibrium Problem

Let K be the set of shippers and M the set of carriers. Each carrier, $m \in M$, operates a network represented by the directed graph $G(N_m, A_m)$, where N_m is a set of nodes and A_m a set of links (arcs) representing carrier m's physical system. Carriers may represent different transport modes and/or competitors within a mode. Interline or transfer connections between carriers and modes are included as links in one of the carrier networks. The total transportation system is described by the directed graph $G(N,A)$, where $N = \bigcup N_m$ and $A = \bigcup A_m$. For $a \in A$, let x_a be the total arc flow, and $\boldsymbol{x} = (\ldots, x_a, \ldots)$, the vector of total arc flows. The total cost of shipping over arc $a \in A$ is $c_a(\boldsymbol{x})$, and depends on the vector of flows on the network. We assume that $c_a(\boldsymbol{x})$, is convex in $\boldsymbol{x}$, but do not require that it be separable in arc flows, nor do we make any assumptions about the symmetry of the Jacobean. The function $c_a(\boldsymbol{x})$ reflects the congestion on the link, and can include the effect of flows in the reverse direction or other service classes.

Shipper $k \in K$ controls the set W_k of origin-to-destination (OD) shipments. Shipment $w \in W_k$ moves from origin r to destination s $(r,s \in N)$, with flow q_w, and has delivered inverse demand function $p_w(q_w)$. We assume that the marginal revenue is decreasing, which ensures that the revenue from each shipment is strictly concave. Let $\boldsymbol{q}^k = (\ldots, q_w, \ldots)$ be the vector of OD shipments for shipper k, and $\boldsymbol{q} = (\ldots, \boldsymbol{q}^k, \ldots)$ the total shipment vector. The cost of production for shipper $k \in K$ is $c_o^k(\boldsymbol{q}^k)$, and is assumed to be convex in the vector of shipments from shipper k. A shipper may control an individual OD shipment, or an extensive set of shipments (e.g. General Motors). The set of all OD shipments is $W = \bigcup W_w$.

The standard path-flow formulation is used. Associated with each OD pair w, the index $p \in P_w$ will refer to a path through the network for the OD shipment w, with path flow f_p^w. In this formulation, the path-flows f_p^w are the decision variables, and we let $f_w = (\ldots, f_p^w, \ldots)$ be the vector of path flows for shipment w, and $f = (\ldots, f_w, \ldots)$ the vector of all path flows.

We define the strategy space for carrier $m \in M$ to be the set of tariff schedules or contracts R_m^w that the carrier charges for each shipment $w \in W$, with vector $R_m = (R_m^w)$. Each carrier has profit

$$\pi_m(R_m) = \sum_{w \in W} R_m^w - \sum_{a \in A_m} c_a(x).$$

For each shipper $k \in K$, the strategy space for each shipment $w \in W_k$ is the output q_w and the path flows f_w. The set of carriers used to move shipment w is

$$C_w = \left\{ m | a \in A_m, f_p^w \delta_{a,p}^w > 0 \right\},$$

where $\delta_{a,p}^w = 1$ if arc a is part of path p for OD pair w, and 0 otherwise. The profit for shipper k is

$$\pi_0^k(q^k, f^k) = \sum_{w \in W_k} \left\{ q_w p_w(q_w) - \sum_{m \in C_w} R_m^w \right\} - c_0^k(q^k).$$

Each agent has a strategy space with a payoff function defined over that strategy space. The Cartesian product of the individual strategy spaces is the strategy space of the problem. An equilibrium point is a point in the strategy space where each agent maximizes its payoff with respect to its strategy space, holding the optimal strategy choices of the other agents fixed.

3. Integrated System Problem

Before solving for an equilibrium for the FNEP, we consider the problem if the freight system is integrated or owned by a single agent. The integrated system (vertically between shipper and carriers, and horizontally amongst carriers) profit or surplus is

$$\Gamma(\boldsymbol{q},\boldsymbol{x}) = \sum_{m\in M} \pi_m(R_m^w) + \sum_{k\in K} \pi_O^k(\boldsymbol{q}^k,\boldsymbol{f}^k)$$
$$= \sum_{k\in K}\sum_{w\in W_k} q_w p_w(q_w) - \sum_{k\in K} c_O^k(\boldsymbol{q}^k) - \sum_{a\in A} c_a(\boldsymbol{x}).$$

Optimal flows, which we shall call the vertically efficient flows, are found by solving the Integrated System Optimization problem (ISP):

$$\begin{aligned}
\max \quad & \Gamma(\boldsymbol{q},\boldsymbol{x}) = \sum_{k\in K}\sum_{w\in W_k} q_w p_w(q_w) - \sum_{k\in K} c_O^k(\boldsymbol{q}^k) - \sum_{a\in A} c_a(\boldsymbol{x}) \\
s.t. \quad & q_w = \sum_{p\in P_w} f_p^w \\
& x_a = \sum_{k\in K}\sum_{w\in W_k}\sum_{p\in P_w} f_p^w \delta_{a,p}^w \\
& f_p^w \geq 0, \qquad p\in P_w, w\in W_k, k\in K.
\end{aligned} \tag{ISP}$$

This concave optimization problem is in a form that is easily solved for the vertically efficient flows $\boldsymbol{q}^*, \boldsymbol{x}^*, \Gamma^* = \Gamma(\boldsymbol{q}^*,\boldsymbol{x}^*)$, and marginal cost, u_w, for shipment $w \in W$, using available traffic assignment algorithms.

Clearly, there is an incentive for the carrier and shipper to move $\boldsymbol{q}^*$, with network flows $\boldsymbol{x}^*$, if the agents can find a mechanism to divide the additional profit. That is, if the shipper and carrier were vertically integrated, then in general total profit would be greater. In general, full economic integration of vertical firms is not necessary to achieve vertical efficiency if the two are able to contract at the vertically efficient output. The carrier's problem, then, is to find a tariff schedule that induces the shipper to choose $\boldsymbol{q}^*$, and at the same time, allow the carrier to make as high a profit as possible. Tariffs that induce the shipper to behave as if the system were vertically integrated are said to be vertically efficient.

4. Profitability of Each Shipment

To set tariffs we need to know the joints profit generated by each shipment. While the revenue is a function of output, both production and transportation costs may be joint, and a cost allocation method is required. We shall use the arbitrary rule that prorates joint cost based on the vertically efficient flows. Other allocation methods could be used.

Following Friedman[2], we develop a notation for strategy combinations that allow the strategy for one of the players to be varied while the strategies for the remaining players is fixed. Let $(x \backslash q_w)$ denote the vector x with only changes due to the single variable q_w. Thus $c_a(x \backslash q_w)$ is the cost on arc a as a function of the single variable q_w, given the remaining variables x are fixed. Following this convention, $(x_a^* \backslash q_w)$ is the total flow on arc a as a function of q_w, given all other flows are fixed at their vertically efficient values.

Consider first the production cost. Let $c_{ow}(q_w)$ be the cost of production for shipment w. If the shipper's production costs are separable in output, then this is just the cost of producing output q_w. If the shipper's production cost are joint, let

$$c_{ow}(q_w) = c_o^k(q^{k*} \backslash q_w) - \frac{\sum_{v \in W_k, v \neq w} q_v^*}{\sum_{v \in W_k} q_v^*} c_o^k(q^{k*}).$$

This allocation ensures that

$$c_o^k(q^{k*}) = \sum_{w \in W_k} c_{ow}(q_w^*).$$

To calculate the cost to transport shipment w, define

$$x_a^w(q_w) = \sum_{p \in P_w} f_p^w \delta_{a,p}^w,$$

the flow on arc a due to shipment w. If transportation costs are also prorated based on the vertically efficient flows, then the total carrier cost without shipment w is

$$c^{-w} = \sum_{a \in A, x_a^* > 0} \frac{(x_a^* - x_a^w(q_w^*))}{x_a^*} c_a(\boldsymbol{x}^*)$$

while the cost of moving q_w^* is

$$c_w = \sum_{a \in A, x_a^* > 0} \frac{x_a^w(q_w^*)}{x_a^*} c_a(\boldsymbol{x}^*).$$

The joint profit without shipment w is

$$\hat{\Gamma}_w = \Gamma^* - q_w^* p_w(q_w^*) + c_w + c_{ow}(q_w^*)$$

and the profit generated by shipment w is $\Gamma^* - \hat{\Gamma}_w$. Note that

$$\sum_{w \in W} \left(\Gamma^* - \hat{\Gamma}_w\right) = \Gamma^*,$$

so that the profit from the individual shipments aggregate to Γ^*.

5. Noncooperative Equilibrium

The essential feature of a noncooperative game between players (shippers and carriers) is that each player has a strategy space with a payoff function defined over that strategy space. The Cartesian product of the individual strategy spaces is the strategy space of the game.

An equilibrium point is a point in the strategy space of the game where each player maximizes his payoff with respect to his strategy space, given the optimal strategy choices of the other players.

We now calculate the share of the joint profit, generated by shipment w, that each agent can claim. The vertically efficient coalition is

$$C_w^* = \left\{m \middle| a \in A_m, f_p^{w*}\delta_{a,p}^w > 0\right\},$$

which is the set of carriers that moves the vertically efficient flow q_w^*. The shipper can
use the vertically efficient set of carriers C_w^*, or possibly some other set of carriers. A subset y of carriers in C_w^* is feasible to excluded, if the remaining carriers $\{C_w^* \setminus \gamma\}$ plus possibly other carriers not in C_w^* can feasibly move the shipment. Let L_w be the set of all possible sub-sets of C_w^* that can feasibly be excluded. For each $\gamma \in L_w$, let the joint profit be $\Gamma_w^{-\gamma}$.

Let π_m^w be carrier m's profit from shipment w. If the shipper selects C_w^*, its profit from the shipment is

$$\pi_o^w = \Gamma^* - \hat{\Gamma}_w - \sum_{m \in C_w^*} \pi_m^w.$$

If she selects $\{C_w^* \setminus \gamma\}$, then

$$\pi_o^w = \Gamma_w^{-\gamma} - \hat{\Gamma}_w - \sum_{m \in \{C_w^* \setminus \gamma\}} \pi_m^w, \qquad \gamma \in L_w.$$

Carrier profit, $\pi_m = 0,\ m \notin C_w^*$, since these carriers have no claim.

In the noncooperative game, each carrier in the vertically efficient set will maximize its profit π_m^w subject to the conditions that the shipper is free to select the carriers used for shipment w. Each carrier $m \in C_w^*$ simultaneously solves

$$\begin{aligned} \max \quad & \pi_m^w \\ s.t. \quad & \pi_o^w = \Gamma^* - \hat{\Gamma}_w - \sum_{m \in C_w^*} \pi_m^w \\ & \pi_o^w = \Gamma_w^{-\gamma} - \hat{\Gamma}_w - \sum_{m \in \{C_w^* \setminus \gamma\}} \pi_m^w, \qquad \gamma \in L_w \\ & \pi_o^w, \pi_m^w \geq 0, m \in C_w \end{aligned} \qquad (LP_m^w)$$

given the value for the other carriers π's. Note that the above constraint set defines the core of the cooperative game between shipper and carriers, however, we define it based on the noncooperative actions of the shipper. To obtain a solution to this problem, we use the following lemma:

Lemma Let D be a convex set in $\pi = (\pi_m), m \in C$, and suppose the vector π^* solves

$$\begin{aligned} \max \quad & \sum_{m \in C} \pi_m \\ s.t. \quad & \pi \in D. \end{aligned}$$

Then for each $m \in C$, π_m^* solves the optimization problem

$$\begin{aligned} max \quad & \pi_m \\ s.t. \quad & \pi \in D, \text{ given } \pi_k^*, k \in C, k \neq m \end{aligned}$$

Proof Since $\sum_{m \in C} \pi_m$ is convex, and D a convex set in π, then,

$$\sum_{m \in C} \pi_m^* \geq \max \pi_m + \sum_{k \in C, k \neq m} \pi_m^* \quad s.t. \quad \pi_m, \pi_k^* \in D, k \in C, k \neq m.$$

Rewriting, we have

$$\pi_m^* \geq max\, \pi_m \quad s.t. \quad \pi \in D, \text{ given } \pi_k^*, k \in C, k \neq m,$$

which completes the proof.

Thus, in the noncooperative game between shippers and carriers, for each shipment $w \in W$, an equilibrium division of profit can be found by solving the linear programming problem

$$\pi_w^* = \max \sum_{m \in C_w^*} \pi_m^w$$
$$s.t. \quad \pi_o^w = \Gamma^* - \hat{\Gamma}_w - \sum_{m \in C_w^*} \pi_m^w. \qquad (LP_w)$$
$$\pi_o^w = \Gamma_w^{-\gamma} - \hat{\Gamma}_w - \sum_{m \in \{C_w^* \setminus \gamma\}} \pi_m^w, \qquad \gamma \in L_w.$$
$$\pi_o^w, \pi_m^w \geq 0, m \in C_w$$

We now state the main results of the paper:

Theorem 1. Let (q^*, x^*) solve (ISP). For each shipment $w \in W$, let $\pi_w = (\pi_m^w)$ solve (LP_w), and suppose each carrier in C_w^* offers the shipper the tariff

$$R_m^w = \left\{ \pi_m^w + \sum_{a \in A_m, x_a^* > 0} \left\{ c_a(x^* \setminus q_w) - \frac{\left(x_a^* - x_a^w(0)\right)}{x_a^*} c_a(x^*) \right\} \right\} \delta_m^w,$$

to move shipment w, where $\delta_m^w = 1$ if $x_a(q_w) > 0, a \in A_m$, and 0 otherwise. Then $\{R_m, m \in M \text{ and } (q^*, f^*)\}$ is a noncooperative equilibrium for the K-shipper, M-carrier problem, that is vertically efficient.

Note that the tariff offered by each carriers is a cost-plus tariff, where π_m^w is a fixed fee equal to the profit carrier m receives if the shipment w uses any part of carrier m's network. The terms in the summation is the cost carrier m incurs moving q_w, if the costs $c_a(x^*)$ are prorated on a flow basis.

Proof Shipper k's profit from shipment w is

$$\pi_{ow}^k(q_w,f_w)=q_w p_w(q_w)-\sum_{m\in C_w} R_m^w - c_{ow}(q_w)$$

$$=q_w p_w(q_w)-\left\{\sum_{m\in C_w}\pi_m^w+\sum_{a\in A_m, x_a^*>0}\left\{c_a(x^*\setminus q_w)-\frac{\left(x_a^*-x_a^w(0)\right)}{x_a^*}c_a(x^*)\right\}\right\}\delta_m^w$$

$$-\left\{c_o^k(q^{k*}\setminus q_w)-c_{ow}(q_w)\right\}.$$

If the shipper uses the vertically efficient carriers, then

$$\pi_{ow}^k(q_w,f_w)=\Gamma(q^*,x^*\setminus q_w,f_w)+constant,$$

with maximum at q_w^*, f_w^*, the vertically efficient output and path flows. Suppose the shipper selects another set of carriers $\{C_w^*\setminus\gamma\}$, $\gamma\in L_w$, then the change in shipper profit is

$$\Gamma_w^{\sim\gamma}-\sum_{m\in\{C_w\setminus\gamma\}}\pi_m-\Gamma^*+\sum_{m\in C_w}\pi_m=\sum_{m\in\gamma}\pi_m-\left(\Gamma^*-\Gamma_w^{\sim\gamma}\right)\le 0.$$

Therefore there is no incentive for the shipper to switch from using the vertically efficient carriers C_w^* to move shipment w, and the set of tariffs is vertically efficient.

Now consider the carriers. Suppose the shippers send the vertically efficient shipments q^*, routed according to f^*. Shipment $w\in W$ is moved by carriers C_w^*, defined by the vertically efficient path flows f_w^*. If each carrier receives profit from shipment w that satisfy LP_w, then by the lemma, no carrier is able to increase its profit from that shipment, and the proof is complete.

6. Cooperative Equilibrium

In the cooperative formulation, carriers are able to form coalitions. Carriers, individually or in coalition, set a tariff for each shipment, and shippers then select output and the carrier or coalition of carriers used, given these freight rates. Hurley and Petersen[9,10] show the following:

Theorem 2. Let (x^*,u^*) solve (ISP). For each shipment $w \in W$, let π_w^* solve (LP_w). Suppose the coalition of carriers C_w^* offers the shipper a tariff of one of the following forms:

cost-plus

$$R_w(q_w) = \pi_w^* + c(x^* \backslash q_w) - c^{-w}$$

two-part

$$R_w(q_w) = b_w + a_w q_w$$

where

$$a_w = u_w,$$
$$b_w = \pi_w^* + \mu_w(q_w^*), \quad \text{and}$$
$$\mu_w(q_w) = c_w - u_w q_w$$

quantity-discount

$$R_w(q_w) = \begin{cases} p_l q_w & q_w < q_w^* \\ p_h q_w & q_w \geq q_w^* \end{cases}$$

where $p_h = (\pi_w^* + c_w)/q_w^*$ and $p_l > p_h$ is sufficiently large such that if $q_w < q_w^*$, the shipper's profit from shipment w is less than π_o^w.

minimum-charge

$$R_w(q_w) = \begin{cases} R_o & q_w \leq q_w^* \\ R_o + u_w(q_w - q_w^*) & q_w \geq q_w^* \end{cases}$$

where $R_o = \pi_w^* + c_w$.

Then $\{R_w, w \in W \text{ and } (q^*, x^*)\}$ is an equilibrium for the K-shipper, cooperative M-carrier problem, that is vertically efficient.

The cooperative formulation has appeal in that it shows that cooperation amongst carriers does not lead to any inefficiencies the transportation system. The noncooperative formulation is appealing in that it shows that no arrangement between carriers is required to achieve a vertically efficient equilibrium, and price supports are obtained.

Both formulations are equivalent, in the sense that an equilibrium exists for each that has identical shipper outputs and network flows. This result can be generalized. We have structured the problem so that the carrier's strategy is to set tariffs, while the shipper's strategy is to select output and the routing for each shipment. This give the carriers the "lead" role with the shipper "shopping" for the lowest cost alternative. This formulation seems to best describe practice in the freight industry, however, we note that there may be exceptions, such as the auto industry, where the shipper is the leader and can force carriers to accept minimal profits by maximizing π_o^w subject to the constraints in LP_w. We can even go one step further in generality. The constraints in LP_w define the core of the cooperative game between shipper and carriers, which von Neumann and Morgenstern argue to be reasonable solutions for such games. The integrated system is the grand coalition. A necessary condition for a division of profit to be in the core is that shipper outputs and network flows be vertically efficient. The power of vertically efficient tariffs is that group rationality can be attained by individual agents acting in their best own interest.

7. Examples

In this section we present simple examples to illustrate the theory. We begin with the case where two carriers with similar costs compete to move a shippers product to market.

Example 1. One Shipper-Two Carriers, Traffic on Both Carriers

Consider two carriers competing to move a shipper's freight as shown is Figure 1. Let carrier a be the lower cost carrier with cost function $c_a(x_a)=4x_a+0.05x_a^2$, and carrier b, the higher cost carrier, has cost $c_b(x_b)=5x_b+0.05x_b^2$. Let $p_o(q)=20-0.2q$ and $c_o(q)=5q$. Carrier a is the low cost carrier, but costs are close enough that the

Figure 1. Competing Carriers

vertically efficient flow is split over both carriers with $x_a^*=16.67$, $x_b^*=6.67$ and $\Gamma^*=125$. If carrier a were not available, then the traffic would move over carrier b, with $x_b^{\sim a}=20$ and $\Gamma^{\sim a}=100$. Similarly, if carrier b were not available then all the traffic would move over carrier a, with $x_a^{\sim b}=22$ and $\Gamma^{\sim b}=121$. The share of the surplus each player can claim is obtained by solving

$$
\begin{aligned}
\pi^* &= \max \pi_a + \pi_b \\
s.t.\quad & \pi_o + \pi_a + \pi_b = \Gamma^* = 125 \\
& \pi_o + \pi_a \qquad \geq \Gamma^{\sim b} = 121 \\
& \pi_o + \qquad \pi_b \geq \Gamma^{\sim a} = 100 \\
& \pi_o, \pi_a, \pi_b \geq 0
\end{aligned} \tag{Ex1}
$$

with solution $\pi_o=96, \pi_a=25$, and $\pi_b=4$. The constraint set for (Ex1) defines the core of the 3-person game and is shown in Figure 2.

The following tariffs are vertically efficient:

Individual carrier pricing:

Each carrier offers the following tariff schedules:

$$R_a(x_a) = 25 + 4x_a + 0.05x_a^2$$
$$R_b(x_b) = 4 + 5x_b + 0.05x_b^2.$$

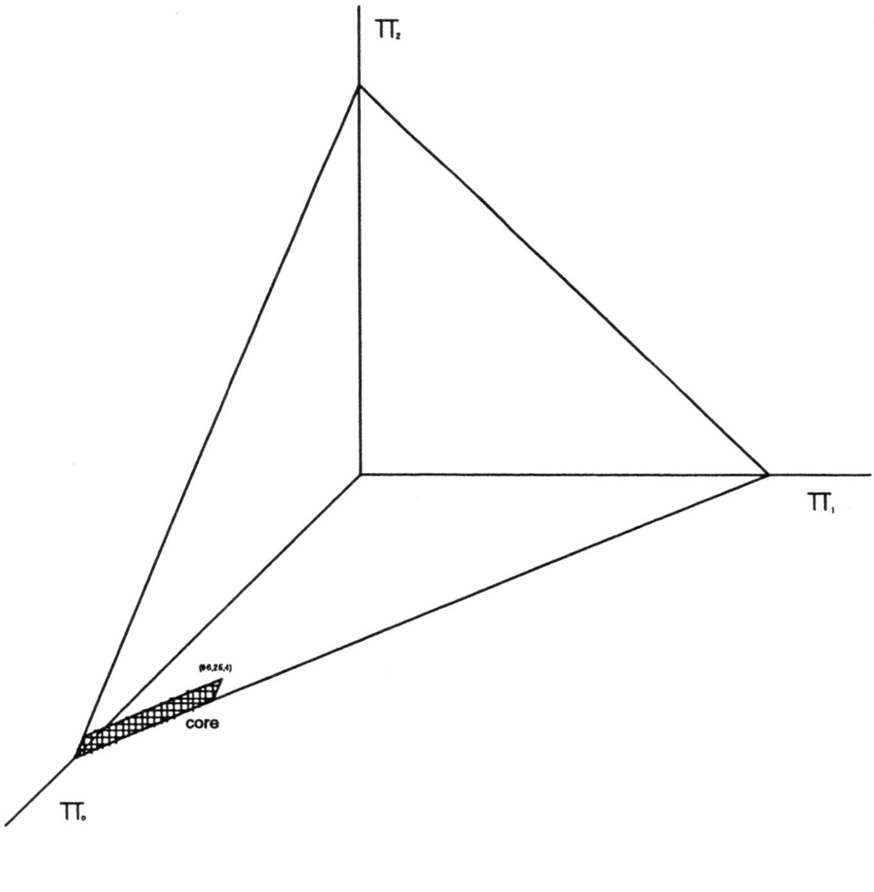

Figure 2. The Core of the Game

Coalition pricing:

The coalition $\{a,b\}$ offers the shipper one of the following tariffs:

cost-plus

$$R(q)=\begin{cases}29+4q+0.05q^2 & q \le 23.33\\ 126.5-5.5q+0.025q^2 & q \ge 23.33\end{cases}$$

two-part

$$R(q)= 12.89 + 5.67q$$

quantity-discount

$$R(q)=\begin{cases}6.3q & q \le 23.33\\ 6.219q & q \ge 23.33\end{cases}$$

minimum-charge

$$R(q)=\begin{cases}145.11 & q < 23.33\\ 145.11+5.67(q-23.33) & q \ge 23.33\end{cases}$$

For each tariff, $q^* = 23.33$, $x_a = 16.67$, $x_b = 6.67$, $\pi_o = 96$, $\pi_a = 25$, and $\pi_b = 4$.

Example 2. Two Shippers, Two Carriers

Consider the problem in Example 1 with two shippers. The data is the same as in that example, except that shipper 2 has inverse demand function $p_2(q_2) = 20 - 0.3q_2$ and production cost $c_{o2}(q_2) = 6q_2$. Solving (ISP) we obtain:

$$q_1^* = 21.90, f_a^1 = 11.00, f_b^1 = 10.90,$$
$$q_2^* = 12.93, f_a^2 = 11.41, f_b^2 = 1.52,$$
$$x_a = 22.41,\ x_b = 12.42 \text{ and } \Gamma^* = 178.88.$$

The joint profit without shipment 1 is $\hat{\Gamma}_1 = 63.90$ and without shipment 2 is $\hat{\Gamma}_2 = 114.98$.

For shipment 1, if carrier a were not available, then $\Gamma_1^{-a} = 167.99$, while if b were not available then $\Gamma_1^{-b} = 168.19$. The profit distribution for shipment 1 is obtained by solving

$$
\begin{aligned}
& \pi_1^* = \max \pi_a^1 + \pi_b^1 \\
& s.t. \quad \pi_o^1 + \pi_a^1 + \pi_b^1 = \Gamma^* - \hat{\Gamma}_1 = 114.98 \\
& \qquad \pi_o^1 + \pi_a^1 \qquad \geq \Gamma_1^{-b} - \hat{\Gamma}_1 = 104.29 \\
& \qquad \pi_o^1 + \qquad \pi_b^1 \geq \Gamma_1^{-a} - \hat{\Gamma}_1 = 104.10 \\
& \qquad \pi_o^1, \pi_a^1, \pi_b^1 \geq 0
\end{aligned}
$$

with solution $\pi_o^1 = 93.41$, $\pi_a^1 = 10.88$, and $\pi_b^1 = 10.69$.

For shipment 2, if carrier a were not available, then $\Gamma_2^{-a} = 167.99$, while if b were not available then $\Gamma_2^{-b} = 168.19$. The profit distribution for shipment 2 is obtained by solving

$$
\begin{aligned}
& \pi_2^* = \max \pi_a^2 + \pi_b^2 \\
& s.t. \quad \pi_o^2 + \pi_a^2 + \pi_b^2 = \Gamma^* - \hat{\Gamma}_2 = 63.90 \\
& \qquad \pi_o^2 + \pi_a^2 \qquad \geq \Gamma_2^{-b} - \hat{\Gamma}_2 = 63.68 \\
& \qquad \pi_o^2 + \qquad \pi_b^2 \geq \Gamma_2^{-a} - \hat{\Gamma}_2 = 51.80 \\
& \qquad \pi_o^2, \pi_a^2, \pi_b^2 \geq 0
\end{aligned}
$$

with solution $\pi_o^2 = 51.58$, $\pi_a^2 = 12.10$, and $\pi_b^2 = 0.22$.

It should be noted that there are multiple equilibria. In this example, the solution to (ISP), (q,x), is unique, however, the flow allocation f is not. Each flow allocation corresponds to a different equilibrium.

Example 3. Interlining

Consider the network shown in Figure 3, where one path involves interlining the shipment over arcs a and b. Suppose the demand and shipper cost function data is the same as example 1, with arc costs

$$c_a(x_a) = 4x_a + .05x_a^2,$$
$$c_b(x_b) = 5x_b,$$
$$c_c(x_c) = 9x_c + .06x_c^2, \text{ and}$$
$$c_d(x_d) = 12x_d.$$

The integrated system optimization problem has solution

$$q^* = 13.2, x_a^* = x_b^* = 7.2,\ x_c^* = 6.0,\ x_d^* = 0, \text{ and } \Gamma^* = 39.6.$$

The equilibrium pricing will depend on the carrier ownership of the transportation arcs. We consider two alternatives.

A. Different Ownership of Each Arc

Assume each arc is a different carrier. Then $C_w^* = \{a,b,c\}$. The set L_w is the sub-set of C_w^* that can leave the coalition and the remaining carriers can feasibly move the shipment. In this problem $L_w = \{a,b,c,ab,ac,bc,abc\}$, with

$$\Gamma^{\sim a} = \Gamma^{\sim b} = \Gamma^{\sim ab} = 34.615,$$
$$\Gamma^{\sim c} = 36,\ \Gamma^{\sim ac} = \Gamma^{\sim bc} = \Gamma^{\sim abc} = 11.25.$$

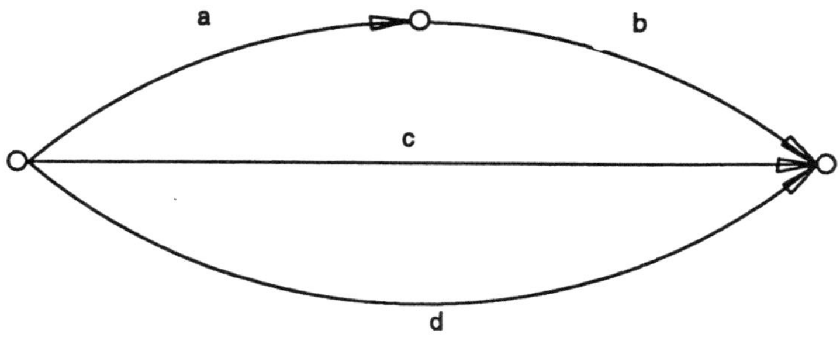

Figure 3. Interlined Carriers With Competition

The sharing of the surplus is determined by solving

$$
\begin{aligned}
&\pi^* = max\ \pi_a + \pi_b + \pi_c \\
&s.t.\quad \pi_o + \pi_a + \pi_b + \pi_c = \Gamma^* = 39.6 \\
&\qquad \pi_o \qquad\quad \pi_b + \pi_c \geq \Gamma^{\sim a} = 34.615 \\
&\qquad \pi_o + \pi_a \qquad + \pi_c \geq \Gamma^{\sim b} = 34.615 \\
&\qquad \pi_o \qquad\qquad + \pi_c \geq \Gamma^{\sim ab} = 34.615 \\
&\qquad \pi_o + \pi_a + \pi_b \qquad \geq \Gamma^{\sim c} = 36 \\
&\qquad \pi_o + \qquad \pi_b \qquad \geq \Gamma^{\sim ac} = 11.25 \\
&\qquad \pi_o + \pi_a \qquad\qquad \geq \Gamma^{\sim bc} = 11.25 \\
&\qquad \pi_o \qquad\qquad\qquad \geq \Gamma^{\sim abc} = 11.25 \\
&\qquad \pi_O, \pi_a, \pi_b, \pi_c \geq 0
\end{aligned}
$$

with a solution

$$
\begin{aligned}
&\pi_o = 31.015 \\
&\pi_a = 4.985\alpha, \pi_b = 4.985(1-\alpha), \quad 0 \leq \alpha \leq 1 \\
&\pi_c = 3.6.
\end{aligned}
$$

Carrier d is not in the vertically efficient coalition and has no claim to any of the joint surplus. The division of profit between carrier a and carrier b is not determined uniquely. Any division in the above range results is an equilibrium. In practice the division of benefits between interlining carriers is usually governed by industry rules.

B. Common Ownership of Arcs a and c

Now suppose that carrier 1 owns both transportation arc a and c, carrier 2 owns arc b, and carrier 3 owns arc d. Now $C_w^* = \{1,2\}$ and $L_w = \{1,2,1\text{-}2\}$, with $\Gamma^{-1} = \Gamma^{-1\text{-}2} = 11.25$, and $\Gamma^{-2} = 34.615$. The sharing of the surplus is obtained by solving the LP

$$
\begin{aligned}
\pi^* &= max\, \pi_1 + \pi_2 \\
s.t.\quad & \pi_o + \pi_1 + \pi_2 = \Gamma^* = 39.6 \\
& \pi_o + \pi_1 \qquad \geq \Gamma^{-2} = 34.615 \\
& \pi_o + \qquad \pi_2 \geq \Gamma^{-1} = 11.25 \\
& \pi_o, \pi_1, \pi_2 \geq 0
\end{aligned}
$$

with multiple solutions

$$
\begin{aligned}
&\pi_o = 11.25 \\
&\pi_1 = 23.365 + 4.985\alpha, \quad \pi_2 = 4.985(1-\alpha),\ 0 \leq \alpha \leq 1.
\end{aligned}
$$

Note that carrier 1 has much stronger market power, and the shipper's profit drops from 31.015 to 11.25.

This example clearly demonstrates that the level of freight rates for a particular shipper will depend on the ownership structure of the network. If a shipper has at least two carrier routing options which are independently owned, and the costs of these alternatives are relatively close, then the shipper's rates will be low. On the other hand, rates will be high if there is only one option or the second best route has a high cost relative to the low cost route. This example also illustrates that the efficiency of the transportation system is not affected by the pattern of carrier ownership, since a vertically efficient tariffs is always used.

8. Discussion

Equilibrium flows for the FNEP are obtained by solving ISP, a concave traffic assignment problem. For each shipment, the division of profit in the cooperative formulation is obtained by solving a linear program. This division of profit is also a solution for the noncooperative formulation. Equilibrium tariffs for the noncooperative game must be of the cost-plus form, while if carriers cooperate then two-part, quantity-discount and minimum-charge tariffs may also be used.

One point that should be emphasized is that there may be multiple equilibria. In solving ISP, the shipper outputs and arc flows may be unique, however, the path flows are not, a well known property of the traffic assignment problem. Each different set of path flows can result in a different division of benefits amongst the agents. Clearly, each agent must select its policy based on the same path flows, which would be the particular vertically efficient path flows that are realized. At equilibrium, this would be the flows observed by each agent, and would be used by each to set its strategy.

To calculate tariffs, each carrier must know the shipper demand function and the competing carrier cost functions. The total flow on each arc together with the flow due to each shipment can be observed. With this data, the carrier can calculate the value of the shipment and the share each agent can claim. Thus each carrier can individually post a tariff of the cost-plus form which results in a vertically efficient equilibrium.

As the examples show, carrier profit depend on the closeness of the competitor's cost functions. Depending on the pattern of ownership of the transportation system, profit can vary substantially. However, ownership has no impact on the efficiency of the freight transportation system, which implies that the strategic role for carriers is limited.

References

[1] Caroline S. Fisk, "A Conceptual Framework for Optimal Transportation Systems Planning with Integrated Supply and Demand Models", *Trans. Science*, 20, pp. 37-47, 1986.

[2] James W. Friedman, "Game Theory with Applications to Economics", *Oxford University Press*, 1986.

[3] Terry L. Friesz, "Transportation Network Equilibrium, Design and Aggregation: Key Developments and Research Opportunities", *Trans. Research A* 19, pp. 413-427, 1985.

[4] Terry L. Friesz and Patrick T. Harker, "Freight Network Equilibrium: a Review of the State of the Art", Chapter 7, *Analytical Studies in Transport Economics*, Andrew Daugherty editor, Cambridge University Press, 1985.

[5] Terry L. Friesz, Joel A. Gottfried and Edward K. Morlok, "A Sequential Shipper-Carrier Network Model for Predicting Freight Flows", *Trans. Science* 20, pp. 80-91, 1986.

[6] P.T. Harker, "Predicting Intercity Freight Flows", *VNU Science Press*, Utrecht, The Netherlands, 1987.

[7] P.T. Harker, "Multiple Equilibrium Behaviors on Networks", *Trans. Science* 22, pp. 39-46, 1988.

[8] W. J. Hurley, "A Theory of Rail Freight Pricing", unpublished Ph.D. *dissertation, School of Business, Queen's University*, Kingston, 1988.

[9] W. J. Hurley and E. R. Petersen, "Nonlinear Tariffs and Freight Network Equilibrium", *Trans. Science* 28, pp. 236-245, 1994.

[10] W. J. Hurley and E. R. Petersen, "Freight Network Equilibrium and Vertically Efficient Tariffs", *School of Business Working Paper* No. 93-17, Queen's University at Kingston, 1993.

[11] E. R. Petersen and H. V. Fullerton, "The Railcar Network Model", *Canadian Institute of Guided Ground Transport*, No. 75-11, Queen's University, Kingston, 1975.

[12] P. Taborga and E. R. Petersen, "RAIL User's Manual", *The World Bank*, Washington, D.C., (1982).

[13] P. Rey and J. Tirole, "The Logic of Vertical Restraint", *American Economic Review* 76, pp. 921-39 (1986).

[14] Jean Tirole, "A Theory of Industrial Organization", *MIT Press* (1989).

THE IMPACT OF PREDICTIVE INFORMATION ON GUIDANCE EFFICIENCY: AN ANALYTICAL APPROACH

Moshe Ben-Akiva

Professor of Civil and Environmental Engineering
Massachusetts Institute of Technology, Cambridge MA 02139, USA

André de Palma

Professor of Economic Sciences - Université de Cergy Pontoise
33, Bd. du Port - F-95011 Cergy Pontoise, France

Isam Kaysi

Assistant Professor of Civil and Environmental Engineering
Faculty of Engineering and Architecture, American University of Beirut
PO Box 11-0236 Beirut, Lebanon

The acceptance of route guidance advice by motorists is expected to occur only whenever such advice is experienced to be valid and reliable. Three major factors may cause route guidance systems to provide motorists with unreliable advice:

(1) the traffic information constituting the basis for guidance advice is inaccurate;
(2) the impact of a large fraction of motorists responding to the guidance, and the subsequent overreaction that occurs, is ignored whenever the guidance advice is being set; and,
(3) concentration effects may occur and induce a higher level of (unpriced) congestion when drivers are provided with better information.

In this paper, the impact of the above factors is investigated for a small prototypical network. We provide numerical and analytical results in order to identify the critical factors involved in the design of driver information systems. The analysis presented in the paper involves (a) estimating the benefits of accurate predictions of traffic conditions; (b) assessing the sensitivity of guidance provision to inaccuracies and imperfections in the traffic predictions; and (c) comparing the behavior of uninformed drivers with that of drivers receiving myopic or predictive guidance. Finally, conclusions and policy implications regarding the design features of advanced traveler information systems are presented.

1. Introduction

Advanced Traveler Information Systems (ATIS) based on modern information technology are expected to play an important role in reducing traffic congestion and improving traffic flows and safety. ATIS have the potential to reduce delays due to both incident and recurrent congestion by providing information to motorists with respect to alternative paths to their destinations and to actual traffic conditions on links of interest using a combination of roadside signals and onboard systems.

Potential Adverse Impacts of Guidance

In considering the potential benefits of alternative traveler information systems, it is also essential to evaluate the potential adverse effects that information may induce. Several behavioral phenomena may negate some of the beneficial effects of improved information, prime among them being overreaction and concentration [2].

ATIS Design Issues

The actual benefits realized from traveler information systems will depend heavily on the quality of the traffic information or guidance provided to drivers [3]. For instance, adverse impacts may not be avoided if the driver information system is not well-designed (see [5]). Moreover, inaccurate traffic information may result in poor guidance directives being provided to drivers. The likely impact would be a gradual loss of confidence in the system and its eventual abandonment.

Desirable Properties of Route Guidance

Guidance provided to drivers has to possess a number of *desirable properties* if it is to be useful for dynamic traffic management and if it is to be adopted by drivers [4]. To be useful for dynamic traffic management, guidance has to be able to *respond in real-time* to emerging traffic conditions and *anticipated* demand levels in such a way that it *reduces the overall extent of congestion*. In addition, since traveler information may prove to be counterproductive in some cases, guidance should be provided in such a way that it *reduces the potential adverse impacts of information.*

From the drivers' point of view, guidance should be *valid* in the sense that the travel time a driver experiences on the route he was guided to should be shorter than that on alternate routes. Moreover, guidance has to improve the *reliability* of travel times experienced by drivers by reducing variability and eliminating unusually long travel times. If drivers receive guidance which is not characterized by these desirable properties, they will not have confidence in the information system. As a result, drivers may not comply with the guidance being provided, in which case the dynamic traffic management potential of the information system would not materialize.

The objective of this paper is to conduct prototypical analyses of ATIS design elements using analytical approaches. A previous paper by the authors [6] provided a similar evaluation analysis which was, however, based on simulation tools.

2. The Sample Network

Network Structure

The network used in the analytical study was chosen to be simple but at the same time to include features that will enable an analysis of the factors involved in the design of traveler information systems. The network is presented in Figure 1 and consists of 4 links (labeled 1 to 4). Each of these four links has a potential bottleneck at its end.

O-D Flows

Associated with the network are 3 O-D flows all of which are destined to node G, as follows:

- $\Phi(t)$ rate of flow entering at node C at time t
- $\rho_2(t)$ rate of flow entering at node E at time t
- $\rho_4(t)$ rate of flow entering at node F at time t

Flow $\Phi(t)$ has two possible routes to the destination. Flows $\rho_2(t)$ and $\rho_4(t)$ have no route choice as they have to follow links 2 and 4, respectively.

Guidance Provision

The network consists of two routes in parallel. Drivers receive route guidance at point C concerning which of the two routes they should follow to their destination, point G.

$\Phi(t)$ represents the incoming flow which is to be guided. $\Phi_1(t)$ and $\Phi_3(t)$ are the flows departing from C at time t and following routes A (links 1 and 2) and B (links 3 and 4), respectively ($\Phi(t) = \Phi_1(t) + \Phi_3(t)$).

Figure 1. Prototypical Network

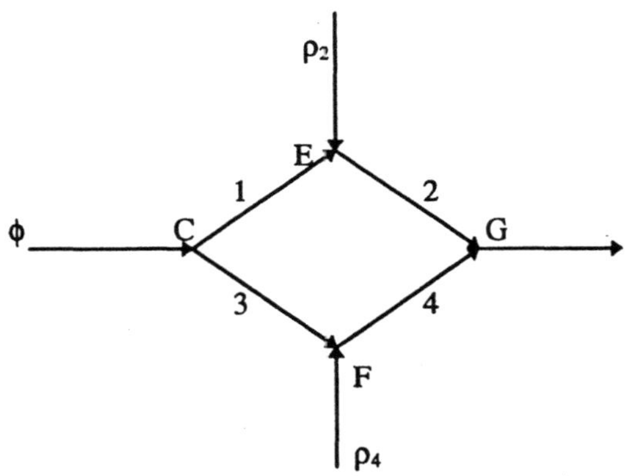

If traffic to be guided constituted the only flow on the two routes (no side flows; i.e., $\rho_2(t)=\rho_4(t)=0$) then information on the current traffic conditions would be sufficient to compute future traffic conditions and to guide traffic properly, in the absence of accidents or any other factors that could vary route capacity. However, for the network adopted here the side flows need to be taken into consideration when providing guidance at C since future traffic conditions will depend on the future arrival of side flows in the network. These latter flows represent "external demands" that are loaded onto the network and affect its traffic conditions but are outside the scope of the guidance system at point C in the sense that they can not be controlled or diverted.

Network Performance

The link performance functions used to determine the travel times on links, or the time a vehicle spends on a specific link, assume that each link consists of a running section that has a constant free flow travel time and a bottleneck at the end. Without any loss of generality, the travel time on the running sections of the four links is set to zero. As such, we will consider delays at bottlenecks as the only measure of travel disutility faced by traffic. Moreover, each bottleneck is modeled as a deterministic queue which possesses a specific discharge capacity. The network has four bottlenecks in total (of capacities s_1, s_2, s_3, and s_4), two on each of the parallel routes. Queues in the bottlenecks build up or dissipate based on the level of the incoming flows relative to the bottleneck discharge capacity.

Network Topology

The adopted network represents an extreme case of network topology as it relates to the spatial frequency of guidance points. This network represents the case in which a unique decision point exists, and traffic that is guided at point C cannot switch from the route selected at C. As such, *no possibilities* exist for adapting to evolving traffic conditions or for recovering from mistakes that were made when guiding vehicles at C. Cases whereby drivers obtain updated guidance advice at other points along their route are not addressed here, and the conclusions reached in analyzing the network at hand may not necessarily apply to such cases.

3. Guidance System and Information Regimes

Route Distributive Guidance

The objective of the route guidance system is to split traffic at C between routes A and B such that drivers are guided to the minimum travel time route. If both routes A and B are used at time t, then traffic has to be *distributed* between the two routes in such a way that their travel times are equal (the "user equilibrium" objective). We refer to guidance which splits traffic among the two routes A and B as route distributive guidance. As discussed in [5], the implementation of route distributive guidance in reality may not be straightforward. For instance, if in-vehicle units are used to transmit guidance advice to drivers, it is technically possible to provide different drivers with different route directives at the same time to split them among alternate routes. However, such a procedure may not be politically or legally acceptable due to its inequity implications. An alternate scheme for achieving route distributive guidance is likely to be required. One possibility, which can also be implemented with variable message signs (VMS), is to change the guidance message or sign repetitively so that the time average of the route directives would correspond to the splitting fractions to be achieved for each route. In either case, all drivers are considered to be "informed" and to follow guidance advice so that the intended splitting fractions are actually achieved.

The Information Regimes

Given the network elements described above, the analysis considers three information systems, namely, instantaneous information, predictive information, and no information. Each of the information regimes is discussed next.

3.1 Instantaneous Information

The first information system uses *instantaneous* information to determine how many drivers should be diverted to routes A and B.

Travel time on a route at time t is the sum of the (instantaneous) travel time on different links of that route as computed at time t and given the prevailing traffic conditions. The traffic control system aims here to achieve user equilibrium in the sense of equating the travel times on routes A and B as determined using instantaneous link travel times. This is, of course, an ex-ante user equilibrium objective since the travel time experienced by the drivers will be different in general from the instantaneous or *myopic* travel time.

Let $Q_i(t)$ be the queue length at time t at the bottleneck entrance i, i=1...4. The travel time for a driver joining the queue before bottleneck i at time t is $Q_i(t)/s_i$. The instantaneous travel time for a driver arriving at C at time t and following route A is:

$$T^M{}_A(t) = Q_1(t)/s_1 + Q_2(t)/s_2 \tag{1a}$$

and likewise for route B it is:

$$T^M{}_B(t) = Q_3(t)/s_3 + Q_4(t)/s_4 \tag{1b}$$

Equations (1a) and (1b) reflect the fact that, with the instantaneous information regime, the travel time on each link of the network is calculated at time t, and not at the time the link will be used, as should be the case.

The evolution of the queue length is provided by the following flow conservation relationships (assuming that all the bottlenecks are used at full capacity):

$$dQ_1(t)/dt = \Phi_1(t) - s_1 \tag{2a}$$
$$dQ_2(t)/dt = s_1 + \rho_2(t) - s_2 \tag{2b}$$
$$dQ_3(t)/dt = \Phi_3(t) - s_3 \tag{2c}$$
$$dQ_4(t)/dt = s_3 + \rho_4(t) - s_4 \tag{2d}$$

The user equilibrium objective based on instantaneous travel times is $T_A(t) = T_B(t)$. Differentiating this equation and using (1) and (2) one gets

$$\Phi_1(t) = \frac{s_1 s_3}{s_1 + s_3}\left[\frac{\Phi(t)}{s_3} + \frac{s_3 + \rho_4(t)}{s_4} - \frac{s_1 + \rho_2(t)}{s_2}\right] \tag{3}$$

All the comparative statistics are obvious and intuitively sound except for the effects of s_1 and s_3. In the absence of side flows ($\rho_2(t) = \rho_4(t) = 0$) the capacities s_1 and s_3 should not play any role in determining the relative split of traffic between the two parallel routes. The reason is that, considering route A, when the two bottlenecks 1 and 2 are in series with $s_1 > s_2$ (see equation (2b)), s_1 should not affect travel time on route A. In the case of equation (3), the splitting fraction depends on s_1 and s_3 even if the side flows are zero. This is a consequence of the fact that instantaneous rather than actual travel times have been used to achieve the user equilibrium objective. Comparative statistics as a function of s_1 and s_3 are thus ambiguous and the result of a mathematical artifact.

It is a matter of simple algebra to show that in general the travel time experienced by drivers will not be the same on the two routes. The exact travel times expressions are complicated even for simple patterns of arrivals and are not discussed here.

3.2 Predictive Information

The second information system is based on *predictive information*. Here it is recognized that the time to clear bottleneck 2 on route A is not $Q_2(t)/s_2$ (the travel time computed at time t) but rather that Q_2 has to be computed at time $t+Q_1(t)/s_1$ which represents the arrival time at the second bottleneck of a car joining the queue of bottleneck 1 at time t. This compensates for the fact that a vehicle will not reach the second bottleneck instantaneously - at time t - but will do so only after all cars in front of it in bottleneck 1 have cleared this bottleneck. Therefore, in the case of perfect foresight, the travel times for a driver arriving at C at time t are computed using the correct expressions, as follows:

$$T^P{}_A(t) = Q_1(t)/s_1 + Q_2[t+Q_1(t)/s_1]/s_2 \quad \textbf{(4a)}$$

$$T^P{}_B(t) = Q_3(t)/s_3 + Q_4[t+Q_3(t)/s_3]/s_4 \quad \textbf{(4b)}$$

In practice, Q_1 and Q_3 will be based on the latest measurements of traffic conditions and Q_2 and Q_4 need to be predicted. The user equilibrium objective, based on foresight, is achieved by requiring that $T^P{}_A(t) = T^P{}_B(t)$. Again, differentiating this equation and using (2) and (4), one obtains:

$$\Phi_1(t) = \Phi(t) \quad \frac{\Psi_2(t)^{-1}}{\Psi_2(t)^{-1} + \Psi_4(t)^{-1}}$$

where

$$\Psi_2(t) = \frac{s_1 + \rho_2^F(t)}{s_1 s_2}; \quad \Psi_4(t) = \frac{s_3 + \rho_4^F(t)}{s_3 s_4} \quad \textbf{(5)}$$

$$\rho_2^F(t) = \rho_2\left[t + \frac{Q_1(t)}{s_1}\right]; \quad \rho_4^F(t) = \rho_4\left[t + \frac{Q_3(t)}{s_3}\right]$$

Note that the splitting fraction at time t depends on the observed queues on links 1 and 3 at time t and on the side flows at later times. Thus, the implementation of this route guidance objective requires measurements of the queues at 1 and 3 and predictions of the side flows. If traffic patterns are repetitive, information concerning future traffic conditions may be approximated from data on historical travel patterns.

In this case all the comparative statistics give intuitive results. It can also be checked that when the side flow on either (or both) route is null, only the capacity of the downstream bottleneck matters. The splitting pattern defined by equation (5) guarantees that the travel time experienced by the drivers are the same at each instant on the two routes (although, of course, they generally change as a function of t).

3.3 No Information

The above two cases reflect the situation whereby drivers are split between the two routes based on either myopic or predictive information. The split is implemented by providing drivers with information regarding the route to be followed to their destination. The information will either reflect instantaneous traffic conditions at the time drivers reach point C or traffic conditions predicted to materialize by the time drivers reach the downstream bottlenecks.

The travel times experienced based on the two information regimes also need to be compared with the case where drivers are uninformed. In such a case, the behavior of drivers is assumed to reflect their response to the "average" traffic conditions they are likely to experience on the downstream bottlenecks. That is, such drivers formulate a "predictive" view of the traffic conditions on an average day and split themselves in such a manner that the travel time on the two routes will be equal if "average" traffic conditions materialize. Obviously, on any given day, traffic conditions differ from the average, and therefore uninformed drivers will be unable to adjust their behavior since they have no information regarding actual or future traffic conditions when they make their route choice at C. In general, their route choices will be suboptimal.

4. Evaluation of Myopic and Predictive Information Regimes

In this section two basic cases will be addressed, namely, constant flows and day-to-day flow variability.

4.1 Constant Flows

In what follows, the situation whereby all travelers receive either myopic or predictive information is considered and the effectiveness of these two information regimes is compared.

4.1.1 Formulation

Here we discuss the case whereby all the arrival flows (upstream and side flows) are constant as a function of time. In such a case the splitting factors are constant over time and can be computed explicitly from equations (3) and (5). The actual travel times on the two routes can be computed using equations (4a) and (4b) and by integrating equations (2a)-(2d). It can be checked that when using the instantaneous information splitting factor, the actual travel times on the two routes are generally different and consequently that a fraction of users have ex-post been misguided.

Alternatively, when using the perfect foresight splitting factors, the travel time on the two routes are the same and are given by:

$$T_A(t) = T_B(t) = \left(\phi \frac{\Psi_2 \Psi_4}{\Psi_2 + \Psi_4} - 1 \right) t \tag{6}$$

4.1.2 Equivalence Between Information Regimes

An obvious question concerns the conditions under which the two information regimes would provide the same splitting ratios and, consequently, the same guidance. In other words, it may be the case that traffic evolves in such a way that errors in the instantaneous case wash out and the two formulas (3) and (5) become equivalent. This can happen in two situations: (i) if the numerical values of the factors included in the two expressions just happen to produce the same splitting factors; and, (ii) for a more systematic case which occurs under the following conditions:

$$\frac{s_1 + \rho_2^F(t)}{s_1 s_2} = \frac{s_3 + \rho_4^F(t)}{s_3 s_4} \tag{7a}$$

and

$$\frac{s_1 + \rho_2(t)}{s_1 s_2} = \frac{s_3 + \rho_4(t)}{s_3 s_4} \tag{7b}$$

For the current case of constant flows, the above two conditions simplify to

$$\frac{s_1 + \rho_2}{s_1 s_2} = \frac{s_3 + \rho_4}{s_3 s_4} \tag{7c}$$

A sufficient condition for the above two equalities (i.e., (7a) and (7b)) to hold simultaneously is that the downstream queue length computed by the instantaneous case (at time t) and by the predictive case (taking into account the expected delays at the upstream bottlenecks) be the same, even though these may be time dependent. Another interpretation is to assume a "naive" prediction method such that the predicted value for $\rho_i^F(t)$ is $\rho_i(t)$, i=2, 4. In this case, although the instantaneous model does not provide correct estimates of travel times, the prediction errors on the two routes are such that the splitting ratios are unaffected. Note finally that when the downstream capacities are "infinite" (that is, practically, when they are not binding), the two information approaches are equivalent as expected.

4.1.3 Ranking Predictive and Myopic Information Regimes

For the network under consideration, and with constant flows, it is possible to make a definite statement concerning the benefits of predictive information compared to instantaneous information. For this purpose, we first compute the total travel time experienced by flow incoming to C at time t. The expression for this total travel time can be written as:

$$TT(t) = T_A(t).\Phi_1 + T_B(t).\Phi_3 \tag{8}$$

TT is computed for the predictive case based on equations (4) and (5). For the myopic case, the splitting ratio is obtained from equation (3) while $T_A(t)$ and $T_B(t)$ to be used in equation (8) relate to the *actual* travel times experienced on the two routes, based on the designated splitting ratio. Comparing the total travel time for flow incoming at time t for the two cases of predictive (P) and myopic (M) information, the following expression is obtained:

$$\frac{TT^M(t) - TT^P(t)}{t} = \frac{\left(\Phi_1^M - \Phi_1^P\right)^2}{\Psi_2 + \Psi_4} \tag{9}$$

In other words, it can be concluded that $TT^M \geq TT^P$; that is, the total travel time experienced by flow incoming to C and split based on predictive information is at all times less than or equal to the value experienced if traffic is split according to myopic information. The equality holds for the condition stated in equation (7c). In addition, point equalities occur whenever $\Phi_1^M = \Phi_1^P$. It should be noted, however, that the result presented above is not obvious since perfect information could leave traffic conditions worse off than with imperfect information. For instance, Arnott, de Palma, and Lindsey [1] concluded that instances exist whereby zero information can be more desirable from the social point of view than perfect information.

4.1.4 Sample Results

To illustrate some of the results for the network being considered, a simple numerical example will be considered.

The following values of the input parameters will be utilized: $\Phi = 10$; $s_1 = 2$; $s_2 = s_3 = 4$; $s_4 = 6$; $\rho_2 = 2$; $\rho_4 = 5$. It should be noted that the basic assumption that all bottlenecks are used up to capacity at all times introduces restrictions on the feasible combinations of these input parameters. Although the results presented below are associated with specific input parameters, the conclusions are robust with respect to variations in the values of these parameters.

Level of Incoming Flows - Figure 2 indicates that the total travel time incurred with instantaneous information regarding traffic conditions is somewhat larger than that incurred by traffic which is split based on perfect foresight information. However, a more critical conclusion relates to the observation that the ratio of travel time on the two alternate routes when traffic is split according to instantaneous information becomes quite different from 1.00 as Φ increases. This indicates that confidence in such information, and therefore a high level of compliance, are unlikely to be sustained as drivers using the longer route are bound to find out that they are being misguided ex-post. In other words, while a slightly lower guidance efficiency is to be expected from using instantaneous information, the compliance rate is likely to be quite low. The figure also shows that the inefficiency introduced by the use of myopic information increases as the incoming flow at C becomes larger since the inaccuracy introduced by the use of such information increases.

Level of Side Flows - Figure 3 illustrates the variation of TT^M/TT^P and T_A/T_B for the myopic case as the side flow ρ_2 varies between 2 and 8 units, representing the feasible interval for ρ_2. This figure indicates that the ratio of TT^M/TT^P changes from 1.006 (at ρ_2=2), to 1 (at ρ_2=4 and ρ_2=5), and to 1.008 (at ρ_2=8). As such, this ratio is always greater than or equal to 1, and predictive information is always more effective than instantaneous information, as guaranteed by equation (9). Moreover, the ratio of the travel times on the two alternate routes for the myopic case, T_A/T_B, varies from a low of 0.75 (at ρ_2=8) to a high of 1.01 (at ρ_2=4.5) while T_A=T_B occurs at the two values of ρ_2=4 and 5 since the expression for T_A=T_B is quadratic in ρ_2. These last results indicate that certain values of the side flows can generate wide discrepancies in the travel times on the two alternate routes for the myopic case. As discussed above, this can potentially have adverse implications in terms of compliance with and confidence in the information system being adopted.

Figure 2. Effect of Incoming Flow Level

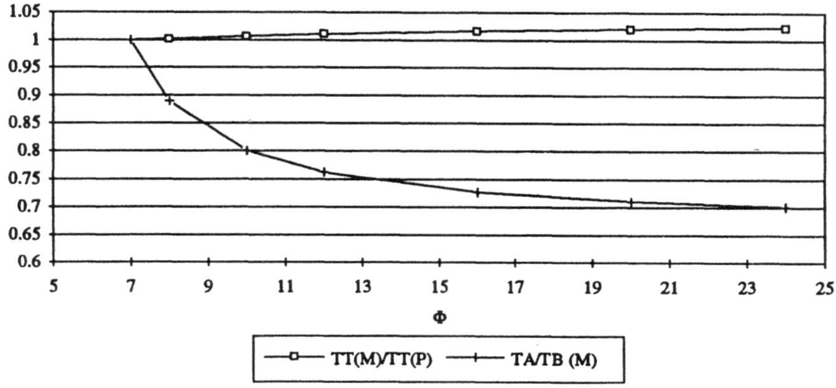

Figure 3. Effect of Side Flow Levels

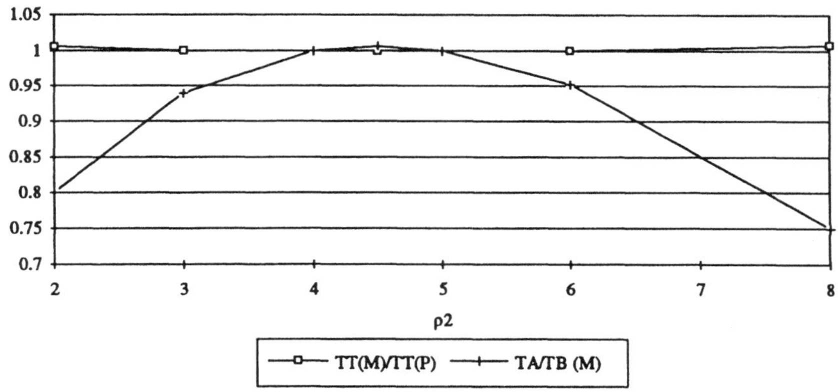

Prediction Accuracy - It is clear that perfect foresight constitutes an ideal situation given that it requires totally accurate predictions of the levels of the future side flows. Predictions of future side flows can be obtained in general but are unlikely to be perfect. Figure 4 illustrates how the ratio of the total travel times as well as the ratio of travel times on the two alternate routes for the predictive and myopic cases vary as the (imperfectly) predicted level of ρ_2 varies around the actual level which is equal to 5 units. It should be noted that the splitting ratio for the myopic case, as determined by equation (3), is based on the correct value of ρ_2 which is observed, and as such no inaccuracies are introduced in the evaluation of the ratio. The results presented below are associated with a value $\Phi=20$.

The results indicate that the superiority of predictive information over instantaneous information is quite robust with respect to the quality of flow prediction. Even though the total travel cost for the predictive case TT^P increases as the quality of the flow prediction deteriorates, predictive information remains superior to instantaneous information even with a prediction error in ρ_2 as large as 40% (that is, with the imperfectly predicted ρ_2 varying between the values of 3 and 7). Moreover, even with quite inaccurate predictions of ρ_2, predictive information still does a much better job at achieving almost equal travel times on the two alternate routes, and thus user equilibrium is almost maintained.

Figure 4. Effect of Prediction Accuracy
(actual ρ_2 = 5)

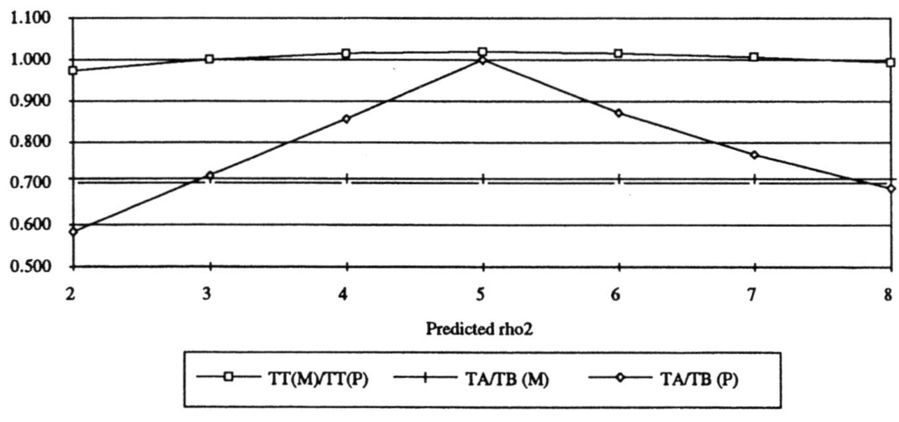

4.2 Day-to-Day Flow Variability

In the previous section we considered a case whereby flow levels are constant over the course of the period of analysis. We now expand our analysis to consider situations in which flow levels remain constant over any day but vary from day to day. In specific, we consider three different types of days, each with a different combination of (constant) flow levels and a probability of occurring, as follows:

Day Type	Probability	Φ Level	ρ_2 Level	ρ_4 Level
1	0.25	Φ	$\theta\rho_2$	ρ_4
2 ("normal")	0.50	Φ	ρ_2	ρ_4
3	0.25	Φ	ρ_2	$\theta\rho_4$

The parameter θ reflects the extent to which side flows vary on days with heavy side traffic from levels experienced on "normal" days. In the above table, a "normal" day has a 0.5 probability of occurring while days whereby each of the side flows is heavy ($\theta \geq 1$) have a probability 0.25 each of occurring. Figure 5 presents average values of the total travel times (TT) over the different day types for drivers affected by each of three different information availability scenarios: predictive information, instantaneous information, and no information. The first two cases are familiar by now with the only difference being that splitting ratios are set based on traffic information that will reflect (in a myopic or predictive sense) the type of day being experienced. On the other hand, the splitting ratios for the uninformed drivers remain the same from day to day and are simply based on achieving equal travel times on the two routes if a normal day is experienced. That is, such drivers are assumed to adopt a behavior which responds to normal traffic conditions and which is not expected to perform well if there is wide variability in traffic conditions.

Figure 5 indicates that for the day types being considered, and for values of θ ranging from 1 to 2, total travel time for uninformed drivers is less than that experienced by drivers responding to myopic information. On the other hand, predictive information provides benefits, especially as θ increases, reflecting greater variability in traffic conditions. The implication here is that in case traffic conditions do not display large variability from day to day ($\theta < 1.8$) *and* uninformed drivers are able to respond to "normal" traffic conditions properly, then uninformed drivers are better off than drivers provided with guidance based on myopic information. Moreover, for values of $\theta < 1.2$, indicating very little variability in traffic conditions from day to day, predictive information offers only marginal improvement over the no information case.

Figure 5. Day-to-Day Flow Variability

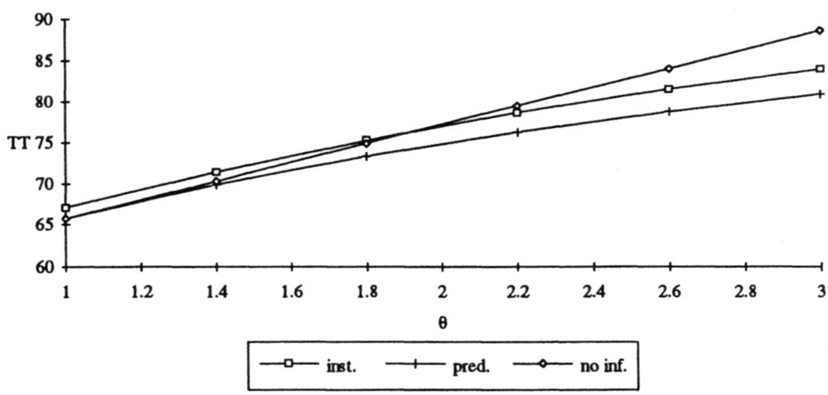

5. Conclusions and Policy Implications

Previous Research

Previous research by the authors (see [6]) addressed ATIS design issues using simulation tools and concluded that one of the critical characteristics of successful guidance system designs consists of basing guidance on predictive traffic information which accounts for driver reaction. Simulation results indicated that guidance based on very good flow predictions provide minor reductions in average delay but significant improvements in guidance validity levels. These same conclusions were confirmed by the analytical derivations and numerical results presented in this paper, as described below.

Conclusions and Policy Implications

The analytical study described in this paper has demonstrated, for a small prototypical network, the value of predictive information as a basis of route guidance. Three main conclusions can be made, namely: (1) While total travel time is only reduced slightly by basing guidance on predictive rather than instantaneous information, the latter type of information results in large discrepancies in travel time between the two alternate routes; (2) The superiority of predictive information over instantaneous information is quite robust with respect to the quality of required flow prediction; and, (3) When day-to-day variations in traffic flows are not large, uninformed drivers who behave according to "normal" traffic conditions may be better off than drivers who receive guidance based on myopic or instantaneous information. Moreover, if the day to day variability is very limited, guidance based on predictive information is not called for as it offers only marginal improvements over the no information case. Since the success of any ATIS implementation will clearly depend on the quality of the guidance it provides to drivers, guidance validity represents one of the major considerations at the ATIS design stage. The first conclusion described above indicates that a guidance system based on instantaneous information is not likely to sustain operation due to the large discrepancies in travel times between the two alternate routes which result in low guidance validity levels and, as a result, a loss of confidence in the system.

Further Research

Ongoing research is addressing the issues of within-day flow variability as well as the impact of market penetration and value of information using analytical techniques. The results of this research will be compared with conclusions reached by investigating similar issues based on simulation tools, as was undertaken in previous research. Other extensions which will be considered include the issues of endogenous flow arrival rates and the possible coordination between traffic control measures and route guidance systems.

References

[1] Arnott, R., A. de Palma, and R. Lindsey (1994) "Dynamic Model of Peak Period Congestion with Heterogeneous Commuters", *Journal of Transportation Economics and Policy*, forthcoming.

[2] Ben-Akiva, M., A. de Palma, and I. Kaysi (1991) "Dynamic Network Models and Driver Information Systems", *Transportation Research*, Vol. 25A, No. 5.

[3] Bonsall, P.W. and A.D. May (1989) "Evaluation of In-Vehicle Route Guidance Systems", paper presented at the *5th WCTR*, Yokohama, Japan.

[4] Kaysi, I. (1991) "Framework and Models for the Provision of Real-Time Driver Information", Unpublished Ph.D. *Dissertation, Department of Civil Engineering, MIT*, November 1991.

[5] Kaysi, I., M. Ben-Akiva, and H. Koutsopoulos (1993) "An Integrated Approach to Vehicle Routing and Congestion Prediction for Real-Time Driver Guidance", *Transportation Research Record*, No. 1408.

[6] Kaysi, I., M. Ben-Akiva, and A. de Palma (1994) "Design Aspects of Advanced Traveler Information Systems", *Compendium of Papers*, 1992 Capri Seminar on Urban Traffic Networks.

DYNAMIC TRAFFIC PREDICTION FOR MOTORWAY NETWORKS

M. Ben-Akiva
Massachussets Institute of Technology, USA

E. Cascetta
University of Naples, Italy

H. Gunn
Hague Consulting Group, Netherlands

D. Inaudi
Centro Studi sui Sistemi di Trasporto S.p.A. of Turin, Italy

J. Whittaker
University of Lancaster, England

A dynamic prediction system for an inter-urban motorway network which forecasts traffic conditions on the network in real time, and provides information to a motorway traffic control centre is described.
The model receives traffic data such as traffic flow, average speed and average detector occupancy in real-time from a motorway monitoring system. Additionally, a database of historic traffic information will be used.
The model system contains several submodels: a statistical state space model for filtering noisy traffic data, and for very short term predictions (prediction horizon: 5-15 min.); a dynamic traffic assignment model for short term predictions (prediction horizon: 15-60 min.); a real-time O/D matrix prediction model to provide the input flows to the statistical and assignment models. Predictions from the two models are merged where their respective time horizons overlap.

1. Introduction

Many advanced industrial societies are feeling capacity constraints on their motorway networks in the form of increasing congestion and lengthening travel times. Improved management of the existing network is one of the few options available to ameliorate the traffic conditions. Dynamic traffic models transform on-line roadside measurements of traffic flow into explicit assessments about the current and future state of an inter-urban network and thus provide transport management with a tool for monitoring, prediction and potentially control.

Two somewhat different methodological approaches are combined in the prediction system described herein, which take advantage of the two rather different sources of information available to a traffic control centre. One is essentially behavioural, in which O/D flows are mapped to entry/exit points on the network and then dynamically assigned to the links. The other is essentially statistical, in which the observed correlations between traffic variables, together with traditional ideas of mathematical traffic modelling, are mapped into the framework of a state space model.

The apparently large quantity of data (both in time and space) is mainly uninformative and is simplified by investigation of the conditional independence structure of the process. This structure in part reflects the given spatial topology of the network, and is incorporated in the state transition equations. The transition and observation equations lead to an associated Kalman filter which provides the relative discounting of past data in proportion to the underlying variability of the traffic process. Prediction ahead is an instance of Bayesian forecasting. Basic ideas of traffic flow, such as conservation of vehicles and the fundamental diagram are built into the state transition equations. However some complications ensue because of non-linearities in speed-flow relationships and resulting non-linearity of the equations.

The paper describes project work underway in the EC/DRIVE II initiative, and in cooperation with partners from the Hague Consulting Group, DVK (from the Dutch Ministry of Transport), CSST and ELASIS (company part of the FIAT group), and from the Universities of Lancaster and Napoli. The overall aims of the project entitled Dynamic Traffic Modelling (DYNA) are to develop practical forecasting tools and to validate these in a full-scale pilot project on an inter-urban motorway network centred on Rotterdam. The emphasis of this work is to provide practical working methods based on existing technology, but preliminary pilot studies have naturally raised some interesting statistical and theoretical questions.

The O/D module is described in Section 2, dynamic traffic assignment in Section 3, and the state space prediction system in Section 4.

2. O/D Estimation and Prediction

An overview is presented of how the O/D model is set up; after this the underlying theory is presented.

2.1 Schematic View

The diagram below shows the O/D subsystem. Traffic counts and historical O/D matrices are made available through the Data Interface. The most recent assignment matrix is made available by the Dynamic Traffic Assignment (DTA) module. This assignment matrix maps the (unobservable) O/D flows into the observable link flows, and constitutes a key input to the subsystem. The O/D estimation module takes the counts, the historic O/D matrices, and the most recent assignment matrix to produce an estimate of O/D flows which are on the network. This estimated O/D matrix is used to obtain a predicted O/D matrix for the subsequent time-intervals.

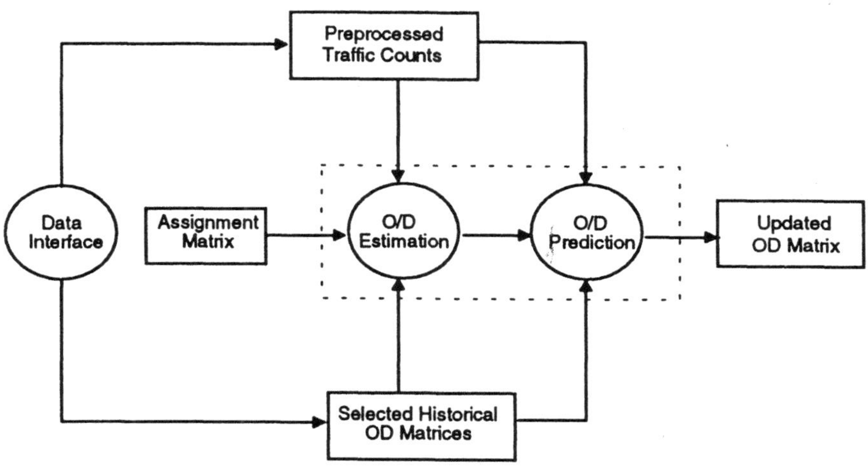

2.2 Theoretical Framework

The estimation of dynamic O/D matrices from traffic counts can be approached in two different ways. *Simultaneous* estimation can be used to obtain the O/D matrix for multiple time intervals. *Sequential* estimators iteratively produce the O/D matrix per time interval using as input the traffic counts for that time interval and the estimated O/D matrices of previous time intervals. The latter approach, which is computationally less demanding is the only possible option for real-time applications. Discrete time intervals are used throughout.

Definitions and notation

Let d_{ij} be the demand for O/D pair i which departs from the origin during time interval j; f_{lh} be the flow at measuring point l during interval h; a_{lh}^{ij} be the fraction of d_{ij} crossing measuring point l during interval h (called in the following *crossing fractions*) then

$$f_{lh} = \sum_{i} \sum_{j \leq h} a_{lh}^{ij} d_{ij} \tag{1}$$

or in matrix formulation

$$f_h = \sum_{j \leq h} A_h^j d_j \tag{2}$$

where: d_j denotes the demand vector that departs during interval j; f_h denotes the flow vector during interval h; and A_h^j denotes the assignment matrix for traffic during interval h which departs from the origin during interval j.

Problem formulation

Sequential estimation can be performed by expressing traffic counts for the current interval as a linear stochastic function of the unknown demand for the same interval, starting from relationship (2) between demand and traffic flows:

$$f_h = \sum_{j \leq h-1} A_h^j d_j^* + A_h^h d_h + e_h \tag{3}$$

where: f_h is the traffic count vector for the current interval h; d^*_j is the estimated demand vector for interval $j<h$; d_h is the unknown demand vector for the current interval h; e_h is a random error. Usually j is greater than $h-n$, where n identifies a "past horizon" in terms of number of intervals which demand could contribute to the composition of f_h.

The crossing fractions contained in the assignment matrices should be computed through a dynamic network loading procedure that is dependent on the demand vector, leading to a fixed-point problem. For real-time applications an O/D estimation for interval h is performed at the end of interval h, based on the estimate for previous time intervals up to $h-1$ and the prediction for time interval h, which can be used to estimate crossing fraction matrices for interval h. Through relationship (3) the dynamic estimation problem can be formulated as a minimisation problem. Let d^0_h be the a-priori estimate of the demand vector for interval h (d^*_{j-1} can be used if nothing else is available). The estimator of the demand vector for interval h is:

$$d_h^* = \arg\min\left[\phi_1\left(d_h, d_h^0\right) + \phi_2\left(d_h, f_h, d_1^*, \ldots, d_{h-1}^*\right)\right] \tag{4}$$

where ϕ_1 is the error function for the a-priori demand vector and ϕ_2 is the error function for the available traffic counts during interval h. Several formulations can be adopted for ϕ_1 and ϕ_2 based on different statistical assumptions about the random errors. For more detail on the estimation of dynamic O/D matrices see Cascetta et al. (1993).

O/D prediction will be based on a simplification of the approach developed in Ashok and Ben-Akiva (1993). It is based on an autoregressive process describing the pattern of deviations between the a-priori matrix d^0_h and the estimated matrix d^*_h. Given the calculated deviations the a-priori matrices for intervals $h+1$, $h+2$, ... are adjusted using the coefficients of the previously estimated autoregressive process. For more details on the O/D prediction see Cantarella et al. (1993).

3. Dynamic Traffic Assignment

For a given O/D matrix, a Dynamic Traffic Assignment (DTA) model predicts the flows and travel times on the links of the network. A schematic overview of the structure of DYNA's DTA is given below, followed by a brief outline of the theoretical framework underlying the dynamic assignment approach.

3.1 Schematic View

The diagram below shows a detailed view of the DTA subsystem. Central to this subsystem is a module that keeps track of the past, current, and latest predicted 'State of the Network'. The data interface supplies traffic measurements to this process, which enables it to update current travel times. Up-to date infrastructure data is used, reflecting any changes made by the traffic operator through the operator interface. Directly from the operator interface is an arrow, indicating that the traffic operator may access the state of the network directly.

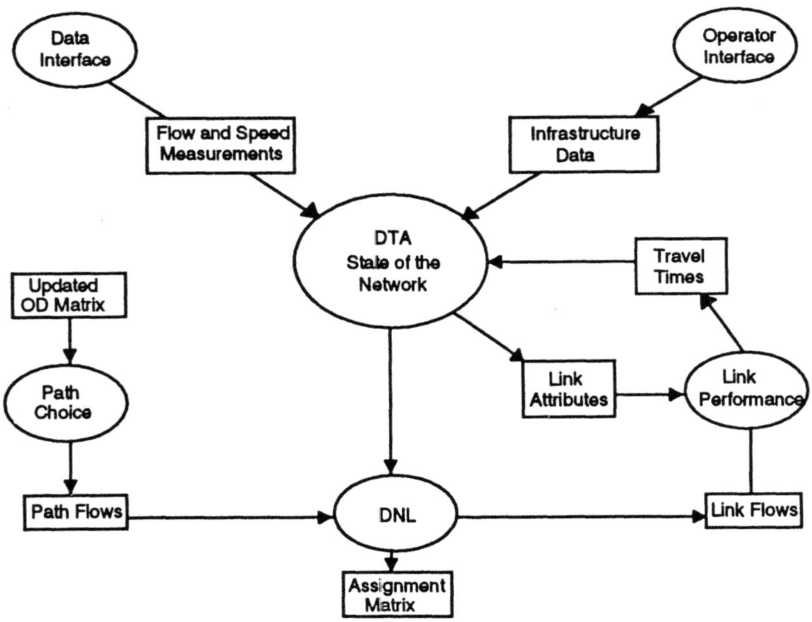

The previously described O/D prediction subsystem produces an updated dynamic O/D matrix. This O/D matrix enters the path choice module, where given the observed (past and current) and projected (future) States of the Network a set of Path Flows is produced. The path flows and the state of the network are used by the dynamic network loading (DNL) module to derive the link flows. The link performance module uses the link flows calculated in this way by the DNL module, and the current link attributes (e.g. speed, density, etc.), to calculate the travel times of all links. These, in turn, are used to update the state of the network.

The DNL module is based on a "continuous packet" approach, where the dynamic O/D flows are represented during the simulation by a number of "packets" of cars. Separate packets are generated for each combination of: O/D pair, user-type (e.g. whether or not equipped with on-board information systems), selected path through the network, and departure time. Continuous packets have variable length with explicit location of the front and back ends of each packet on the network. For more detail see Cascetta and Cantarella (1991) and Di Gangi (1992).

Link Performance will be simulated using link-specific speed-flow curves calibrated on observed flows. A two-regime model will be used, with speed-flow curves fitted in off-line mode being used during uncongested operation, and additional waiting times being modelled using a queuing model once congested operation has been reached (see also Toorenburg (1989) and Transpute (1992)).

The path choice module simulates path choice behaviour of the drivers using a procedure which can be summarised as follows:

(i) drivers entering the network have an initial choice set, containing their feasible paths given their entry and exit points;

(ii) on the basis of the information available to them at that moment, their path choice at network entry is modelled using a probabilistic discrete choice model;

(iii) at each diversion point in the network, the drivers (may) given up-to-date information about the network conditions, consider the path-alternatives relevant for them at that point in the network, and (if relevant) reconsider their choice for the remaining part of their trip on the network. Similarly they may be advised at the diversion node to use a certain route; the model then simulates their compliance with the given advice, and produces a modified prediction of their route choice.

This procedure simulates the en-route diversion that might occur in response to new information or route-advice.

In the path choice process different classes of drivers may be simulated, for instance frequent drivers (commuters, with past experience about the network conditions) and infrequent drivers (those travelling for leisure, relying mainly on external information), who may have different perceptions of the network conditions and who may have different preferences in choosing their paths.

Recent research has been carried out into route-choice behaviour (see Ben-Akiva et al. (1984)) yielding models which could be used in conjunction with expectations of journey times to estimate proportions of drivers using the different paths. In addition, a large-scale market-research exercise has been undertaken to investigate 'values-of-travel-time-savings', differentiating between user groups and contexts (see HCG (1990)).

For the prototype DYNA system, simple models will be assembled from these sources. When the control systems are actually operational, it will be feasible to sample travellers and obtain data on the basis of actual decisions (compliance rates with advice, evaluation and use of information). This will be the most useful basis for operational models, but will not be available for some time. It is to be expected that such a sampling process will be needed on a regular basis, at least in the early years of a new information system, as familiarity with the potential of the system grows.

3.2 Theoretical Framework

Traffic assignment models are used to simulate link flows on transportation networks and the resulting link performances, such as travel times, congestion, pollution and energy consumption. Traffic assignment models are based on a demand model predicting user behaviour and a supply model representing network performance.

If travel demand and supply are assumed to be uniform over the simulated period (e.g. the morning peak), the assignment model is called within-day or intra-periodic static. Time dependent demand (for instance during the build-up of the rush hour) and/or changes in supply (due for instance to weather conditions or accidents) generally lead to time-dependent flows and queues which can only be simulated through within-day or intra-periodic dynamic models. Such models can also take account of the effects of real-time control measures (such as variable message signs or traffic information broadcasts).

Demand model

Users are grouped into n_i classes that are homogeneous in all behavioural characteristics. In particular they travel between the same O/D pair (possibly for the same purpose), and they have a common set K_i of available paths. It is assumed that on each day, d_i users of class i travel between the O/D pair. Elasticity of demand can be dealt with through the introduction of a fictitious alternative (e.g. a dummy path for within-day static models, and a dummy departure time for within-day dynamic models) corresponding to the choice of not travelling at all. In this case d_i is the number of potential travellers in user class i, and d is the corresponding O/D demand vector.

The reference period is divided into n_j intervals, which can be assumed, without loss of generality, to have equal length T; F_{jk} denotes the path flow, i.e. the number of users that depart during interval j and embark on path k (where k denotes the actual path followed to the final destination when all en-route diversions have been taken into account). Denote the corresponding path cost by C_{jk}, the path flow vector by F and the path cost vector by C.

The demand model describes user behaviour by expressing path flows as a function of path cost. Let $p_{i,jk}$ denotes the probability that a user of class i selects departure interval j and path k, and denotes the corresponding $n_i*(n_j*n_k)$ choice matrix (usually modelled through random-utility discrete-choice models) by P. Assuming that P directly depends on the path cost vector C (see Cascetta Cantarella (1992a), Cascetta Cantarella (1992b) for more general models) leads to:

$$F=P(C)d \tag{5}$$

Supply model

The supply model describes the relationship between path costs and path flows. In the within-day static case the supply model can be easily formulated using the link-path incidence matrix. In the within-day dynamic case a more complex approach is needed, which is described below.

In traffic assignment models the performance of a link in the network is related to a variable associated with each link, usually called 'link flow'. Unlike the static case, in a within-day dynamic network the definition of link-flow is not unique since time and space averages need not coincide. Different definitions can be adopted corresponding to different modelling approaches. Denoting the flow (however defined) on link l during time interval h by f_{lh}, and the corresponding link flow vector by f, the relationship between link and path flows can be formally expressed through the Dynamic Network Loading (DNL) map as follows:

$$f=\Phi(F) \tag{6}$$

In general this map is non-linear, and cannot be expressed analytically as in the within- day static case.

Let r denotes the link performance (such as running speed or queuing delay) vector, which in congested networks generally is a function of the link flow vector:

$$r=r(f) \tag{7}$$

In the within-day dynamic case both the time needed to traverse a link and the generalised link cost (if different from link travel time) depend on link-performance as defined by the link performance function and the flow. Since this flow can be time-varying, the cost of traversing a link is determined by arrival time at the link. Once the link flows are known through the DNL map, the link performance vector can be determined and the arrival times at each of the links of the path can be computed for any departure time. Then, the corresponding link costs can be defined as functions of both link performances and arrival times.

As above, the generalised transportation cost on path k, averaged across all users leaving during interval j and following path k is denoted by C_{jk}, and the corresponding path cost vector by C. Path costs for a given departure interval j can be computed from the corresponding link costs, so that path costs depend on the link performance vector:

$$C=\Gamma(r) \tag{8}$$

A within-day dynamic supply model is obtained once both the DNL maps and the link- performance functions are specified:

$$C=\Gamma(r(\Phi(F))). \tag{9}$$

Dynamic Network Loading

In the following the DNL map is examined in more detail. A straightforward generalisation of the within-day static formulations is not possible since, as mentioned before, the definition of link flow is not unique. Therefore link and path flows must be redefined in a consistent way. Additionally a fixed point problem arises in the definition of the DNL map since the time-dependent link flows depend on the time-dependent link performance, which in turn depend on the links flows.

Denoting the fraction of the path flow F_{jk} contributing to link flow f_{lh} by crossing fraction b_{lh}^{jk}, the flow on link l can be expressed as:

$$f_{lh} = \sum_{j \le h} \sum_{k} b_{lh}^{jk} F_{jk} \tag{10}$$

Note that the crossing fraction is equal to zero when link l does not belong to path k, the departure interval j is later than h ($j>h$), or when path flow F_{jk} does not cross link l during interval h. The crossing matrix, denoted by B, is a lower triangular $(n_l * n_j)*(n_k * n_j)$ block matrix, and the following holds:

$$f=BF. \tag{11}$$

Crossing fractions b_{lh}^{jk} consistent with the definition adopted for link flows depend on the network topology and the time needed to reach link l, travelling on path k and departing during interval j. Therefore they depend on the travel times on all links preceding link l on path k, which are functions of the link performance vector r, so that:

$$B=B(r) \tag{12}$$

Therefore the relationship between link flows and path flows can be expressed in a general way as:

$$f=B(r)F \tag{13}$$

On the other hand, as stated above, link performance in congested networks depends on the link flows so that by combining (7) and (13) a fixed point problem arises in the definition of the DNL map (6):

$$f=B(r(f^*))F \tag{14}$$

Once link-performance functions (7) have been specified, a dynamic network loading method essentially reduces to the computation of the crossing fractions, i.e. the $b_{lh}^{jk}=b_{lh}^{jk}(f)$ that satisfy (14). It is worth noting that in non-congested networks where travel times are constant and independent of link flows, the link-path crossing matrix does not depend on link flows and the fixed point problem reduces to $f^*=BF$, formally analogous to the within-day static case.

For on-line application with relatively long time intervals the most effective approach towards the solution of the DNL problem seems to be the continuous packet method. Several approaches can be adopted for defining link flows f_{lh} and for solving the fixed-point problem (14). Real-time applications may require the use of a simple link-flow definition (e.g. the number of vehicles crossing a given point of the link per time unit) and of a heuristic algorithm for solving (14).

4. The State Space Prediction System

The Kalman filter, Kalman (1960), is an important algorithmic device in the theory of signal detection and automatic control, see for example Ljung and Söderstrom (1983) or Young (1984). It has many fields of application, including statistics, see West and Harrison (1989), and economic time series modelling, see Harvey (1990).

Underlying the filter is the state space model, a multivariate stochastic process, for which the filter is an optimal Bayes solution to a prediction problem. The state space model supposes that at each discrete time point t, a (vector-valued) observation y_t is related to x_t, the state, through an observation equation. The state process is an unobserved first order Markov process and is specified by a transition equation. On the basis of past observations $y^t=(y_1, y_2, ..., y_t)$ and the current observation y_{t+1} it is required to best predict x_{t+1}. It turns out that the filter depends on the observations only through the previous estimate of the state m_t and the current prediction error, and so allows a recursive computation.

The basis of the model are the transition and observation equations

$$\begin{aligned} x_{t+1} &= A_t x_t + Bu_t + q_t, \qquad && \mathrm{var}(q_t) = Q \\ y_t &= Cx_t + r_t, && \mathrm{var}(r_t) = R. \end{aligned} \tag{15}$$

where x_t is the state vector, y_t the observation vector and u_t the exogenous input. The matrix A_t is the transition matrix, C is the observation matrix and B represents the manner in which the input affects the states. The vectors q_t and r_t are independent errors in the state process and in the measurement process.

The structure of these equations is held in the conditional independence graph, or influence diagram,

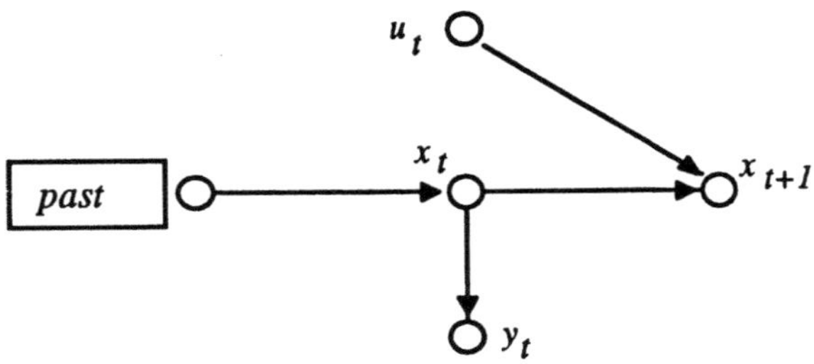

where it is the missing edges that indicate independences. See Dempster (1990), Smith (1990) for the relationship between graphical modelling and Kalman filtering. The *past* represents past values of the variables. That there is no edge to x_{t+1} from the past indicates that enough information is held in the current state x_t to render past information redundant in predicting its evolution.

4.1 Estimation (filtering) and Prediction

There are many ways to write the updating formula for estimating the current state of the process. Here we choose one expressed in terms of the conditional mean m_t and conditional variance p_t of the value of the state vector x_t at time t given all past and present observations y^t and u^t. We use the fairly standard notation that the past data vector is $y^t=(y_1, y_2, \ldots, y_t)$; $y^0=(\cdot)$. Put $m_t=\mathrm{E}(x_t| y^t, u^t)$ and $p_t=\mathrm{var}(x_t| y^t, u^t)$.

The recursion that determines the Kalman filter is

- initialise: choose m_0, p_0
- recurse:

$$K_{t+1} = (A_t p_t A'_t + Q) C' [C(A_t p_t A'_t + Q)C' + R]^{-1} \quad (16)$$
$$m_{t+1} = A_t m_t + Bu_t + K_{t+1}(y_{t+1} - C(A_t m_t + Bu_t)) \quad (17)$$
$$p_{t+1} = (I - K_{t+1} C)(A_t p_t A'_t + Q) \quad (18)$$

The Kalman gain matrix is K_t.

The recursion for m_t in (17) can be re-expressed in three steps: the first is to calculate the best 1 step ahead predictor of x_{t+1} from the best predictor of x_t alone, $m_t^1 = A_t m_t + Bu_t$. The second is to calculate the prediction error of y_{t+1} from this, $y_{t+1} - C\, m_t^1$; and the third is to error correct the 1 step ahead predictor using the updated Kalman gain and so obtain the updated filtered estimate of the new state. As A_t depends on the states this is an instance of the extended Kalman filter. The k-step ahead predictor at time t is obtained by iterating this sequence through

$$m_t^k = A_{t+k-1}\, m_t^{k-1} + Bu_{t+k-1} \quad (19)$$

which depends on knowledge or forecasts of $u_{t+1}, u_{t+2}, \ldots, u_{t+k-1}$ and of $A_{t+1}, \ldots, A_{t+k-1}$.

4.2 State Space Models for Traffic Networks

The features of a state space model that make it appropriate for traffic network applications are that it is multivariate, stochastic (both inherently random and including measurement error) and above all, from the manner by which structure may be imposed.

The salient features of structure are:

1. Traffic variables may be represented at different conceptual levels: observed, latent, derived, so for example, link flow is observed, link density is latent, link travel time is derived, and on-ramp flow is exogenous.
2. Physical understanding, the traffic dynamics, is incorporated through the transition equation: density is predictable (but not observed) through an accounting equation; derived variables such as velocity are consequent on density via the fundamental traffic

diagram; and flow (which is observed) is dependent on upstream and downstream links, and is a form of spatial interaction.

3. The temporal and spatial Markov structure of the physical process is mapped onto the Markov structure of the state space model; this both allows recursive processing and decomposition into subnetworks.
4. Flexible observation equations allow for the vagaries of the measurement process, for instance, malfunctioning detectors, asynchronised monitors, locally smoothed input measures.

The state space model under development consists of the following components: a latent state vector x_t holding traffic variables density and link outflow (r, q) for all links on the network, and (r, q_1, q_2) at forks; an observation vector y_t of traffic measurements from conductance loops at monitoring stations; and a vector of exogenous inputs, the on-ramp flows, u_t, from the monitoring stations.

The lack of an edge between y_t and the past in the independence graph above indicates that given the state of the network x_t any further variation is due to independent measurement error. Importantly there is no arrow from the past or from x_t to the input u_t. This assumes that traffic demand is not dependent on past state of the network (and of course this may be unrealistic).

4.3 The Transition Equations

The transition equations and the transition matrix A_t carry both the network topology (which are time invariant) and the traffic dynamics.

Network topology

The network is viewed as a set of stations, resultant links and blocks composed of links. A link is the section of motorway between two adjacent stations. Motorway networks are relatively simple because there are at most two upstream and two downstream links to any station. Consequently the sections of the motorway fit together in the following way

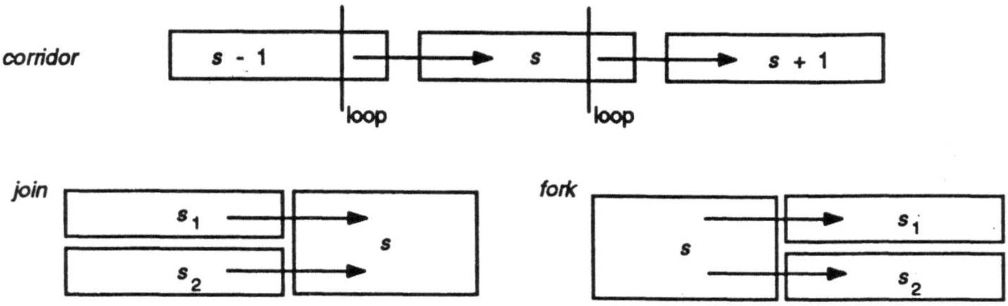

A block then is a combination of links: either a single section, or the two sections that form a join, or the two sections that form a fork. An on-ramp is a join block, an off-ramp is a fork block; intersections are combinations of forks and joins. Roundabouts and cross-roads are prohibited. The whole motorway network may be decomposed into these 3 types of blocks.

If the state space vector holds just the density and the exit flow(s) for each block then the section and join blocks have dimension 2, while the fork blocks have dimension 3 as there are two exiting flows. The block adjacency graph in part determines the spatial structure of the influence diagram of the state space blocks.

Transitions: time varying traffic dynamics

The transition matrix $A_t=A(x_t)$ depends on traffic dynamics. Classically these follow differential equations analogous to those of fluid dynamics, and for a review see the articles in Papageorgiou (1991). There are three types of equations:

-a balance or accounting equation

-a flow equation based on the identity $q=vr$, and

-an equation determining velocity from the fundamental traffic diagram $v=V(r)$.

There are three types of transitions to consider section to section, sections to join and fork to sections. The independence graph of a section to section transition is constructed by replicating the graph

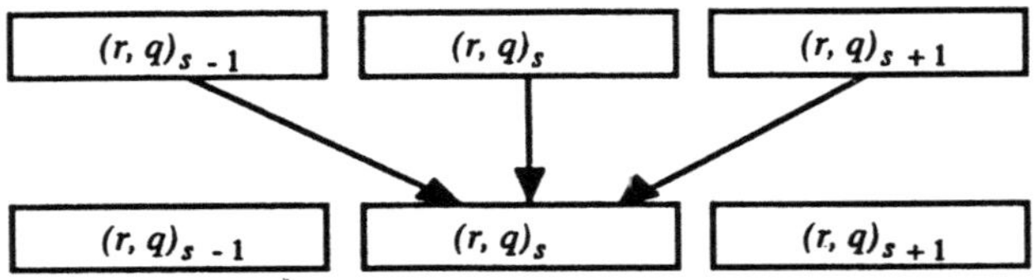

over s. There are no other directed arrows into block s.

The maths that lies behind these arrows is a balance equation of the form

$$r_{s,t+1} = r_{s,t} + k_s \, (q_{s-1,t} - q_{s,t})$$

and one of several possible flow equations. The simplest is

$$q_{s,t+1} = \min(v_{s,t}, v_{s+1,t}) \, r_{s+1,t}.$$

The flow equations incorporate velocity, and so give a non-linear function of the state space, and reflect the physical reality of congestion.

Similar equations hold for sections to joins; fork to sections are slightly more complicated and need to introduce a state variable representing the turning fraction between the two downstream choices. The unknown parameters require estimation, which may be performed on-line from current measurements or off-line from an historical data base.

The observation equations for measured flow are straightforward, and graphically result in a single directed arrow from each traffic block to an independent observation y_s that holds the observed flow from the monitoring station exiting the block. Hence the C matrix is block diagonal. There are two observed flows in a fork block. The observation equation that incorporates measured velocity is slightly more involved but the graph is the same.

4.4 Sparse Multiplication and Sub-Networks

In principal, the state space is typically of order of several thousand states, and with several hundred monitoring stations computation of the time varying Kalman gain requires inversion of huge matrices. Evaluation, storage, multiplication let alone inversion are neither sufficiently accurate nor feasible in real time.

The conditional independence graph of (x_t, x_{t+1}, y_t) derivable from the figures above when exploded over all blocks s is not decomposable and so does not lead to any simple exact computational scheme. However there are relatively few edges in this and sparse multiplication is often possible, a reflection of the sparse connectivity of the block adjacency graph and of the state transitions. Furthermore since the important edges are the ones relating $(x_{s-1,t}, x_{s,t}, x_{s+1,t})$ to $x_{s,t+1}$, it turns out that adequate approximations can be derived by decomposing the full network into overlapping subnetworks, as for example in the following diagram.

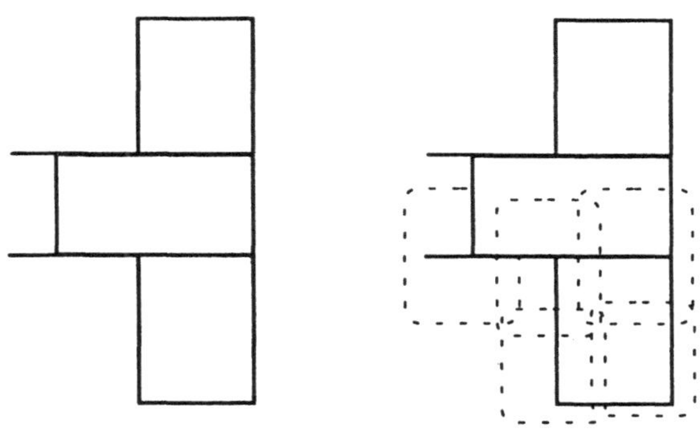

The consequence of having a transition matrix A_t which is a function of the state space and so of time is that the Kalman gains needs to be recomputed at each time point. Ways of simplifying and speeding up this calculation are under research.

5. Concluding Remarks

We have outlined above a partial description of the monitoring and prediction system that is under development in the DYNA project. There are many issues not dealt with here, and specific topics of investigation include: self calibration for parameter estimation; forecasting network entry flows using historical data base methodology; assessment of system performance from diagnostic routines based on one step ahead forecasts; generalisation of traffic flow models at a point to encompass a whole network; assessment of non-linearities in the state transitions required to model traffic flow and their effect on longer term predictions; methods of handling large matrix multiplications and inversions; decomposition techniques for simplifying networks, especially cycles in which path choice plays a part and which correspond to non-decomposable graphical model; representations within this model framework of queues where spill back blocks upstream off-ramps.

Work is in progress on detailed specification of the algorithms, the detailed design of the computer software, and the testing of the models with actual measurements.

Acknowledgements

The work reported in this paper was undertaken as part of the DYNA project supported by the European Commission within the DRIVE II programme. The partners in the project are Hague Consulting Group, Centro Studi sui Sistemi di Trasporto, ELASIS, the University of Lancaster, the University of Napoli, the Université Libre de Bruxelles and the Netherlands Rijkswaterstaat. In preparing this paper we have benefited from the contribution of: Giulio Cantarella, Massimo Di Gangi, Simon Garside, Eric Kroes, Karel Lindveld, Simonetta Manfredi, Andre de Palma, Stef Smulders, Silvana Toffolo.

References

[1] Ashok, K., Ben-Akiva, M. (1993), "Dynamic origin-destination matrix estimation for real-time traffic management systems", *Proceedings of the ISTTT*, Univ. of California, Berkeley, pp. 465-484.

[2] Ben-Akiva, M., Bergman, M., Daly, A., Ramaswamy, R. (1984), "Modelling inter-urban route choice behaviour", *Proceedings Ninth International Symposium on Transportation and Traffic Theory*, VNU Science Press.

[3] Cantarella, G.E., Cascetta, E., Inaudi, D. (1993), "Real time Estimation and Prediction of Origin-Destination Matrices from traffic counts", *Proceedings of the PTRC Conference*, 1993 Manchester.

[4] Cascetta, E., Cantarella, G.E. (1991), "A day-to-day and within-day dynamic stochastic assignment model", *Transportation Research*. 25A,5,277-291.

[5] Cascetta, E., Cantarella, G.E. (1992a), "Non linear dynamics and equilibrium in transportation networks", Submitted to *Advances in Equilibrium modelling, analysis and computation* - Annals of operation research, A. Nagurney Ed.

[6] Cascetta, E., Cantarella, G.E. (1992b), "Modelling dynamics in transportation networks: part I the within-day static case", Presented at *Triennial Symposium on Transportation Engineering*, Univ. Naples, Italy.

[7] Cascetta, E., Inaudi, D., Marquis, G.(1993), "Dynamic Etimators of Origin-Destination Matrices using traffic counts", *Transportation Science*, vol. 7, N. 4, pp. 363-374, November 1993.

[8] Dempster, A.P. (1990), In Oliver, R.M. and Smith, J.Q. (eds) "Influence diagrams, belief nets and decision analysis", *Wiley*: Chichester.

[9] Di Gangi, M. (1992), "An algorithm for continuous-flow dynamic network loading", presented at the *6th. World Conference on Transportation Research*, Lyon, France.

[10] HCG (1990), "The Netherlands Value of Time Study: Final Report", Study commissioned by the Dienst Verkeerskunde, Rijkswaterstaat to Hague Consulting Group, *Final report February 1990.*

[11] Harvey, A.C. (1989), "Forecasting, structural time series models and the Kalman filter", *C.U.P.: Cambridge.*

[12] Kalman, R.E. (1960), "A new approach to linear filtering and prediction problems. Journal of Basic Engineering", *Transactions ASME. Series D*, 35-45.

[13] Kitagawa, G. (1987), "Non-Gaussian state space modeling of nonstationary time series [with discussion]", *Journal of the American Statistical Association* , 1032-63.

[14] Ljung, L., and T. Söderstrom (1983), "Theory and Practice of Recursive Identification", *Cambridge, Mass: MIT Press.*

[15] Ministry of Transport (1992), "Dynamic Traffic Management in the Netherlands Transportation and Traffic Research Division", *Directorate-General for Public Works and Water Management, Ministry of Transport, Public Works and Water Management.*

[16] Papageorgiou, M. (1991), "Concise Encyclopedia of Traffic and Transportation Systems", *Pergamon Press: Oxford.*

[17] Smith, J.Q. (1990), "Statistical principles on graphs", In Oliver, R.M. and Smith, J.Q. (eds) *Influence diagrams, belief nets and decision analysis*. Wiley: Chichester.

[18] Toorenburg, J.v. (1989), "Dynamisch schatten van vertragingstijden en file-omvang", *Transpute report to Dienst Verkeerskunde*, Rijkswaterstaat.

[19] Transpute (1992), "Vervolgonderzoek algorithmen voor schatten van vertragingstijden en file-omvang bij verkeersmonitoring", *Transpute report to Dienst Verkeerskunde*, Rijkswaterstaat CXT91068.RAP

[20] Vlist, M.v.d., Arem, B. v., Smulders, S. (1992), "An on-line procedure for estimating capacity under prevailing roadway and traffic conditions".

[21] West, M., and P.J. Harrison (1989), "Bayesian forecasting and the dynamic linear model", *Springer-Verlag.*

[22] Whittaker, J. (1990), "Graphical Models in Applied Multivariate Statistics", *Wiley: Chichester.*

[23] Whittaker, J. and Garside, S. (1993), "State space models for dynamic traffic networks", in Gunn, H. *DYNA-DRIVE II project V2036 Annual Project Review Report* - Part A Section 2. Submitted to EC R&D Program Telematic Systems in the Area of Transport.

[24] Young, P. (1984), "Recursive Estimation and Time-Series Analysis", *Berlin: Springer-Verlag.*

A PARKING SIMULATION MODEL FOR EVALUATING AVAILABILITY INFORMATION SERVICE

Yasuo Asakura

Associate Professor of Department of Civil and Ocean Engineering, Ehime University, Matsuyama, 790, JAPAN

Parking guidance and information (PGI) systems aim to provide information to drivers concerning the location, direction and availability of parking spaces. The PGI systems are expected to redistribute parking demand among car parks and to reduce queues at the most popular car parks. This paper aims to evaluate the effects of PGI systems, in particular availability information service, on drivers' parking choice behavior and system performance. A simulation model is developed which consists of demand, performance and information service sub-models. The demand model describes a driver's car park choice behavior on the network. A combined model of disaggregate logit and Bayesian statistics model is formulated explicitly incorporating the variable on availability information and the refining process of the expected utility function and the choice probability. The performance sub-model presents the interaction between congestion and level of service in a network and car parks. In the information service sub-model, the outputs of the performance sub-model are transferred to the availability information such as FULL or SPACES. Numerical examples are then calculated for evaluating the effects of different types of availability information service. Changing both the congestion level of car parks and the percentages of informed drivers, different availability information is given to informed drivers. It is shown that the amount of user's time savings by PGI system depends on the ratio of numbers of informed and non-informed drivers.

1. Introduction

Small towns as well as big cities have suffered parking congestion and consequent problems in road traffic. In a highly developed area of the city center, sufficient supply of car parks seems difficult and existing parking stock should be efficiently used. It is sometimes observed in peak hours that some car parks still have vacant spaces when popular car parks are full. Parking guidance and information (PGI) systems aim to provide information to drivers concerning the location, direction and availability of parking spaces. The locational or direction information is usually static information which intends to guide unfamiliar drivers to car parks. The availability information shows the degree of congestion in a car park such as FULL, CONGESTED, SPACES or CLOSED. The PGI systems are expected to redistribute parking demand among car parks and to reduce queues at the most popular car parks.

It is probably true that the PGI systems can contribute to solving parking problems. However, it is not easy to find the practical and/or theoretical reports/papers on the evaluation of the effects of PGI systems. According to Polak et al.[7], there is little comprehensive before and after study of the effectiveness of PGI systems. This is partly because the study of driver's behavior with regards to parking information has only recently begun and behavioral models available for evaluating a PGI system have not been completed yet. Young et al.[8], Muromachi et al.[6] and Cascetta et al.[3] had a similar research motivation and studied the evaluation of parking systems. However, the parking information is not explicitly involved within the models.

This paper aims to develop a simulation model for evaluating the effects of PGI systems on drivers' parking choice behavior and resulting performance of the system. In particular, we focus on the differences between types of availability information, such as FULL/SPACES, expected queuing time and the number of vacant spaces.

2. PGI Systems in Matsuyama

According to Polak et al.[7], more than 50 PGI systems are in operation worldwide. The Ministry of Construction of Japan reported that almost 30 cities had established intelligent PGI systems supported by high technology by the end of 1994, and another 10 systems were under construction or planned.

The PGI system in Matsuyama City with a population of 450,000 has been working since April 1994. It covers 100 hectares of the central area of the city. The system involves 35 car parks and 4,874 parking lots which amount to 40% of car parks and 70% of off-street charged parking lots in the area, respectively. The system provides direction and availability information through three types of LED information boards. Figure 1 shows the location of information boards and car parks involved. A driver travelling from suburbs to the city center will find a block information board located on the entrance road of the city. The block information board displays the availability of the zones. The area is divided into 6 zones and the Kanji characters meaning full, congested or spaces are depicted on a schematic map. A detailed area information board is on a main street in the area and provides the availability information of car parks which are possibly approached from the street. When a driver enters a block, he/she will find a board which lists the names and the availability or the direction of individual car parks near by. The number of listed car parks are about 5 to 7. Then, he/she can choose a car park with spaces or may follow the queue of a popular and congested car park.

The system is expected to have the following effects.

- Users will save time spent searching for a vacant car park, time in the queue of car parks will be reduced and the alternative choices of unknown car parks will be widened.
- Mobility and safety of road traffic will be improved by reducing mazing and/or queuing vehicles and illegally on-street parked vehicles.
- Owners of car parks will be benefited through raising the efficiency of unpopular car parks.

However, the system has just started and it is not known how much it will contribute to solving parking problems in the city. For studying driver's recognition of the system, a questionnaire survey was made in May 1994 just after the installation of the system. According to the provisional aggregation of 391 samples, it is found that 79 % of car drivers have recognized the existence of the system. However, just 10 % of drivers who are aware of it utilized the information. The reasons drivers did not use the information were "did not intend to use (58%)", "did not notice (13%)", "was not legible or not understandable (10%)" and "miscellaneous(19%)". Although detailed analysis must be made, these results imply that it may take time until the PGI system becomes popular, or the information shown by the PGI system might be necessarily reconsidered.

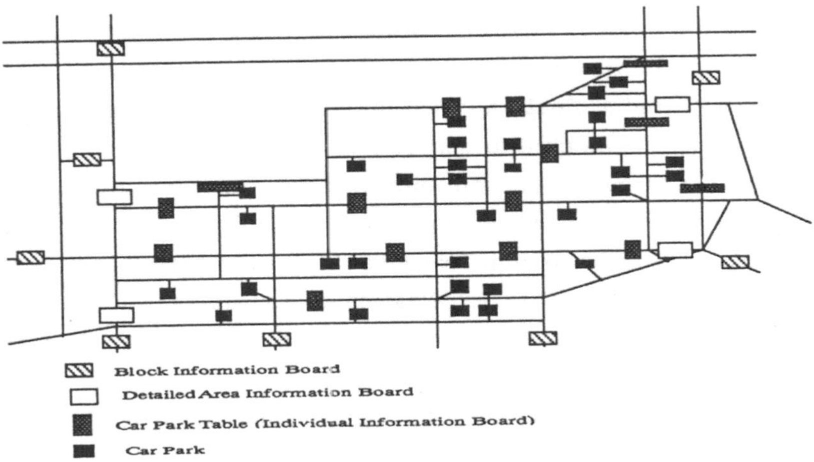

Figure 1 Location of Information Boards and Car Parks in Matsuyama City

3. Model Framework

3.1 Necessity Conditions

The effects of the PGI system will be dependent on the traffic conditions, the type of information and so on. It is desirable to build up a model so that it could consider the following aspects.

(a) The differences of traffic conditions, in particular, that the congestion level of the network and car parks may cause different results.

(b) When a lot of drivers are informed, a specific car park or route may become congested and the system performance might deteriorate. The relation of the system effects with respect to the percentages of the number of informed car drivers must be examined.

(c) If all of the car parks in the system are occupied, FULL/SPACES information is no longer informative. The type of information must be compared in relation to the degree of congestion.

(d) The location of car parks and the road network must be composed in a model, and a driver's route and car park choice should be described.

(e) Parking demand is usually time-dependent and this must be considered.

(f) The system control authority monitors the condition of the system and provides the information to drivers. The difference in decision criteria of the authority might be identified.

3.2 Model Framework

A simulation model is developed which consists of demand, performance and information service sub-models. Figure 2 shows the structure of the model, which is similar to a simulation type model of dynamic traffic assignment incorporating route choice sub-model and traffic flow sub-model. In addition to those sub-models, the proposed model involves an information service sub-model.

The demand model describes the sequential choice behavior of a driver on a road network including car parks. The disaggregate modeling approach is applied to a driver's car park choice and route choice behavior. The choice a behavior under uncertainty is formulated using a multinominal logit model combined with the Bayesian approach. Behavioral differences between informed drivers and non-informed drivers are explicitly described. A non-informed driver is assumed to have prior information on the attributes of a parking, for example, parking charge and walking distance from his/her final destination. An informed driver is assumed to be served with availability information as well as prior experience. His/her subjective probability on the system availability is up-dated using presented availability information.

The performance sub-model presents the interaction between the degree of congestion and level of service in a car park and that of road network. Car park performance model and network performance model are separated. The car park model is available to describe the condition in which time dependent arrival rate sometimes excesses service rate. In the information service sub model, the outputs of the car park performance model are transferred to the availability information such as FULL or SPACES. More detailed information such as queuing time and number of vacant spaces are also produced. These variables are used as the inputs for the demand model mentioned above.

The fundamental structure of the model was developed by Asakura and Kashiwadani[2], in which the road network was not considered. The proposed framework explicitly incorporates the network in the model.

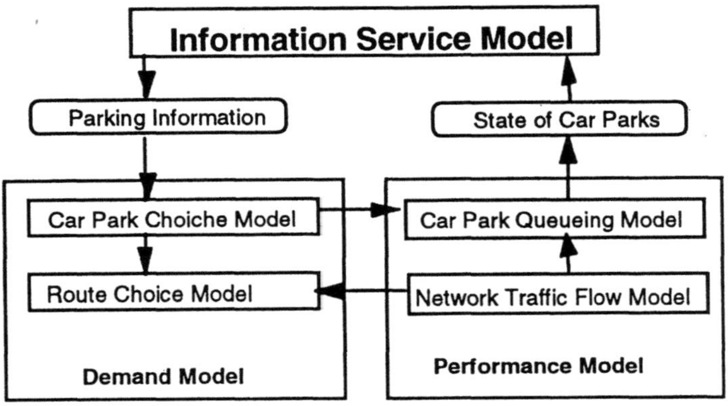

Figure 2 Simulation Model Framework

4. Demand Model

The demand model describes the behavior of a driver on a network including car parks. A combined modelling of route choice and car park choice has been developed by Cascetta et al. [3]. The model is capable of simulating user's choices relative to path, parking type and location through a Probit random utility model. It is well defined model, however, the parking information was not explicitly included. Thus, we intend to identify a driver's choice behavior model with and without parking information.

4.1 Assumptions on Route and Car Park Choice

In order to simplify the model, the following assumptions are employed.

(a) Every car park is an off-street charging car park since existing PGI systems deal with this type of car park.

(b) The origin, destination and departure time of a driver are given.

(c) A driver first determines a car park and next chooses the route to the car park.

(d) Once a driver determines a car park and a route, he/she never changes the decision on the network. Since the area covered by the usual PGI system is not so wide (about 1 km square) and the traffic conditions and the parking information will not change dramatically, a driver does not intend to change his/her decision within the area.

(e) A driver is welcomed to the tail of the queue, when it exists, of the chosen car park. He/she is never rejected and never escapes.

(f) Even if the information is not provided, a driver has knowledge of the traffic congestion in a network and the attributes of car parks except for availability. For example, he/she knows the travel time in the network, the location, the direction, the walking distance to the final destination and the parking fee.

The assumption (c) does not always neglect the effects of the attributes of the route on the car park choice behavior, since it is possible to include route attribute into the car park choice model. The assumption (d) is valid when the parking information is provided just outside of the area covered by the PGI system. When drivers obtain the parking information within the area and the information is frequently varying, the assumption (d) may be altered by the assumption of which drivers may change their alternative enroute. The assumption (e) is possibly relaxed as follows. If a driver finds a long queue in front of the chosen car park, he/she may give up the car park and start to search for another car park. The choice behavior of whether to follow the queue or to start searching can be described. If a driver determines to search for another car park, it is regarded as if a new trip is produced from the present car park.

The model framework is capable of applying these modification. However, they require another unknown factors in the model. Thus, we employ simpler assumptions on driver's behaviour as is stated above.

4.2 Car Park Choice Model (Asakura and Morikawa[1])

Even if parking information is not shown, a driver has a prior subjective probability for availability of car parks. The probability would be refined when the driver obtains parking information. The choice probability would be the conditional probability of the information, and be derived from the expected utility function which is weighted utilities with the subjective probability. Such choice behavior under uncertain situations could be well described by the Bayesian statistics theory. In this section, we apply the Bayesian theory and combine it with the multinominal logit (MNL) formulation.

The following sets[1] are used in model formulation.

S={s} : the set of discrete states, in which s_j denotes the j-th state. For example, s_1 and s_2 mean FULL and SPACES, respectively. Since a state is uncertain, sj is a random variable.

A={a} : the set of behavior, in which a_i denotes the event that car park i is chosen.

M={m} : the set of information, in which m_j denotes the information of state j.

A driver has a prior subjective probability $p_i(s)$ for car park i. If a PGI system does not exist, he/she estimates the availability of a car park through previous experience. Such prejudice on availability is reflected in the value of $p_i(s)$. Let us show a simple example of $p_i(s)$. There are two car parks and two states, FULL or SPACES. When a driver does not have any experience, he/she gives uniform probability on each car park in both states. A matrix of a driver's prior subjective probability $p_i(s)$ is represented in Table 1.1.

State	Car Park - 1	Car Park - 2
FULL	0.5	0.5
SPACES	0.5	0.5

Table 1.1 Example of Prior Subjective Probabilitywithout Information

[1] To simplify the following discussion, we consider the case in which the state and information are discrete. However, it is possible to assume a continuous space for both.

State	Car Park - 1	Car Park - 2
FULL	1.0	0.0
SPACES	0.0	1.0

Table 1.2 Example of Prior Subjective Probability with Complete Information

When a driver has rather complete information or strong beliefs, an example of $p_i(s)$ is represented in Table 1.2

Let V_{ij} and ε_{ij} represent the deterministic and random part of utility function of car park i at state j, respectively. V_{ij} is explained by the variables of the attributes of car park i such as parking charge, walking distance and the 0-1 variable which shows the state of the car park, for example, FULL(=1) or SPACES(=0). The attributes of the route to car park i, for example the route travel time, could be reflected in V_{ij}.

The value of utility U_i for choosing car park i under the uncertain situation of whether the car park is available or not is defined as the expected utility of every U_{ij} weighted by subjective probability $p_i(s_j)$. Assuming that ε_{ij} is independent from state, the net utility U_i is written as;

$$U_i = \sum p_i(s_j)\, U_{ij} = \sum p_i(s_j)\,(V_{ij} + \varepsilon_{ij}) = \sum p_i(s_j)\, V_{ij} + \varepsilon_i \qquad (1)$$

Assuming IID Gumbel for the distribution of random residuals, the choice probability of car park i without information (m=0) is then formulated as the following MNL form of,

$$p(a_i \mid m=0) = \exp\left(\sum p_i(s_j)\, V_{ij}\right) / \sum \exp\left(\sum p_k(s_j) V_{kj}\right) \qquad (2)$$

A decision maker of the parking information service monitors the state of car parks and gives availability information to drivers. When a car park is currently not so congested, but expected to be occupied in near the future, the message of FULL may be shown. Thus, the conditional probability f(m|s) is introduced, which shows the appearance of information m when state s is true. This represents the judgment or control strategy of the parking service authority.

The value of f(m|s) is defined for a combination of state and information of a car park. For car park i, the value of $f_i(m_j|a_k)$ represents the probability in which state j is informed in the condition that state k is true. If the strategy of the information service authority, i.e. the judgment of the state, differs in each car park, the value of f(m|s) depends on each car park. Otherwise, subscript i could be dropped. As an example of this case, Table 2 shows values of $f_i(m_j|a_k)$ for two availability states.

State	Information "Full"	Information
Full	0.8	0.2
Spaces	0.1	0.9

Table 2 Example of f(m/s)

When a driver finds availability information m, he/she will refine the subjective probability. The Bayesian theory can be applied to describing the renewal process of a driver's subjective probability. The posterior subjective probability is consequently obtained as;

$$p_i(s_k|m_j) = f_i(m_j|s_k)\, p_i(s_k) \,/ \sum f_i(m_j|s_h)\, p_i(s_h) \qquad (3)$$

Table 3 shows the calculated value of p(s|m) through Eq.3 using Tables 1.1 and 2. Note that the values of p(s) and f(m|s) are common for car park 1 and 2.

State	Information "Full"	Information
Full	0.89	0.18
Spaces	0.11	0.82

Table 3 Example of p(s/m)

The expected utility function is then calculated using p(s|m) and V_{ij} as;

$$V_i(m) = \sum p_i(s_k|m) \, V_{ik} \tag{4}$$

where $V_i(m)$ denotes the utility of car park i when the information m is shown. The posterior choice probability P(a|m) is finally formulated in Eq.5, which shows a driver's choice probability for given information m.

$$p(a_i \,|m) = \exp\,(\sum V_i(m) \,/ \sum \exp\,(V_l(m)) \tag{5}$$

For example, a driver finds the information that "Car park 1 has SPACES(s_2) and car park 2 is FULL(s_1)". His/her expected utility functions and posterior choice probabilities are;

$$V_1(m_2) = p_1(s_1|m_2)\, V_{11} + p_1(s_2|m_2)\, V_{12} \tag{6.1}$$

$$V_2(m_1) = p_2(s_1|m_1)\, V_{21} + p_2(s_2|m_1)\, V_{22} \tag{6.2}$$

$$P(a_1|m) = \exp(\, V_1(m_2)) \,/\, (\exp(\, V_1(m_2))+\exp(\, V_2(m_1))) \tag{7.1}$$

$$P(a_2|m) = \exp(\, V_2(m_1)) \,/\, (\exp(\, V_1(m_2))+\exp(\, V_2(m_1))) \tag{7.2}$$

Assume that the prior subjective probability is indifferent such as in Table 1.1 and the information service authority observes the condition of car parks and provides the information on the situation of car parks without any modification such as in Table 4.

State	Information	Information
Full	1.0	0.0
Spaces	0.0	1.0

Table 4 The value of f(m/s)

In this case, the model yields to the conventional MNL model. Let us show an example. Suppose that V_{i1} and V_{i2} are explained as the following forms;

$$V_{i1} = \beta_1 WD_i + \beta_2 PC_i + \beta_3 \tag{8.1}$$

$$V_{i2} = \beta_1 WD_i + \beta_2 PC_i + \beta_4 \tag{8.2}$$

where WD_i and PC_i are the walking distance to the final destination and the parking charge of car park i, and β_3 and β_4 denote the contribution of the state 1 and 2, respectively. Since $p_i(s_j)$ is equal to 0.5 for each car park, the deterministic part of the prior expected utility function is calculated from Eq.1 as,

$$V_i = 0.5\, V_{i1} + 0.5\, V_{i2} = \beta_1 WD_i + \beta_2 PC_i + 0.5(\beta_3 + \beta_4) \tag{9}$$

When a driver finds the message that "Car park 1 has SPACES(state 2) and car park 2 is FULL(state 1)", the information is true since there is no modification as in Table 4. Thus, the value posterior subjective probability is calculated using Eq.3 as in Table 5.

State	Information	Information “SPACES”
Full	1.0	0.0
Spaces	0.0	1.0

Table 5 The value of p(s/m)

His/her posterior expected utility functions are,

$$V_1(m_2) = p_1(s_1|m_2)\, V_{11} + p_1(s_2|m_2) V_{12} = V_{12} = \beta_1 WD_1 + \beta_2\, PC_1 + \beta_4 \tag{10.1}$$

$$V_2(m_1) = p_2(s_1|m_1)\, V_{21} + p_2(s_2|m_1) V_{22} = V_{21} = \beta_1 WD_2 + \beta_2\, PC_2 + \beta_3 \tag{10.2}$$

Both prior and posterior situations, the choice probabilities are given by the conventional logit equation. Then, the conventional MNL model is a specific case of the combined Bayesian statistics and MNL model.

4.3 Route Choice Model

The modelling approach for route choice is the same as the above mentioned car park choice model. If a traveller information system is not installed, a driver has to choose a route to the predetermined car park using his/her prior experience. The prior route choice probability is formulated as the same form of Eq.(2). When a driver obtains the availability information on the route in a network, for example route travel time or degree of congestion, he/she will refine the subjective probability through the Bayesian process shown in Eq.(3). The route choice probability with route information is then calculated as Eq.(5). If a driver has perfect information without any uncertainties, the most attractive route will be chosen. This corresponds to the shortest route principle in the conventional traffic assignment. Therefore, a detailed explanation of the route choice model is not presented here[2] .

5. Performance Model

The performance model consists of a car park performance model and network traffic flow model which describe the time-to-time traffic situation in car parks and in a road network, respectively. It has the same function as the traffic flow sub-model in a dynamic assignment simulation model like a CONTRAM (Leonard et al.[5]). Due to the assumption on car park choice and route choice behavior, a driver does not change his/her chosen car park and route until he/she arrives at the car park. It means that driver's decisions in the demand model are used as the inputs of the performance model. The outputs of the performance

---

[2] It is not necessary to enumerate all the alternative paths for large network. The choice set of paths will be predetermined using the value of the prior subjective probability.

model are the degree of congestion in car parks and of a network. Those are used as the inputs of the information service model and the demand (route choice) model, respectively.

5.1 Assumptions

To simplify the model, the following assumptions are employed. These are closely related to the demand model.

(a) A vehicle is identified in the network and in the car parks. The chosen car park and the route to the car park for the individual vehicle has been determined by the demand model.

(b) A car is never rejected from following the queue of a car park. This means that there is no constraint on the queue length of a car park.

(c) The queue of a car park is made vertically. Queuing vehicles do not directly reduce the capacity of the road links connected to the car park.

The assumption (b) corresponds to the assumption (e) of the demand model. Thus, this could be relaxed so that the model incorporates the constraint of the number of queuing vehicles. The model framework is capable of considering this point. However, it is necessary to describe the behaviour of rejected car drivers and this may cause another complexity.

5.2 Network Representation

A network including some car parks are represented as Figure 3. Models like this have already been used by several authors like Eldin et al. [4] and Cascetta et al.[3]. Using the network, a user on the road network runs on road links and then arrives at a car park. He/she parks the car and walks to the final destination. When the most popular car park is full and the driver prefers to walk rather than to follow the queue, he/she may choose a car park located in a neighboring zone.The network flow model and the car park model are sequentially calculated. The movement of a vehicle from an origin to the entrance node of a car park is traced in the network flow model. After arriving at the car park, the queuing and parking service for the vehicle is simulated in the car park model.

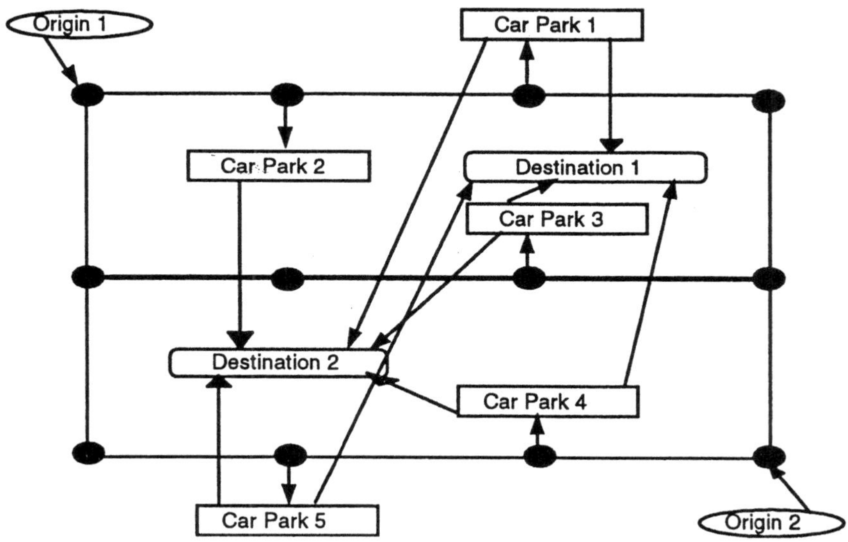

Figure 3. Network Configuration Including Some Car Parks

5.3 Network Flow Model

The network flow model corresponds to a traffic flow simulation model in a network. Several previous micro-scopic simulation models could be used for the network flow model. In this paper, the model is designed as simply as possible. A directed link has a single lane and the first-in-first-out principle is applied for the movement of vehicles. The difficulty of right turn or left turn is not considered. A vehicle is moved from a link to the next on the predetermined route. The link travel time consists of the running time on that link and the queuing time at the exit of the link. The running time of a vehicle is calculated using the function of the number of vehicles on that link when the vehicle enters the link. If the preceding link is saturated, the vehicle has to wait in the present link. The queuing time is determined by the degree of congestion in a preceding link. The calculated link travel time is used in the demand model as one of the explanatory variables of driver's route choice.

5.4 Car Park Model

Two variables are used to describe the states of car parks. NQ_i and NS_i denote the number of queuing vehicles and the number of parked vehicles in car park i, respectively. Figure 4 shows the flow chart of the car park model. The model consists of several different branches.

(1) When a vehicle is just allocated to car park i during time period t ~ (t +dt), the vehicle has to follow the queue ($NQ_i = NQ_i + 1$) if the car park is full. Otherwise, the vehicle can enter the car park ($NS_i = NS_i + 1$), and the time that the vehicle leaves the car park is calculated. The exponential distribution is assumed to represent the parking service time distribution.

(2) If a vehicle finishes the service during time period t ~ (t +dt), such a vehicle leaves the parking facility ($NS_i = NS_i - 1$). If there is a queue at the car park, the first vehicle of the queue can enter the car park ($NQ_i = NQ_i - 1$, $NS_i = NS_i + 1$) and the time that the vehicle leaves the service is calculated. When another vehicle finishes its service in the same time period, this process is iterated again.

(3) If no car finishes parking service, the state at time t continues at next time interval.

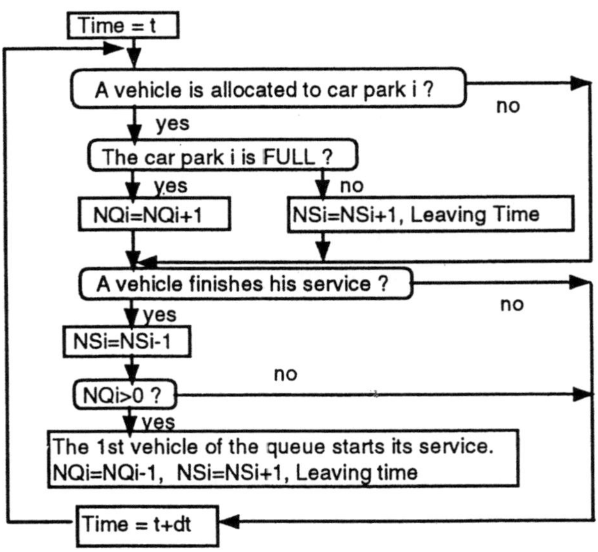

Figure 4 Performance Model Structure

6. Information Service Model

When we monitor the situation of car parks, it is possible to produce the following availability information: FULL/SPACES, the number of vacant spaces and the expected waiting time. For simplicity, the strategy of the parking authority is assumed to show the current states of car parks to drivers without any modification and prediction. One of the following three types of availability information is provided to drivers.

(1) When $NQ_i > 0$, the message FULL in car park i is displayed. Otherwise, the message SPACES in car park i is displayed.

(2) The number of vacant spaces in car park i is calculated as $C_i - NS_i$ where C_i denotes the capacity of car park i.

(3) The expected waiting time of a car park means the expected value of the waiting time when a vehicle arrives at the tail of the queue of the car park. The value of waiting time is equal to zero if there are no queuing vehicles. If $NQ_i > 0$, the expected waiting time is calculated as,

$$WT_i = (NQ_i - 0.5)/(\mu C_i) \quad \textbf{(11)}$$

where μ denotes the service rate corresponding to the reciprocal of the average service time[3] .

7. Numerical Examples

7.1 Parameter Estimation of Demand Model

A questionnaire survey was made for identifying the demand model. An individual's parking preference was asked under hypothetical situations in which the walking distance and parking charge of car

---

[3] The vehicle at the head of the queue is expected to wait $0.5/(\mu C_i)$ and each vehicle in the queue is expected to spend $1.0/(\mu C_i)$ to proceed. The vehicle at the tail of the queue is expected to wait $(NQ_i - 0.5)/(\mu C_i)$ to enter the car park.

parks were shown to all drivers. Respondents were required to choose one from five alternative car parks in this condition. Then, one of three types of availability information was additionally provided and the respondents had to make another decision. Since the parameter estimation process for the combined Bayesian and MNL models has not been completed yet, the conventional MNL model is used as the demand model of car park choice. 1500 samples including non-informed and informed drivers were pooled and used for parameter estimation. This is because the parameter values of the variables representing prior knowledge should be common between non-informed drivers and informed drivers.

For non-informed drivers, the deterministic part of the utility function of car park i was estimated as,

$$VO_i = \underset{(-18.99)}{-0.01857\, WD_i} - \underset{(-21.75)}{0.02104\, PC_i} \qquad (12)$$

here WD_i and PC_i denote walking distance (meters) and car park charge (yen/hour), respectively. Figures in the parenthesis shows t-statistics of the parameters. For informed drivers, the following utility functions were estimated.

$$VI_i = VO_i + \underset{(17.93)}{5.52917\, FS_i} \qquad (13.1)$$

$$VI_i = VO_i + \underset{(5.81)}{0.08191\, NV_i} \qquad (13.2)$$

$$VI_i = VO_i - \underset{(-13.25)}{0.34511\, WT_i} \qquad (13.3)$$

here FS_i, NV_i and WT_i denote full and space information (FS_i,=1 if the information "car park i has spaces" is given), number of vacant spaces information (vehicles) and waiting time information (minutes), respectively. Note that a driver is provided one of three types of availability information. This means that one of three functions above is applied to each informed driver. The signs of the estimated parameters are consistent and the values of t-statistics are sufficient. The value of χ^2 was 0.1835 x 10^4 and the percentages of hit was 79.5%. Thus, the estimated function is statistically valid.

7.2 Case Studies

Numerical examples were calculated using the network shown in Figure 3. The network has 2 origins, 2 destinations and 5 car parks. Three factors were controlled in the simulation; the types of information, the percentages of informed drivers and the value of arrival rate for controlling the degree of congestion.

Figures 5.a and 5.b depict the values of the averaged waiting time of two cases of availability information for the different proportion of informed drivers. Figures 5.a and 5.b correspond lower and higher congestion cases, respectively. Number of vacant spaces information was the most inferior among three types of information in any cases of simulation, so it is omitted from the figures.

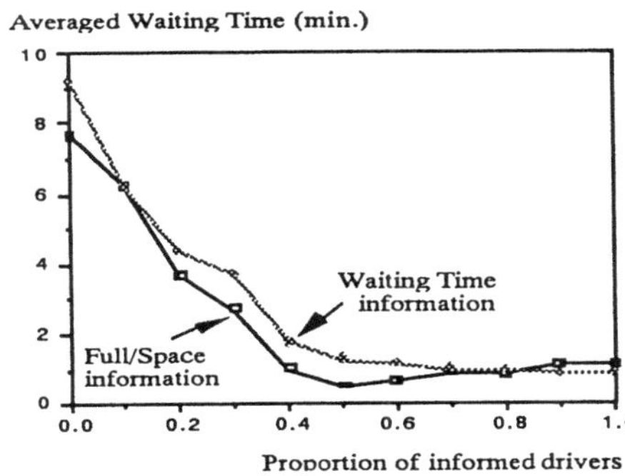

Figure 5.a Effects of Information at Lower Level of Congestion

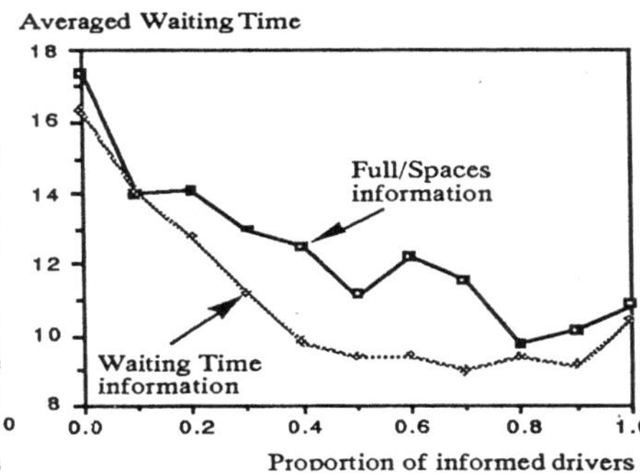

Figure 5.b Effects of Information at Higher Level of Congestion

It is commonly observed that the averaged waiting time reduces when the proportion of informed drivers becomes higher. In particular, the curves steeply decreases at lower percentages of informed drivers. This is because informed driver disperse to less congested car parks. However, such effects of providing information deteriorates when almost of all car drivers are informed. At the lower percentages of informed drivers, providing availability information to drivers is effective in reducing congestion. However, the effects may deteriorate when a large proportion of drivers are informed. This may be because of the time difference between obtaining information and arriving at a car park. The informed drivers will choose a less congested car park before entering the network and traveling on the network. When they arrive at the car park, the situation of the car park will change and they may experience congestion. This suggests the concentration of informed drivers to the car parks which were informed less congested might cause another congestion at those car parks. Such negative effects could be eliminated if the information is up-dated as fast as possible.

Figure 5.a shows that the effect of Full/Spaces information and that of waiting time information are not so different. However, Figure 5.b points out that the waiting time information is more effective than Full/Spaces information in reducing congestion. It is valid that Full/Spaces information may have no meaning if all of the car parks are full. These results imply that Full/Spaces information will be effective when some car parks are full and others are vacant, and waiting time information should be displayed when all car parks are occupied.

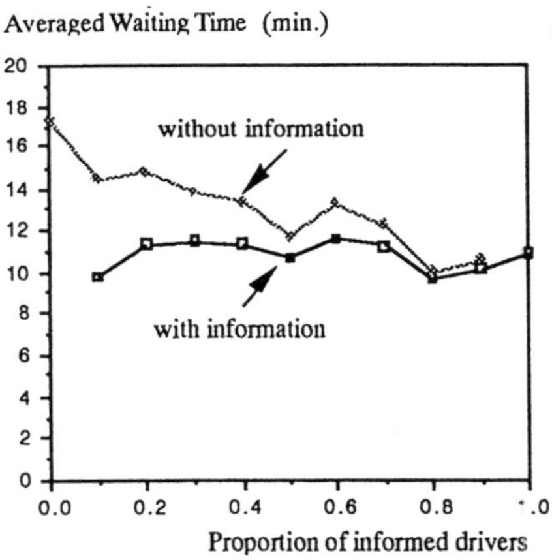

Figure 6.a Comparison between Two Groups (Full/Spaces Information)

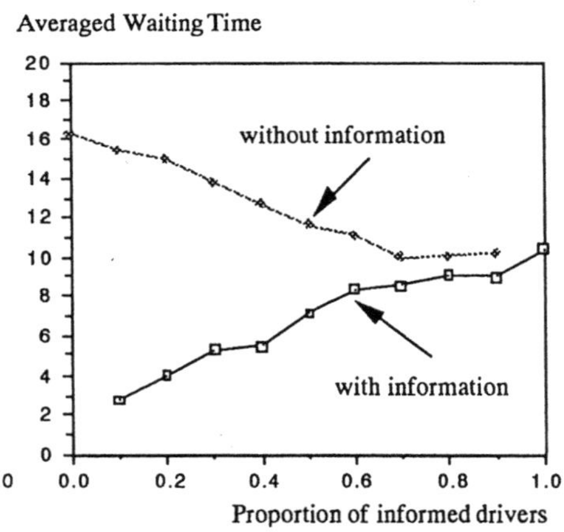

Figure 6.b Comparison between Two Groups (Waiting Time Information)

Figures 6.a and 6.b show the difference of the averaged waiting time between two groups of drivers; with information and without information. These two figures correspond to Full/Spaces information (Figure 6.a) and waiting time information (Figure 6.b), respectively. Congestion level is fixed at higher level.

It is shown that the non-informed drivers can reduce their waiting time when the percentages of informed drivers becomes higher. The informed drivers, on the other hand, are superior that the non-informed drivers. However, they will lose their advantages at higher percentages of themselves. The difference between two groups of drivers are not too much for Full/Space information, while the differences are distinguished for waiting time information, in particular at lower percentages of informed drivers. This proves the value of information is relatively higher under the condition that a few drivers are properly informed at congested situation.

8. Discussion

In this paper, we proposed a simulation model for evaluating the effects of parking availability information service. The entire simulation model was calculated for different combinations of the types of information, congestion level and the proportion of information. The results of the model were consistent with our experience in actual cities. Almost all current PGI systems show only FULL/SPACES information to drivers, and such information may loose its benefit in congested conditions. The provisional study for analyzing the driver's acceptance of the PGI system in Matsuyama city suggests that the type of information might be reconsidered. The results of the simulation and the findings in the empirical study must be compared in detail.

The proposed model is still in the development stage and further improvement is necessary. In particular, the demand model should be improved so that it could consider the parking search behaviour and the heterogeneity of drivers. For example, drivers are disaggregated into segments based on the attitude for the provided information. The parameter estimation process for the demand model must be also developed. It is necessary to modify the conventional procedure for the MNL model.

The performance model is now under improvement, for example the background traffic including vehicles just departed from car parks is incorporated in the model. The findings in the dynamic traffic assignment models will be applied for improving the performance model.

Another aspect to consider is evaluation of the "benefit to the individual driver" as well as the system performance indexes. The economics of information seems promising in evaluating those aspects of PGI systems.

Acknowledgements

The author would like to express his appreciation to Professor Masuo Kashiwadani for his constructive comments at early stage of this research. The gratitude is also given to Mr. Shiro Sakamoto for his computational work and to Mr. Satoshi Mori for his analysis of questionnaire survey.

References

[1] Asakura. Y. and Morikawa. T. (1993). "Evaluation of Parking Information Systems using Behavioral Choice Models", *Proc. 26th ISATA* in Aachen, September 13-17, pp.218-226.

[2] Asakura. Y. and Kashiwadani. M. (1994). "Evaluation of Availability Information Service by Parking Choice Simulation Model", Proc. the Int'l Conf. on A3TM in Singapore, pp.335-342.

[3] Cascetta E., Nuzzolo A. and Bifulco G.N. (1991) "Some Application of a Stochastic Assignment Model for the Evaluation of Parking Policy", *Proc. of the 19th PTRC* Summer Annual Meeting Seminar G, pp.17-29.

[4] Eldin M.S.N., El-Reedy T.Y., and Ismail H.K. (1981) "A Combined Parking and Traffic Assignment Model", *Traffic Eng. and Control* , Vol.22, No.10, pp.524-530.

[5] Leonard, D.R., Tough, J.B. and Baguley, P.C. (1987) "CONTRAM: A Traffic Assign-ment Model for Predicting Flows and Queues During Peak Periods", *TRRL LR841*.

[6] Muromachi Y., Harata N. and Ohta K. (1991) "A Study on Parking Choice Model Considering Parking Guidance and Information System", *Proceedings of Infrastructure Planning*, No.14(1), pp.139-146.

[7] Polak J.W., Hilton I.C. Axhausen K.W. and Young W. (1990). "Parking Guidance and Information Systems: Performance and Capability", *Traffic Eng. & Control*, Vol.31, No.10, pp.519-524.

[8] Young W., Taylor M.A.P., Thompson R.G., Ker I. and Foster J. (1991) "CENCIMM, a software package for the evaluation of parking systems in central city areas", *Traffic Eng. & Control*, Vol.32, No.4, pp.186-193.

FLEXIBLE DISPATCHING CONTROL TOOLS IN PUBLIC TRANSPORT

Andrzej Adamski

Institute of Automatics, University of Mining and Metallurgy, 30-059, Cracow, Poland.

In the public transport transit on-time service has been a chronic operational problem created by the existence of unpredictable operational random events and high complexity of the public transport processes. The potential of new information and communication technologies offers great opportunities for the development of effective management and control tools for improvement of transit service reliability. This paper focuses on the distributed dispatching control problem, embedded into the follow-up direct control layer in the hierarchical control system. Autonomous dispatching control actions are practically realized by vehicles equipped with on-board computers and some communication media. In the paper wide spectrum of flexible single and multicriteria dispatching control tools are presented. Different control criteria described in terms of waiting time and control random variables moments are used as service irregularity and control cost measures. The analytical solutions for various types of non-linear regulators (half-wave rectifier, relay, polynomial, flexible with insensitivity zone) and various types of dispatching control strategies are presented and numerically illustrated.

1. Introduction

It is generally agreed that maintaining on-time transit service is a key operational problem which affects both travellers (reliability of service) as well as operators (productivity and efficiency). Although many researchers have studied various aspects of this problem and recommended some remedial headway-based or schedule-based control actions, adequate and operational control tools are still lacking.

The new information and communication technologies (wireless communication systems) and hardware capabilities (new AVLM systems, low-cost highly powered on-board computers, new generations of intelligent traffic selective detectors) offer great opportunities for the development of effective management and control tools for public transport.

This paper focuses on the optimal dispatching control problem [6][15] Previous research related to this control problem was devoted to:

- Creation of analytical [3, 7, 16, 18, 25-26], simulation [17, 19, 22, 24, 28], probabilistic [27] and empirical [1, 2] models for the description of the phenomena observed on the bus lines and studying the sources of service unreliability. The analytical models are frequently based on highly simplified assumptions (e.g. uniformity of route sections, infinite capacity of buses, continuity of passengers arrivals). The complex simulation models, on the other hand, pose statistical problems in the analysis of the output and, similarly as empirical models have limited transferability (these are in essence models for a given place and given set of input data). The sophisticated probabilistic model is numerically demanding and poses also the representativity problems.
- Application of the above models to developing and evaluation (in some cases experimentally) the efficiency of the simple static control actions (e.g. holding, skipping, turning, rescheduling, reserve actuation) realized on the bus lines [1-3, 10-11, 18, 22, 25-26, 29-31]. Most of these approaches were concentrated on simple holding threshold type (checkpoint) strategies and analytical [11-12, 18, 25-26], or simulation [1, 24, 29, 31] attempts to optimize threshold value parameter with respect to average passenger waiting time. The analytical solutions for more sophisticated holding strategies (e.g. flexible or synchronizing type) and for multicriteria problems were presented in [10-11, 13-15]. The general conclusion concerning these works may be as follows: The simple control actions which are realized at

appropriate points along the route may be to a high degree profitable even when vehicle locations are roughly estimated.

- Development of the dynamical dispatching control tools [3, 6, 10] realized in the hierarchical structures [2-4, 6, 8, 10, 15] or expert systems [11, 14-15]. This approach makes it possible to integrate sensibly in time and space simple control actions into an efficient dynamic control strategy [3].
- Public transport priority control at traffic signalized intersections [9, 10, 21, 23]
- Development of hardware for vehicle location and monitoring, signal pre-emption, management of peripheral on-board passengers information facilities, cashless fare collection, efficient communication means between vehicles and dispatching centres etc [7 and survey cited there].

In this paper the solution of on-line real-time adaptive, distributed dispatching control problem, embedded into follow-up direct control layer in hierarchical public transport management and control system, is presented. The main purpose of this control is the compensation of off-schedule transit service deviations [3, 10, 13]. It suppresses various operating disturbances being supported by effective real-time vehicle location and communication feedback. In consequence the dispatching control layer for moderate and fast disturbances may act autonomously because the stabilization feedback ensures then proper accommodation. This motivates the applicability of the distributed control idea [3], in which the autonomous control actions are realized by local control units i.e. vehicles equipped with on-board computers and some communication media or by human operators equipped with computer support tools [11]. At this point it should be noted that both local control equipment units have been practically realized on the basis of microcomputer system called ZPD and that this support system has controlled 15 bus lines in Cracow in normal exploitation conditions for several years. In the paper a wide spectrum of optimal dispatching control analytical solutions (punctuality, regularity, synchronizing type) for different control criteria which represent the service irregularity and control costs are provided. The second section deals with the background of the dispatching control. The analytical solutions for a family of flexible holding strategies, punctuality and regularity type, are derived in the next section and illustrated by numerical examples. These strategies minimize the weighted linear combination of waiting and holding times measures.

The extensions of these holding strategies to multicriteria dispatching control problems with the real-time adaptive adjustment of domination structure are also presented.

2. Background of the Dispatching Control

The planned passengers service determinants in public transport are described in terms of the number of operating vehicles and service schedules: in space (routing, stops locations) and in time (given arrival/ departure times to/from timing points). The original premise for dispatching control application is the discrepancy between the planned and realized service. However, the complexity of the public transport system cause that the broad spectrum of decision-making and control functions (e.g. process control, scheduling [20], planning) play essential role in the effective operation of this system with respect to its planning transportation goals. The important control premise is concerned with the total integration of the control plant in the framework of the hierarchical control structure [5][15](horizontally from supply resources to final product i.e. passenger service, vertically from single vehicle control to global system management). The new control proposals properly integrate the new control philosophy (i.e. multilayered and multilevel hierarchical structure, multicriteria management approach, on-line real-time adaptive and distributed control actions) with the presently existing hardware capabilities [15]. In the multilayered structure (see Fig. 1) the concept of the functional decomposition of the original problem into a set of different functionally ordered layers tasks is introduced. The layers are working at different time intervals (i.e. intervention times T_i) and use differently aggregated models so that the decision-making horizons and models aggregation levels tends to increase as we proceed up the hierarchy. Layers are mutually integrated via interlayer interface e.g. information feedback from the operating system and data base. The detailed description of this multilayer structure was presented in [10,15]. At this point we present the optimisation layer which is directly connected with the dispatching control actions.

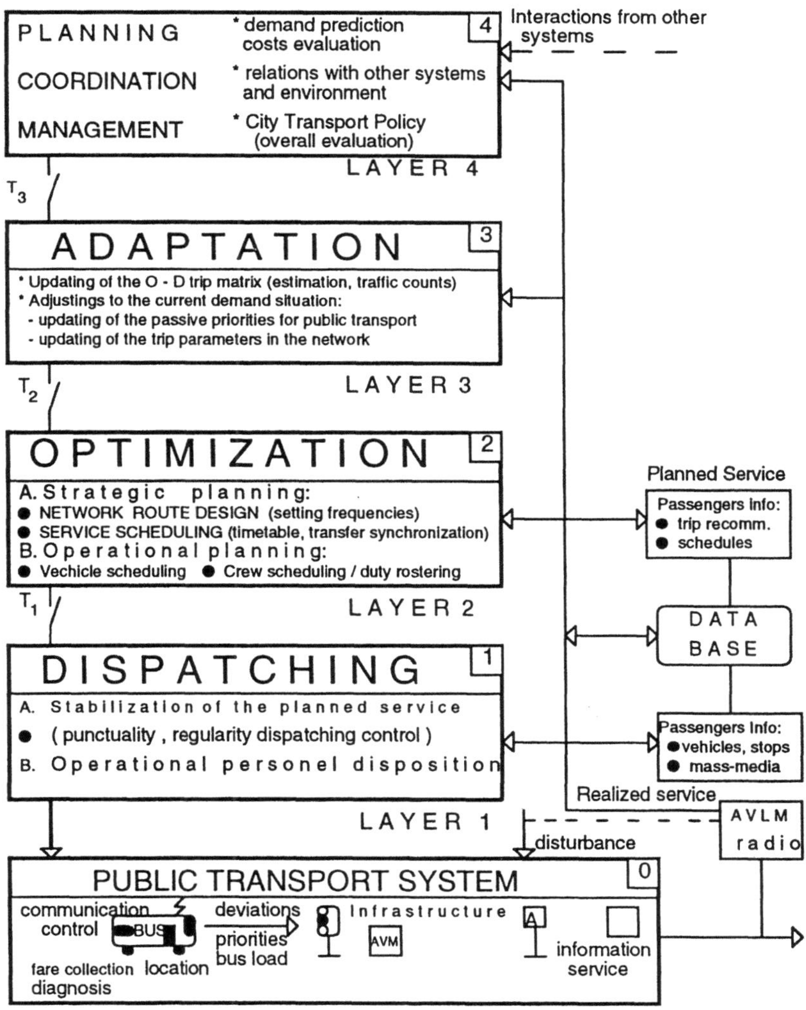

Fig. 1 Hierarchical Multilayer Public Transport Control and Management System

The optimisation layer solves the transit planning optimisation problem aimed at efficient (from the user and operator viewpoint) transport of O-D transit passengers with O-D updating at adaptive layer. It includes in general two main interrelated elements i.e. Network Route Design (setting frequencies) and Service Schedule (timetable development, bus and driver scheduling). At this layer a wide spectrum of operational requirements have to be fulfilled by the optimal solution: high operating and productivity efficiency, high temporal and spatial service flexibility e.g. different operating modes (line-haul, route deviation, demand-actuated), different service variants throughout the day, real-time schedule and route modifications, robust features. To cope with these requirements several mechanisms were used at this layer. First of all, the solution task for this problem is split into the direct dispatching control layer task (described below) which realise regulatory (schedule follow-up) control on the lines and provides a stable steady-state operation of the lines, and optimisation layer task which realise optimising (transit planning set-point) control of the steady-state operated lines. This causes that in the transit planning optimisation problem the steady-state approach and a simplified (to essential nonlinearities and dominant disturbance effects) models are well motivated. Additionally, this makes it possible to effectively use of feedback to increase the optimal solution robust features (relative dispatching control stabilization task independence on the bus line model accuracy and errors of the optimizer) and operating flexibility e.g. real-time on-line repetitive scheduling [4]. At this point it should be emphasised that without the existence of adaptive (O-D matrix updating) and dispatching control (stable, steady-state guarantor) layers it is very difficult to formulate a reasonable transit planning optimisation problem (e.g. due to uncertainty in fixed demand assumption and in using the steady-state trip cost estimates). To our knowledge no approach tackles this problem explicitly. The output from optimisation layer includes the set of optimal bus routes and synchronized schedules for lines. This output for a given line may be interpreted in time-space context as scheduled vehicles trajectories.

At the direct control layer the optimal dispatching control problem is solved. The main purpose of this control is stabilization (follow-up) of real vehicles trajectories (punctuality control) or headway's (regularity control) around schedule trajectories/headway's and efficient counteraction to the off-schedule deviations [3][4][15]. It suppresses various disturbances and transient effects on the basis of effective real-time vehicle location feedback [7], real-time data base and passengers information service tools [Fig. 1] Especially important advantage resulting from the action of this dispatching control layer is the fact that for moderate and fast disturbances this layer may act autonomously because in these cases the appropriate stabilization feedback ensures proper accommodation. The feedback efficiency however is conditioned to a high degree by the operating appropriateness as well as time and location adequacy of the implemented solution of corresponding dispatching control problem. In general this control problem is a very complex one due to large scale network character, occurrence of functional non-homogeneity with diversified stochastic interactions from other vehicles, passengers demand and system environment. To cope with this control task complexity the multilevel hierarchical approach supplemented by "distance and times of control influence" decomposition mechanisms was proposed. It makes full use of the system structure i.e. partition (decompose) of this control task into a collection of smaller local control problems and integrate theirs (via inter-local problems co-ordination instruments) in order to achieve the overall control objective. The direct benefits result from the fact that the labour demand increases non linearly with the problem dimension. In this case the system may be decomposed not only according to its natural (geographical or lines) structure but additionally along lines of weak interactions (determined by "distances and times of control influence"), which gives significant reduction in the required frequency of co-ordination actions. In accordance with multilevel control idea the whole system consist of independent controlled subsystems (e.g. bus lines on which vehicles tracking the scheduled trajectories determined for each line at the optimizing layer). To represent various inter-line requirements connected for example with inter-line passengers transfers or operation of different lines on common sections, the integrating co-ordination tools e.g. arrival windows, offset times are introduced into the schedules. The lines control problems are further spatio-temporally decomposed into local control problems in accordance with the line homogeneity features (e.g. into line zones), "distance of control influence" (measured spatially by part of line or by the number of vehicles to be influenced) or "time duration of control influence" (e.g. rush hour period). To fulfil high control plant description flexibility in the sense of various degree of aggregation: in space (stop, part of line, line) in the population of vehicles (one vehicle, several vehicles, all vehicles on the line) and in time the public transport line was represented by means of two standard operating modules BRMI/BRM (see Fig. 2 and 3).

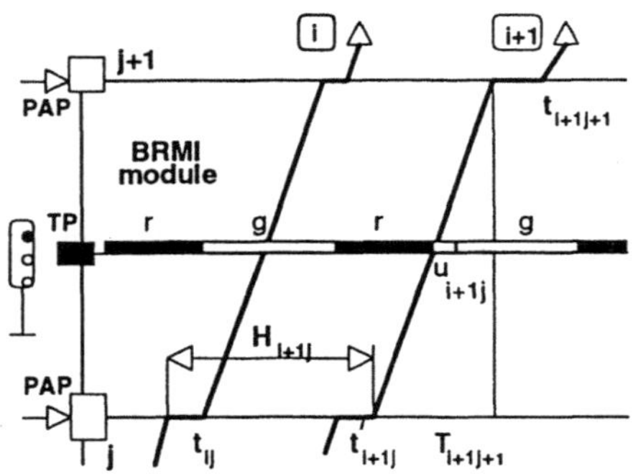

Fig. 2 BRMI standard module

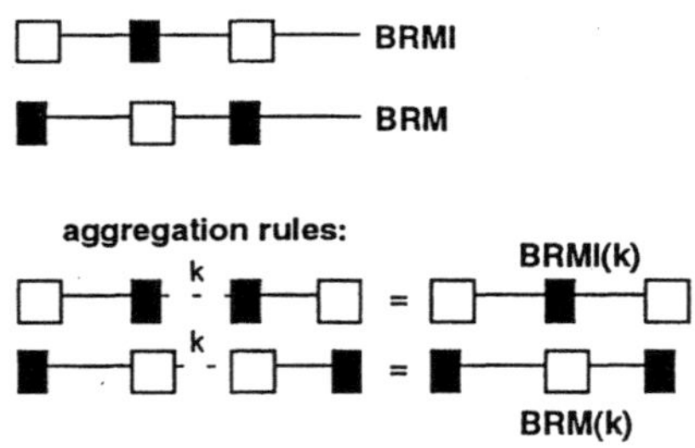

Fig. 3 BRMI and BRM standard modules

These modules consist of (Fig. 3): the inter-stop line segment and two type points denoted in Figs. 2 and 3 by white and black boxes respectively: first one the service point (e.g. bus stop) and the second timing or disturbance point at which control actions (e.g. at signalized traffic intersection) or disturbances can be introduced. Modules create the base for local dispatching control actions and may be described on various levels of aggregation. The whole line operation, being aggregation of passengers and vehicles processes can be flexibly represented by a connection of such modules (see Fig. 4). As can be seen the appropriate parallel connection of modules can represent the transfers or overlapping segments of various lines.

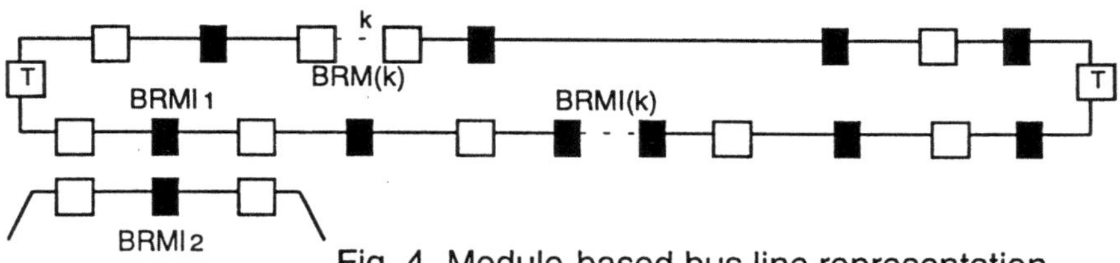

Fig. 4. Module-based bus line representation

The control plant module-based representation together with the simple connection algebra is advantageous because it is a flexible computer-oriented tool for the representation of various network problems and makes it possible to concentrate on the solutions of the distributed "small" control problems for modules [10].

This motivates the usefulness of distributed flexible control tools developed in this paper which may be practically used in the form of autonomous dispatching control actions realized by vehicles equipped with on-board computers.

3. Flexible Dispatching Control Tools

Traditionally, the dispatching control and supervision is run by human dispatchers (located at control centres, terminals, control points and equipped with some hadware and communication media).

Experience has shown that control decisions realized by human operators are characterized by:

- limited range and repertory depending on individual skills, professional experience and intuition.

- considerable amount of subjective heuristic logic
- diversified level of efficiency, violently reduced with elapsed working hours or superposition of several disturbances at the same time (e.g. often at multibus line terminals and complex peak hour traffic situation the selection of appropriate decision exceeds the capabilities of human dispatchers).

In the light of high sensitivity of control plant with respect to irrational control decisions, the improvement dispatching control actions efficiency by computer support tools are an important step to increase service quality.

3.1. Flexible Holding Strategies

Let us consider a bus line realizing service with high/medium frequency so that the random passengers arrival pattern is well motivated [1, 7, 22, 25, 31]. We assume that the dispatching holding control strategy may be used to an early vehicle at any point along the route and is mainly directed to compensate vehicle headway's (in regularity control) or scheduled times (in punctuality control) variability. This means that control impacts on the mean scheduled values can be negligible. The control actions u(e) depend on off-schedule deviations (e) which are random variables having normal type $N(0,\sigma)$ distribution [1, 28] with parameters recursively estimated in real-time [7]. The idea of the flexible holding strategy with the half-wave rectifier type of regulator is presented in Fig. 5.

The flexibility of the presented holding strategies result from the fact that for a wide family of control criteria and non-linear u(e) dependencies (types of regulators, see Table 1) the optimal analytical solutions are derived, in which parameters can be adaptively adjusted.

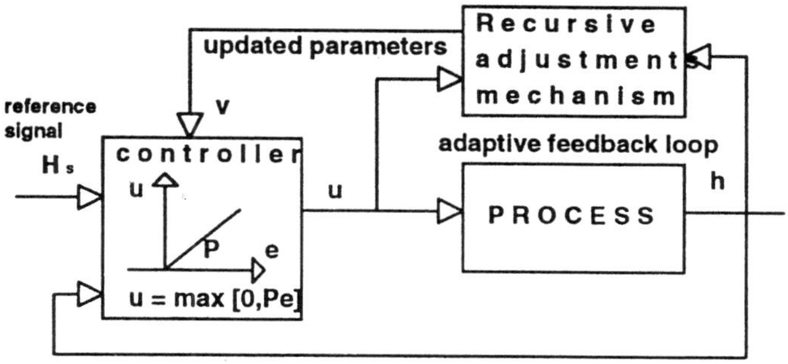

Fig. 5. Block diagram of Flexible Holding strategy

The optimal dispatching holding problem may be stated as follows:

$$O_{min} \quad u \quad S(u) = \gamma \; Q_1(u) + (1-\gamma) \; Q_2(u) \quad | \; u \geq 0 \qquad \textbf{(1)}$$

where weighting by constant $\gamma \in [0,1]$, $Q_1(u)$ and $Q_2(u)$ components represent in general holding costs (i.e. disutility measure of on-board passengers in the held bus) and disutility measure of arriving passengers in control zone respectively. In the papers [2, 10, 18, 28, 30] for example, the average amount of time for a bus to be held $Q_1(u)=E(u)$ and average wait time expressed by the Welding formula $E(w)=E(h)/2 + Var(h)/2\,E(h)$ were considered.

Table 1. Optimal parameters of flexible control strategies

	CONTROLLER		OPTIMAL STRATEGY	
	Formula	**Type of ch-cs**	**Regularity: R/H**	**Punctuality: P/H**
1	**u = max [0,Pe]** $P \in [0,1]$	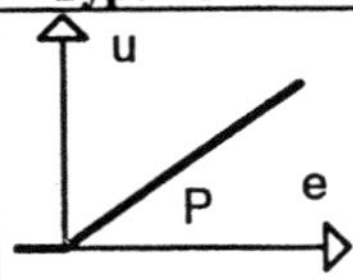 Half-wave rectifier	$P^* = 1 - d\,v$ $d=\sqrt{2/\pi}$; $v=\gamma H_s/(1-\gamma)\sigma$	$P^* = (2-dv)/(2-d^2)$ $d=\sqrt{2/\pi}$; $v=\gamma H_s/(1-\gamma)\sigma$
2	$u=a\,e^n 1(e-e_m)$ 1 ( ) Heaviside unit function	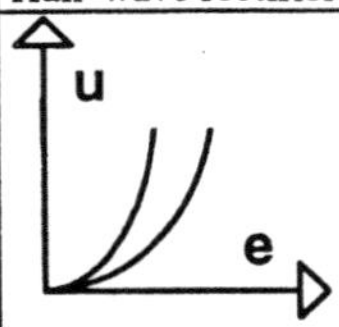 Polynomial	$a^*=[A(n)-\sigma v A(n-1)]/A(2n-1)$ $A(j) = d^{i(j)}\,j!!\,\sigma^j/2$ $i(j) = [(-1)^j -1]/2$	$a^*=[A(n)-\sigma v A(n-1)/2]/[A(2n-1)-\sigma d A^2(n-1)]$ $A(j) = d^{i(j)}\,j!!\,\sigma^j/2$ $i(j) = [(-1)^j -1]/2$
3	$u = a\,1(e-e_m)$ 1 ( ) Heaviside unit function	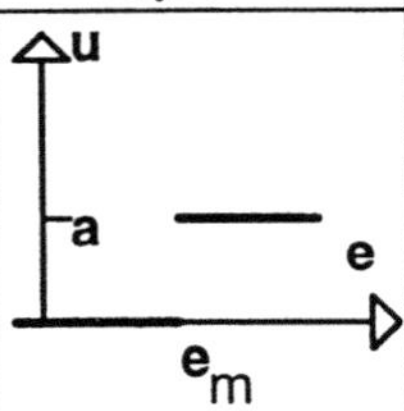 Relay	**1.** $a=e_m$ $a=\sigma[-0.5+2.15\sqrt{x}]$ $x=(v-0.2)^2+0.4$ for $v\geq 0.9$; $a=2.14v+0.5$ **2.** $a\neq e_m$ $e_m=0.7\sigma v(x/2+1)$ $a = 0.7\sigma v(x-0.86)$ $x=\sqrt{(1+3.3/v^2)}$	**1.** $a=e_m$ $a=\sigma[0.06(v+0.8)^2+0.8]$ for $v \in [0, 3]$ $a=\sigma[v-1/v]$ for $v>3$ **2.** $a\neq e_m$ $e_m=\sigma(0.6+0.4v)$ $a = 2\sigma(0.6-0.1v)$
4	$u=P(e-e_m)1(e-e_m)$	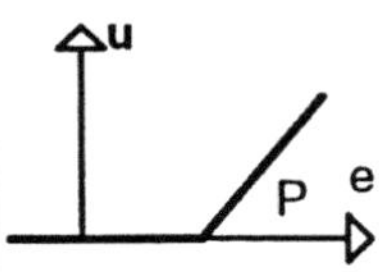 Flexible with insensitivity zone	1. Minimal interval: $P^*=1$; $e_m=\sigma v$ $h_{min} = H_s - \sigma v$ 2. Elastic interval: $P^*=8.8-3v$; $e_m=2.5\sigma$ special cases: PREFOL strategy [29] $P^*=0.5$; $e_m=\sigma E(v)$ Holdback/Pushout[22] $\gamma=0$; $P^*=0.5$; $e_m=0$	1 Special cases: ($\gamma=0$) $P^* = 1.$; $e_m = 0$ RUNON strategy [22] $P^*=1$; $e_m=\sigma v/2$ 2 Elastic strategy: $P^*=8.75-1.55v$ $e_m=2.5\sigma$

3.2. The Half-Wave Rectifier Type Regulator

It is convenient to start the presentation of the flexible holding strategies from the half-wave rectifier type regulator $u(e)=\max[0,Pe]$ where e is the amount the vehicle is early and the control parameter $P\in[0,1]$ is the fraction of the time by which the vehicle to be held (Fig. 5, Tables 2.1 - 2.2).

In this case the optimal regularity control problem may be stated as:

$$PO_{min} \quad P \quad S(P) = \gamma\ E(u) + (1-\gamma)\ E(w) \quad |\ 0 \le P \le 1 \qquad (2)$$

and its optimal solution has the simple form [Tables 2.1 - 2.2 and 4]

$$P^* = 1 - d\ v\ /\ \sigma \qquad (3)$$

where $v=\gamma H_S/(1-\gamma)\sigma$, H_S- is scheduled headway, σ^2- is the variance of off-schedule headway's deviations e, and $d = E|e| = \sigma\sqrt{2/\pi}$

Remarks:

1. For $\gamma=1/\gamma=0$ obvious intuitive solutions $P^*=0$ / $P^*=1$ are obtained.
2. The special case ($P=1$, $e=h_{min}-h$) of this control strategy is the well-known "minimal-headway" threshold holding strategy (in which the threshold value $h_{min}\in[U, H_S]$ (in practice $U=(0.65-1)H_S$). The optimal value of the threshold parameter is equal to $h_{min}= \max[U, (1-2\gamma)H_S/(1-\gamma)]$ and confirms the results presented in [28, 30]
3. The robust features of the solution may be conveniently illustrated by the analytical solution of bicriterial Min-Min problem with the Pareto preference ordering [15]. The set of compromise solutions AB, in the criteria space is shown in Figs. 7.1 -7.2. For a fixed value of the σ parameter the relative sensitivity of Q1 and Q2 criteria with respect to the variability of the γ parameter is well illustrated by the tangent to the compromise set at the point P*.

As can be seen, the increase of scheduled headway H_S improves the sensitivity feature. For a fixed H_S value and decreasing σ parameter the criteria tend to be more co-operative (i.e. from random service with $\sigma = H_S$ to full co-operation for regular service $\sigma=0$) (Fig. 7.2).

4. The adaptive implementation of this strategy with recursive adjustment of regulator parameter is presented in Fig. 5. For real-time recursive estimation of the time-varying parameter v several RLS methods may be used (e.g. RLS with exponential forgetting) [10]

5. The family of flexible polynomial types of regulators $u(e)=ae^n 1(e)$ is a natural extension of the above control strategy. It can adjust control parameters to particular lines, control points, time periods. In this case $Q_1=E(u^k)=a^k\sigma A(nk-1)$ where $A(m)=[d^{i(m)}m!!\, \sigma^m /2]$ and $i=[(-1)^m-1]/2$. The optimal value of the control parameter "a" is the solution of the algebraic equation: $v\sigma kA(nk-1)a^{k-1}+A(2n-1)-A(n)=0$. The solution for k=1 is presented in Table 1. In the special cases: $\gamma=0$: $a=A(n)B(n)/\sigma^n$; $\gamma=1$: $a=0$. As can be checked for k=n=1, the solution (3) is obtained.

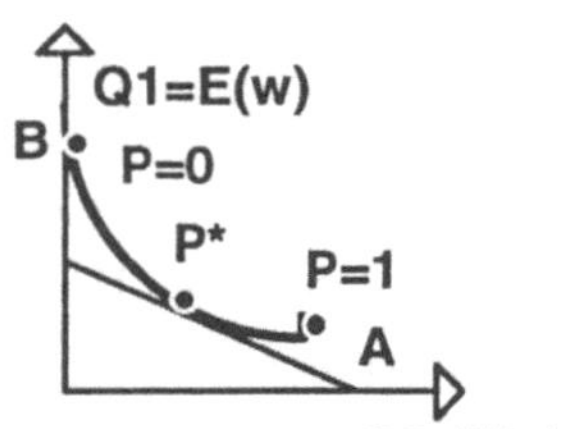

Fig.7.1 Compromise Set AB in criteria space

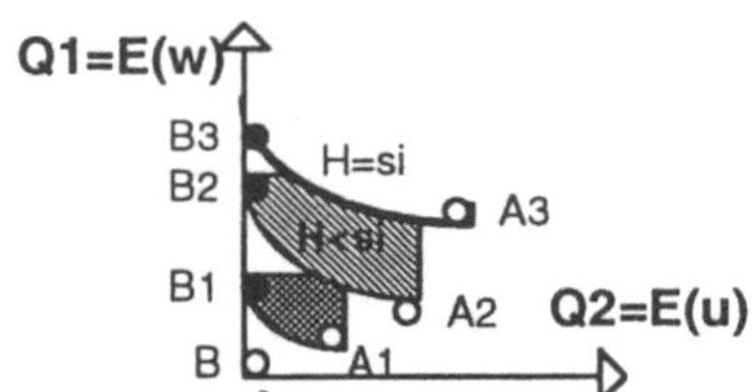

Fig. 7.2 Evolution of Compromise Set AB with standard deviation si

The possible extensions of the above mentioned approach may be conducted in a few (not necessary disjoint) directions for other types of: **performance criteria** (Tables 2), **regulators** (Table 1) and **strategies** e.g. punctuality, synchronizing type (see Table 1, Fig. 6).

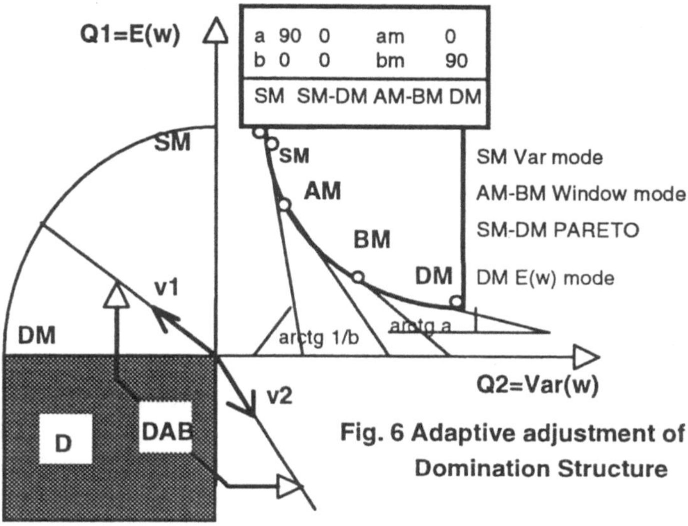

Fig. 6 Adaptive adjustment of Domination Structure

Various passengers disutility functions may be used as objective functions in our optimisation problems. The common choice in literature is the mean waiting time. The attractiveness of such a measure however can be misleading because it penalizes long and short waiting times with the same weight and does not take into account service reliability aspects. This is why the variance and second order moment of control and wait time random variables (which more penalizes long than short waiting times) may be more attractive disutility measures. In particular, the optimal analytical solutions of problem (1) for very general cases i.e. Q1 equal to $E(u^k) = a_k\ (k-1)!!\ (P\sigma)^k / 2$ where ($a_k=1$ for even and $a_k=d/\sigma$ for uneven k-th), $E^2(u)$, and Var(u) and Q2 equal to E(w), $E(w^2)$, $I=Var(h)/2H_s$ and $I=Var(w) / \sigma^4$ are also presented in Table 2.1.

To illustrate these results the optimal P* values for each of four cases in Table 2.1 and various parameters from four bus lines A to D (as in Example 1) are also presented in Table 2.2.

$S(u) = \gamma\, Q_1 + (1-\gamma)\, Q_2$ **Table 2.1**

	Q_1	Q_2	$u = \max\{0, P^* e\}$, $P^* \in [0,1]$; $v = \gamma/(1-\gamma)\, C_H$; $C_H = \sigma/E(H)$; $d = \sigma\sqrt{2/\pi}$		
	$E(u^k)$	$E(w)$			
1	$k = 1$		$P^* = 1 - d\, v/\sigma$	$v \le \sigma/d$	
	$k = 2$				$B=1$
	$E^2(u)$		$P^* = 1/(1 + 2B\sigma v)$;	$v \ge 0$	$B=1/\pi$
	$Var(u)$				$B=1-1/\pi$
	$k = 3$		$P^* = 2/\left[1+\sqrt{1+24 d v \sigma}\right]$		
	$k \to \infty$		$P^* \to 1$		
	$E(u^k)$	$I = Var(h)/2H_s$		$A = \pi/(\pi-1)$	
2	$k = 1$		$P^* = A(1 - 2v)$	$(v \in [1/2\pi, 1/2])$	
	$k = 2$				$B=A$
	$E^2(u)$		$P^* = A/(1 + 2B\sigma v)$	$(v \ge 0)$	$B=A-1$
	$Var(u)$				$B=1$
	$k = 3$		$P^* = 2A/\left[1+\sqrt{1+24 d\sigma v A}\right]^2$		
	$k \to \infty$		$P^* \to A$	$(P^* = 1)$	
	$E(u^k)$	$E(w^2)$		$A = (H_s - d)/(H_s - 2d)$	
3	$k = 1$		$P^* = A\{1 - v(A-1)/(2A\sigma)\}$	$v \in [2\sigma, 2A\sigma/(A-1)]$	
	$k = 2$				$B=1$
	$E^2(u)$		$P^* = A/[1 + B\sigma v(A-1)/d]$	$v \ge d/B\sigma$	$B=1/\pi$
	$Var(u)$				$B=1-1/\pi$
	$k = 3$		$P^* = 2A/\left[1+\sqrt{1+8\sigma v A(A-1)}\right]$		
	$k \to \infty$		$P^* \to 1$		
	$E(u^k)$	$I = Var(w)/\sigma^4$		$A = (1 - C_H^2)/(1 - 2C_H^2)$	
4	$k=1$		$P^* = A[1 - d v \sigma H_s (A-1)/2A]$	$v \le 2A/(d\sigma H_s(A-1)$	
	$k=2$				$B=1$
	$E^2(u)$		$P^* = A/[1 + 2B\sigma^3 v H_s (A-1)]$;	$v \ge 1/2B\sigma^3 H_s$	$B=1/\pi$
	$Var(u)$				$B=1-1/\pi$
	$k = 3$		$P^* = 2A/\left[1+\sqrt{1+16 d v H A (A-1)\sigma^3}\right]$		
	$k \to \infty$		$P^* \to 1$		

Table 2.2.

γ\P	k = 1	k = 2	k = 3	$E^2(u)$	Var (u)	k = 1	k = 3	k = 2	E^2 (u)	Var (u)
case 1	A: H = 10 (min), σ = 1(min)					B: H=10, σ=2.5		C: H=5 (min), σ=1 (min)		
0	1	1	1	1	1	1	1	1	1	1
0.1	0.11	0.31	0.70	0.60	0.40	0.76	0.80	0.47	0.73	0.56
0.5	0	0.05	0.09	0.13	0.07	0	0.29	0.09	0.24	0.13
0.8	0	0.01	0.02	0.04	0.02	0	0.15	0.02	0.07	0.04
1	0	0	0	0	0	0	0	0	0	0
γ_{max}	0.11	1	1	1	1	0.24	1	1	1	1
case 2	B: H = 10 (min), σ = 2.5 (min)					D: H= 5 (min) σ = 1.5 (min)				
0.1	0.16	0.34	0.25	0.72	0.46	0.38	0.40	0.55	0.96	0.69
0.3	0	0.18	0.13	0.29	0.15	0	0.22	0.20	0.49	0.28
0.5	0	0.05	0.09	0.14	0.07	0	0.15	0.10	0.26	0.13
0.8	0	0.01	0.05	0.04	0.02	0	0.08	0.02	0.07	0.04
1	0	0	0	0	0	0	0	0	0	0
γ_{max}	0.11	1	1	1	1	0.13	1	1	1	1
case 3	B: H = 10 (min), σ = 2.5 (min)					D: H= 5 (min) σ = 1.5 (min)				
0.1	1	1	0.82	1	1	1	0.98	1	1	1
0.3	1	0.78	0.53	1	0.90	1	0.66	0.80	1	0.94
0.5	1	0.50	0.37	0.83	0.62	0.69	0.47	0.50	0.90	0.63
0.8	0.27	0.18	0.20	0.43	0.24	0	0.25	0.17	0.42	0.23
1	0	0	0	0	0	0	0	0	0	0
γ_{max}	0.55-083	≥0.166	≥0.05	≥0.39	≥0.21	0.47-074	≥0.01	≥0.193	≥0.43	≥0.23
case 4	B: H = 10 (min), σ = 2.5 (min)					A: H= 10 (min) σ = 1 (min)				
0.1	0.29	0.100	0.15	0.27	0.140	0.97	0.78	0.826	0.94	0.88
0.3	0	0.030	0.08	0.08	0.040	0.68	0.57	0.540	0.78	0.63
0.5	0	0.012	0.05	0.03	0.018	0.60	0.42	0.330	0.60	0.42
0.8	0	0.003	0.02	0.01	0.004	0	0.245	0.110	0.29	0.15
1	0	0	0	0	0	0	0	0	0	0
γ_{max}	0.01-013	≥0.0008	≥0.0002	≥0.0003	≥0.002	0.02-071	≥0.003	≥0.005	≥0.004	≥0.007

Example 1:

Let us consider the bus lines with scheduled intervals H_S and intervals standard deviations σ :

A. H_S = 10 (min), σ = 1 (min); B. H_S = 10 (min), σ = 2.5 (min);

C. H_S = 5 (min), σ = 1 (min); D. H_S = 5 (min), σ = 1.5 (min) ;

Remarks: (to Table 2.2)

1. For the second case, when k increases then $P^* \rightarrow A$ ($A=\pi/(\pi-1)$). Hence delaying the vehicles for a longer period then the supplement to H_S (i.e. for $e=H_S$ -h) seems to be beneficial.

2. From the 4-th case in the table it can be easily seen that when we take into account the higher control moments then the longer holding is beneficial due to lower importance of these moments with respect to the irregularity service measure.
3. The practicall importance of these results is concerned with the wide admissible range and small magnitudes (costs) of beneficial control actions. This makes it possible to realize very profitable (in the sense of present and future quality of service measures) distributed control actions in the "control by opportunity" mode [15].

The control experiments on the real bus lines in Cracow proved the practical importance of such type control mode.

At this point it should be noted that all considerations concerning the multicriteria solutions and adaptive implementations may be applied to these very general cases. In Fig.7.3. for illustration the Sets of Compromises for various criteria (second item in Table 2.1) are presented. As can be seen, for fixed σ and H_S the Compromise Set range increases for higher order moments.

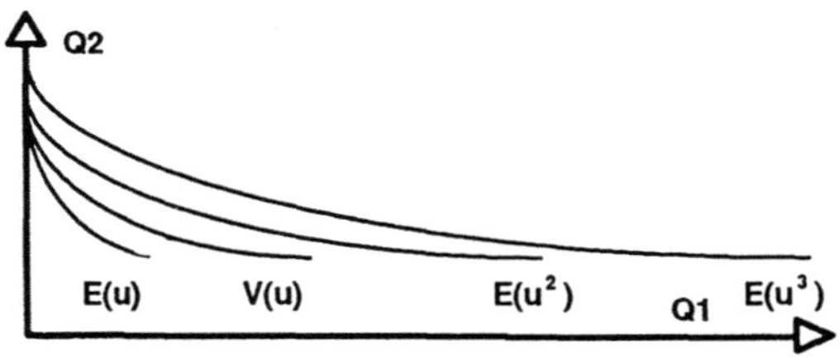

Fig. 7.3 Compromise Set for various criteria

3.3. The Relay Type of Regulator

The relay type regulator is described by $\mathbf{u(e)=a1(e\text{-}e_m)}$ where $\mathbf{1(\bullet)}$ is Heaviside unit function and two control parameters **a** and $\mathbf{e_m}$ represent the magnitude of the holding time (a) and threshold value (e_m) of the regulator insensitivity zone respectively. The two types of this regulator: A. $\mathbf{a \neq e_m}$ and B. $\mathbf{a=e_m}$ will be considered in this holding strategy. Denoting the p.d.f. and cumulative distribution functions of the normal distribution N(0,1) by f(u) and F(u) respectively we define **C(u)=f(u)/[1-F(u)]** function. We can describe $Q_1 = E(u^k) = a^k f(u)/C(u)$ in terms of C(u) (where $\mathbf{u=e_m/\sigma}$) and assuming Q_2=E(w) [10] to solve the optimal control problem (1). The optimal parameter values e_m and a for two types A and B relay regulator and approximation of the C(u) function are presented in Table 3.1 and numerically illustrated by two bus lines A and D (see Example 1) in Table 3.2.

$S(u)=\gamma Q_1 + (1-\gamma) Q_2$ **Table 3.1**

	Q_1	Q_2	$u=a1(e-e_m); e_m \in [0,H_s]; a \in [0,H_s]; v=\gamma/(1-\gamma)C_H; C_H=\sigma/E(H)$
A	$E(u^k)$ k=1	E(w)	**A: $a \neq e_m$ Relay type of regulator** $a = \sigma v\,[0.7A - 0.6]$ where $A=\sqrt{1+3.3/v^2}$ $e_m = 0.35\,\sigma v\,[2+A]$
	k=2		$a = 1.3\sigma/(1+2\sigma v)$; $e_m\ 0.65\,\sigma$; special cases: $\gamma=0$: $a=1.3\sigma$, $e_m=0.65\sigma$; $\gamma=1$: $e_m \to \infty$, $a=0$
B	$E(u^k)$	E(w)	**B: $a = e_m$** :special cases: $\gamma=0$: $a=0.886\,\sigma$, $\gamma=1$: $a \to \infty$,
	k=1		$a=\left\{2.15\sqrt{(v-0.2)^2+0.4}-0.5\right\}\sigma$ for $v \le 1$ $a=\sigma[2.14v-0.5]$ for $v \ge 0.9$
	k=2		$a=\sigma\sqrt{B+\sqrt{B^2+2}}$; $B=1/(2v\sigma-1)$
	Q_1	Q_2	$u=P(e-e_m)1(e-e_m); e_m=H-h_{min} \in [0,H]; v=\gamma/(1-\gamma)C_H; C_H=\sigma/E(H)$
A	$E(u^k)$	E(w)	$P^* = 1$, $h_{min} = \max[0, h_{min}^*]$
	k=1		$h_{min}^*=(1-2\gamma)H_s/(1-\gamma)$ $\gamma \in [0, 0.5]$
	k=2		$h_{min}^*=H_s - \sigma^2 v/(0.62+0.341\,\sigma v)$ $\sigma v \le 0.62/(C_H-0.341)$
	$E^2(u)$		$h_{min}^*=H_s - \alpha\sigma[A-\sqrt{A^2-1}]$, $\sigma v \in [0.0, 1.22]$; $\alpha=2$ $A=1.1+1.25/\sigma v$ $\sigma v \in [1.22, 6.0]$; $\alpha=1.25$, $A=0.92+1.08/\sigma v$; $\sigma v \in [6.0, 26.0]$; $\alpha=1.6$, $A=0.97+0.85/\sigma v$
	k=3		$h^*_{min} = H_s - \sqrt{A^2+3.67\,\sigma^2} + A$; $A=1.4627\sigma+0.611/v\sigma$
	k=1	$E(w^2)$	$h^*_{min} = \sqrt{H_s^2 - \sigma v}$
B	$E(u^k)$ k=1	E(w)	**$P^* \ge 0$ Elastic minimal headway strategy** **1. $P^* = 1$ $e_m = \sigma v$** **2. $P^* = 11.9 - 3.5\,v$; $e_m = 3.16\,\sigma$** special cases: $\gamma=0$: $P=1$, $e_m=0$; $\gamma=1$: $P \to \infty$, $e_m=3.7\sigma$

Table 3.2.

RELAY	A: H= 10 (min), σ= 1 (min)					D: H= 5 (min) σ= 1.5 (min)				
type:	A : $a \neq e_m$ k = 1		k=2 e_m=0.65	B k=1	k=2	A: $a \neq e_m$ k = 1		k=2 e_m=0.98	B k=1	k=2
	e_m	a	a	$a=e_m$	$a=e_m$	e_m	a	a	$a=e_m$	$a=e_m$
0.1	1.52	0.81	0.40	1.11	1.56	4.27	1.62	0.92	0.9	5
0.2	2.6	0.66	0.22	2.5	1.30	2.31	1.34	0.56	1.42	2.28
0.3	4.6	0.69	0.14	4.28	1.24	1.8	1.14	0.36	2.46	1.98
0.5	10 -	1.2	0.06	10	1.21	3.66	0.99	0.18	5	1.83
0.6	-	-	0.04	-	1.20	5.34	1.06	0.12	-	1.81
0.7	-	-	0.02	-	1.20	8.2	1.328	0.04	-	1.80

3.4. Flexible Regulator with Insensitivity Zone

The flexible minimal headway type strategy can be realized by the regulator $u(e)=P(e-e_m)\ 1(e-e_m)$ where similarly as in relay regulator the control parameters $P \geq 0$ and $e_m=H_s-h_{min}$ represent the regulator gain (P) and insensitivity zone (e_m) coefficients respectively. As the special case (P=1) of this strategy corresponds to the well-known minimal headway strategy we consider this case first and in the sequel generalize it to $P \geq 0$. The general optimisation problem is formulated as: $PO_{min}\ S(u)= \gamma\ E(u^k)+(1-\gamma)\ E(w)$ where

$$E\ (\ u^k\ ) = f\ (u)\ \ C_{k+1}\ /\ \sigma \tag{4}$$

$C_1=\sigma/C(u);C_2=\sigma^2[1-u/C(u)]$ while for $k \geq 3$ the following recurrence hold

$$C_k = \ \sigma^2\ (k-2)\ C_{k-2} - \sigma\ u\ C_{k-1} \tag{5}$$

Remarks:

1. The following property for C_k can be inductively proved:

$$dC_k\ /du = -\ C_{k+1}\ /\sigma \tag{6}$$

Proof: For k=1 and k=2 this property can be checked by differentiation of C_1 and C_2. We assume the validity of (6) for k= n-1 and n and prove it for k= n+1. From the recurrent relation (5) for C_k it follows that $dC_{n+1}/du = \sigma^2$ (n-2) $dC_{n-1}/du - \sigma C_n - \sigma u\, dC_n/du$. Replacing the both derivatives in the right hand side according to (6) and using (5) we obtain $- C_{n+2}/\sigma$. Thus , the proof is completed.

2. For $\gamma=0$ (i.e. Waiting Mode (WM)) the optimal solution is $e_m=0$ i.e. $h_{min}=H_s$. For $\gamma=1$ (i.e. Control Mode (CM)) the optimal solution $e_m = - C_{k+1}/C_k$ can be derived from the $C_{k-1}=0$ condition.

3. For $\gamma\neq1$ the general optimal solution is equal to $e_m=\sigma v k C_k/C_1$ where $v=\gamma.H/(1-\gamma)\sigma$.

Proof. $dS(u)/du = (-v/\sigma^2)[C_{k+2}+\sigma u\, C_{k+1}] + u\, C_1= 0$ and from (5) the expression in square brackets can be replaced by $\sigma^2 k\, C_k$ so that the solution of this equation is equal to $u = e_m / \sigma = vk\, C_k / C_1$

The analytical closed form solutions for minimal headway strategy (with analytical representation of the C(u) function derived in [10]) are presented (as case A1: Q_2=E(w); case A2: $Q_2=E(w^2)$) and numerically illustrated in Tables 3. The general case with P≥0 can be considered in a similar way [10]. The analytical solutions and illustrative examples (case B) are also presented in Tables 3.

Table 3

γ\P	k=1	k = 2	k = 3	$E^2(u)$		k = 1	k=3	k=2	E^2 (u)	
Case A 1	**A: H = 10 (min), σ = 1 (min)**					**D: H= 5 (min) σ = 1.5 (min)**				
0	10	10	10	10		5	5	5	5	
0.1	8.9	8.84	9.23	9.52		4.4	3.96	4.48	4.7	
0.3	5.7	7.93	9.10	9.3		2.86	2.62	4.34	4.25	
0.5	0	7.51	9.05	8.86		0	1.77	4.316	3.88	
0.8	0	7.19	9. 0	7.41		0	0.96	4.29	2.6	
1	0	0	7.1	7.1		0	0	0.61	0.65	
Case A 2	**A: H = 10 (min), σ = 1 (min)**					**D: H= 5 (min) σ = 1.5 (min)**				
γ→	0	0.1	0.3	0.5	0.8	0	0.1	0.3	0.5	0.8
k=1	10	9.94	9.76	9.49	0	5	4.91	4.78	4.48	2.25
Case B	**A: H = 10 (min), σ = 1 (min)**					**D: H= 5 (min) σ = 1.5 (min)**				
k=1 γ→	0.1	0.2	0.3	0.4	0.5	0.1	0.2	0.3	0.4	0.5
P=1 e_m	1.11	2.5	4.28	6.67	10	0.55	1.25	2.14	2.22	5
P / e_m	8/3.16	3.16/3.2				10/ 4.7	9/ 4.7	7/ 4.7	4.1/ 4.7	0.25/4.7

3.5. Punctuality Strategies

The examples of equivalents of punctuality and synchronizing types of strategies are presented in Table 1 and Table 4. Consider a simple holding punctuality strategy for transfer synchronization [13] between C and N buses: buses depart immediately if they are late (with respect to scheduled departure times t^S) and at their scheduled departure time otherwise.

Table 4

Δ	E(w)	E_0(w)	Gain%	E(w^2)	E_0(w^2)	Gain %
4	4.38	4.00	- 9.5	21.20	18.00	- 17.70
3	3.50	3.33	- 5	16.70	19.38	+ 13.70
2	3.00	3.49	+ 14	23.72	77.80	+ 69.50
1	3.53	5.00	+29.4	52.80	90.90	+ 41.90
0	5.91	10.00	+40.9	116.80	201.79	+ 42.11

Δ	V(w)	V_0(w)	Gain%	F(w*)	F_0(w*)	Gain%
4	2.06	1.90	- 8.42	0.1735	0.2100	+17.40
3	4.31	8.32	+48.19	0.0176	0.0814	+78.40
2	14.63	52.90	+72.34	0.1455	0.164	+11.35
1	40.40	67.67	+40.30	0.2000	0.200	+ 0.00
0	81.98	101.90	+19.50	0.3696	0.500	+26.00

For this case the transfer wait function can be compactly expressed as:

$w(x,y)=x+\max\{0,v_C\}$ if $x+\max\{0,v_C\}>0$ or $w(x,y)=y+\max\{0,v_N\}$ otherwise, where $v_C=t_C{}^S-t_C$, $v_N=t_N{}^S-t_N$ are off-schedule deviations of C and N buses at the transfer point, and $u_1=\max\{0,v_C\}$, $u_2=\max\{0,v_N\}$ are control variables. Analytical formulas for different disutility functions of the transfer waiting time and assumed Shifted Truncated Exponential distribution of bus arrivals times to the transfer point are derived in [13]. The comparison of control efficiency (relative to no-control case) in terms of values of various disutility functions and offset times control variables Δ, are presented in Table 4 . It is easy to see that the application of control at the transfer point is highly profitable.

The idea of the multicriteria approach in this case with adaptive adjustment of the domination structure according to dynamically evolving "arrivals windows" on both lines. is presented in Fig. 6. The preference ordering is introduced by means of convex dominated cones e.g. Pareto preference is introduced by $D=R^2$ cone. To obtain the adaptive control actions we relate the domination structure with the currently detected transfer situations. To achieve this, it is proposed to represent the domination structure in an adaptive self-organizing way by dominated cones generated in real-time by preference directions v^i of the "arrival windows " modes which are most suitable for the current transfer situation. In Fig. 6. the domination cone $D_{\alpha\beta}$ dependent evolution of Compromise Set is illustrated. It is easy to see that for the D_{BA} domination cone the compromise set corresponds to the set of points AM-BM, from which final solution will be selected. The polyhedral D_{BA} cone can be considered as the convex cone generated by preference directions $\{v^1,v^2\}$ connected with arrival windows modes $D_{BA}=\{\sum_{i=1}^{2} a_i v^i \mid a_i \geq 0\}$. In a similar way the other operational modes may be represented.

4. Conclusions

1. New hardware capabilities make it possible to design distributed hierarchical management and control systems which have highly profitable features for application in the traffic control.
2. The attractiveness of the presented control idea is that many criteria, types of regulators and control strategies can be implemented and that it is a computer-oriented and network tool, applicable to wide spectrum of demands met at real traffic situations. It makes it also possible to work in not exactly known

traffic environment and to start with a simple traffic model which is adjusted during the system operation to the real traffic situations (i.e. it has on-line real time dynamical modelling and model-tracking features).

3. The presented results constitute an important step to the generation of highly efficient robust adaptive multicriterial solutions for the networks problems. The closed form of analytical solutions enables us to deal with the full representation of the solution which is crucial for the solution sensitivity (robustness) analysis, as well as for representation of the Set of Compromise solutions in the multicriterial optimisation.
4. The high improvement in the quality of service may be obtained in the synchronization way, by the proper integration of scheduling and dispatching control actions.

The future work will concentrate on the development of network related analytical and numerical results for a wide range of alternative control actions, and network criteria.

References

[1] M. Abkowitz., A. Eiger and I. Engelstein "Optimal control of headway variation on transit routes", *Journal of Advan. Transp.*, 20/1: 73-88, 1986.

[2] A. Adamski and A.,Rudnicki, "Public Transport as a Control Plant", *Proc. VII Conf. of Automatics Rzeszów, POLAND, 1977*, 1: 722-730, 1977. (in Polish)

[3] A. Adamski, "Optimal Dispatching Control of Bus Lines", *Proc. 9th IFIP, Conference on Optimization Techniques, Lecture Notes in Control and Information Science Springer Verlag*, No. 23, Part 2: 334-344, 1980

[4] A. Adamski, "Multilayer control system for public transport", *ZNAGH*, 131:101-11, 1980.

[5] A. Adamski and A. Korytowski, "Optimal Structures of Hierarchical Traffic Control Systems", *System Science,* 7/1: 110-116, 1981.

[6] A. Adamski, "Optimal Dispatching Control of Bus Lines", *Proc.of the 4th IFAC/IFIP/IFORS Inter. Conf. on CONTROL in TRANSPORTATION SYSTEMS Baden-Baden:* 67-72, 1983.

[7] A. Adamski and A.Rudnicki, "Control systems in the public transport", (*Chapter 6 in a book: Computer Traffic Control Systems: Application Examples,*): 104-148, WKi £ 1984.

[8] A. Adamski, "Optimal dispatching control of bus lines in hierarchical system", *Proc. 4th Inter. Conf. "Science and Practice of Transport" Jachranka, POLAND.* 4: 1-11, 1985.

[9] A. Adamski, "Polyoptimal integrated control of the traffic on an intersection", *Proc. 5th IFAC/ IFIP/IFORS Inter. Conf. on CONTROL in TRANSPORT. SYSTEMS July 8-11, Vienna 1986.*

[10] A. Adamski, "Optimal Dispatching Control in Public Transport", *Hab. Thesis. Scien. Bulletins of AGH, AUTOMATICS, Bulletin no. 50,* 1989.

[11] A. Adamski, "Expert system for dispatching control of the public transport vehicles at terminals" *Proc. International Conference TRANSSYSTEM-89,* 2, 1989.

[12] A. Adamski, "Probabilistic Models of Passengers Service Processes at Bus Stops", *Transpn Research.,* 26B/4: 253-259, 1992.

[13] A. Adamski, "Transfer Optimization in Public Transport", *Sixt Inter Workshop on Computer-Aided Scheduling in Public Transport Lisboa,* 1993.(in press)

[14] A. Adamski, "Real-time Computer-Aided Control in Public Transport from the point of view of Schedule Reliability", *Sixt Inter. Workshop on Computer-Aided Scheduling in Public Transport, Lisboa,* 1993. (in press)

[15] A. Adamski, "Optimal adaptive dispatching control in an integrated public transport management system", *Sec. Meeting of the EURO Working Group on Urban Traffic and Transportation 15-17 Sept., Paris ,FRANCE:* 1993.

[16] O. Adebisi, "A Mathematical Model for Headway Variance of Fixed Route Buses", *Tranpn. Res.* 20B: 59-70, 1986

[17] P Anderson, A. Hermansson, E. Tengvald and G. Scalia-Tomba, "Analysis and simulation of an urban bus route", *Transpn. Research,* 13A: 439-466, 1979.

[18] A. Barnett, "On controlling randomness in transit operations", *Transp. Sci.*, 8/2:102-116, 1974.

[19] P.H. Bly, R.L. Jackson, "Evaluation of Bus Control Strategies by simulation", TRRL 637, 1974.

[20] A. Ceder and N.H.M. Wilson, "Bus Network Design", *Transp. Research,* 20B/4: 331-44, 1986.

[21] R.L. Gordon, "Control of bus headways by traffic signal timing method", *IEEE Trans. Veh. Tech.*, 27/2: 69-76, 1978 .

[22] R. Jackson, "Evaluation by simulation of control strategies for a high frequency bus service", *TRRL- 807*, 1977

[23] R. Jacobson and Y. Sheffi, "Analytical Model of Traffic Delays under Bus Signal Preemption:Theory and Application", *Transpn. Research,* 15B/2: 127-38, 1981

[24] D. Koffman, "A simulation study of alternative realtime bus headway control strategies", *Trans. Res. Record,* 663: 41-46, 1978.

[25] G.F. Newell, "Control of Pairing of Vehicles on Public Transportation Route, Two Vehicles, One Control Point", *Transportation Science,* 8/3, 1974.

[26] E.E Osuna and G.F. Newell, "Control Strategies for an Idealized Public Transportation System", *Transp. Science,* 6: 52-72, 1972.

[27] W.B. Powell and W.B. Sheffi, "A Probabilistic Model of Bus Route Performance", *Transp. Sci.*, 17/4: 376-404, 1983

[28] M.Turnquist and L.Bowman, "Control of Service Reliability in Transit Networks", *U.S.-DOT Rep. 5/ 1979*

[29] M. Turnquist, "Strategies for Improving Bus Transit Service", *Final. U.S. DOT Rep. 27/ 1982.*

[30] M. Turnquist and S.W. BLUME, "Evaluation the Potential Effectiveness of Headway Control Strategies for Transit Systems", *Transp. Res. Record,* 746: 25-29, 1980.

[31] U. Vanderbona and A.J.,Richardson, "Effect of checkpoint control strategies in a simulated transit operation", *Transp. Research,* 20A/6: 429-36,1986.

QUEUING OPTIMIZATION OF SIGNALIZED INTERSECTIONS

Avi Ceder

Chief Scientist, Ministry of Transport Israel and Professor at the Civil Engineery Department. Technion-Israel Institute of Technology, Haifa, Israel 32000.

Israel Reshetnik

Chief Engineer, Haifa Traffic Control Center, Haifa Municipality, P.O.B. 4811, Haifa, Israel 31047.

Previous research regarding optimization of isolated intersections was carried out in the past as well as in recent years, yet, optimization techniques implemented on the signal controller's level are very primitive and unsatisfying. Three major obstacles prevent the controller from adapting programs to varying traffic conditions :

1. The necessity to deal with several constraints, such as intergreen times, minimum green times, phase sequence, green wave constraints etc.
2. Complex non-polynomial optimization problems, especially for saturated and oversaturated conditions.
3. The difficulty to collect the data required to control an intersection at oversaturated conditions.

In this paper, a self-tuning system for controlling signalized intersections in undersaturated and oversaturated conditions is depicted. The system was successfully tested in one of Haifa's most congested intersections. The system tailors some well-known control techniques together with some specially developed methods and algorithms to overcome the difficulties.

The philosophy of system is based on a very well known actuated traffic control which was proved to be very efficient at unsaturated intersections and is very simple to implement. A method of constructing a basic program that meets the required constraints and can be modified by means of traffic detectors is also briefly described. The paper formulates several optimization problems that can be solved in polynomial time like Simplex algorithms (virtually in some fractions of a second) for oversaturated as well as undersaturated conditions. Finally, at the end of the paper, a method for collecting the required data at reasonable cost is presented at the end of the article.

1. Introduction

The creation of a signalization program for an intersection generally requires consideration of constraints such as: phase sequence, minimal green times for every movement, inter-green times, internal green-waves (i.e. for adjacent intersections which are operated by one traffic controller, etc.). The construction of an optimization problem which is intended to ensure optimal green for all directions and also to maintain all the planning constraints is a difficult and highly complex process. A possible method which permits control of a signalized intersection is control by means of a stopping point.

According to the above mentioned method a basic program, which satisfies all the constraints, is inserted into the controller. Points are established, in the basic program, at which the program stops and waits for a certain condition to be fulfilled. This condition can be a lack of demand or the presence of a queue at a particular detector, the arrival of a releasing pulse from an adjacent intersection or from a control center, or a manual release provided by a policeman. As a result of the controller's pause at the stopping point the green light is prolonged for the movements "stuck" at this point and the waiting times at all the stopping points create the overall picture of the cycle.

The basic program is constructed, not in real time, manually or by means of algorithms (a type of Gantt algorithm). The program is constructed to establish activities (green phases) in the timetable, subject to the constraints of minimum green and inter-green times and other constraints. The stopping points described above are established in the basic program. These points are generally established as close as possible to the conclusion of the phase which will become prolonged at this point. At the stopping point a transition can occur to another basic program as a result of lack of demand at a particular detector. An example of a signalization program for an intersection is described in Figures 1 to 3.

Constructing a signalization program in this way permits optimization only of the duration of waiting time at the stopping point without consideration for the constraints expressed when in outline program. The number of decision variables is less than 10 in this case and only at a few intersections with a complex geometry does the number of stopping points exceed 6.

Problems of this order of magnitude can be resolved in real time for the time ranges required by traffic light control systems when the optimization problem is not linear. The applicability of various optimization problems which are suitable for operating a signalized intersection will be examined in the next section.

2. Conversion Matrix of Waiting Times to Green Times

The various optimization problems are mostly expressed in respect of the green times which are allotted to the different movements at an intersection. In order to express the various mathematical problems in respect of the waiting times at the stopping points it is necessary to find a way of presenting the green times by means of the waiting times at the stopping point.

Definition: A conversion matrix **W** is a matrix in which the number of rows **m** is equal to the number of movements on which the optimization is done and the number of columns **n** is the same as the number of stopping points which can be prolonged. The value of the term **Wij** becomes **1** if movement **i** is prolonged as a result of waiting at stopping point **j**, otherwise the value of **Wij** is **0**.

Let **g** be the green time vector of the different movements and let g_m be the minimal green times according to the minimal (skeleton) program; one can then write :

$$g = W \cdot y + g_m \tag{1}$$

where y is the vector of waiting times at the various stopping points.

By substituting the function of **y** for **g** one can formulate various objective functions and constraints in accordance with the operating strategy.

3. Unsaturated Traffic Conditions

The allocation of times at unsaturated intersections does not constitute a special problem and there are a number of objective functions which can be optimized relatively rapidly. The various objective functions are mostly based on minimization of the delay caused to an isolated vehicle or to all the vehicles at an intersection. The first delay function was developed by Webster[18] in 1958 and since then was improved many times by other researchers (Miller[16], Newell[17], McNeil[13], et al). It is presently customary to use the delay function presented at HCM 1985[11] and which can be minimized by expressing the optimization problem quoted below:

$$\text{Find } \min \sum_i d_i \tag{2}$$

s.t. $$\langle \mathbf{1}, \mathbf{g} \rangle = c$$

where the delay $\mathbf{d}_i$ for every movement i, is defined by the formula:

$$d_i = \left\{ \frac{0.38 \cdot c \cdot (1 - g_i / c)^2}{1 - g_i / c \cdot X_i} + 173 \cdot X_i^2 \left[(X_i - 1) + \sqrt{(X_i - 1)^2 + \frac{16 \cdot X_i}{g_i \cdot S_i}} \right] \right\} \cdot Fr \tag{3}$$

and c - the cycle time (in seconds).

X - the degree of the traffic saturation $(X = (c \cdot V)/(g \cdot S))$.

V - hourly traffic volume (vehicles/hour).

S - saturation flow of the traffic (vehicles/hour).

Fr - correction coefficient due to non-random arrival (in a green or red wave).

Minimization of the objective function proceeds quickly over the only constraint when the green times **g** are represented by the waiting times **y**, as described in Equation 1. In this case the optimization problem will appear as follows:

$$\text{Find } \min \sum_i d_i(\mathbf{y}) \tag{4}$$

s.t. $$\langle \mathbf{1}, \mathbf{W} \cdot \mathbf{g} + \mathbf{g_m} \rangle = c$$

By carrying out a linear search over the range of the logical cycle times one can deduce the duration c of the optimal cycle although this is not critical since, at intersections operated by detectors the cycle time can be modified in real time by the intersection detector.

Another method which is simpler and as logical is based on the desire to clear all the vehicles of all movements at the intersection during the course of the cycle. The green time allotted by this method suffices for the discharge of the volume accumulating during the cycle. In this case the green allocation problem becomes a linear programming problem as follows:

$$\text{Find } \min \quad c = \sum_i g_i \tag{5}$$

s.t. 1) $$\frac{V_i \cdot c}{S_i \cdot g_i} \leq 1 \Rightarrow S_i \cdot g_i - V_i \cdot \sum_i g_i \geq 0 \qquad \forall i = 1..m$$

2) $$\mathbf{g} \geq \mathbf{g_m}$$

As the traffic volumes vary from one cycle to another it is usual to slightly increase the volumes arriving (for example to calculate the 70th or the 80th percentile of the volume in accordance with the Normal or Poisson distribution formula), when the local detectors are supposed to shorten the green times for the actual traffic volumes.

Here also the optimization problem can be simplified by substituting waiting times at the stopping point for the green times (according to Equation 1). The new optimization problem becomes :

$$\text{Find min} \quad \sum_i y_i \tag{6}$$

s.t. 1) $S_i \cdot \langle \mathbf{W}^i, \mathbf{y} \rangle - V_i \cdot (\mathbf{1Wy} + \mathbf{c}_m) \geq 0 \qquad \forall i = 1..m$

2) $y \geq 0$

This problem is also a linear programming problem and the complexity of the green time allocation problem under unsaturated conditions is therefore a polynomial problem at the stopping points.

4. Traffic Control for Congested Conditions

The operation of signalized intersections under conditions of congestion cannot be carried out based on counting the number of vehicles arriving at the intersection. The reasons are as follows:

- The accepted delay models (Webster, HMC) for steady state traffic flow are not quite suitable for describing delay during congestion.
- The use of the method for clearing an intersection is not practicable, since during congestion the intersection cannot be cleared at the end of the cycle.
- The demand of traffic aiming at passing the intersection cannot be assessed since the detector can gauge only vehicles which passed over it on one hand while on the other hand this quantity becomes smaller as congestion increases (because of the reduction in speed).

These limitations have led to the development of methods, based on data from a queue detector, for the purpose of allocating green-times to the traffic lights. Models based on an optimization of queue lengths have been suggested in the past by a number of researchers

(Michalopoulos & Stephanopoulos[15], Church & Revelle[6], et al), although these methods required prior knowledge at the time on the mode of development of the queue and were not appropriate to the method of control which adapts itself in real time to traffic volumes.

The first adaptive method was proposed by Gartner[9] in 1983. The method is based on the measurement of the delay by integrating the length of the queue over a certain interval of time when the decision variables are the points in time when the green light has to change. The foregoing objective function is described in general terms as follows:

$$\text{Find min} \quad \sum_i \Phi_i(x_1, x_2 \dots x_p) \tag{7}$$

$$\Phi_i(x_1, x_2 \dots x_p) = \int_0^T (Q_0 + A(t) - D(x,t))$$

where :

x1, x2, ..., xp	- decision points when green light has to change.
Qo	- The queue at the start of the time interval T being examined.
A(t)	- The vehicles arrival function.
D(x,t)	- The vehicles discharge function.

However these methods were based on integer programming or on dynamic programming and were unsuitable for optimization which has to be executed in real time. This paper suggests a number of alternative methods for controlling a signalized intersection in a congested condition.

5. Queuing Optimization Model

The principle of the proposed method is that the optimization has to be done on a restricted number of stopping points (Ceder & Reshetnik, 1995) and that the objective function has to be calculated in real time in accordance with data arriving once per cycle and not to be based on data for a longer interval of time (apart from slight corrections in accordance with past observations). Since the accumulating queue cannot be discharged in all traffic directions during a single cycle one of the following strategies can be employed :

- The increment to the accumulating queue of vehicles during a single cycle in all the critical movements will be minimal.
- The maximum accumulating queue in all the critical movements will be minimal.

If one does not take into account the number of lanes or the differences in the saturation flow between the various movements, the use of both strategies will produce the same outcome which is a queue of more or less the same length in all the critical directions. Otherwise priority will be given to movements in which the total saturation flow is higher.

The following is the mathematical expression of this optimization problem regarding the first strategy:

$$\text{Find min } \sum_i \Delta Q_i \tag{8}$$

s.t.

1) $Q_i^0 + \Delta Q_i \le Q_i^{max}$ $\quad \forall i = 1...m$

2) $Q_i^0 + \Delta Q_i \ge 0$ $\quad \forall i = 1...m$

3) $\sum_i y_i + c_{min} \le c_{max}$

4) $y_i \ge 0$ $\quad \forall i = 1...m$

where :

ΔQ_i - increment to vehicle queue in movement i at the end of the cycle.

Q_i^0 - vehicle queue in movement i at the end of the green light.

y_i - duration of prolongation at the stopping point of movement.

c_{min} - duration of the minimal cycle according to the minimal (skeleton) program.

c_{max} - the maximal desired cycle duration.

The queue increment to the movement at the traffic light can be expressed by means of the green times which all the movements get at the traffic light or equivalently by prolongation of the stopping points.

$$\Delta Q_i = \alpha_i \cdot (\langle \mathbf{1y} \rangle + c_{\min}) - s_i \cdot \langle \mathbf{W}^i \mathbf{y} \rangle \tag{9}$$

where :

α_j - average arrival rate of vehicles at the intersection during the cycle.

s_j - saturation flow of traffic.

The first constraint limits the maximal length of the queue which a particular movement is allowed to accumulate. Without the said bound a trivial solution will be obtained according to which the movement with the largest saturation flow will receive most of the green time (see also the results of research by P.G.Michalopoulos & G.Stephanopoulos[14]). The second constraint is to prevent the allocation of green time to a movement in which the entire queue has been discharged and to avoid wasting time. The third constraint restricts the length of the cycle at the intersection and the last constraint dictates only constructive solutions to the problem.

Notwithstanding that the problem appears to be a linear programming problem, it is not possible to use the objective function described in Equation 8 as it stands because the problem is liable to converge into very short cycles due to the queue increment getting smaller as the measuring time shortens, i.e. the cycle duration. In order to obtain a reliable index the objective function has to be normalized in relation to the length of the cycle at the intersection. The final optimization problem thus acquires the following form:

$$\text{Find min } \frac{\sum_i \Delta Q_i}{\sum_i y_i + c_{\min}} \Leftrightarrow \text{ Find max } \frac{\sum_i s_i \cdot \langle \mathbf{W}^i \mathbf{y} \rangle}{\sum_i y_i + c_{\min}} \tag{10}$$

s.t.

1) $Q_i^0 + \Delta Q_i \le Q_i^{\max} \quad \forall i = 1 \ldots m$

2) $Q_i^0 + \Delta Q_i \ge 0 \quad \forall i = 1 \ldots m$

3) $\sum_i y_i + c_{\min} \le c_{\max}$

4) $y_i \ge 0 \quad \forall i = 1 \ldots m$

One can see that the problem of minimizing a queue is commensurate with the problem of maximizing the normalized saturation flow in relation to the length of the cycle.

In spite of the defined optimization problem not being a classic linear programming problem one can easily see that the level sets of the objective function are hyperplanes having the form:

$$\sum_i s_i \langle \mathbf{W}^i \mathbf{y} \rangle = Const. \cdot (\sum_i y_i + c_{\min}) \Rightarrow \sum_i (s_i \langle \mathbf{W}^i \mathbf{y} \rangle - Const. \cdot y_i) = Const. \tag{11}$$

The significance of this is that the solution to the problem is obtained in one of the apexes of the linear constraint polyhedron. In the scope of our research a computer program was developed which implements an algorithm to search the polyhedron apexes, similarly to the Simplex algorithm, in which the transit direction from apex to apex is determined by the gradient size in the direction of the apex to which the transit is made.

6. An Implementation Example

Figures **1** to **3** present a test case system was built and implemented in order to examine the theoretical and practical problems in the proposed method of control at the Check Post intersection in Haifa.

A plan of the intersection with the location of the detectors is given on Figure **1**. Figure **2** describes the phases and the flow diagram according to which the traffic light controller operates. Figure **3** shows an example of some skeleton programs with the stopping points at which the optimization is done, for the branches which were presented in the flow diagram.

At certain points in the flow diagram the optimization program can be read. For test purposes this program run on an external 386 PC computer. The maximal green times for each phase have been directly written to the memory of the traffic light controller via the communications port. The data for volumes and queues were processed by the traffic light controller and transmitted to the external computer. It was further determined that if the intersection does not reach the calibration points within three cycles a self-calibration cycle will be carried. The description of the complete Queue Control System (QCS) is presented in Figure **4**.

Running the system during peak hours showed a drastic improvement in the length of the queues in the directions of congested traffic as opposed to only a slight worsening in the other directions. The overall state of the traffic at the intersection improved considerably and the duration of the rush hour was shortened. Figure5 illustrates the results obtained.

References

[1] R.Akcelik, "The highway capacity manual delay formula for signalized intersections", *ITE Journal* Vol. 58, No 3, pp.23-27, 1988.

[2] R.E. Allsop, "A diagram showing practicable settings for a fixed time traffic and their effect on delay", *Transp. Research*, Vol.5, no 1, pp.59-65, 1971.

[3] K.L. Bang, "Optimal control of isolated traffic signals", *Traffic Engineering and Control*, Vol.17, pp.288-292, 1976.

[4] M.S. Bazaraa and C.M Shetty, "Nonlinear programming", *John Wiley & Sons*, 1979.

[5] I. Catling, "A time dependent approach to junction delays", *Traffic Engineering and Control*, Nov. 1977.

[6] R. Church and C. Revelle, "Modeling an oversaturated intersection", *Transp. Research*, Vol.12, No. 3, pp. 185-189, 1978.

[7] G.C.D'ans and D.C. Gazis, "Optimal control of oversaturated store-and-forward transportation networks", *Transp. Science*, Vol.10, No.1, pp.1-19, 1976.

[8] N.H. Gartner and J.D.C. Little and H.Gabbay "Optimization of traffic signal settings by mixed-integer linear programming. Part - II : the network synchronization problem", *Transp. Science*, Vol.9, No.4, pp.344-363, 1975.

[9] N.H. Gartner, "OPAC: A demand-responsive strategy for traffic signal control", *Transp. Research Record*, No.906, pp.75-81, 1983.

[10] Highway Capacity Manual, *Transp. Research Board*, Special Report, No. 209, 1985.

[11] A.J.Mayne, "Traffic signals for congested traffic", *Proc., Engineering Foundation* Conference on Research Directions in Computer Control of Urban Traffic Systems, Calif., Feb. 1979.

[12] D.R.McNiel, "A solution to the fixed-cycle traffic light problem for compound poisson arrivals", *J. Appl. Prob.*, No.5, pp.624-635, 1968.

[13] P.G.Michalopoulos and G.Stephanopoulos, "Oversaturated signal systems with queue length constraints - I : Single intersection", *Transp. Research*, Vol.11, No.6, pp.413-421, 1977.

[14] P.G. Michalopoulos and G.Stephanopoulos, "Oversaturated signal systems with queue length constraints - II : Systems of intersections", *Transp. Research*, Vol.11, No.6, pp.423-428, 1977.

[15] A.J.Miller, "Settings for fixed-cycle traffic signals", *Opnl. Res. Quart.* Vol.14, pp.373-386, 1963.

[16] G.F.Newell, "Approximation methods for queues with application to the fixed-cycle traffic light", *SIAM Rev.*, No.7, pp.223-239, 965.

[17] F.V.Webster, "Traffic signal settings", *Road Research Technical* Paper, No.39, Her Majesty's Stationery Office, London, 1958.

[18] A.Ceder, I.Reshetnik, "Real-time queuing optimization for traffic signal systems", paper submitted to *Transportation Science Journal*, 1995.

Figure 1. Geometric plan of the experimental intersection

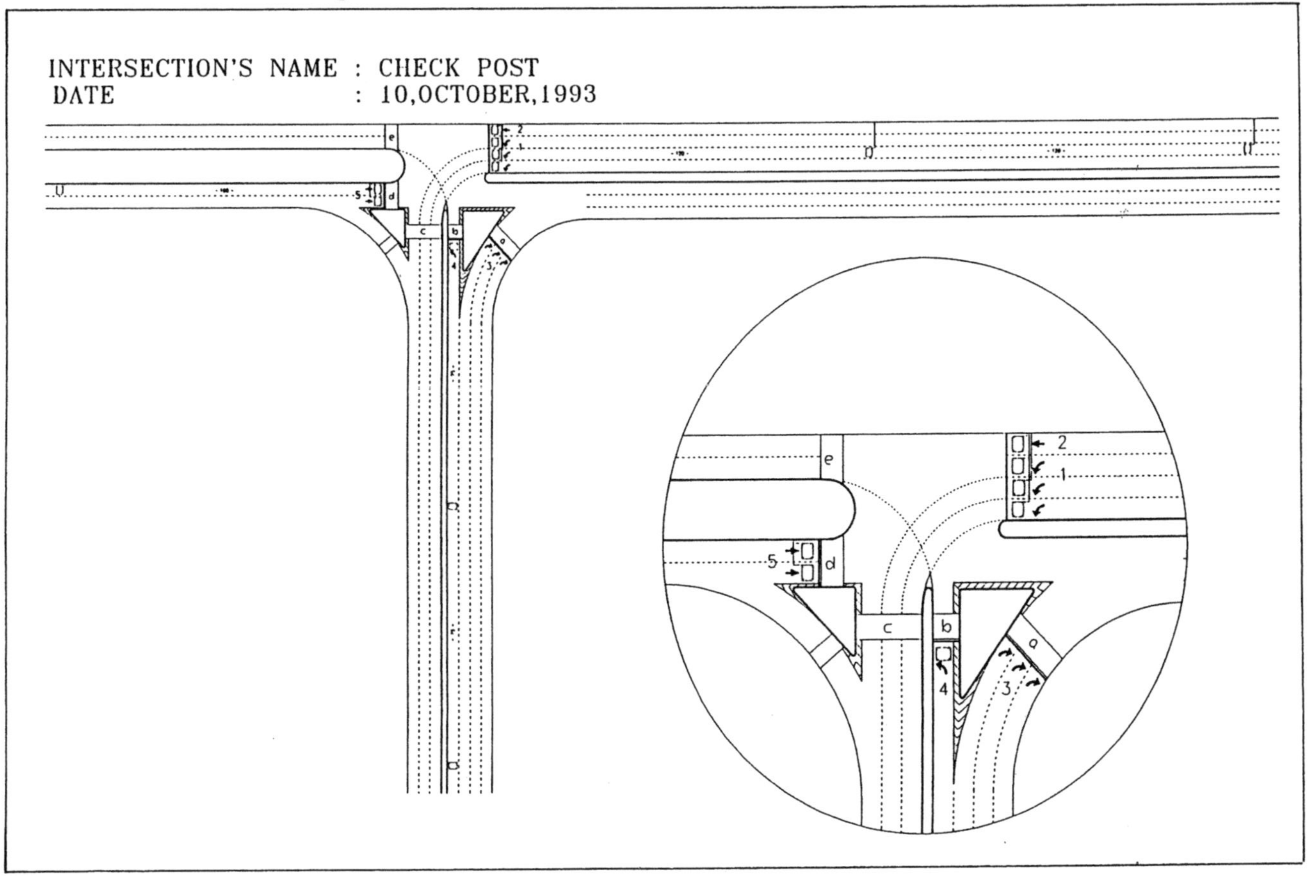

Figure 2. Flow Diagram and Phases for the signal's controller

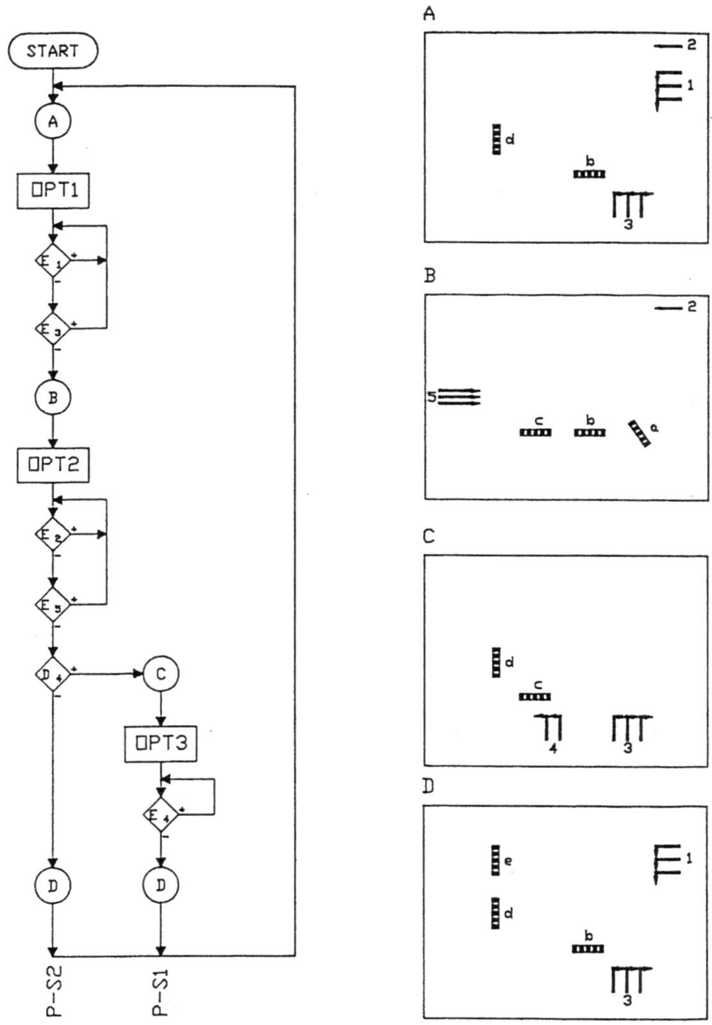

Figure 3. Skeleton Programs and Stopping Points

```
Name         : Check Post
Intersection: KN020101                    Date    : 10/10/93
Program      : P-S1                       Cycle   :  53"
Phases       : A-B-C-D
```

Movement	Ts	Tf	Tt
1	37	5	21
2	52	18	19
3	27	5	31
4	23	32	9
5	11	17	6
a	8	16	8
b	35	16	34
c	13	28	15
d	20	2	35
e	40	46	6
Optimize Extension Demand			

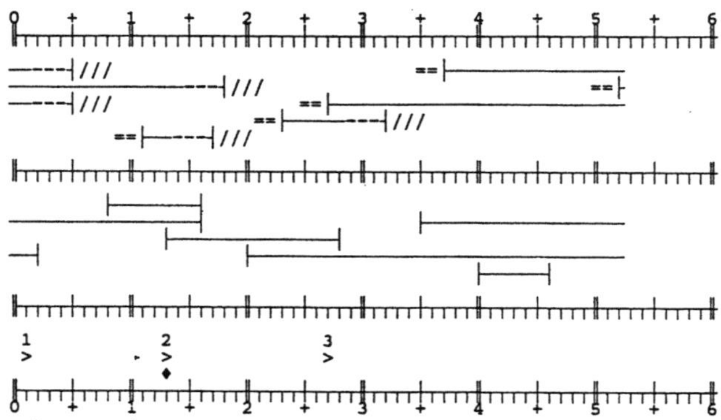


```
Name         : Check Post
Intersection: KN020101                    Date    : 25/12/94
Program      : P-S2                       Cycle   :  40"
Phases       : A-B-D
```

Movement	Ts	Tf	Tt
1	28	5	17
2	39	20	21
3	28	5	17
4	0	0	0
5	11	20	9
a	8	17	9
b	0	40	40
c	13	19	6
d	23	2	19
e	27	33	6
Optimize Extension Demand			

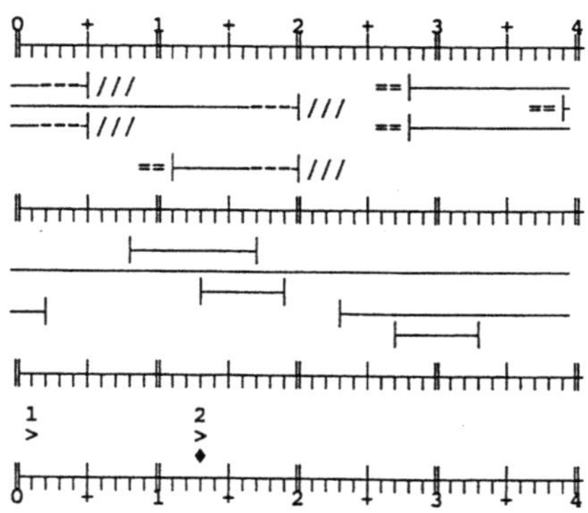

Figure 4. The main components of the Queue Control System

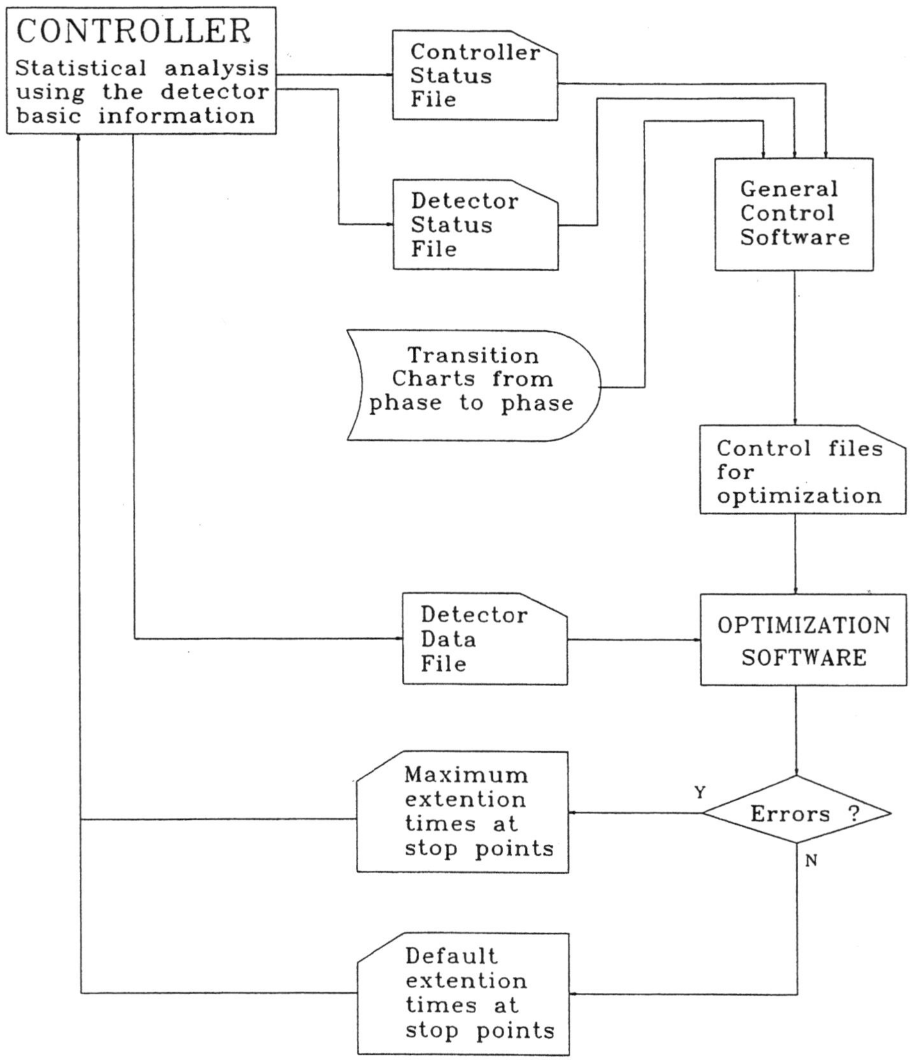

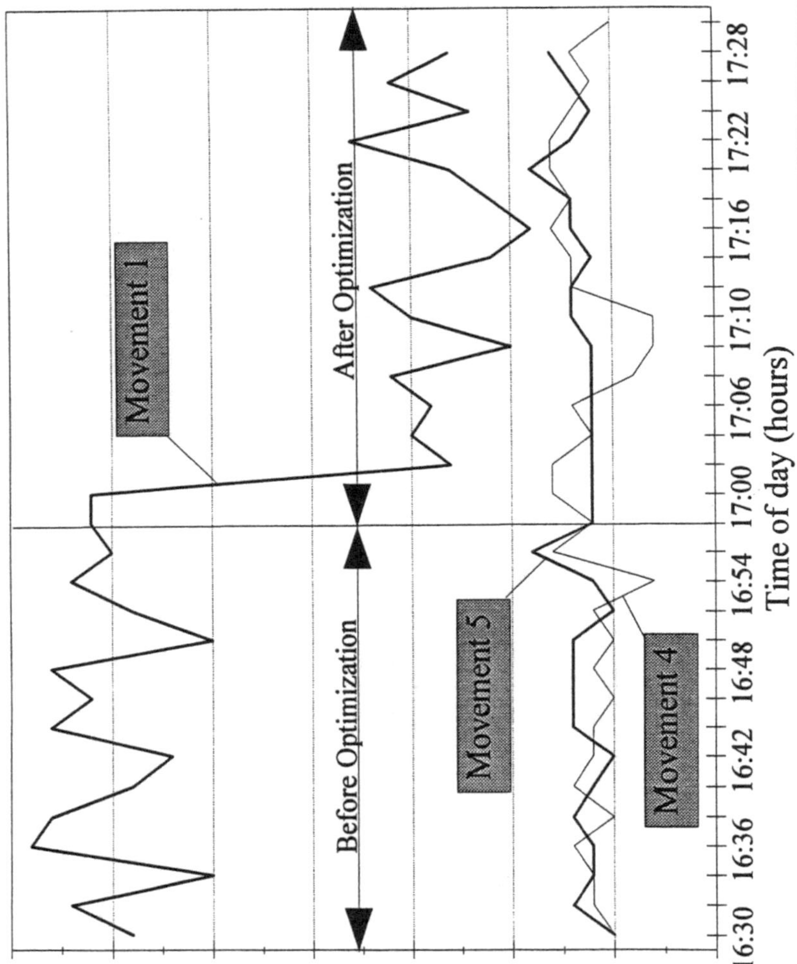

Figure 5. Queues on Intersection Approaches

A MODEL FOR REAL-TIME TRAFFIC COORDINATION USING SIMULATION BASED OPTIMIZATION

Paolo Dell'Olmo

Electronic Eng. Dpt. University of Rome "Tor Vergata"
Viale della Ricerca Scientifica, 00133 Roma, Italy

Pitu B. Mirchandani

Systems and Industrial Engineering Department
University of Arizona, Tucson, Arizona 85721, USA

An approximate traffic model, *APRES-NET*, is proposed for a simulation-based optimization approach for real-time network coordination of traffic flows. The objective of *APRES-NET* is twofold: to provide network-wide prediction of traffic movements and to evaluate, approximately but quickly, the performance of any signal timing strategy. To make a prediction, it uses the observed network vehicular traffic and signal data for a period of time, a future signal plan, and statistical information on vehicle movement, to predict the traffic for some future time period. The *APRES-NET* simulation consists of propagating into the future the movement of the vehicles detected in the network during the "last detection period". To quickly, and approximately, simulate the vehicle movement through the whole network, a simplified queuing mechanism is implemented at the intersections. At an intersection, a turn is assigned to each vehicle based on turning probabilities. Performance measures are calculated continuously as the simulation run is being conducted. *APRES-NET* is being used for real-time coordination of signal phase timings by making it behave as a "function evaluator" in an iterative optimization scheme. The paper discusses the basic structure and the algorithms used by *APRES-NET*.

1. Introduction and Background

Traffic simulation models are used extensively for transportation network planning and traffic operations management. As in other applications, the main purposes of these simulation models are to predict response to planning and/or management decisions, to compute the resultant performance of the system, and to evaluate whether a decision is acceptable (or optimal) based on some performance criterion. Common specific usage of traffic simulation models include:

(1) prediction of link traffic volumes due to potential network planning decisions,
(2) evaluation of highway design and geometrics (turns, merges, etc.) in terms of traffic flow and traffic safety, and
(3) prediction of traffic performance (delays, stops, air quality, etc.) for a given signal timing plan.

In the first case, vehicle traffic is represented in the models in an aggregate fashion (e.g., as number of vehicles per hour), and aggregate traffic flow characteristics (e.g., volume-speed relationships) are used to simulate traffic flows. These models are sometimes referred to as aggregate or *macroscopic* models. In case (2), vehicles are represented as discrete entities but travel times and queue delays are calculated using macroscopic flow characteristics and queuing models so that transients and smooth/turbulence flow characteristics can be predicted. Such models are sometimes referred to a *"mesoscopic"* simulations.

For case (3), to predict traffic performance for a given signal timing plan, two commonly used models are TRANSYT [Wallace et al., 1981] and TRAF-NETSIM [Rathi and Santiago, 1989]. TRANSYT is actually a signal optimization package that uses a *macroscopic* simulation model for performance evaluation and "optimizes" on timing parameters until a good traffic performance is obtained. On the other hand, TRAF-NETSIM is a "microscopic" simulation model, where each vehicle is separately represented and the modeling of its movement in time and space include vehicle interactions, car-following and driver-behavior characteristics. TRAF-NETSIM is used by traffic engineers for *"what-if"* evaluations - that is, <u>what</u> is the traffic performance <u>if</u> a given timing plan is implemented.

A common feature in all of the above applications is the off-line usage of the models. In other words, a discrete number of planning or management decisions are evaluated off-line with respect to given measures of performance and a decision is chosen based on an optimization or cost criterion. The simulation application that is described in this paper requires an on-line performance evaluator for real-time traffic control. Here we need to continuously observe traffic via detectors and surveillance equipment, quickly predict the traffic performance for the next few minutes and develop optimal signal timings. It is for this simulation-based optimization approach that the APRES-NET *(Approximate Prediction in Response to a Signal Network)* simulation model reported here has been developed.

A detailed simulation such as TRAF-NETSIM becomes computationally unwieldy when one uses the simulation model as a function evaluator (i.e., for evaluating the performance function for each candidate signal setting) in an optimization routine. One may use TRANSYT in an on-line fashion and compute signal settings every few minutes and download those settings to the field. In a way, this is exactly what SCOOT [Hunt et al., 1981] does. Recent versions of SCOOT do consider real-time measurements of link traffic – through *link profile units* – but assume that future flows are the same in the computation of optimal signal settings, which reduces the optimization to a single cycle. On the other hand, our simulation-based approach REALBAND [Dell'Olmo and Mirchandani, 1994], briefly described below, measures, in real-time, platoons and their speeds in the network (or sub-network) and performs optimization over several future "cycles" for determining optimal green phases or green bands.

The function of the APRES-NET is to quickly (1) provide a network-wide prediction of traffic movement and (2) evaluate the performance of a signal timing strategy in terms of given performance measures. APRES-NET serves both as a traffic predictor and an objective function evaluator/estimator. The performance measures currently being estimated are (1) the number of stops, (2) total delay and (3) throughput. The quality of different timing plans can be evaluated by comparing their corresponding APRES-NET objective function estimates. Thus, APRES-NET, using any given signal control strategy that generates signal timings, such as REALBAND, can be utilized to quickly estimate the performance of a control strategy with consideration of network-wide traffic dynamics.

In making a prediction, APRES-NET uses observed network vehicular traffic and signal data for a period of time (say 5 minutes), a future signal plan, and statistical information on vehicle movement to predict future traffic (say, for another 5 minutes). Past network traffic and signal data is provided to the APRES-NET module thorough an area-wide surveillance database interface which stores the surveillance/detector information. A future signal plan is an external input to the module. Parametric, or statistical, input to the module consist of information such as turning probabilities and average waiting times in the queues. Currently, this statistical information is given to the module explicitly, but an objective of this research is to derive this information using historical conditions, updated by current traffic data and trends. APRES-NET also requires topological information of the network which includes parameters such as link lengths, position of the detectors, number of lanes for each link, and other link/node attributes of the network.

2. Modeling for Real-Time Network Coordination

To further motivate the development of APRES-NET model, it may be useful to describe REALBAND and the role of the simulation model in REALBAND.

Unlike prototypical off-line approaches to network coordination such as TRANSYT [e.g., Wallace et al., 1981], MAXBAND [Little et al., 1981; Chang et al., 1988], and PASSER II-IV [e.g., Chang et al., 1987; Chaudhary and Messer, 1993], REALBAND explicitly considers available real-time information for computing signal timings. It first identifies platoons and predicts their movement in the network – their arrival times at intersections, their sizes and their speeds – by fusing and filtering the traffic data obtained, from various sources, in the last few minutes. APRES-NET is used to propagate the predicted platoons through the network, for a given time horizon. The signals are set so that the predicted platoons are provided appropriate green times to optimize a given performance criterion. It is obvious that two platoons demanding conflicting movements may arrive at an intersection at the same time. In that case, then either one or the other will be given priority on the green time, or it may be necessary to split one of the platoons, to maximize the given measure of performance. Optimally resolving such conflicts in real time is the main objective of the REALBAND algorithm.

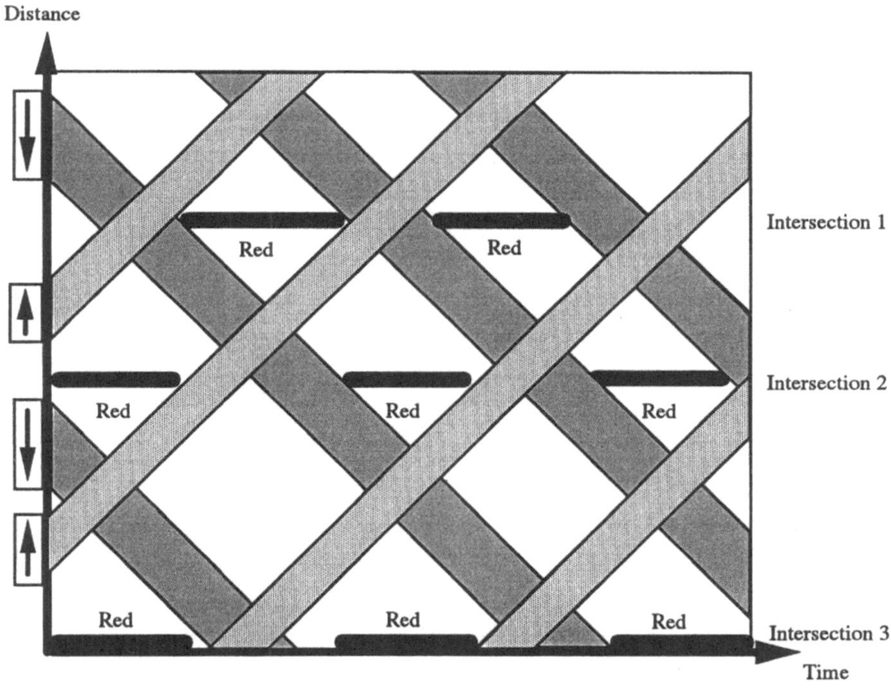

Figure 1. The MAXBAND Concept

Consider the time-distance diagram on a single arterial as shown in Figure 1. The goal of off-line arterial progression algorithms, such as MAXBAND and PASSER II, is to set the signal timings so that the number of vehicles that can traverse the arterial in either direction without stopping (other similar criteria may be incorporated) is maximized. The figure shows these bands of green times. Note the following drawbacks: it is assumed (1) that platoons are of equal size, and (2) that platoons travel at the same constant speed.

Now consider the time-distance diagram as shown in Figure 2. Here we have platoons of different sizes and different speeds. The green times required for these platoons are different than those required for the uniform case shown in Figure 1, and therefore the smooth anticipated progression is disrupted.

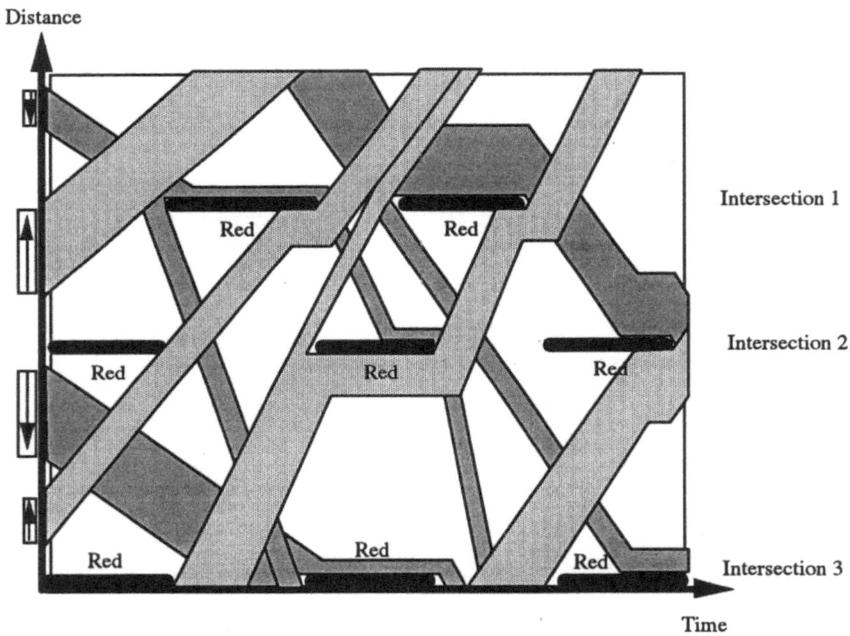

Figure 2. Actual MAXBAND Performance

By slightly adjusting the red times, we now reinstate the green bands for the given platoons with their own sizes and speeds, as shown in Figure 3.

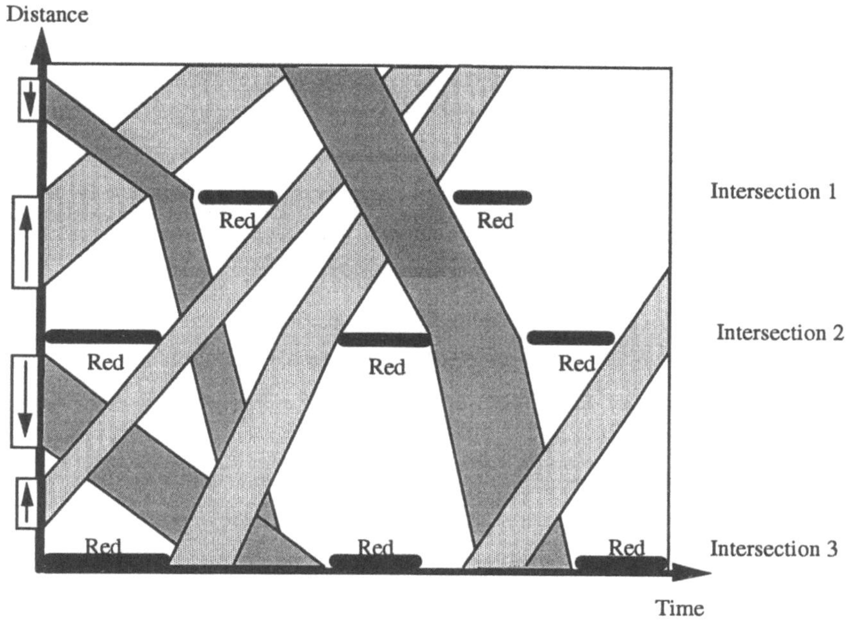

Figure 3. The REALBAND Concept for Single Arterial

In this manner we can also include platoon dispersion and compression but for convenience we haven't shown this in the figure.

The flow chart for REALBAND optimization is given in Figure 4. Note that to begin the recursion it is suggested that an initial signal plan be given. This is so that the network flow control optimization itself begins in the right "ball park". The initial plan may be obtained from an off-line method such as TRANSYT using historical data.

We note that in relation to the use of APRES-NET for function evaluation, the spatial region for the simulation model needs to be bigger than the area of control for network coordination, so that one can predict the movements of all real-time platoons within the region of control for several minutes in the future. As an illustration, Figure 5 shows the region of control for REALBAND (left in Figure 5) and the area simulated using the APRES-NET simulator (right).

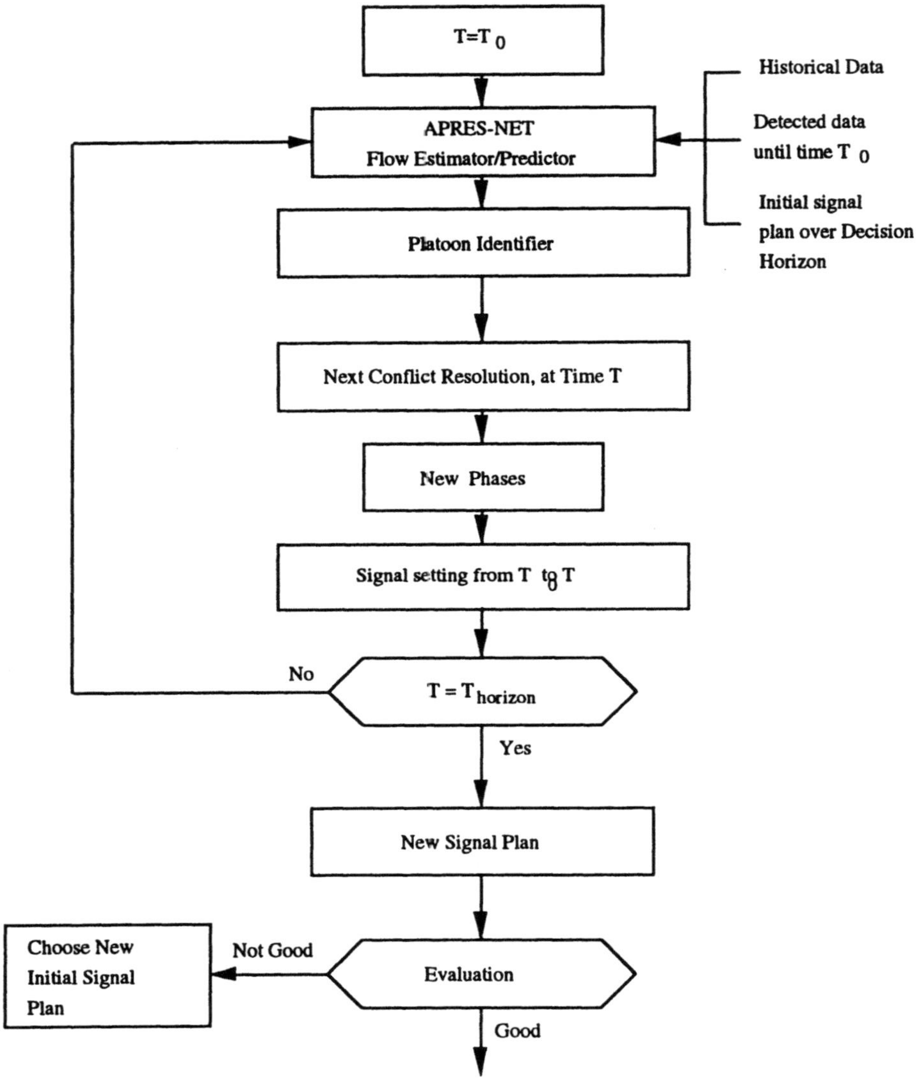

Figure 4. Flow Chart for *REALBAND*

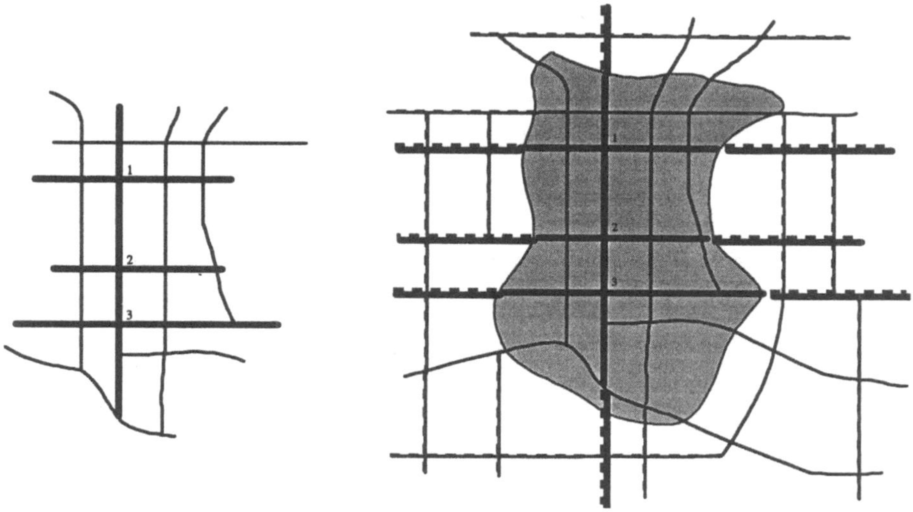

Figure 5. Region for Network Flow Control (left)
Region for Simulation (right)

Although Figure 4 shows a rather simple flow-chart, there are two issues that must be considered to make the code both, be effective in providing good if not optimal solutions and be efficient for real-time applications: (1) filtering detector data for identifying platoons, and (2) propagating REALBAND through time using APRES-NET.

The definition of platoons for traffic coordination depends on the level of traffic in the network. In low congestion, with vehicles traveling at high speeds, a platoon may be defined with as little as three cars having average headway of two seconds. Whereas, on a more heavily congested road, a platoon may consists of many more cars with average headway of a second. In any case, if a traffic engineer looks at vehicle detector data, he can easily recognize "platoons". Any platoon identification algorithm, where the goal of the algorithm is to filter out individual cars and identify platoons, should be able to imitate the traffic engineer. For the purpose of developing the REALBAND algorithm, we have used a *platoon identifier* based on user-specified threshold parameters, maximum headway between two vehicles in the same platoon and the minimum number of vehicles that constitute a platoon, to define a platoon.

The *platoon identifier* then filters the detector data to identify platoons. Examination of the platoon data from the initial run of APRES-NET will indicate that at times a platoon will be stopped or split so that another platoon is allowed to pass through a conflicting movement. REALBAND will identify the first time, say at time T_0+DT, this occurs. This is the first conflict to be resolved and hence the time has been propagated by DT. APRES-NET will then be run using the signals corresponding to having the stopped platoon to go through and the other (conflicting) platoon stopped. Then another scenario is obtained along with a new set of identified platoons (the platoons may have changed due to splitting and/or combination of platoons). For the two sets of scenarios formed, the next conflict is identified and the process is repeated. In this manner, as REALBAND propagates through time, it develops a decision tree, keeping at each node a feasible UB (performance measure for the corresponding scenario) for the performance. The algorithm terminates when REALBAND propagates to the planning horizon; the phase timings corresponding to the leaf node with minimum UB becomes the timings sent to the lower (intersection) level controllers.

In summary, the REALBAND procedure for network flow control exploits the availability of real-time traffic data to control vehicular traffic through a network to optimize a given performance measure. It is being used for real-time network coordination within a real-time hierarchical traffic management system referred to as *RHODES* [Head et al., 1992].

3. The APRES-NET Model

APRES-NET simulation consist of propagating into the future the movement of the vehicles detected in the network during the "last" detection period. The individual vehicles in the network are identified when they cross (detector crossings are obtained through the database interface). During the simulation, the detected vehicles are first propagated to the downstream detectors at the intersections using simple travel and delay time expressions. These predicted vehicle detections, in turn, are propagated to the succeeding downstream detectors and so on. To quickly, and approximately, simulate the vehicle movement through the whole network, a simplified queuing mechanism is implemented at the intersections. Depending on the signal status, given through any future signal plan, a vehicle may or may not be delayed by a stop or a queue at an intersection. At an intersection, a turn is assigned to each vehicle based on the turning probabilities.

A "run" of APRES-NET consists of propagating the detected vehicles in a recursive manner until the prediction horizon is reached. Performance measures are calculated continuously as the simulation run is being executed.

Currently, vehicles are continuously generated at "sources" to ensure that the area being simulated is populated with vehicles. The generation rates are based on the observed rates of vehicle arriving to the point in the simulated area. In the envisioned implementation, a large network can be thought of as consisting of several adjacent sub networks or cells. In that setting, the information about the vehicles entering a given cell from neighboring cells will be communicated from the latter cell.

APRES-NET vehicle movement propagation is used by REALBAND to identify platoons and their movement in the network. We are currently using a platoon identifier based on user-specified threshold parameters, maximum headway between two vehicles in the same platoon and the minimum number of vehicles that constitute a platoon, to define a platoon.

In real traffic networks, vehicles are distributed all over the network. They move continuously along network links, change lanes, overtake other vehicles, cross intersections, stop at traffic lights and wait in queues. Good quality traffic simulators try to emulate this vehicle behavior. Basically, such simulators, when working in regime, explicitly represent the position of each single vehicle in the network (on link or at an intersection) and update the position at discrete time intervals (say, every one second in simulation time). For each update, the new position of each vehicle is computed taking into account several parameters, such as vehicle occupancy in the vicinity, position in a queue, signal status, and acceleration/dispersion of vehicles in the vicinity.

The "logical" representation of the network in such a simulation model is composed of a set of links such as that shown in Figure 6, where vehicle positions are referenced with respect to a two dimensional space.

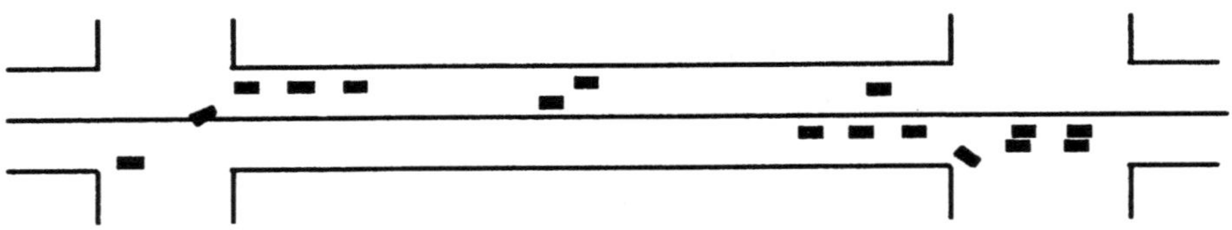

Figure 6. Accurate representation for a network link

Network-wide simulation with such a detailed network representation is not applicable for real-time purposes. Therefore, we consider only the minimal quantity of information necessary for the approximate propagation of vehicle movement for evaluating real-time control strategies. This minimal information is focused at representing vehicles at intersections. More precisely, we would like a model that is capable of representing vehicles that either go through the intersection without stopping, or stop at the intersection and wait in queue. Therefore, in APRES-NET, a link is limited by two mechanisms: one to represent vehicles moving through an intersection and the other to represent the stops (and waiting time) of vehicles. The former can be thought of as a *passage detector* and the latter as a *simplified queue*. Thus, we approximate a traffic network in APRES-NET by a logical representation which includes a set of links like that shown in Figure 7.

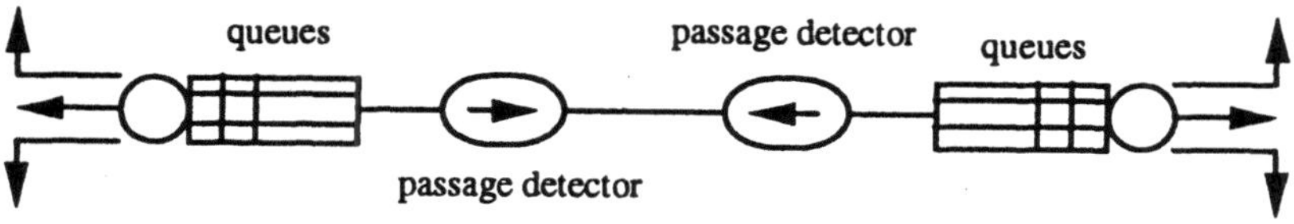

Figure 7. Approximate representation for a network link

In APRES-NET, possible positions for a vehicle are limited to link queues or link passages at detector positions. That is, a vehicle exists in the network when only its passage is detected or predicted, or when it is predicted to be in a queue. Thus, we drastically reduce, in comparison with the requirements for Figure 6 representation, the number of possible vehicle positions in the network and corresponding update operations.

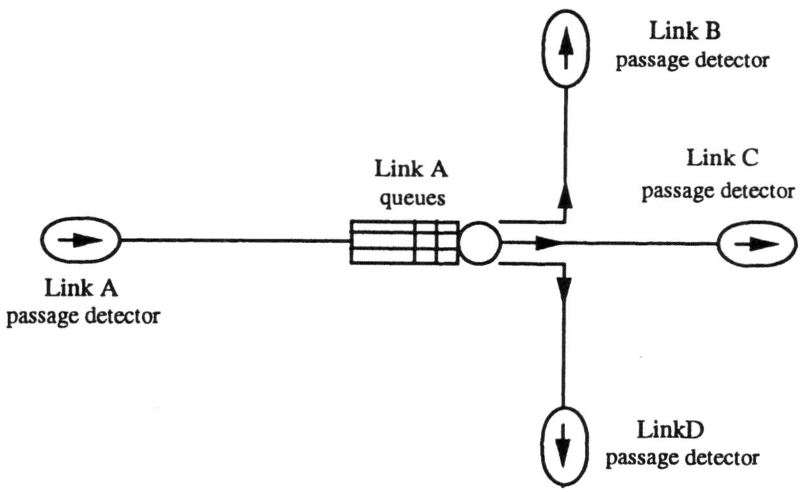

Figure 8. An Example of Intersection in the Approximate Network

The individual vehicles in the network are identified when they are at detectors (obtained through the database interface). During the simulation, the detected vehicles are first propagated to downstream detectors at the intersections using simple travel time/delay expressions. These predicted vehicle detections, in turn, are propagated to succeeding downstream detectors and so on. A simplified queuing mechanism is implemented at the intersections, (see Figures 8 - 10).

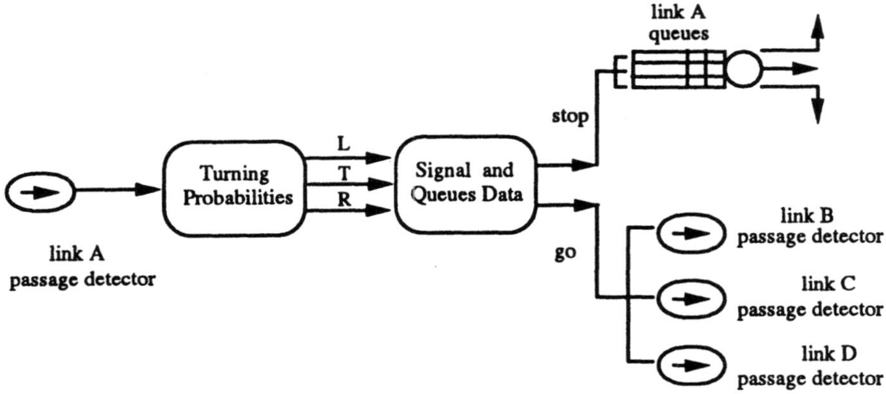

Figure 9. Propagation from a detector

Depending on the signal status corresponding to the candidate signal phase timings, a vehicle may or may not be delayed by a stop or queue at an intersection. Subsequently, a turn is assigned to each vehicle based on turning probabilities.

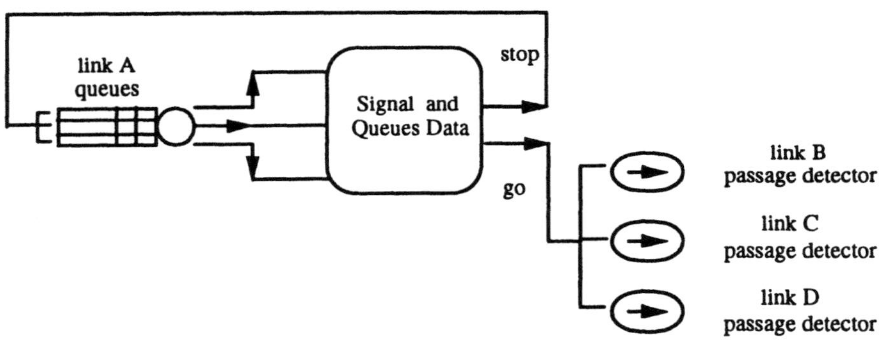

Figure 10. Propagation from a queue

An APRES-NET "run" consists of propagating the detected vehicles in a recursive manner until the prediction horizon is reached. Performance measures are calculated continuously as the simulation run is executed.

3.1 Traffic Propagation Procedure

Basically, the propagation procedure predicts the future location of each detected vehicle using the turning probabilities, the lengths of links, detector positions on the links, the link free flow speed, and approximate relationships to predict arrival times at the downstream detectors or queues.

Figure 11 shows a flow chart of the simulation algorithm. The outer loop runs the simulation from its starting time to the planning horizon. The inner loop propagates vehicles on the links. At every "simulation instance", all vehicles are propagated from detectors/ intersections to downstream detectors/intersections.

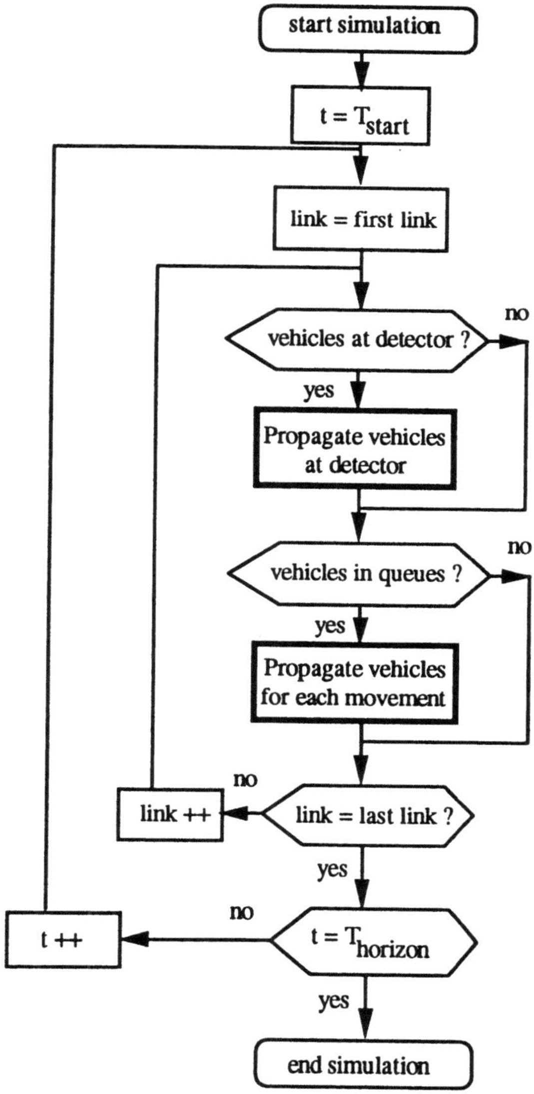

Figure 11. Flow Chart of the Simulation Procedure

In order to quickly perform a simulation run, vehicles are examined only when they cross a detector passage or when they are in a queue. Hence, there are two basic prediction algorithms which use different propagation formulae for vehicles depending on whether a vehicle is detected at a passage or it is in a queue. The parameters (such as acceleration for left or right turning vehicles, turning probability, etc.) and formulae (such as for estimating arrival time at an intersection in case of a queue) can be adjusted to specific traffic network conditions. Most crucial in the propagation of vehicles are the turning probabilities. In real-time settings, turning probabilities may be estimated by a "rolling horizon" statistical scheme using detector counts [e.g., Maher 1984].

We need a direct access data structure to fast load the "network status" via the database interface. Such a data structure must also allow fast update of the network condition in our approximate model, that is, the update of vehicle passages and status of queues.

As we discretized time, we can record the vehicle passages at detectors every time unit (say one second). Moreover, since the number of detectors in the network is of the same order of magnitude as the number of links, we can use for each detector a one-dimensional single array in which each entry is the number of vehicle passages for each time interval. In this way, vehicle passages over the detector are simulated is discrete time. Similarly, for queues at intersections, the queue status for each movement (left-turn, right-turn and through movement) is simulated by a one-dimensional array, where each entry denotes the queue length at each time instance. These simple sequential direct access data structures that can be scanned very quickly. Such a representation consumes significant memory space. However, for real-time application, saving computational time for reading and updating entries is more crucial than the size of memory being used.

The evolution of the approximate network status can be simulated and recorded every time interval (every one second in our case). The performance measures related to, for example, the number of stops, throughput, delay time, and queue lengths can be computed for each of these times. Moreover, this data structure also allows for easy platoon identification, as we shall describe shortly.

We remark on the applicability of this approach to a parallel or distributed computer architecture with shared memory. As illustrated in Figure 12, synchronized read-write operation on the queue and detector data structure within the memory allow parallel processors to simultaneously perform simulations for several *sub networks* cells. The memory areas where reading and writing operations are performed by different processors may correspond to detectors of the links defining cell boundaries. In this parallel or distributed simulation process, the reading entry of each processor (for any time instance) is the same. Since the simulation time increments are the same also, a *writing* operation (i.e. a prediction of a vehicle arrival at a detector or in a queue) performed at time t can be read in "future" simulation times by any *reading* processor.

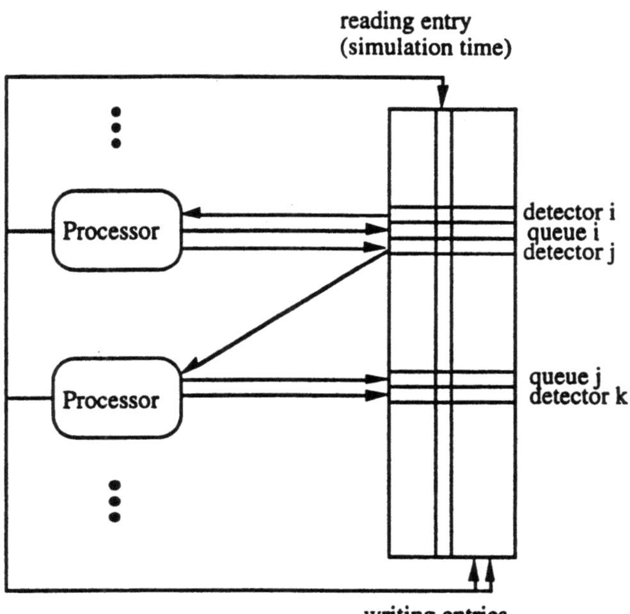

Figure 12. Scheme of the Simulation Process on a Parallel Architecture

4. Experimental Evaluation and Conclusions

We evaluated the APRES-NET model in a laboratory setting. In the experiments we performed, the "actual data" are taken from a detailed simulation model (we are using TRAF-NETSIM for this purpose). As an example of the performances, we report in Figure 13 typical plots comparing actual data and that predicted by APRES-NET. The number of vehicles reported are counted by detectors located on the network links, for every five-second period. The plots from time T=0 to T=300 (current time) are identical because this is the data which APRES-NET used to predict flows from T=300 to T=500, which represents the decision horizon for the experiments. The simulation interval from time T=0 to T=300 is required by APRES-NET to build up queues, if any. Turning probabilities assumed in APRES-NET were the same as those used in TRAF-NETSIM. Since in operational situations, turning probabilities must be estimated from detector counts [e.g., Maher 1984], which of course would result in additional estimation errors, predicted flows in the field may not be as accurate as that shown in Figure 13. Fortunately, in this application for REALBAND, only approximate predictions are needed.

As we expected, predicted data do not match exactly with actual. On the other hand, looking at the vehicle flow from a network coordination point of view, (i.e. considering only platoons and not single vehicles) we find the prediction more reliable. Figures 14 and 15 show the same flows of Figure 13 but now individual vehicles have been filtered using the *platoon identifier* of REALBAND (see Figure 4). In order to assign bands and control progression in the network, what we need to know is the approximate time interval for a platoon approaching an intersection and its approximate size. In fact, in the RHODES control system [Head et al., 1992], these approximate time intervals and sizes is downloaded to an *intersection controller* which subsequently fine tunes the phase timings to allow for the detected platoons and other vehicles not identified or ignored by the *platoon identifier* of REALBAND. From this view point, the prediction seems very good.

Running time of the APRES-NET depends on the volume of traffic. In our experiments it was always less than 0.2 seconds for a 41-node, 42-link actual network (representing a section of Tucson) and a simulation time of 300 seconds on a SUN workstation.

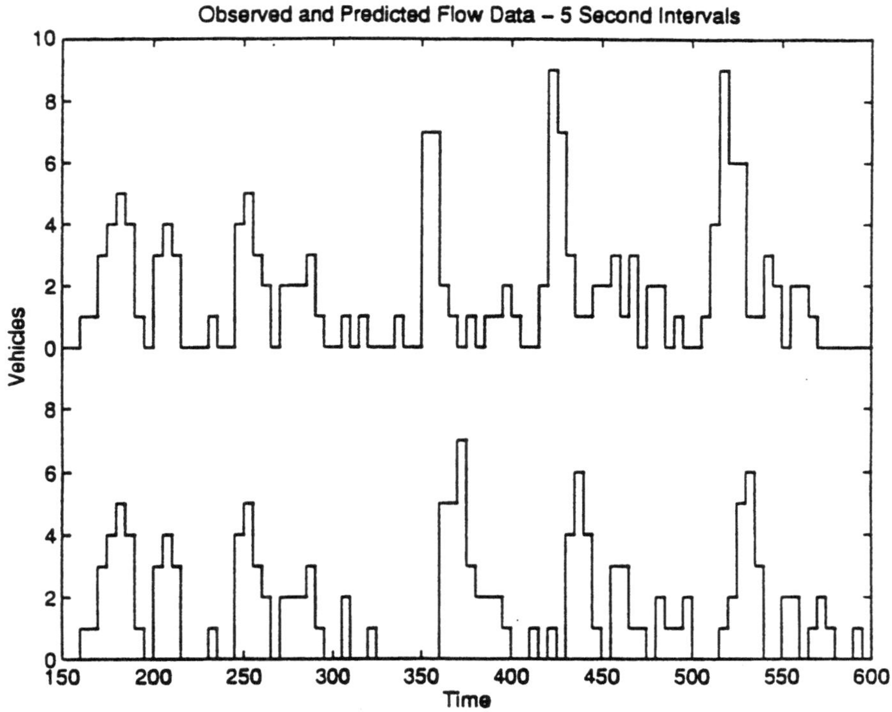

Figure 13. Comparison of "actual" and Predicted Flows

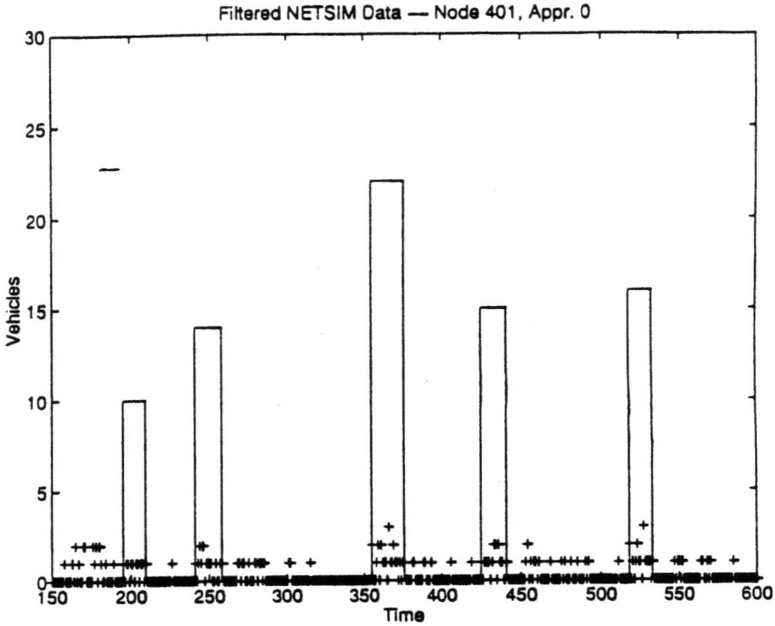

Figure 14. Identified Platoons in the "actual" Flows

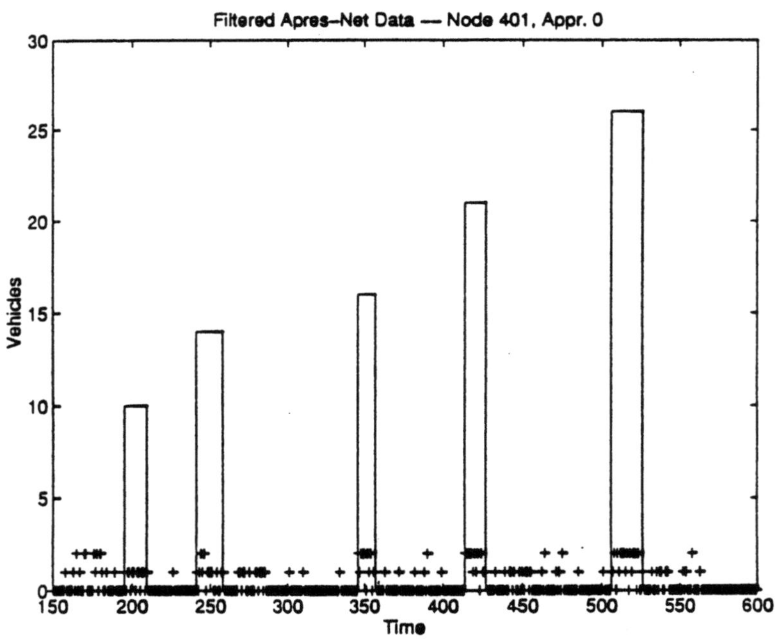

Figure 15. Identified Platoons in the "predicted" Flows

REALBAND, using APRES-NET, is currently being evaluated in a laboratory setting. In this evaluation a detailed simulation model (we are using TRAF-NETSIM for this purpose) is used to generate realistic vehicular traffic. This traffic is detected through simulated inductive loops; and the detector outputs are sent to the APRES-NET/REALBAND control scheme. Preliminary results indicate that the traffic performance is significantly improved.

Acknowledgments

The first author acknowledges Consiglio Nazionale delle Ricerche, Progetto Finalizzato Trasporti 2 for partially supporting this research; the second author acknowledges the support of Arizona Department of Transportation (A. G. Contract KR93 1973TRN) and the Federal Highway Administration (Contract DTFH61-94-C-00010). The authors are also very appreciative of the assistance and helpful comments of Larry Head and Michael Whalen. Their help in the laboratory experiments with TRAF-NETSIM was invaluable in the evaluation of APRES-NET.

References

[1] Chang, E.C., B.G. Marsden, and R. Derr [1987], "PASSER II-84 Microcomputer Environment System - Practical Signal-Timing Tool", *Journal of Transportation Engineering, 113*, 1987, 625-641.

[2] Chang, E.C., S.L. Cohen, C. Liu, C.J. Messer, and N.A. Chaudhary [1988], "MAXBAND-86: Program for Optimizing Left-Turn Phase Sequences in Multiarterial Closed Networks", *Transportation Research Record 1181*.

[3] Chaudhary, N.A., and C.J. Messer [1993], "PASSER IV, A Program for Optimizing Signal Timing in Grid Networks", Paper No. 930825, 72nd Annual TRB Meeting, Washington D.C., January.

[4] Dell'Olmo, P. and P.B. Mirchandani [1994]. "REALBAND: An Approach for Real-time Coordination of Traffic Flows on Networks", Technical Report, M.I.S. Dpt., University of Arizona.

[5] Gartner, N.H., S.F. Assmann, F. Lasaga, and D.L. Hou [1990], "MULTIBAND - A Variable-Bandwidth Arterial Progression Scheme", *Transportation Research Record 1287.*

[6] Hadi, M.A., and C.E. Wallace[1992], Improved Optimization Efficiency in TRANSYT-7F," Transportation Research Center, University of Florida, Gainesville, Florida, for *presentation at the ORSA/TIMS Meeting*, Orlando, FL, April.

[7] Head, K.L., P.B. Mirchandani and D. Sheppard [1992], "A Hierarchical Framework for Real-Time Traffic Control", *Transportation Research Record,* 1360,82-88.

[8] Hunt, P.B., D.I. Robertson, R.D. Bretherton, and R.I. Winton [1981], SCOOT - A Traffic Responsive Method of Coordinating Signals, *Report No. LR 1014*, Transport and Road Research Laboratory, Crowthorne, Berkshire, England.

[9] Little, J.D.C., M.D. Kelson, and N.H. Gartner [1981], "MAXBAND; A Program for Setting Signals on Arteries and Triangular Networks", *Transportation Research Record 795,* 40-46.

[10] Maher, M.J. [1984], "Estimating the Turning Flows at a Junction," *Traffic Engineering and Control*, May .

[11] Rathi, A. K. and A.J. Santiago [1989], "The New NETSIM: TRAF-NETSIM Version 2.00 Simulation Program", presented at the Transportation Research Board's 68th Annual Meeting, Washington, D.C., January 1989.

[12] Wallace, C.E., K. Courage, D.R. Reaves, G.W. Schoene, and G.W. Euler [1981], *TRANSYT-7F User's Manual*, Federal Highway Administration.

AN APPROXIMATE LABELLING ALGORITHM FOR THE DYNAMIC ASSIGNMENT PROBLEM

Warren B. Powell and Derek H. Gittoes

Department of Civil Engineering and Operations Research
Princeton University - Princeton, NJ 08544, USA.

We develop a solution strategy for the dynamic assignment problem, an important class of routing and scheduling problems with many applications in transportation and logistics. A label setting rolling horizon algorithm (LASER) was developed for the solution of the dynamic assignment problem. An extensive numerical analysis on a variety of randomly generated and real-world problems was conducted in order to evaluate the algorithm. The breadth-first LASER algorithm was compared with a depth-first tour construction algorithm and an LP-optimal Dantzig-Wolfe column generation procedure. The LASER algorithm proved to be effective when compared to the other solution procedures from a solution quality and computational efficiency standpoint for most problems.

1. Introduction

We address the problem of solving a static scheduling problem where resources must be assigned to cover a known set of tasks (see figure 1). The assignment must obey complex work rules, covering as many tasks as possible. Each time a task is covered, we receive a revenue (or reward), but we are not required to cover all tasks. Although we address the static problem in this paper, our solution approach is motivated by the need to solve this problem in real-time, under demanding time restrictions on the performance of the algorithm. The driver assignment problem is complicated in practice by the mixture of wide and narrow time windows, and the fact that we often cannot cover all the tasks. In this case, we must not only determine the routing and scheduling of the drivers, we must also determine which tasks should be "refused".

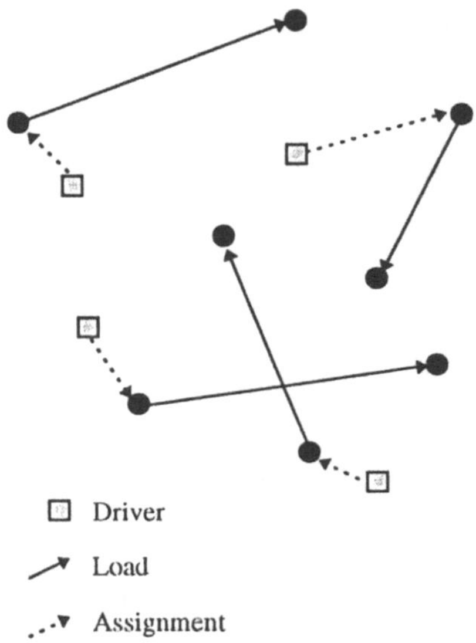

Figure 1. Driver Assignment Problem

The complexity of the problem is further amplified by the work rules that govern the construction of tours and the costs associated with the tours. For example, there are regulations regarding the number of hours on the road and on duty a driver can accumulate in a given day. Collective union agreements may require guarantees with regard to hours worked and driver preferences for certain routes. There are also constraints regarding the service of the tasks. For example, if the task is comprised of moving freight between an O/D pair, there may be time window constraints for both the origin and destination, equipment requirements, and customer preferences with regard to the driver and vehicle type. The large number and intricacy of the driver work rules, complicated cost functions, and the size of most operations culminates to make the dynamic assignment problem very difficult.

The DAP is a general framework that can be used to model many transportation and logistics problems. Obvious motor carrier applications include long-haul full-truckload trucking and short-haul intermodal drayage operations. Many inventory distribution problems may also be described with the DAP. For example, bulk chemical manufacturers faced with the problem of maintaining customer inventories of various products must efficiently manage their fleet of delivery vehicles and trailers. Operations managers must design vehicle routes that distribute the necessary amounts of products to their customers while adhering to the many operational constraints regarding the construction of the tours. Dynamic crew scheduling problems provide another possible application for the DAP.

This paper introduces an optimization-based heuristic, LASER, which is designed for solving dynamic assignment problems with complex work rules. An extensive computational study was undertaken to provide a basis for evaluating the performance of the LASER algorithm. A column-generation procedure using Dantzig-Wolfe decomposition was used to generate continuous optimal solutions to provide rigorous bounds for the problem. The LASER algorithm proved to be effective from both a computational and solution quality standpoint. The algorithm was capable of finding solutions within 1% of the LP-optimal objective function value for a number of randomly generated and real-world test problems. For problems that were tightly constrained with respect to the number of available vehicles, the algorithm's performance was poor; the LASER solutions were within 10% of the LP-optimal solution. The poor performance of the algorithm for the constrained problems was primarily due to its inability to efficiently utilize the available vehicle duty time.

2. Dynamic Assignment Model

There are several dynamic routing and scheduling problems that can be described by the dynamic assignment model. Section 2.1 describes the dynamic assignment problem from the context of a motor carrier application. Section 2.2 then provides a mathematical formulation for the dynamic assignment model.

2.1 Problem Description

Assume we have a fleet of vehicles that are to be scheduled to service a set of known tasks. Each vehicle starts at a known location with a known time of availability. The set of tasks (loads to be transported between specified origins and destinations), and all pertinent task information, is assumed to be known at the beginning of the planning horizon. The goal is to construct minimum cost (maximum profit) tours for each vehicle subject to a number of constraints.

Every task is assumed to be a full trailer with every vehicle having the capacity of one trailer (no freight consolidation). Each task has a specified time window $[a_i, b_i]$ and revenue r_i; the company receives the revenue if the task is serviced within the time window. Otherwise, the task is assumed "lost" with a corresponding loss in potential revenue. A vehicle is allowed to arrive before the start of the time window but must wait until a_i before starting service, and a vehicle is never allowed to arrive and start service after b_i. In the soft time window case vehicles would be allowed to arrive outside of the time window with a corresponding early/late penalty. The service time for each task is a combination of the load/unload times and the loaded travel time between the origin and destination. The length of every tour is limited by a known maximum duty time for each vehicle. Every tour must return the vehicle to its domicile without violating the duty time restrictions. A vehicle's initial location and domicile are not assumed to be the same physical location. It is not necessary to satisfy all customer demands; uncovered tasks may be viewed as refused loads or loads given to a common carrier.

For the purposes of our experimental study, we have made a number of simplifying assumptions that are not necessarily true for most real-world applications. For example, we have assumed a homogeneous fleet of vehicles and human operators. In addition, we have chosen to ignore equipment requirements for tasks. These and other simplifying assumptions were made to keep the model as clean as possible from a mathematical standpoint. However, these features are trivial to incorporate into LASER.

For most real-world applications the assumption of a static problem is incorrect. Customers call in new tasks during the course of the planning horizon and the vehicle tours must be reconstructed. One cannot simply insert new tasks into existing tours and guarantee that the solution will remain optimal. The definition of an optimal solution is not even clearly defined in a real-time setting. For the purposes of this paper we assume all problem information is known at the beginning of the planning horizon. The LASER algorithm is amenable to reoptimizing as new data becomes available for a real-time implementation. The next section formulates the DAP described above as a mixed integer program.

2.2 Formulation

The dynamic assignment problem can be modelled as a network with complicating constraints. Consider the following notation for the DAP described above:

- D - set of vehicle nodes (source nodes)
- L - set of task nodes (transshipment nodes)
- $\{H\}$ - vehicle domicile node (sink node)
- $V = D \cup L \cup \{H\}$ - complete set of nodes
- A_k - set of directed arcs from vehicle node $k \in D$ to nodes $L \cup \{H\}$
- $A = \cup_{k \in D} A_k$ - complete set of vehicle arcs
- B_L - set of directed arcs between task nodes
- B_H - set of directed arcs between task nodes and the domicile node
- $B = B_L \cup B_H$ - complete set of task arcs
- $E = A \cup B$ - the complete set of arcs for the network
- c_{ij} - cost for arc $(i,j) \in E$. Denote $c_{k,H}$, $k \in D$ as the cost of an unused vehicle. All costs are assumed to be strictly positive
- r_j = net contribution associated with task j (where $r_j \leq 0$, since we are minimizing costs)
- t_{ij} - travel time for arc $(i,j) \in E$
- s_i - service time for node $i \in V$

- $[a_i, b_i]$ - service time window for node $i \in V$.

Note:

1. For arcs (i,j): $j \in L$, the cost c_{ij} is equal to total operating costs (travel cost, wages, etc.). For arcs (i,H), the cost is equal to operating costs of returning the driver home.
2. Service time s_i is equal to the combination of the load time at the origin, unload time at the destination, and loaded travel time between the origin and destination for $i \in L$, and zero otherwise.
3. For $k \in D$ the service time window is equal to $[t°, t°]$, where $t°$ is the time that the vehicle is first available. For domicile node H the service time window is equal to $[t°, t° + \delta]$, where δ is the maximum duty time for a vehicle.

The variables are binary flow variables and continuous time variables:

$$x_{ij}^k = \begin{cases} 1 & \text{of vehicle } k \in V \text{ flows on arck } (i,j) \\ 0 & \text{otherwise} \end{cases} \tag{1}$$

$$z_j = \begin{cases} 1 & \text{if task } j \in L \text{ is covered} \\ 0 & \text{otherwise} \end{cases} \tag{2}$$

$$t_i = \text{time that a vehicle arrives at node } i \tag{3}$$

Given the above problem information and notation the dynamic assignment problem may be formulated as the following mixed integer program:

$$\max \sum_{j \in T} r_j z_j - \sum_{k \in D} \sum_{(i,j) \in A \cup B} c_{ij} x_{ij}^k \tag{4}$$

subject to:

$$\sum_{k \in D} \sum_{i \in D \cup L} x_{ij}^k = 1 \qquad \forall j \in L \tag{5}$$

$$\sum_{(i,j)\in A_k} x_{ij}^k = 1 \qquad \forall k \in D \qquad (6)$$

$$\sum_{(i,H)\in A_k \cup B_H} x_{iH}^k = 1 \qquad \forall k \in D \qquad (7)$$

$$\sum_{k\in D}\sum_{i\in D\cup L_j} x_{ij}^k - \sum_{k\in D}\sum_{i\in \{H\}\cup L_j} x_{ij}^k = 0 \qquad \forall j \in L \qquad (8)$$

$$a_i \le t_i \le b_i \qquad \forall i \in V \qquad (9)$$

$$t_i + s_i + t_{ij} - t_j \le (1 - x_{ij}^k)\, M \qquad \forall (i,j) \in E,\ k \in D \qquad (10)$$

$$x_{ij}^k,\ z_j \in \{0,1\} \qquad \forall (i,j) \in E,\ k \in D \qquad (11)$$

Note:

- M = large constant
- $t_i = t°$ for $i \in D$

The objective is to maximize total profits over the entire network. Constraint set (5) ensures that every task is covered by at most one vehicle. Constraint sets (6) and (7) ensures that every vehicle tour starts at the vehicle's initial location and terminates at the domicile node. Flow conservation at the task nodes is captured by constraints (8). The hard time window constraints are enforced by constraint set (9). The logical relationship between the flow variables and the time variables is represented by (10). The final set of constraints are binary constraints on the link flow variables. Constraints (9) and (10) eliminate subtours provided $t_{ij} + s_i > 0$.

The problem exhibits a potentially large number of integer variables. A problem with 100 vehicles and 400 tasks would result in as many as 20,000,000 binary variables. The problem can be solved in principle using set partitioning formulations used for airline crew scheduling, but the wide time windows will likely make such techniques very slow. Local search heuristics would apply, but we believe (but do not prove) that our approach, though its successive application of global solution methods, will provide an overall higher quality solution. Finally, our notational framework is extremely flexible with respect to incorporating complex rules governing operating strategies and policies.

3. LASER Development

The dynamic assignment model presented above generally cannot be solved directly with the level of efficiency required for real-world implementations. Section 3.1 describes a dynamic decomposition for the DAP that reduces it to a pure network problem. Then section 3.2 presents the development of the LASER algorithm which uses the decomposed network to solve the original DAP.

3.1 Dynamic Decomposition

One heuristic approach to solve the DAP is the repeated use of the classical assignment model. For each vehicle a set of assignment arcs to customer demands can be generated. Each assignment arc represents the repositioning of the vehicle from its current location to the load's origin. Only those arcs which are feasible are considered. For example, given that a vehicle is available at time t at location i, all arcs must satisfy the condition $t + t_{ij} \leq b$, where j is the task's origin and b is the latest time at which the load can be picked up. After solving the assignment model each vehicle's time and location of availability and remaining duty hours can be updated. Those demands which were covered in the first assignment can be removed from the problem. A new assignment problem can then be generated and solved based on the updated information. This process can be repeated until there are no more tasks to be covered. In this manner a complete vehicle tour is built over the planning horizon.

The greedy algorithm described above is myopic and tends to perform poorly when the planning horizon is long. The dynamic decomposition approach attempts to perform multiple assignments by implicitly updating the vehicle information and generating a complete vehicle tour in one iteration. In order to accomplish the above goal the concept of a potential resource is introduced.

We define a potential resource in the context of the DAP as a vehicle at the end of a task. Upon completion of every task a vehicle will become available to possibly be assigned to another task. However, the state of the potential resource is not completely known. For the DAP, the state of the potential resource would include time and location of availability along with remaining duty time. The location of availability is the destination of the task associated with the potential resource. Each task has a time window so one does not know when the potential resource associated with the task will become available. The amount of remaining duty time for the potential resource is also unknown. Note that the state of a vehicle and a potential vehicle contain the same elements.

One approach is to set the state of each potential resource to its most optimistic value. For example, for a potential resource associated with task i with time window $[a_i, b_i]$ and service time s_i, the optimistic time of availability would be $a_i + s_i$. Similar optimistic values could be used for other elements of the state variable. One could then generate an assignment problem with vehicles and potential vehicles and complete vehicle tours could be produced in one iteration. By tracing the solution to the decomposed dynamic assignment problem one can determine the identities of the potential vehicles. For example, if vehicle 1 is assigned to task A and potential vehicle A is assigned to task B, then the tour for vehicle 1 consists of task A followed by task B.

Consider the following notation for the decomposed DAP:

- $V =$ set of vehicle nodes, $|V| = n$.
- $L =$ set of task nodes, $|L| = m$.
- $P =$ set of potential vehicle nodes, $|P| = m$.
- $\{D\} =$ dummy task (sink) node.

- $L' =$ $L \cup \{D\}$, total set of task nodes.
- $A =$ set of vehicle to task assignment arcs (includes feasible arcs only).
- $B =$ set of potential vehicle to task assignment arcs (includes feasible arcs only).
- $C =$ set of task to potential vehicle arcs (demand arcs).
- $c_{ij} =$ cost of assigning vehicle i to task j.
- $d_{ij} =$ cost of assigning potential vehicle i to task j.
- $r_j =$ net contribution associated with task j (where $r_j \leq 0$, since we are minimizing costs).
- $P(x,y,z) =$ total profits derived from the solution.

The cost of assigning a vehicle or potential vehicle to a task is the total operating costs associated with relocating the vehicle from its current position to the origin of the task. The net contribution for a task is equal to revenue minus operating costs associated with covering the task (usually a negative quantity). The arcs from the real/potential vehicles to tasks represent empty movements while task to potential vehicle arcs represent loaded movements. Arcs into the dummy task node represent vehicles returning to their domicile. Feasible potential vehicle to task arcs are generated based on the most optimistic state for the potential vehicle.

It is important to emphasize that our formulation does not require that all tasks be covered. If we cover a task, then we receive a revenue r_j for covering task j. The net contribution can be interpreted as lost revenue if the load is not covered. Alternatively, r_j can be interpreted as the cost of having the load covered by an outside vendor (this was the case in the examples we were testing).

There are three types of integer flow variables for the decomposed DAP:

$$x_{ij} = \begin{cases} 1 & \text{if vehicle } i \in V \text{ is assigned to task } j \in L' \\ 0 & \text{otherwise} \end{cases} \quad (12)$$

$$y_{ij} = \begin{cases} 1 & \text{if potential vehicle } i \in P \text{ is assigned to task } j \in L' \\ 0 & \text{otherwise} \end{cases} \tag{13}$$

$$z_j = \begin{cases} 1 & \text{if task } j \in L \text{ is covered} \\ 0 & \text{otherwise} \end{cases} \tag{14}$$

The decomposed DAP can then be formulated as the following pure network problem:

$$\max P(x,y,z) = \sum_{j \in L} r_j z_j - \sum_{(i,j) \in A} c_{ij} x_{ij} - \sum_{(i,j) \in B} d_{ij} y_{ij} \tag{15}$$

subject to:

$$\sum_{j \in L'} x_{ij} = 1 \qquad \forall i \in V \tag{16}$$

$$\sum_{j \in L'} y_{ij} = 1 \qquad \forall i \in P \tag{17}$$

$$\sum_{i \in V} x_{ij} + \sum_{i \in P} y_{ij} - z_j = 0 \qquad \forall j \in L \tag{18}$$

$$\sum_{i \in V} x_{ij} + \sum_{i \in P} y_{ij} = n \qquad j = D \tag{19}$$

$$x_{ij} \geq 0, \text{ integer} \qquad \forall (i,j) \in A \tag{20}$$

$$y_{ij} \geq 0, \text{ integer} \qquad \forall (i,j) \in B \tag{21}$$

$$z_j \text{ binary} \qquad \forall j \in L \tag{22}$$

The objective function (15) seeks to maximize total profits, given by total revenues minus total costs. Constraint sets (16) and (17) ensure that real or potential vehicles are assigned to exactly one task (including the dummy task). Constraints (18) restrict each task to be covered by at most one real/potential vehicle. The last three sets of constraints are the bounds and integrality constraints for the flow variables. The decomposed DAP is a pure network problem that can be solved efficiently by using a network simplex code.

By assuming the potential resources have the most optimistic states, the time window and length of tour constraints have been implicitly relaxed. Therefore, the solution to the decomposed DAP may result in tours which are infeasible. For example, if the time window for task *A* is between 11:00 am and 4:00 pm with a service time of 60 minutes, the potential vehicle is expected to be available at 12:00. When the entire vehicle tour is traced it may result in the real vehicle assigned to task *A* not completing the task until 3:00. The fact that the vehicle was available at 3:00, not 12:00, may make the next leg in the tour infeasible, i.e. the vehicle would have to leave sometime before 3:00 to satisfy the time window constraints for the next task in the tour. The same type of scenario can happen with respect to length of tour constraints.

For problems with perfectly tight time windows and no tour length limits, the decomposed dynamic assignment model and the original DAP are equivalent. The presence of time windows and finite tour length limits results in the decomposed DAP being a relaxation of the original DAP. Another drawback of the decomposed DAP is the possible existence of subtours in the solution. Depending on the structure of the problem it is possible for a solution to contain cycles (e.g. potential vehicle A assigned to task B and potential vehicle B assigned to task A). However, if the solution to the decomposed DAP does not violate any of the time windows, tour length constraints, or contain subtours, then the solution is optimal for the DAP.

In most applications the use of the optimistic state variables for the potential vehicles will result in a solution with infeasible tours. The question of how to use the concept of potential resources to construct good feasible tours then arises. The following section describes the LASER algorithm which uses the potential resource concept to construct vehicle tours.

3.2 LASER Algorithm

The decomposed DAP is a pure network and can be solved efficiently using a standard network simplex code. The primary disadvantage is that the uncertainty in estimating the potential resource's state can lead to highly infeasible solutions. The LASER algorithm is a tour construction procedure that builds tours based on a rolling horizon using the concept of potential resources. The following section describes the label setting rolling horizon algorithm.

For every potential resource in the network there is an associated label which contains the following elements:

1. expected time of availability (ETA)
2. ETA location
3. available duty time (ADT)
4. status
5. other attributes

The first three elements constitute the state variable for the potential vehicle. The status of the label may be any one of three values: undefined, temporary, or permanent. A potential vehicle with an undefined status cannot be assigned to any tasks because the elements of the state variable are too uncertain. A potential vehicle with temporary status can be assigned to tasks, but the state variable elements are subject to change. Finally, a potential vehicle label has permanent status if all the elements of the state variable have been fixed. A potential vehicle with permanent status is equivalent to a real vehicle. For a potential vehicle's state variable to be fixed the path (tour) that includes that task must be fixed. We now formally present the LASER algorithm.

- *Step 0:* INITIALIZE
  - Get T° and T^{f}, the start and end of the planning horizon respectively and set $T=T^{\circ}$.
  - Get $\delta > 0$ and $\varepsilon > 0$ with $\delta \geq \varepsilon$.

- Initialize all potential vehicle labels by setting the state variable elements to the optimistic values and the status to UNDF (undefined).
- Build the initial decomposed DAP network.
- NUMTEMP = 0

• *Step 1:* SOLVE THE NETWORK
  - Solve the decomposed DAP network.
  - Note: A potential vehicle cannot be assigned to a task if its label status = UNDF.

• *Step 2:* UPDATE POTENTIAL VEHICLE LABELS
  - Check for cycles: if a task is covered on a subtour set the label status of its corresponding potential vehicle to UNDF.
  - Set $T' = T + \delta$
  - Simulate each acyclic vehicle tour, calculate NEW-ETA and NEW-ADT for each potential vehicle.
  - For each potential vehicle with label status $\neq$ PERM do:
    * If NEW-ETA $\leq T'$ and the tour preceding the PV is feasible, then set status = TEMP, ETA = NEW-ETA, and ADT = NEW-ADT.
    * Else if status = TEMP and NEW-ETA $> T'$, then set status = UNDF.
  - Update NUMTEMP.

• *Step 3:* UPDATE PERMANENT LABELS
  - *Step 3.1:* Set T_{min} = minimum ETA over all potential vehicles with label status = TEMP ($T_{min} = \infty$ if NUMTEMP = 0).
  - *Step 3.2:* If $T_{min} \leq T + \varepsilon$, then $\bar{T} = T_{min}$, for all potential vehicles with TEMP label status and ETA $\leq \bar{T}$ set label status = PERM, update NUMTEMP and go to Step 3.1.
  - If NO labels were set to PERM above and $T < T^f$, then set $T \leftarrow T + \varepsilon$. Go to Step 4. Else stop.
  - If $\bar{T} < T^f$ then set $T = \bar{T}$. Go to Step 4. Else stop.

• *Step 4:* ADJUST NETWORK

- For potential vehicles with status = PERM fix flow on the associated path.
- For potential vehicles with status = TEMP recalculate the feasibility and cost for arcs to task nodes given the potential vehicle's current state variable.
- For potential vehicles with status = UNDF, restrict flow to the domicile node.
- Go to Step 1.

The initial network is equivalent to the network in the decomposed DAP with the potential vehicle state variables set to their optimistic values. However, initially the potential vehicle labels have an undefined status so that flow is restricted to the domicile node only (i.e. $y_{ij} = 0$, $\forall j \in T$). Therefore, the first network that is solved is identical to solving a simple vehicle to task assignment problem.

Step 2 examines the current solution and determines which potential vehicles will have their status changed from undefined to temporary or vice-versa. Every vehicle tour is simulated to determine the state of the potential vehicle associated with every task on the tour. Suppose $T' = 11{:}00$ and vehicle 1 has the following tour $\{A \rightarrow D \rightarrow C$ domicile$\}$, with the following label status values PV_A = TEMP, PV_D = TEMP, and PV_C = UNDF. Simulation of the vehicle tour results in the following data:

- PV_A: NEW-ETA = 10:00 NEW-ADT = 10 hrs. tour = FEASIBLE
- PV_D: NEW-ETA = 12:00 NEW-ADT = 8 hrs. tour = FEASIBLE
- PV_C: NEW-ETA = 16:00 NEW-ADT = 4 hrs. tour = INFEASIBLE

Given the results of the tour simulation and the current value of T', PV_A would have its state variable updated to the new values and its status would remain temporary. The label for PV_C would have its status changed from temporary to undefined because its NEW-ETA is greater than T'. The label for PV_D would remain unchanged because its NEW-ETA is greater than T' and the tour is infeasible at this point.

The next step determines which potential vehicles will have their label status changed to permanent. Continuing with the example above, suppose $\bar{T}$ = 10:00. The status for PV_A would be changed to permanent resulting in the assignment of vehicle 1 to task A being fixed. If the end of the planning horizon has not been reached then the network is adjusted. The algorithm may be described as a window that is rolled forward at each iteration, with δ being the width of the window and ε the minimum advancement (see figure 2). Potential vehicle labels with ETAs beyond the window have an undefined label status. Labels of potential vehicles within the window may have either temporary or undefined status depending on the feasibility of the tour associated with the potential vehicle. As the window rolls forward potential vehicles with temporary label status are set to permanent.

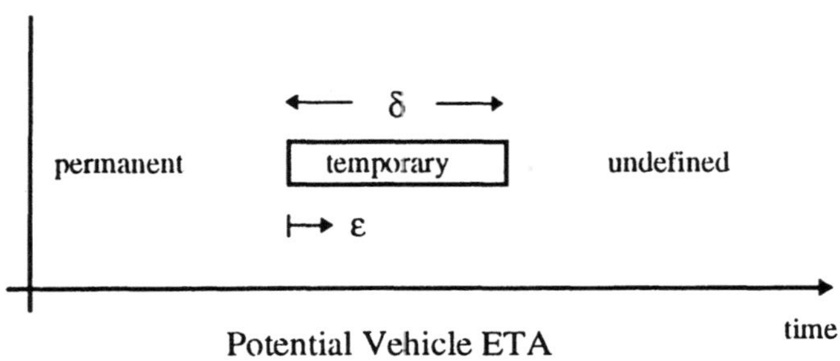

Figure 2. Rolling Window

The adjustment of the network is conducted based on the new values for the potential vehicle labels. If the status has been set to permanent then the path that preceded the potential vehicle is fixed. In the above example this would result in x_{1A} being set to one. Note that when a potential vehicle node has its label status set to permanent, the label status for all preceding potential vehicles nodes on that path have permanent label status. For potential vehicle nodes with temporary label status the arcs to task nodes are re-evaluated. If the arc (i,j) is infeasible based on the new state variable for PV_i, then y_{ij} is set to zero. Otherwise the value of d_{ij} is updated based on the new state at

PV_i. Those potential vehicle nodes with undefined label status have $y_{ij} = 0, \forall j \in T$. The network is then resolved based on the recent adjustments and the above procedure is repeated.

The end of the planning horizon is reached in a finite number of steps; the maximum number of steps required is $(T^f - T°)/\varepsilon$. At the end of the planning horizon all of the potential vehicle labels will be either permanent or undefined. The tasks associated with potential vehicles with undefined label status are uncovered tasks. Depending on the specific application, these tasks may be viewed as refused loads or loads that will be given to a common carrier. Tasks may be refused because there were not enough vehicles or because the revenue associated with the load was not enough to offset the operating expenses of covering the task. The other tasks are covered by feasible vehicle tours.

The LASER algorithm described above is very flexible with respect to incorporating additional constraints. Suppose there are two types of vehicles, V_1 and V_2, and each task specifies whether it can be serviced by V_1, V_2, or both. We would then add a vehicle type to the state variable for the potential vehicles. The vehicle type would be uncertain for potential vehicles corresponding to tasks that could be serviced be either V_1 or V_2. We could use the same label updating mechanism described above to address equipment requirement constraints.

The LASER algorithm is also well suited for real-time implementations which are particularly important for most real-world applications. Information updates can be incorporated by modifying the existing network and re-executing the LASER algorithm. For example, if a new task is called in by a customer it would simply be appended to the current network by adding the necessary task and potential vehicle nodes and the appropriate arcs. If an equipment failure occurs and results in a vehicle being available at a later time than expected, the driver to task arcs would then be re-evaluated in terms of feasibility and cost. Information from the previous solution can be used when resolving the problem. For example, parts of a entire tour can be fixed by keeping the appropriate potential vehicle status labels on the path set to permanent.

4. Computational Study

The following section contains a computational study and evaluation of the LASER algorithm. The algorithm is compared to a shortest path tour construction procedure (DPGA) and a Dantzig-Wolfe decomposition column generation procedure (DWCG). The DWCG solution method is used to provide a lower bound on the optimal objective function and the DPGA algorithm provides an upper bound. The size of the rolling window and the rate at which it advances are also studied to see how they impact on the algorithm's performance. The algorithms are tested on randomly generated problems and problems from a real-world motor carrier. The algorithm analysis uses a number of evaluation criteria including cpu time, and different measures of solution quality.

4.1 Algorithm Implementation

The following section describes issues that arose in implementing the LASER algorithm. The dynamic programming-based algorithm for the column generation procedure is based on the methods described in Desrosiers et al. [4].

LASER There are a number of implementation issues that arise in the use of LASER. The first major issue is the generation of the network. For large problems explicitly enumerating all the possible real/potential vehicle to task arcs would be too time consuming. We implemented a spacefilling curve procedure for finding a candidate set of arcs for the real/potential vehicle nodes (see Bartholdi and Platzman [1]). The procedure added arcs to the network by first finding the n closest tasks for each vehicle and then finding the n closest vehicles for each task. The closeness of a task to a vehicle is determined by their relative positions on the spacefilling curve.

The optimal solution to the network was determined by using SIMPNET, a network simplex code (see [7] and [6]). SIMPNET is capable of solving relatively large pure network problems in a few seconds on a typical computer workstation. All of the algorithms were coded in ANSI C and compiled and executed on a Silicon Graphics Indy workstation.

Dynamic Programming with Greedy Assignment (DPGA) In order to evaluate the performance of LASER we implemented a greedy dynamic programming-based procedure. The DPGA procedure determined vehicle tours by finding the least cost path from a vehicle node to its domicile node using the same network generated for LASER. The algorithm is based on the shortest path label correcting algorithm of Glover et al. [5] and the shortest path with time windows algorithm of Desrochers and Soumis [4], [2]. The algorithm determines the least cost path from the vehicle node to the domicile node while satisfying the time window and length of tour constraints (see [8]).

The DPGA algorithm is as follows:

- *Step 0:* INITIALIZE
  - Generate the decomposed DAP network.
  - $k = 1$
- *Step 1:* CONSTRUCT VEHICLE TOUR
  - Determine the least cost path for vehicle k.
- *Step 2:* UPDATE
  - Remove the tasks covered on the path from the network.
  - If there are still tasks to be covered and k is less than the number of vehicles then $k \Leftarrow k + 1$, Go to Step 1. Otherwise stop.

The DPGA algorithm outlined above is dependent on the order in which the vehicles are treated. For example, if there are ten vehicles with ADT = 20 and ten with ADT = 40, the solution will vary based on the order in which the vehicle tours are constructed. Nonetheless, the DPGA algorithm provides a good solution to the DAP.

Dantzig-Wolfe Decomposition One possible solution approach for the DAP is to enumerate different feasible paths through the network to develop a set of candidate routes for each vehicle. One could then choose the best route for each vehicle. After choosing the best routes one might ask if there are better routes that were not considered in the candidate set. This is precisely the framework in which Dantzig-Wolfe decomposition with column generation operates. The DWCG procedure implemented for the DAP is based on the vehicle routing problem with time windows work of Desrosiers et al. [4] and [3].

The DAP can be easily transformed into a problem which is well suited for a Dantzig-Wolfe decomposition column generation scheme. The restricted master problem considers a set of feasible routes for each vehicle and selects the optimal subset of paths such that each vehicle is assigned to exactly one path.

The restricted master can be formulated as:

$$min \sum_{k\in D}\sum_{p\in\Omega_k} c_{pk}\theta_{pk} + \sum_{i\in T} r_i s_i \tag{23}$$

subject to:

$$\sum_{k\in D}\sum_{p\in\Omega_k} a_{ipk}\theta_{pk} + s_i = 1 \qquad \forall i \in T \tag{24}$$

$$\sum_{p\in\Omega_k}\theta_{pk} = 1 \qquad \forall k \in D \tag{25}$$

$$s_i \text{ binary} \qquad \forall i \in T \tag{26}$$

$$\theta_{pk} \text{ binary} \qquad \forall p \in \Omega_k, \forall k \in D \tag{27}$$

where:

- Ω_k - set of candidate paths (tours) for vehicle k
- c_{pk} - cost of path p for vehicle k
- r_i - net contribution for task i
- a_{ipk} - 1 if task i is on path p for vehicle k, 0 otherwise
- s_i - slack variable
- θ_{pk} - 1 if path p is utilized for vehicle k, 0 otherwise

The above formulation is a set partitioning problem with additional generalized upper bound (GUB) constraints for each vehicle. The binary constraints can be relaxed and replaced with linear constraints $0 \leq \theta_{pk} \leq 1$ to form a linear relaxation. Let α_i be the dual variables associated with the set partitioning constraints, and let β_k be the dual variable associated with the resource constraint.

The subproblem for this formulation is the same shortest path problem solved in the DPGA algorithm. Note that a shortest path problem must be solved for each vehicle and that these problems are independent. The dual variables from the restricted master problem are used to calculate reduced costs on arcs in the network. The marginal cost $\bar{c}_{pk}$ of path p for vehicle k is given by:

$$\bar{c}_{pk} = c_{pk} - \sum_{i \in T} \alpha_i a_{ipk} - \beta_k \qquad \textbf{(28)}$$

The DWCG solution procedure used the same network as the LASER and DPGAalgorithms. The dual information from the master can be used to calculate the reduced costs for arcs in the decomposed DAP network.

The subproblem can be solved efficiently by using the same shortest path algorithm used for the DPGA procedure. The shortest path algorithm can generate several negative reduced cost columns at each iteration. The DWCG procedure can be accelerated by adding several columns at once for each vehicle subproblem. In the actual implementation an upper limit was set on the number of columns to add for each subproblem. If no negative reduced cost columns can be found then the solution procedure terminates and the current solution provides a lower bound on the value of the optimal integer solution. Experience has shown that this lower bound is tight.

4.2 Problem Sets

The test problems consisted of two classes of DAPs: randomly generated problems and real-world problems. The main difference between the problems is the spatial distribution of the demand movements. The real-world problems contain a central terminal which is either the origin or destination of all the demands, whereas the O/D pairs are uniformly distributed in the generated problems. The following section describes the methodology used to create the generated problems, and the latter section describes the real-world data sets.

Generated Test Problems In order to evaluate the performance of the LASER algorithm a number of randomly generated problems were created. To create the test problems we used a problem generator to simulate customer demands for the DAP. The problem generator used a non-homogeneous Poisson process described by a weekly demand rate parameter with day-of-week and hour-of-day distributions. In this manner the number of loads generated per week is dependent of the current time and day of week. There are several components needed to describe a load: call-in time, origin/destination, pickup time, and the width of the pickup window.

A non-homogeneous Poisson process was used to generate the call-in times. For our testing purposes we were interested in solving a DAP with a planning horizon of one day, with all the demands known at the start of the day. Therefore, we specified the day-of-week and time-of-day distributions so that all demands would be called-in at the start of our planning horizon. The origin/destination pair for a demand was generated uniformly from a given set of locations. The pickup time for the demand was generated from a specified discrete distribution for the hour of day. The minutes after the hour was then chosen from a uniform distribution. The width of the pickup time window was generated from a discrete distribution and appended to the pickup time.

Using the problem generator outlined above we created nine test problems. The test problems are characterized by the number of tasks and the distribution of the time windows. We generated problems with 50, 75, and 100 tasks with tight, wide, and mixed time windows. The width of a wide time window is equal to the length of the planning horizon. In the mixed time window case we used a distribution where 25% of the windows were tight, 25% were wide, and the remainder were uniform between tight and wide. The set of locations was taken from the customer data base for the drayage problem described in the next section. The costs c_{ij} were calculated as $30plus.80$ per mile.

Drayage Test Problems A specific application of the DAP is the drayage problem. The problem arises in intermodal operations where trains and trucks are used to transport freight. Typically a drayage company truck will pickup the freight (container) from the shipper and transport it to the nearest rail terminal. The container is then loaded onto a train and transported to the rail terminal closest to the load's final destination. Finally the freight is moved by truck to the consignee. An intermodal operation will consist of a few geographically dispersed rail terminals each with a fleet of trucks for the pickup and delivery of freight.

The problem is to develop a set of minimum cost driver tours over a planning horizon which is typically 24 hours. Every load has a time window specifying the earliest and latest time at which the load may be picked up or delivered. The time window constraints are hard (i.e. early arrivals are allowed but service cannot begin until the start of the window, and late arrivals are not allowed). There are other constraints including driver work rules and equipment requirements. In addition to using its own fleet of vehicles, the drayage company may use trucks from other companies (at a premium cost) to cover loads. Due to the nature of the drayage problem dispatchers manually construct tours by finding triangulation opportunities. A typical tour is: deliver freight from the rail terminal, move empty to pickup freight, deliver freight to the rail terminal. The rail terminal behaves as a regeneration point.

The fact that most tasks originate or terminate at the rail terminal differentiates the drayage problem from the randomly generated problem. Other features like multiple loads between the same O/D pair with the same characteristics also make the drayage problem different. The naturally occurring degeneracy of the drayage problem makes it a difficult problem class to solve with the DWCG procedure. We used three different real-world drayage sets. The drayage problems contain tasks with a mixed time window distribution similar to the one used in the problem generator.

4.3 Discussion of Results

The subsequent sections discuss the results of the numerical study conducted to evaluate the LASER algorithm. The first section describes the effects of the time window parameters δ and ε on solution quality and cpu time. The second section contains the comparison and analysis of the three algorithms used to solve the DAP.

LASER Parameters The LASER algorithm has two user defined input parameters that affect the potential vehicle label setting procedure. The width of the rolling window is specified by δ, and ε specifies the minimum advancement of the window at each iteration of the algorithm. The width of the time window relative to the length of the planning horizon affects the number of potential vehicle labels considered for temporary status per iteration. For small values of δ the labels of potential vehicles in the future will remain undefined, whereas for large values of δ more labels will be set to temporary earlier in the algorithm. As mentioned earlier, the number of iterations required for the algorithm is bounded $(t^f - t^{\circ})/\varepsilon$, therefore the amount of cpu time is inversely proportional to ε. The smaller the value of ε the fewer potential vehicles labels set to permanent status per iteration.

The combination of window parameters that will produce the best results is not obvious. The LASER algorithm was tested on a randomly generated problem with 100 tasks (wide time windows) and 30 vehicles with different values for δ and ε. The results of the experiment are contained in table 1. All recorded cpu times include network generation and algorithm execution but do not include I/O time.

100 tasks, 30 vehicles		$\varepsilon = 60$	$\varepsilon = 30$	$\varepsilon = 15$	$\varepsilon = 1$
$\delta = 60$	Profit	77,192	78,181	80,742	77,500
	cpu time	3	3	5	12
$\delta = 120$	Profit	76,416	79,686	81,382	83,324
	cpu time	3	4	6	11
$\delta = 180$	Profit	78,961	82,414	81,682	82,453
	cpu time	3	5	6	11
$\delta = 360$	Profit	80,707	84,912	84,724	84,182
	cpu time	4	7	9	18
$\delta = \infty$	Profit	83,493	86,777	87,411	87,455
	cpu time	5	10	13	25

Table 1. Rolling Window Parameters

As expected the amount of cpu time increased as ε decreased; the cpu time increased by four to five times as the value of ε was decreased from 60 minutes to one minute. The solution quality improved as the step size for the rolling window decreased. The cpu time and objective function (profit) increased as the width of the time window increased. The cpu times doubled as the δ was increased from 60 minutes to ∞ (the length of the planning horizon).

Based on the results in table 1 and other experiments it is clear that the best policy for the LASER algorithm is small ε and large δ. There is however a clear trade off between solution quality and cpu time. The results of these rolling window parameter experiments seem to indicate that even some information about a potential vehicle is better than none, and the tours should be built as slowly as possible to allow for changes in subsequent iterations. Given the above results, all subsequent problems solved with LASER had $\varepsilon = 1$ and $\delta = \infty$.

Comparison of Algorithms The performance of the LASER algorithm was evaluated by comparing its solution characteristics to those of the DPGA and DWCG solution strategies for the test bed of generated and real-world problem described earlier. The following statistics were recorded for every problem:

- objective function value in dollars (profit)
- number of idle drivers (idle)
- number of tasks left uncovered (refusal)
- percentage of total miles travelled empty by the vehicles (% empty)
- percentage of total available vehicle duty time utilized (% util.)
- cpu time in seconds

The profit for a set of tours was calculated as the total revenue from covered loads minus total travel costs (loaded and empty miles). All of the algorithms are profit maximization based and do not directly address the other measures outlined above. The other measures are helpful for evaluating differences in the objective function between solutions. For example, two solutions may cover the same number of tasks but one has fewer empty miles and consequently a higher profit.

Every test problem was solved with two values for the number of vehicles. The two different fleet sizes were used to measure the performance of LASER under excess and constrained scenarios. A problem with excess vehicle is one in which all coverable tasks are assigned and there are idle vehicles. Conversely, a constrained problem is one in which there are no idle vehicles. The two fleet size scenarios combined with the generated and real-world test problems resulted in a total of 24 problems. The results for the generated problems are contained in tables 2, 3, 4 and the real-world drayage problem results are presented in tables 5, 6 and 7. Note the DWCG profit values marked with an * are for the optimal integer solutions and DNR denotes problems for which DWCG did not run. All other DWCG solutions were fractional and represent an upper bound on the maximum profit.

The vehicles used for testing purposes were all assumed to be homogeneous with respect to their characteristics (starting time and location, domicile, and available duty time). A homogeneous vehicle fleet was used for two reasons. Firstly, a homogenous vehicle fleet eliminates the order dependence problems associated with the DPGA procedure. Secondly, in the DWCG procedure it is no longer necessary to solve a shortest path problem for every vehicle at each iteration of the algorithm. The GUB constraints were replaced with one constraint of the form $\sum_{p \in \Omega} \theta_p = n$, where n is the number of vehicles. Consequently the cpu times for the DWCG procedure are much less than they would be in the case of a heterogeneous vehicle fleet. The difference in cpu times may be as high as a factor equal to the number of vehicles.

The results of the experiments for both the generated and real-world problems provides explicit evidence on the ability of LASER to solve the DAP. The LASER algorithm performed well with respect to profit when the problem was loosely constrained with respect to fleet size. However, when there were fewer vehicles available to cover the same set of tasks the algorithm's performance deteriorated. The percentage gaps between the DWCG upper bound and the objective function values of LASER and DPGA for the generated problems are presented in table 8. The average gap for the excess vehicle problem is 0.8% indicating that LASER produced optimal or near-optimal solutions. The average gap increased by an order of magnitude to 4.4% for the constrained versions of the same problems.

The DPGA solution procedure displayed the same behaviour with slightly poorer solution quality than LASER. The average percentage gaps were 1.4% and 5.8% for the excess and constrained problems respectively. The DPGA procedure was superior to LASER with respect to cpu time for 23 of the 24 test problems. The DPGA procedure also used fewer vehicles than LASER to cover the same number of loads when there was excess fleet capacity.

	Excess $n = 20$			Constrained $n = 10$		
WIDE	LASER	DPGA	DWCG	LASER	DPGA	DWCG
Profit	45,708	45,298	45,804*	38,872	38,094	42,294
Idle	3	3	2	0	0	0
Refusal	0	0	0	8	9	4.1
% Empty	18.6	25.7	16.8	22.5	22.1	22.6
% Util.	64.4	69.2	60.4	94.3	88.1	98.0
cpu time	10	3	13	10	2	709
	Excess $n = 40$			Constrained $n = 30$		
TIGHT	LASER	DPGA	DWCG	LASER	DPGA	DWCG
Profit	39,462	39,468	39,548*	38,212	36,840	38,948*
Idle	3	6	4	0	0	0
Refusal	4	4	4	6	8	5
% Empty	38.0	37.9	37.1	37.7	36.2	36.8
% Util.	65.4	62.2	61.1	73.1	71.6	71.0
cpu time	9	2	2	9	2	2
	Excess $n = 30$			Constrained $n = 20$		
MIXED	LASER	DPGA	DWCG	LASER	DPGA	DWCG
Profit	44,152	44,156	44,489*	43,500	41,964	44,479*
Idle	5	7	6	0	0	0
Refusal	1	1	1	2	4	1
% Empty	28.1	28.0	22.7	27.4	25.9	22.9
% Util.	69.4	64.7	62.3	90.0	84.5	87.5
cpu time	11	3	5	10	1	13

Table 2. 50 Task Generated Problems

	Excess $n = 30$			Constrained $n = 20$		
WIDE	LASER	DPGA	DWCG	LASER	DPGA	DWCG
Profit	67,834	67,218	68,087	66,292	63,780	68,092
Idle	5	5	5	0	0	0
Refusal	0	0	0	2	5	0
% Empty	19.2	25.6	16.2	20.0	21.5	16.1
% Util.	66.4	69.8	61.7	93.1	86.2	92.4
cpu time	30	6	49	31	5	317

	Excess $n = 50$			Constrained $n = 40$		
TIGHT	LASER	DPGA	DWCG	LASER	DPGA	DWCG
Profit	60,524	60,496	60,632*	60,488	58,353	60,601*
Idle	5	7	6	0	0	0
Refusal	7	7	7	7	10	7
% Empty	36.8	37.0	35.9	37.1	36.6	36.1
% Util.	65.6	64.0	61.9	76.4	73.5	73.3
cpu time	15	4	2	14	4	3

	Excess $n = 50$			Constrained $n = 30$		
MIXED	LASER	DPGA	DWCG	LASER	DPGA	DWCG
Profit	66,275	66,345	66,964*	61,668	64,882	66,952
Idle	12	18	18	0	0	0
Refusal	1	1	1	7	3	1
% Empty	31.2	30.6	24.5	31.3	30.5	24.6
% Util.	65.4	57.3	54.5	91.0	88.9	87.3
cpu time	31	6	16	33	6	35

Table 3. 75 Task Generated Problems

	Excess $n = 40$			Constrained $n = 25$		
WIDE	LASER	DPGA	DWCG	LASER	DPGA	DWCG
Profit	90,234	89,627	90,655	81,865	81,076	90,496
Idle	11	7	9	0	0	0
Refusal	0	0	0	10	11	0
% Empty	15.8	20.7	12.0	18.4	17.5	13.4
% Util.	66.7	69.3	62.0	92.7	86.2	97.1
cpu time	74	12	118	79	10	4,411
	Excess $n = 65$			Constrained $n = 50$		
TIGHT	LASER	DPGA	DWCG	LASER	DPGA	DWCG
Profit	83,328	82,947	83,349*	77,552	74,188	77,556*
Idle	2	3	1	0	0	0
Refusal	6	6	6	14	18	14
% Empty	35.6	37.8	35.5	33.3	34.3	33.9
% Util.	70.0	69.3	68.0	74.1	72.7	70.9
cpu time	24	8	4	21	6	3
	Excess $n = 50$			Constrained $n = 30$		
MIXED	LASER	DPGA	DWCG	LASER	DPGA	DWCG
Profit	86,896	87,324	87,710	76,751	80,357	86,448
Idle	6	11	6	0	0	0
Refusal	4	4	4	17	13	5
% Empty	32.6	29.3	26	28.1	24.2	30.2
% Util.	76.2	68.8	67.9	91.2	92.6	95.1
cpu time	48	13	48	52	11	455

Table 4. 100 Task Generated Problems

	Excess $n = 50$			Constrained $n = 30$		
MIXED	LASER	DPGA	DWCG	LASER	DPGA	DWCG
Profit	3,789	3,695	3,804*	3,771	3,695	3,797
Idle	8	33	16	0	13	0
Refusal	10	10	10	10	10	10
% Empty	12.9	16.7	12.3	13.7	16.7	12.6
% Util.	28.8	23.1	23.7	44.4	37.8	39.7
cpu time	4	4	57	4	4	57

Table 5. 61 Task Drayage Problem

	Excess $n = 40$			Limited $n = 30$		
MIXED	LASER	DPGA	DWCG	LASER	DPGA	DWCG
Profit	3,814	3,819	3,924	3,767	3,705	3,919
Idle	1	10	6	0	2	0
Refusal	12	12	12	12	14	12
% Empty	32.5	32.4	30.4	33.4	32.7	30.5
% Util.	44.8	42.8	39.5	57.0	55.2	51.6
cpu time	5	4	292	5	4	292

Table 6. 73 Task Drayage Problem

	Excess $n = 50$			Limited $n = 30$		
MIXED	LASER	DPGA	DWCG	LASER	DPGA	DWCG
Profit	12,169	11,905	12,482	10,450	11,482	DNR
Idle	1	14	1	0	0	
Refusal	13	13	13	34	18	
% Empty	16.4	20.7	10.8	19.1	25.0	
% Util.	58.0	54.6	52.9	64.7	79.5	
cpu time	13	3,475	19,887	10	3,809	

Table 7. 80 Task Drayage Problem

To further investigate the performance of the algorithms for fleet size constrained DAP's, a 100 task, mixed time window, randomly generated problem was solved with ten different fleet size values. The objective function values are plotted in figure 3 and the percentage gaps are plotted in figure 4. The deterioration in LASER's performance as the problem became more constrained is clear from the results in the two figures. The objective function percentage gap ranged from less than 1% for the excess problems to 25% for the constrained problems.

The DPGA procedure was able to better utilize the available vehicle duty time when the problem was constrained which resulted in more tasks being covered and thus produced a more profitable solution (see figure 3 and 4). The plots in figure 5 show the vehicle utilization for the algorithms. The figure suggest that LASER does not produce efficient vehicle tours for either the excess or constrained fleet size scenarios. When there were excess vehicles LASER used more of the available vehicle time than the other procedures indicating unnecessary vehicle utilization. In the constrained cases the LASER algorithm had a high vehicle utilization measure but still covered fewer tasks.

To achieve the high degree of efficient vehicle utilization required a lot more computational effort for the DWCG procedure. The degree to which the problem is constrained with respect to fleet size appeared to have an important influence on the DAP's level of difficulty. The sometimes dramatic increase in cpu times for the DWCG procedure when the problem was constrained provides further evidence for the above assertion. For the problem examined above, the cpu time for DWCG ranged from 50 seconds to 1,050 seconds as the fleet size varied from 50 to 5, whereas LASER's cpu times ranged from 50 to 20 seconds.

The results of the computational analysis provides a good insight to the nature of DAPs and the ability of the LASER algorithm to solve them. LASER is clearly limited when the problem is constrained with respect to fleet size. However, to achieve the superior results requires a substantial increase in computational effort. The final section describes the main conclusions from the paper and outlines future research opportunities.

PROBLEM	Excess		Constrained	
	LASER	DPGA	LASER	DPGA
50-wide	0.2	1.1	8.1	9.9
50-tight	0.2	0.2	1.9	5.4
50-mixed	0.8	0.7	2.2	5.7
75-wide	0.4	1.3	2.6	6.3
75-tight	0.2	0.2	0.2	3.7
75-mixed	1.0	0.9	7.9	3.1
100-wide	0.5	1.1	9.5	10.4
100-tight	0.0	0.5	0.0	4.3
100-mixed	0.9	0.4	11.2	7.0
61-dray	0.4	2.9	0.7	2.7
73-dray	2.8	2.7	3.9	5.5
80-dray	2.5	4.6	-	-
Average	0.8	1.4	4.4	5.8
Combined Avg	LASER - 2.5		DPGA - 3.5	

Table 8. Objective Function Gap

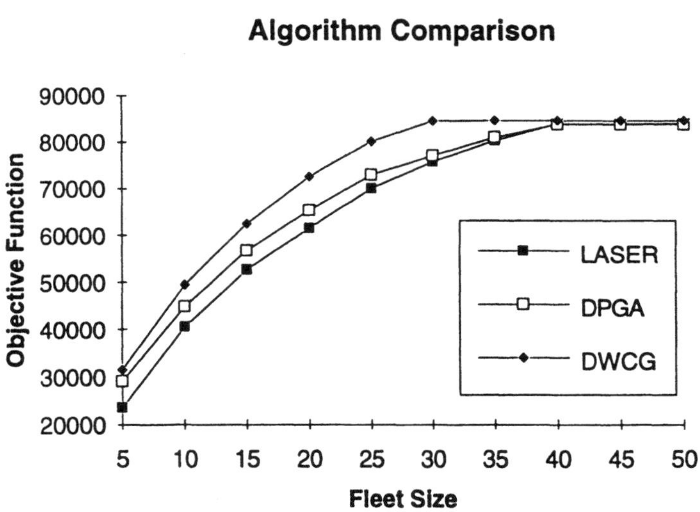

Figure 3. Algorithm Comparison

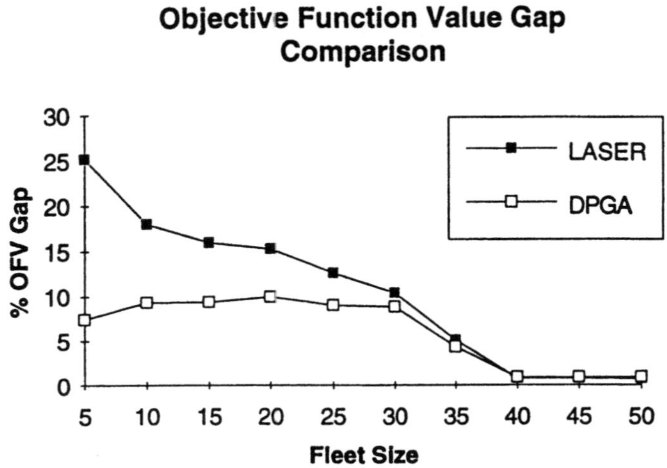

Figure 4. Objective Function Value Gap Comparison

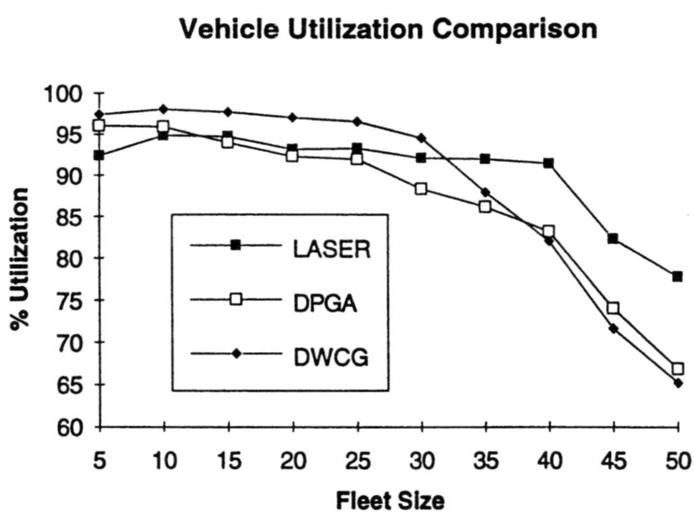

Figure 5. Fleet Utilization Comparison

5. Conclusions

The paper focused on developing a new solution strategy for an important class of dynamic routing and scheduling problems known as the dynamic assignment problem. The DAP has many applications ranging from motor carrier fleet management to airline crew scheduling. Surprisingly, there has been relatively little research done in this area despite the number of existing applications. Our main objective was to design a solution procedure that was able to incorporate real-world problem features, produce high quality solutions, and be amenable to a real-time implementation. To achieve these objectives we developed a generalized label setting rolling horizon algorithm.

To evaluate the performance of LASER we conducted a series of numerical experiments on a test problem set that consisted of randomly generated and real-world drayage problems. A Dantzig-Wolfe column generation scheme was designed to provide a lower bound on the optimal solution, and a shortest path based tour construction heuristic was implemented to provide a good upper bound. The LASER algorithm constructs vehicle tours in parallel while the DPGA procedure constructs vehicle tours sequentially, thus we have a breadth-first versus depth-first comparison for DAP algorithms. There are a number of key findings based on the results of the numerical study.

- LASER produces solutions within 1% of LP-optimal solutions for problems with excess fleet capacity.
- LASER solution gap increases by an order of magnitude when the problems are constrained with respect to fleet capacity.
- LASER produces better solutions than DPGA for problems with wide and tight time windows. The procedures are comparable for problems with mixed time windows.
- LASER generally produces better solution than DPGA for problems constrained with respect to fleet capacity. LASER and DPGA are comparable for problems with excess fleet capacity.
- DPGA uses fewer vehicles than LASER to cover the same number of loads when there is excess fleet capacity.
- DPGA requires less computational effort than LASER, when there is a homogenous fleet. With a heterogenous fleet, LASER is much faster.

- DAP difficulty level is highly dependent on available fleet capacity.

For problems with excess capacity in terms of the number of available vehicles, the LASER algorithm performed well. The LASER algorithm mainly produced results that were within 1% of the DWCG LP-optimal bound. When the fleet size was reduced for the same set of problems the quality of the solutions from LASER deteriorated primarily due to the labelling algorithm's inability to efficiently utilize the available vehicle time. These results were true for both the generated and real-world problems. The fleet capacity constrained DAPs are much more difficult than the excess problems as demonstrated by the large increase in computational effort for the DWCG procedure.

Neither DAP algorithm dominates the other based on the numerical analysis; the DPGA requires less cpu time but the LASER algorithm produces better solutions. The main disadvantage of the depth-first DPGA procedure is its dependence on the order in which the vehicle tours are built when the vehicle fleet is heterogenous.

The LASER algorithm was clearly incapable of producing high quality solutions when the problem became constrained with respect to fleet capacity. Although in many applications there is excess fleet capacity, there are obviously situations in which this is not the case (e.g. end of month effects). The ability to efficiently solve the DAP under these constrained conditions remains an important unresolved issue and requires further investigation. There exists an opportunity to combine elements of the depth-first DPGA procedure with the breadth-first LASER algorithm.

The dynamic assignment model studied in the paper assumed that all of the relevant problem data was deterministic. In many situations this is not the case, thus reducing the effectiveness of the model. For example, when constructing vehicle tours dispatchers may take into account tasks that have yet to be called in, but probably will be called in based on the customer's demand history. The anticipation of future demands can be addressed through forecasting and stochastic methods which opens up new research opportunities.

Acknowledgement

This research was supported in part by grant DDM-9102134 from the National Science Foundation, and by grant AFOSR-F49620-93-1-0098 from the Air Force Office of Scientific Research.

References

[1] John J. Bartholdi III and Loren K. Platzman, "Heuristics based on spacefilling curves for combinatorial problems in euclidean space", *Management Science*, 34:291-305, 1988.

[2] Martin Desrochers and François Soumis, "A generalized permanent labelling algorithm for the shortest path problem with time windows", *INFOR*, 26:191-212, 1988.

[3] J. Desrosiers, Y. Dumas, and F. Soumis, "The multiple vehicles many to many routing problem with time windows", *Publication no. 362, Centre de Recherche sur les transports; Universite de Montreal*, 1984.

[4] J. Desrosiers, M. Solomon, and F. Soumis, "Time constrained routing and scheduling", *Technical report, Groupe d'études et de recherche en analyse des décisions*, 1992.

[5] Fred Glover, Darwin D. Klingman, Nancy V. Phillips, and Robert F. Schneider, "New polynomial shortest path algorithms and their computational attributes", *Management Science*, 31:1106-1128, 1985.

[6] I.J. Lustig, "The influence of computer language on computational comparisons: An example from network optimizationî, *ORSA Journal on Computing*, pages 152-161, 1990.

[7] I.J. Lustig. Simpnet, "A package for the network simplex method in c.", *Technical report, Department of Civil Engineering and Operations Research, Princeton University*, 1990.

[8] Warren B. Powell and Zhi long Chen, "A new algorithm for the shortest path problem wth time windows", *Technical report, Department of Civil Engineering and Operations Research, Princeton University*, 1994.

A HEURISTIC ALGORITHM FOR THE VEHICLE ROUTING PROBLEM WITH BACKHAULS

Paolo Toth and Daniele Vigo

D.E.I.S. - Universita' di Bologna

Viale Risorgimento, 2 - Bologna, Italy

Let us consider the following extension of the Capacitated Vehicle Routing Problem (VRP). We are given a set of customers partitioned in two subsets: Linehaul and Backhaul customers. Each Linehaul customer requires the delivery of a given quantity of product from the depot, while a given quantity of product must be picked up from each Backhaul customer and transported to the depot. This problem is known as the Vehicle Routing Problem with Backhauls (VRPB), or as the Linehaul-Backhaul Problem. VRPB then calls for the determination of a set of vehicle routes visiting once all customers and such that: (i) each vehicle performs one route; (ii) for each route the total load associated with Linehaul and Backhaul customers does not exceed, separately, the vehicle capacity; (iii) in each route no Backhaul customer is visited before that all Linehaul customers have been serviced. The objective is to minimize the total traveled distance.

VRPB is known to be NP-hard in the strong sense, and many heuristic algorithms have been proposed for its approximate solution.

We present a new cluster-first-route-second heuristic algorithm for VRPB which starts from an infeasible solution obtained from a Lagrangian relaxation and tries to build a feasible set of routes through inter-route and intra-route arc exchanges. Extensive computational tests on problem classes proposed in the literature show the effectiveness of the proposed approach.

1. Introduction

Let us consider the following extension of the *Capacitated Vehicle Routing Problem* (VRP). We are given a set of customers partitioned in two subsets: Linehaul and Backhaul customers. Each *Linehaul customer* requires the delivery of a given quantity of product from the depot, while a given quantity of product must be picked up from each *Backhaul customer* and transported to the depot. This problem is known as the *Vehicle Routing Problem with Backhauls* (VRPB), or as the *Linehaul-Backhaul Problem.*

An example of this partition of the customer set, which is extremely frequent in practical situations, is represented by the grocery industry. In this case, supermarkets and shops are the Linehaul customers, and grocery suppliers are the Backhaul customers. In recent years it has been widely recognized that in this mixed distribution/collection context a significant saving in terms of transportation costs can be achieved by visiting Backhaul customers within distribution routes. For example, the Interstate Commerce Commission has estimated in 160 million $ the yearly saving obtained by the U.S.A. grocery industry due to the introduction of Backhauling (see, e.g., Golden, Baker, Alfaro and Schaeffer [13]).

VRPB can be stated as the problem of determining a set of vehicle routes visiting all customers, and such that: (i) each vehicle performs one route; (ii) for each route the total load associated with Linehaul and Backhaul customers does not exceed, separately, the vehicle capacity; (iii) in each route the Backhaul customers, if any, are visited after all Linehaul customers; (iv) the total distance traveled by the vehicles is minimized. The *precedence constraint* (iii) is practically motivated by the fact that, in many applications, Linehaul customers have a higher service priority than Backhaul customers. Another practical reason is that vehicles are often rearloaded, hence the on-board load rearrangement required by a "mixed" service is difficult, or even impossible, to carry out at customer locations.

The problem can be also formulated through the following graph theory model. Let $G=(V_0, A)$ be a complete undirected graph, with vertex set $V_0 := \{0\} \cup L \cup B$. Subsets $L = 1, \ldots, n$, and $B = \{n+1, \ldots, n+m\}$, correspond to Linehaul and Backhaul vertex subsets, respectively. A nonnegative quantity, d_j, of product to be delivered or collected (*demand*) is associated with each vertex $j \in V$, with $V := L \cup B$. Vertex 0 corresponds to the depot (with a fictitious demand $d_0 = 0$), where K identical vehicles with a given *capacity*, D, are stationed. Let c_{ij} be the nonnegative cost associated with arc $(i,j) \in A$, with $c_{ij} = c_{ji}$ for each $i, j \in V_0$ such that $i \neq j$, and $c_{ii} = +\infty$

for each $i \in V_0$. VRPB then consists of finding a min-cost collection of K simple *circuits* (vehicle routes) such that:

i) each vehicle performs one circuit;
ii) each circuit visits vertex 0;
iii) each vertex $j \in V$ is visited exactly once;
iv) the sum of the demands of the Linehaul and Backhaul vertices visited by a circuit does not exceed, separately, the vehicle capacity D;
v) in each circuit the Linehaul vertices precede the Backhaul vertices, if any.

The objective is to minimize the total routing cost, defined as the sum of the costs of the arcs belonging to the circuits.

The "mixed" vehicle routes, i.e. routes which visit both Linehaul and Backhaul vertices, are implicitly oriented due to precedence constraint (v). Let K_L (resp. K_B) denote the minimum number of vehicles needed to serve all the Linehaul (resp. Backhaul) vertices. In order to ensure feasibility, we assume that K (number of available vehicles) is not smaller than the minimum number of vehicles needed to serve all the vertices, i.e. $K \geq \max \{K_L, K_B\}$. The values K_L and K_B can be obtained by solving the Bin Packing Problem (BPP) instances associated to the Linehaul and the Backhaul vertices, respectively. In these instances the bin capacity is equal to D and each item j has weight equal to the demand d_j of the corresponding vertex. In spite of the fact that BPP is NP-hard in the strong sense, many instances with hundreds of items can be optimally solved very effectively (see, e.g., Martello and Toth [16], [17]). Note that, because of requirement (v), it is not allowed to have circuits containing only Backhaul customers.

VRPB is NP-hard in the strong sense, since it generalizes the Capacitated Vehicle Routing Problem, arising when $B = \emptyset$ (see, for example, Fischetti, Toth and Vigo [8]).

Several heuristic algorithms for the solution of VRPB have been presented so far, and some of them will be briefly described in the next section. Deif and Bodin [6] proposed an extension of the well-known

Clarke-Wright [5] VRP heuristic, where the saving of the arcs connecting Linehaul to Backhaul customers is modified so as to delay the formation of "mixed" routes. Different extensions of the Clarke-Wright algorithm have been presented by Casco, Golden and Wasil [3], and Golden, Baker, Alfaro and Schaffer [13]; the latter facing the case where the precedence constraint between Linehaul and Backhaul customers is not present. Goetschalckx and Jacobs-Blecha [11] proposed an algorithm that builds initial routes with customers of the same type by using a spacefilling curves heuristic; these routes are then merged to form the final set of vehicle routes. Goetschalckx and Jacobs-Blecha [12] have also described an extension to VRPB of the Fisher-Jaikumar [10] VRP heuristic.

An exact, set-covering based, algorithm for the special case of VRPB where the number of customers of each type in a circuit is not greater than 4, has been proposed by Yano, Chan, Richter, Cutler, Murty and McGettigan [22]. A branch-and-bound exact algorithm for VRPB has been recently proposed by Toth and Vigo [20].

In this paper we present a new cluster-first-route-second heuristic algorithm for VRPB, which starts from a (possibly infeasible) solution obtained from a Lagrangian relaxation, and tries to improve the current set of routes through a post-optimization phase based on inter-route and intra-route arc exchanges. The paper is organized as follows. A brief survey of the heuristic algorithms proposed in the literature is presented in Section 2. The new heuristic is described in Section 3, while extensive computational tests on problem instances proposed in the literature, showing the effectiveness of the proposed approach, are finally discussed in Section 4.

2. Heuristic Algorithms from the Literature

The heuristic algorithms proposed so far for the solution of VRPB are extensions of algorithms used to successfully solve the VRP.

The first heuristic algorithm for VRPB has been proposed by Deif and Bodin [6] in 1984. The algorithm, called DB hereafter, is an extension of the well-known Clarke-Wright [5] heuristic for the VRP. The Clarke-Wright algorithm starts with an infeasible solution where each vertex is visited by a separate route. Routes are then iteratively combined by considering the "saving", in terms of routing cost, that can be achieved by serving two vertices on the same route instead of leaving them on separate routes. The saving obtained by visiting vertices i and j in sequence in the same route can be expressed as:

$$s_{ij} = c_{0i} + c_{i0} + c_{0j} + c_{j0} - (c_{0i} + c_{ij} + c_{j0}) = c_{i0} + c_{0j} - c_{ij} \quad (1)$$

The results obtained by using the standard Clarke-Wright algorithm for the solution of VRPB are greatly affected by the fact that the precedence constraint substantially reduces the number of feasible mergings. Deif and Bodin have experimentally shown that for VRPB the best results are obtained by delaying the formation of mixed routes. Hence they proposed to modify the saving definition in order to penalize the arcs connecting vertices of different type, thus delaying the union of Linehaul and Backhaul routes. The *Backhaul saving* is defined as

$$s'_{ij} = \begin{cases} s_{ij} - \pi S, & \text{if} \quad i \in L, \quad j \in B \quad \text{or vice-versa} \\ s_{ij} & \text{otherwise,} \end{cases} \quad (2)$$

where S is an estimate of the maximum saving s_{ij}, and π is a real penalty multiplier between 0 and 1.

The Clarke-Wright method, hence the DB algorithm, does not allow for the control of the number of routes of the final solution. Indeed, the solution found for a given instance can require more than K routes to serve all the vertices, thus being infeasible. From a practical point of view, both the routing cost of the solution obtained with the DB algorithm, and the probability that this solution is feasible, are strongly related to the number of route mergings executed. It is then evident that, even if delayed, the route orientation arising in the mixed routes, and the consequent decrease of possible route merging combinations, reduces the effectiveness of this method in facing VRPB, in terms of both the overall routing cost and the number of feasible solutions found.

Deif and Bodin [6] tested their algorithm on randomly generated problem instances with 100 to 300 vertices with a Backhaul percentage between 10% and 50%. Several π values were tested, and the results obtained showed that values of π between 0.05 and 0.20 produced the best solutions. Different extensions of the Clarke-Wright algorithm have been presented by Casco, Golden and Wasil [3], and Golden, Baker, Alfaro and Schaffer [13]; the latter facing the case in which the precedence constraints between Linehaul and Backhaul vertices are not present.

Goetschalckx and Jacobs-Blecha [11] proposed an algorithm, called SF hereafter, for the special case of VRPB arising when the vertices of G are points in the plane, and Euclidean distances are used for the arc costs. The approach is based on the concept of *spacefilling curves*, previously used by Bartholdy and Platzman [1] for the solution of the Euclidean Traveling Salesman Problem. Using the spacefilling curve transformation, Linehaul and Backhaul vertices are, separately, transformed from points in the plane into points along a line. The two separate sequences of points are then partitioned in order to form feasible routes each containing vertices of only one type. Each Linehaul route is, in turn, merged with the nearest Backhaul route (according to the spacefilling mapping), thus obtaining the final set of vehicle routes.

Also in this case the method does not guarantee to build solutions using exactly K routes. Goetschalckx and Jacobs-Blecha tested both DB and SF algorithms on 57 Euclidean instances with 25 to 150 vertices, 20% to 50% of which are Backhauls. The results presented in [11] showed that DB solutions are generally better, while SF is faster for larger problems.

Recently Goetschalckx and Jacobs-Blecha [12] have presented a heuristic algorithm for VRPB, called LHBH, based on the extension of the Generalized Assignment heuristic proposed by Fisher and Jaikumar [10] for the solution of VRP. In the LHBH algorithm the vertices are first grouped together by solving a Generalized Assignment Problem, then the final routes are determined through a modified insertion heuristic algorithm for TSP and post-optimization exchange procedures. The LHBH heuristic has been tested on the problems described in [11], and obtained better results than the previously proposed algorithms.

3. A New Heuristic Algorithm

In this section we describe a new cluster-first-route-second heuristic algorithm for VRPB, called TV. The structure of the proposed method is as follows: TV starts from an initial, possibly infeasible, solution obtained by solving a relaxation of VRPB, and tries to build a final set of feasible routes through vertex movements and arc exchanges.

This approach has been widely used in the literature (see, e.g., Fisher [9] for the symmetric VRP, and Vigo [21] for the asymmetric VRP), and is motivated by the fact that the solutions obtained through good relaxations of an optimization problem are often almost feasible, and contain a high degree of information on the optimal solution structure. The relaxation used to initialize algorithm TV is a Lagrangian approach described in Toth and Vigo [20], and based on a new formulation of VRPB as an asymmetric problem. The structure of a feasible solution to VRPB is illustrated in Figure 1. The underlying basic relaxation requires the determination of:

i) a shortest spanning arborescence on the Linehaul vertex set with fixed outdegree K at the depot;

ii) a shortest spanning anti-arborescence on the Backhaul vertex set with fixed indegree between K_B and K at the depot;

iii) a min-cost collection of K disjoint arcs connecting the Linehaul vertices to either the Backhaul vertices or the depot.

The Lagrangian lower bound is obtained by including into the objective function the degree constraints for each vertex, and a subset of the so-called *capacity-cut constraints*, imposing both the connectivity of the solution and the capacity requirement. Good Lagrangian multipliers are determined through a subgradient optimization procedure, while the subset of the imposed capacity-cut constraints is obtained through an exact *separation algorithm*. This lower bound proved to obtain very good starting solutions, as shown by the computational experiments presented in Section 4. The heuristic is particularly effective if applied in a multi-start fashion to the solutions obtained at different iterations of the subgradient optimization procedure.

In the following we describe algorithm TV by separately examining its main steps.

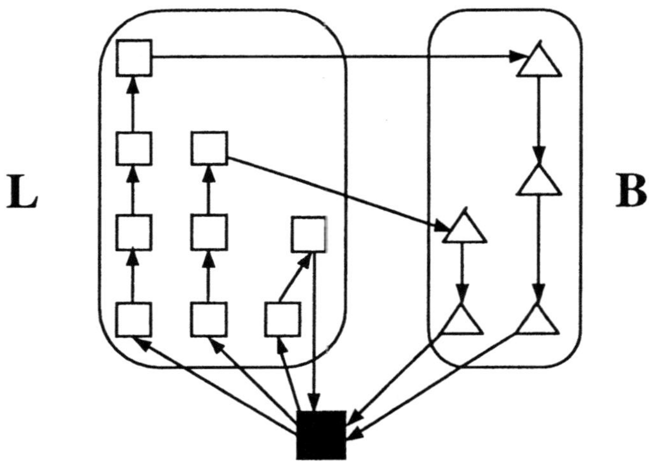

Figure 1. The structure of a feasible solution to VRPB with $n = 9$, $m = 5$, $K = K_L = 3$, and $K_B = 2$. Squares correspond to the Linehaul vertices, triangles to the Backhaul vertices, and the filled square to the depot.

3.1 Clustering

As in general for the cluster-first-route-second approach, the first step of algorithm TV requires the construction of groups of vertices, called *clusters*, containing only Linehaul or Backhaul vertices. These clusters are later combined together to form a basic set of, possibly infeasible, routes. The Lagrangian solution (i.e., the optimal solution to the current Lagrangian relaxation) is used to define K Linehaul clusters and $\hat{K}$ Backhaul clusters, where $\hat{K}$, with $K_B \leq \hat{K} \leq K$, is the number of arcs connecting the Backhaul vertex set to the depot in the Lagrangian solution.

In order to exploit the information on the optimal solution structure embedded in the Lagrangian solution, we choose as clusters the connected components obtained by removing the $2 \infty K$ arcs incident into the depot and the $\hat{K}$ interface arcs, connecting Linehaul to Backhaul vertices. The initial clusters can be determined in 0(n+m) time. Note that some of the clusters obtained at the end of this phase can be infeasible with respect to the capacity constraint.

An example of Lagrangian solution with (n = 9, m = 5, $K = K_L$ = 3 and $\hat{K}$ = 2) is illustrated in Figure 2(a), while the corresponding clusters are shown in Figure 2(b).

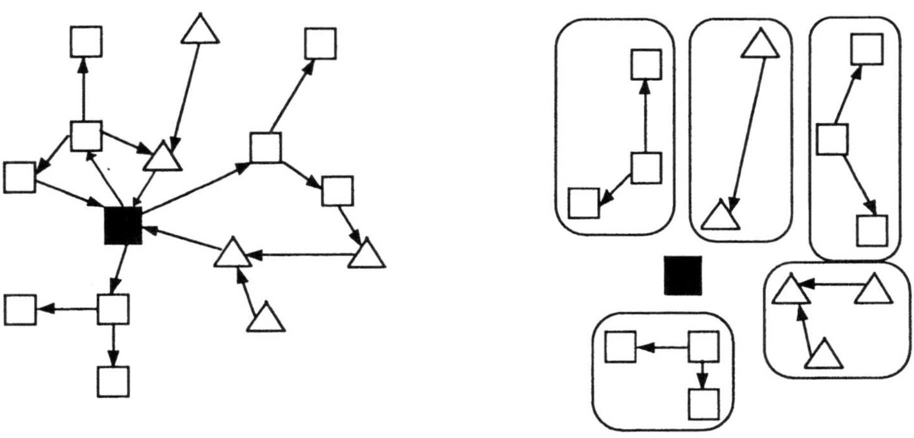

(a) (b)

Figure 2. Example of application of the clustering step for a problem with n = 9, m = 5, $K = K_L$ = 3 and $\hat{K}$ = 2: (a) the Lagrangian solution; (b) the corresponding clusters.

3.2 Matching

The clusters obtained in the previous step must now be combined so as to define the set of vertices associated with the initial routes. In particular, $\hat{K}$ Linehaul clusters must be associated with the $\hat{K}$ Backhaul clusters, hence forming mixed routes, while the remaining $K-\hat{K}$ Linehaul clusters, if any, are associated with the depot.

In order to solve this problem we first consider the removed interface arcs of the Lagrangian solution. If a Linehaul cluster and a Backhaul cluster are both feasible with respect to the capacity constraint and are connected by one or more interface arcs, we combine them. If a feasible Linehaul cluster is connected to more than one feasible Backhaul cluster, we arbitrarily choose one of them and combine it with the Linehaul cluster.

The remaining clusters, say K' and $\hat{K}'$, are then combined by solving an associated Assignment Problem. Let us define a K' x K' matrix, γ, whose rows are associated with the Linehaul clusters. The first $\hat{K}'$ columns are associated with the Backhaul clusters, and the remaining $(K'-\hat{K}')$ columns with the depot. The value of γ_{pq} is an estimation of the cost incurred by serving in the same route the Linehaul vertices of cluster p and either the Backhaul vertices of cluster q (if $q \leq \hat{K}'$), or the depot (if $q > \hat{K}'$). There are many ways to compute this estimation: for example γ_{pq} could be given by the exact or approximate solution value of the TSP associated with clusters p and q (if any), and the depot, when the additional precedence constraint between Linehaul and Backhaul vertices is imposed (this constrained TSP is called TSPB hereafter).

Given any TSPB instance on a complete graph or digraph $\tilde{G} = (\tilde{V}, \tilde{A})$, where the first $\tilde{n}+1$ vertices of $\tilde{V}$ represent the depot and the Linehaul vertices, it is easy to obtain an equivalent asymmetric TSP (ATSP) instance. To this end we remove from $\tilde{A}$ all the arcs connecting either a Backhaul vertex to a Linehaul vertex, or the depot to a Backhaul vertex. Then any exact or heuristic algorithm for ATSP (see, for example, Fischetti and Toth [7]) can be used to solve TSPB as well.

In order to reduce the computational effort, we compute the value γ_{pq} as the cost of the heuristic solution obtained by procedure ROUTING described in subsection 3.3.

The solution of the min-cost Assignment Problem with cost matrix γ, which can be obtained in $0((K')^3)$ time (see, e.g., Carpaneto, Martello and Toth [2]), gives the optimal matching between Linehaul clusters and the depot or Backhaul clusters.

3.3 Routing

The initial, possibly infeasible, routes are now determined for each vertex subset by using a farthest-insertion TSP heuristic, modified so as to take into account the precedence constraints between Linehaul and Backhaul vertices. The algorithm starts with an empty route, and iteratively adds to it the vertex for which the minimum distance from the current route is maximized. The selected vertex is inserted in the best feasible position (i.e., by taking into account the precedence constraint) within the current route.

The pseudo-pascal outline of an efficient implementation of the heuristic procedure ROUTING is given hereafter, where the input parameter S is the set of vertices to be routed. In the procedure U denotes the set of currently unrouted vertices, σ_i the successor of vertex i in the current solution, δ_j the distance from the currently unrouted vertex j to the nearest "compatible" routed vertex, and l the last Linehaul vertex in the current route.

```
procedure ROUTING(n, m, c, S, σ):
begin
1.  U := S; l := 0; σ0 := 0;
    for each j ∈ S do
       begin
          σj := 0;
          if j ≤ n then δj := c0j
             else δj := cj0
       end;
```

```
2.  while U ≠ Ø do
        begin
3.          i := arg max {δj, j ∈ U};
4.          M := +∞; pos := 0;
5.          if i ≤ n then h := 0; last := σl;
            else h := l; last := 0;
6.          repeat
              if chi + ciσh - chσh <M then M:= chi + ciσh - chσh; pos:=h;
                h := σh
            until h = last;
7.          σi := σpos; σpos := i; U := U \ {i};
8.          if (i ≤ n) and (σi > n or σi = 0) then l := i;
9.          for each j ∈ U do
               begin
                  d := + ∞;
                  case
                     (i ≤ n)   and (j ≤ n) : d := min{cij,cji};
                     (i = l)   and (j > n) : d := cij;
                     (i = σl)  and (j ≤ n) : d := cji;
                     (i > n)   and (j > n) : d := min{cij,cji}
                  end
                  if d < δj then δj := d
               end
            end
end.
```

In the main while-do loop, once the new vertex i to be inserted has been selected (step 3), if i is a Linehaul vertex then the insertion point is searched between the depot and the first Backhaul vertex (i.e., the successor or the last Linehaul vertex inserted, l), if any; otherwise, the insertion point is searched between the last Linehaul vertex and the depot.

At steps 7 and 8 the insertion of vertex i in the chosen position (i.e., after vertex pos), is executed, and l is possibly updated. Finally, at step 9, the minimum distances of the unrouted vertices (from the current vertices of the route) are updated by considering only feasible insertions with respect to the precedence constraint.

Procedure ROUTING has an overall time complexity of $0(s^2)$, where $s = |S|$. Indeed, the total number of iterations of the while-do loop to be performed is $O(s)$, and each execution of steps 3, 6 and 9 requires at most $O(s)$ operations.

3.4 Intra-Route Post-Optimization

Each route is improved by applying a post-optimization procedure which considers all the feasible exchanges of two and three arcs belonging to the route (the so-called *intra-route two-exchanges* and *three-exchanges*). The procedure is similar to those described in Lin [14], Lin and Kernighan [15], Christofides and Eilon [4], and Savelsbergh [19].

Given a route, described through the successor and predecessor vectors, σ and π, as well as a pair of distinct vertices, a and b, belonging to the route (with $b \neq \sigma_a$ and $a \neq \sigma_b$), a two-exchange is obtained by replacing arcs (a, σ_a) and (b, σ_b) with arcs (a, b) and (σ_a, σ_b), as illustrated in Figure 3(a). Note that for the symmetric VRP this exchange always produces a feasible solution. If, instead, we consider the directed case, the two-exchange produces a feasible route only if the inversion of all the arcs between vertices b and σ_a is performed, as illustrated in Figure 3(b).

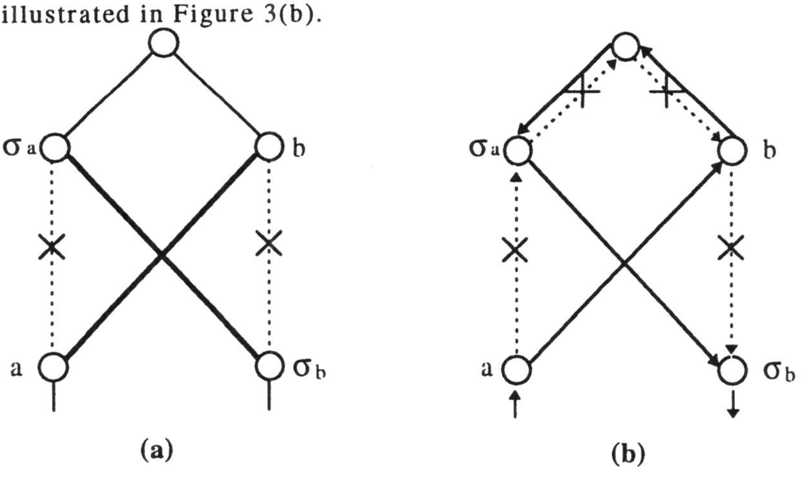

Figure 3. Example of intra-route two-exchanges for VRP and TSP: (a) undirected case, (b) directed case. Removed arcs are indicated with a cross, new arcs are drawn in bold line.

In order to extend the two-exchange to VRPB, we have to take into account that the precedence constraint between Linehaul and Backhaul vertices introduces additional restrictions on the arcs to be considered. In the following we call *Linehaul arc* (resp. *Backhaul arc*) an arc whose endpoints are both Linehaul (resp. Backhaul) vertices. We call *interface arc* an arc starting from a Linehaul and leading either to a Backhaul vertex or to the depot.

If (a, σ_a) is a Linehaul arc, then the arc (b, σ_b) can only be a Linehaul or an interface arc. If, otherwise, (a, σ_a) is an interface or a Backhaul arc, then the arc (b, σ_b) can only be a Backhaul arc. Similar considerations can also be applied to the *three-exchanges*, i.e. to exchanges involving three arcs.

The final solution is obtained by iteratively evaluating the cost of the route produced by each exchange of two or three arcs, and performing the best exchange among all those producing a positive cost reduction (also called *active exchanges*). The procedure is iterated until no active exchanges are found. As usual, we first consider only the two-exchanges, and then we apply the three-exchange procedure.

The evaluation of the cost of the route produced by an exchange can be performed in constant time, through parametric labeling techniques. Hence, the overall computational complexity of one iteration of the three-exchange procedure is $O(\overline{n}^3 + \overline{m}^3)$, where $\overline{n}$ and $\overline{m}$ are the number of Linehaul and Backhaul vertices of the considered route, respectively. Indeed, no exchange of arcs connecting vertices of different type, other than an interface arc, are considered.

3.5 Inter-Route Post-Optimization

The final set of routes is obtained by using further post-optimization local-search procedures where all the feasible movements of a vertex from a route to another, and feasible exchanges of two arcs belonging to different routes are considered (the so-called *inter-route one-exchanges* and *two-exchanges*). We only consider feasible insertions, i.e. insertions that do not either overload the routes or violate the precedence constraint between Linehaul and Backhaul vertices.

For each vertex a, belonging to route ρ_a, the overall "cost", θ_a^b, of its insertion after vertex b, belonging to route ρ_b (with $\rho_a \neq \rho_b$), is given by:

$$\theta_a^b := (c_{ba} + c_{a\sigma_b} + c_{\pi_a\sigma_a} - c_{\pi_a a} - c_{a\sigma_a} - c_{b\sigma_b}) - \alpha\Delta_a, \tag{3}$$

where Δ_a denotes the reduction of the overload of route r_a due to the removal of customer a, and α is an appropriate weighting parameter. Let $D(h)$ be the total demand of Linehaul (if $a \in L$) or Backhaul (if $a \in B$) vertices of route h, then $\Delta_a = \max\{0, \min \{D(\rho_a) - D, d_a\}\}$. At each iteration of the one-exchange procedure, the best insertion (corresponding to the minimum value of θ_a^b) is executed. The procedure is iterated until no feasible insertion exists.

An inter-route two-exchange is illustrated in Figure 4. It is easy to verify that, due to the implicit route orientation introduced by the precedence constraint between Linehaul and Backhaul vertices, this is the only possible inter-route two-exchange for VRPB, both in the directed and in the undirected case. By observing Figure 4 we note that if (a, σ_a) is a Linehaul (resp. Backhaul) arc, then the arc (b, σ_b) can only be a Linehaul (resp. Backhaul) or an interface arc. If, otherwise, (a, σ_a) is an interface arc, then the arc (b, σ_b) can be of any type.

The final solution is obtained by iteratively evaluating the routing cost and the possible infeasibility reduction produced by each exchange of two arcs, and performing the best exchange.

As for the inter-route one-exchange, the score considered is the weighted sum of the routing saving and, if one of the two routes involved in the exchange is infeasible, the overload reduction produced by the exchange. Exchanges that increase the infeasibility of an infeasible route or that produce an infeasible route starting from feasible ones are not considered. The procedure is iterated until no active exchanges are found. The intra-route post-optimization procedure of Section 3.4 is applied to each final route.

At each iteration $O(n^2 + m^2)$ exchanges are considered. The feasibility check and the computation of the score for the new routes can be executed in constant time through parametric labeling techniques. Hence the computational complexity of a single iteration is $O(n^2 + m^2)$.

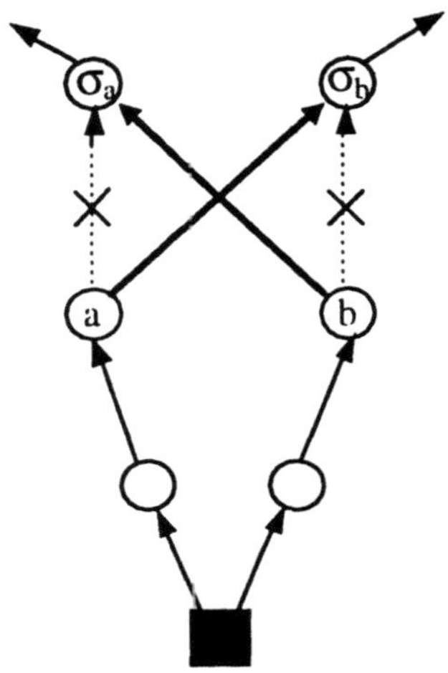

Figure 4. The only possible inter-route two-exchange for VRPB. Removed arcs are indicated with a cross, new arcs are drawn in bold line.

4. Computational Results

The heuristic algorithm TV of the previous Section has been implemented in FORTRAN, and run on an IBM 486/33 personal computer on two classes of test problems proposed in the literature. For the solution of the assignment problems, we used the FORTRAN code APC given in Carpaneto, Martello, and Toth [2]. The results obtained have been compared with the value of the Lagrangian lower bound described in Toth and Vigo [20].

The first class is composed by the 62 randomly generated euclidean instances proposed by Goetschalckx and Jacobs-Blecha [11].

Vertex coordinates are uniformly distributed in the interval [0,24000] for the x values, and in the interval [0,32000] for the y values. The depot is located centrally in (12000,16000). The cost c_{ij} of arc (i, j) is defined as the Euclidean distance between vertex i and j. Vertex demands are generated from a normal distribution with mean value 500 and standard deviation 200. Fourteen values for the total number of vertices, $n + m$, (from 25 to 150) have been considered, with Linehaul percentage equal to 50%, 66%, and 80%. For each value of $n + m$, the vehicle capacity has been chosen so that approximately from 3 up to 10 vehicles are used to serve all the demands. The problem input data have been gently provided by Marc Goetschalckx and Charlotte Jacobs-Blecha.

Table I reports the results obtained by the four algorithms on the instances of this class. For each problem we give the problem name, the problem size (namely the values of n and m), the available number of vehicles K, and the minimum number of vehicles needed to serve the Linehaul and the Backhaul vertices, K_L and K_B. The values of K_L and K_B are determined by solving the associated Bin Packing Problem using the code MTP by Martello and Toth [16]. For each heuristic algorithm the table gives:

i) the percentage deviation of the solution with respect to the lower bound value,
ii) the computing time expressed in IBM 386/20 seconds.

Note that the percentage deviation is an upper bound on the percentage error of the heuristic solution value with respect to the optimal solution value. The solution values and the computing times of algorithms DB, SF and LHBH have been gently provided by Marc Goetschalckx and Charlotte Jacobs-Blecha. As to algorithm TV, each reported result is the best one obtained running it during 100 iterations of the subgradient optimization procedure used for the computation of the Lagrangian lower bound. The heuristic algorithm TV has been applied to the current Lagrangian solution every 5 subgradient iterations, and the computing time reported includes the lower bound computation. We have multiplied the computing times of algorithm TV by four since, according to our experience on this kind of algorithms, an IBM 486/33 is almost four times faster than the IBM 386/20 used by Goetschalckx and Jacobs-Blecha.

The value of the solution obtained by all the algorithms has been computed by using a real-valued cost matrix c, and by rounding the final solution value to the nearest integer. The lower bound values have been, instead, computed by using an integer-valued cost matrix, whose entries are the euclidean distances multiplied by 10 and rounded to the nearest integer. The final lower bound values are then divided by 10 and rounded to the nearest integer.

Table I shows that algorithm TV performed better than the algorithms from the literature, obtaining the best solution in 57 out of 62 test problems within acceptable computing times. The greater effectiveness of algorithm TV can also be seen by noting that, over all the instances of Class I, the average percentage deviation of the solutions obtained by this algorithm, is 105.2% with standard deviation 4.5, while the solutions obtained by DB, SF and LHBH have an average percentage deviation equal to 110%, 126.9% and 107.2%, with standard deviation 5.3, 8 and 4.1, respectively. These results are further illustrated by Figure 5, which reports the trend of the percentage deviation of the solution found by the four algorithms.

Over all the instances of Class I, the average computing time needed by algorithm TV is 256.5 seconds, while DB, SF and LHBH require on average 37.5, 16.6 and 63.6 seconds, respectively.

The second class contains 33 VRPB instances obtained from 11 VRP test problems proposed in the literature, with total number of vertices between 21 and 100. The input data of the VRP test problems are available, through electronic mail, in the TSPLIB library, (see Reinelt [18] for further details). For each VRP test problem we generated three VRPB instances, each corresponding to a Backhaul percentage of 50%, 66%, and 80%. The vertex set has been partitioned by defining as Backhaul one vertex in every two, three or five, respectively, depending on the desired percentage. Vertex demands and vehicle capacity have been set equal to the original VRP values, and we have defined $K = \max\ \{K_L, K_B\}$. We used our implementations of DB and SF heuristics. The results are illustrated in Table II, and confirm the good performance of algorithm TV.

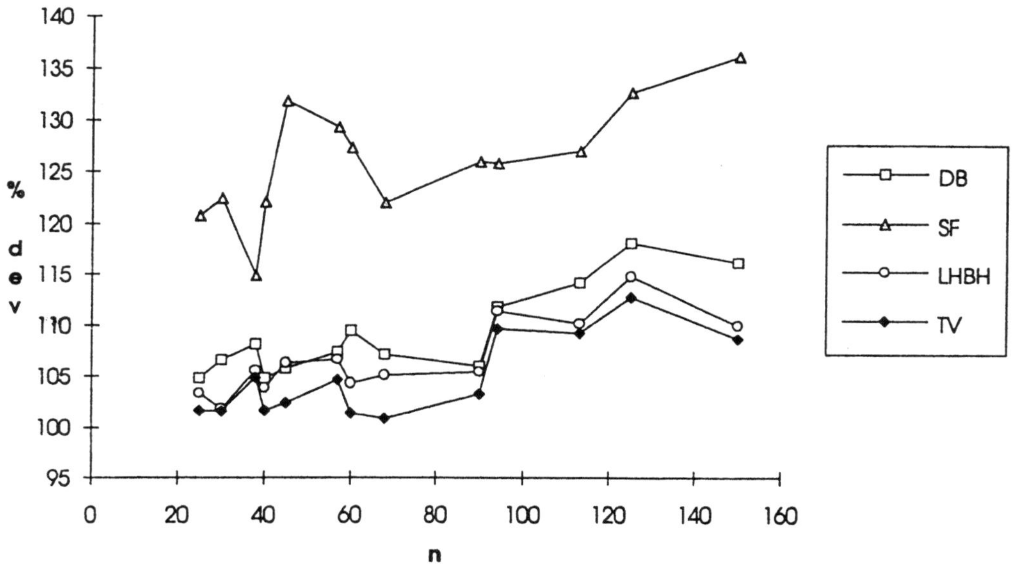

Figure 5. The average percentage deviation of the solutions found by algorithms DB, SF, LHBH and TV on instances of Class I.

Indeed, over all the instances of Class II, the average percentage deviation of the solutions obtained by this algorithm, is 107.4% with standard deviation 5.8, while the solutions obtained by DB and SF have an average percentage deviation equal to 120.1% and 138.1%, with standard deviation 6.9 and 13.2, respectively. Moreover, algorithms DB and SF have not been able to determine a solution with the desired number of routes in 12 and 16 problems, respectively (the number of routes, when different from K, is reported in brackets).

Acknowledgments

This work has been supported by MURST 40%.

References

[1] J. Bartholdy and L. Platzman, "An O(n log n) planar traveling salesman heuristic based on spacefilling curves", *Operations Research Letters*, 1:121-125, 1982.

[2] G. Carpaneto, S. Martello, and P. Toth, "Algorithms and codes for the assignment problem", in B.Simeone, P.Toth, G. Gallo, F. Maffioli, and S. Pallottino, editors, *FORTRAN Codes for Network Optimization*, *volume 13 of Annals of Operations Research*. Baltzer, Basel, 1988.

[3] D. Casco, B. Golden, and E. Wasil, "Vehicle routing with backhauls: Models, algorithms, and case studies", in B. Golden and A. Assad, editors, *Vehicle Routing: Methods and Studies*, pages 127-147. North-Holland, Amsterdam, 1988.

[4] N. Christofides and S. Eilon, "An algorithm for the vehicle dispatching problem", *Operations Research Quarterly*, 20:309-318, 1969.

[5] G. Clarke and J. Wright, "Scheduling of vehicles from a central depot to a number of delivery points", *Operations Research*, 12(4):568-581, 1964.

[6] I. Deif and L. Bodin, "Extension of the clarke and wright algorithm for solving the vehicle routing problem with backhauling", in A. Kidder, editor, *Proceedings of the Babson Conference on Software Uses in Transportation and Logistic Management*, pages 75-96, Babson Park, 1984.

[7] M. Fischetti and P. Toth, "An additive bounding procedure for the asymmetric travelling salesman problem", *Mathematical Programmig*, 53:173-197, 1992.

[8] M. Fischetti, P. Toth, and D. Vigo, "A branch-and-bound algorithm for the capacitated vehicle routing problem on directed graphs" *Operations Research*, 42(5):846-859, 1994.

[9] M. Fisher, "Optimal solution of vehicle routing problems using minimum k-trees", *Operations Research*, 42 (4): 626 - 642, 1994.

[10] M. Fisher and R. Jaikumar, "A generalized assignment heuristic for vehicle routing problem", *Networks*, 11(4):109-124, 1981.

[11] M. Goetschalckx and C. Jacobs-Blecha, "The vehicle routing problem with backhauls" *European Journal of Operational Research*, 42:39-51, 1989.

[12] M. Goetschalckx and C. Jacobs-Blecha, "The vehicle routing problem with backhauls: Properties and solution algorithms", *Technical Report* MHRC-TR-88-13, Georgia Institute of Technology, 1993.

[13] B. Golden, E. Baker, J. Alfaro, and J. Schaffer, "The vehicle routing problem with backhauling: Two approaches", in R. Hammesfahr, editor, *Proceedings of the XXI Annual Meeting of S.E. TIMS*, pages 90-92, Myrtle Beach, 1985.

[14] S. Lin, "Computer solutions to the travelling salesman problem" *Bell System Technical Journal*, 44:2245-2269, 1965.

[15] S. Lin and B. Kernighan, "An effective heuristic algorithm for the travelling salesman problem" *Operations Research*, 21:498-516, 1973.

[16] S. Martello and P. Toth, "Knapsack Problems: Algorithms and Computer Implementations", *John Wiley & Sons*, Chichester, 1990.

[17] S. Martello and P. Toth, "Lower bounds and reduction procedures for the bin packing problem", *Discrete Applied Mathematics*, 28:59-70, 1990.

[18] G. Reinelt, "TSPLIB - a traveling salesman problem library" *ORSA Journal on Computing*, 3:376-384, 1991.

[19] M. Savelsbergh, "Computer Aided Routing" *Ph.D. thesis, Centrum voor Wiskunde en Informatica*, Amsterdam, 1988.

[20] P. Toth and D. Vigo, "An exact algorithm for the vehicle routing problem with backhauls", in *Netflow '93*, Pisa, 1993.

[21] D. Vigo, "A heuristic algorithm for the asymmetric capacitated vehicle routing problem", *European Journal of Operational Research*, to appear.

[22] C. Yano, T. Chan, L. Richter, T. Cutler, K. Murty, and D. McGettigan, "Vehicle routing at quality stores" *Interfaces*, 17:52-63, 1987.

Table I. Test problems proposed by Goetschalckx and Jacobs-Blecha. Computing times in IBM 386/20 seconds.

						DB		SF		LH BH		TV		
Name	n	m	K	K_L	K_B	% dev	time	% dev	time	% dev	time	sol	% dev	time
A1	20	5	8	7	2	104.7	0.7	122.1	6.6	104.7	0.3	229886	104.3	16.8
A2	20	5	5	4	1	105.2	0.6	122.4	5.6	103.0	0.3	180119	102.3	11.4
A3	20	5	4	3	1	103.6	0.5	111.7	5.4	104.0	0.3	163405	100.0	7.8
A4	20	5	3	3	1	105.7	0.6	126.8	4.9	101.7	0.3	155796	100.0	6.2
B1	20	10	7	7	4	110.7	0.9	120.6	7.7	104.4	0.2	239080	104.1	27.6
B2	20	10	5	4	3	107.5	0.7	125.1	5.8	100.9	0.2	198048	100.7	14.6
B3	20	10	3	3	2	101.7	0.7	121.6	4.9	100.0	0.2	169372	100.0	1.8
C1	20	20	7	6	6	110.9	1.6	131.8	8.1	107.0	18.0	253318	105.8	39.0
C2	20	20	5	4	4	100.7	1.4	117.7	6.6	103.9	18.0	215020	100.7	23.2
C3	20	20	5	3	3	102.3	1.4	---	---	102.2	18.0	199346	100.0	11.8
C4	20	20	4	3	3	105.4	1.8	116.7	6.8	102.3	18.0	195367	100.0	12.8
D1	30	8	12	10	3	108.2	2.0	114.9	9.6	104.3	9.8	322705	103.7	49.6
D2	30	8	11	10	3	113.2	2.5	---	---	108.0	9.8	318476	107.1	51.6
D3	30	8	7	6	2	105.9	2.0	115.9	7.2	104.8	9.8	239479	104.6	34.8
D4	30	8	5	5	2	105.4	1.9	113.9	6.6	105.1	9.8	205832	104.0	24.2
E1	30	15	7	6	3	106.4	2.4	128.8	8.0	106.3	17.7	238880	104.0	45.8
E2	30	15	4	4	2	105.3	2.5	135.1	7.4	107.7	17.7	212263	102.2	28.8
E3	30	15	4	3	2	105.6	2.3	131.7	7.9	104.8	17.7	206659	101.0	25.8
F1	30	30	6	5	6	114.4	4.9	---	---	108.6	25.2	263929	103.4	67.8
F2	30	30	7	5	6	110.4	7.3	129.0	9.3	102.4	25.2	265214	100.9	73.6
F3	30	30	5	4	4	107.5	7.1	125.4	9.1	102.1	25.2	241121	100.3	56.0
F4	30	30	4	3	3	105.7	6.2	127.4	9.6	104.3	25.2	233862	100.9	45.0
G1	45	12	10	9	3	112.0	7.5	134.1	9.1	113.4	32.7	306959	110.1	116.8
G2	45	12	6	6	2	107.0	5.2	128.7	8.5	107.6	32.7	245441	106.6	69.2
G3	45	12	5	5	2	108.0	6.4	125.4	9.6	105.0	32.7	230170	104.5	57.2
G4	45	12	6	5	2	107.2	6.4	134.0	9.6	106.6	32.7	232647	103.7	58.2
G5	45	12	5	4	1	107.6	6.0	131.3	9.4	104.8	32.7	221899	102.2	57.0
G6	45	12	4	3	1	102.9	6.0	122.4	15.2	102.7	32.7	213457	100.7	37.0
H1	45	23	6	6	3	108.6	7.6	128.5	10.2	107.2	32.5	270719	103.9	86.6
H2	45	23	5	5	3	107.9	6.4	129.0	13.8	104.6	32.5	253365	100.7	66.2
H3	45	23	4	4	2	106.8	6.2	124.8	13.3	103.8	32.5	247536	100.3	61.2
H4	45	23	5	4	2	108.4	5.9	118.5	12.6	105.2	32.5	250221	100.4	67.4
H5	45	23	4	3	2	104.8	6.4	112.8	13.9	105.0	32.5	246121	100.1	54.0
H6	45	23	5	3	2	106.6	6.0	118.7	12.7	104.9	32.5	249135	100.1	61.4

(follows)

Table I. Test problems proposed by Goetschalckx and Jacobs-Blecha. Computing times in IBM 386/20 seconds.

Name	n	m	K	K_L	K_B	DB % dev	DB time	SF % dev	SF time	LH BH % dev	LH BH time	TV sol	TV % dev	TV time
I1	45	45	10	8	9	107.8	13.2	131.3	13.5	107.0	42.4	354410	105.1	337.6
I2	45	45	7	6	7	105.4	13.8	130.0	13.2	105.6	42.4	315184	103.8	190.4
I3	45	45	5	4	5	108.8	25.2	119.0	32.0	108.4	42.4	298367	104.5	130.0
I4	45	45	6	4	5	106.0	27.6	125.3	25.2	104.1	42.4	295988	101.6	136.4
I5	45	45	7	4	5	102.2	12.7	124.3	19.0	102.4	42.4	302709	101.6	163.0
J1	75	19	10	10	3	114.6	21.4	---	---	116.9	74.3	343476	113.9	290.2
J2	75	19	8	8	2	114.3	25.0	128.2	16.1	111.3	74.3	315644	111.7	245.8
J3	75	19	6	5	2	107.1	23.1	125.2	21.0	108.0	74.3	282447	104.4	187.6
J4	75	19	7	7	2	111.3	23.6	124.1	18.4	109.6	74.3	300548	108.8	228.6
K1	75	38	10	10	5	115.7	52.8	123.8	16.5	112.9	99.8	408303	112.5	479.0
K2	75	38	8	8	4	116.0	67.3	127.4	17.0	109.3	99.8	372423	109.4	330.6
K3	75	38	9	8	4	113.0	59.6	130.9	19.1	111.0	99.8	374417	107.6	380.2
K4	75	38	7	7	3	111.9	77.0	126.0	20.3	107.6	99.8	358308	107.6	331.6
L1	75	75	10	9	9	114.3	86.8	146.3	22.2	110.8	159.0	447422	111.2	1115.6
L2	75	75	8	8	8	120.7	158.6	137.9	27.3	108.5	159.0	411453	106.8	833.8
L3	75	75	9	8	8	111.2	116.6	---	---	108.1	159.0	417990	107.1	878.8
L4	75	75	7	6	6	112.9	122.2	132.5	107.2	107.8	159.0	394133	105.1	709.2
L5	75	75	8	6	6	114.2	128.0	---	---	106.3	159.0	395911	104.1	767.0
M1	100	25	11	10	3	120.8	105.7	137.7	24.0	114.5	157.3	409635	113.6	651.8
M2	100	25	10	10	3	120.3	75.7	---	---	117.6	157.3	410689	116.6	566.8
M3	100	25	9	9	3	115.7	65.7	131.6	28.5	115.0	157.3	387105	113.4	503.0
M4	100	25	7	7	2	116.0	70.8	129.1	44.4	112.3	157.3	352740	107.6	447.8
N1	100	50	11	10	5	119.5	148.2	156.8	5.0	113.5	159.0	430054	112.1	938.8
N2	100	50	10	10	5	121.0	146.3	---	---	116.9	159.0	427593	113.8	866.6
N3	100	50	9	9	4	116.2	146.1	134.8	27.6	109.6	159.0	411543	111.6	787.6
N4	100	50	10	9	4	116.0	111.5	131.6	27.4	110.0	159.0	412704	109.7	848.4
N5	100	50	7	7	3	118.5	145.8	126.2	40.4	110.6	159.0	382861	107.8	547.2
N6	100	50	8	7	3	113.5	128.1	123.7	33.9	108.3	159.0	384287	106.9	543.0

(end)

Table II. Test problems obtained from VRP instances. Computing times in IBM 486/33 seconds.

Name	n	m	K	K_L	K_B	DB			SF			TV		
							% dev	time		% dev	time	sol	% dev	time
eil22_50	11	10	3	3	2		116.3	0.1		133.3	1.1	389	105.4	0.5
eil22_66	14	7	3	3	1	(4)	115.8	0.2	(4)	134.7	1.2	366	100.0	0.6
eil22_80	17	4	3	3	1		103.9	0.2	(4)	145.7	1.1	375	103.9	0.8
eil23_50	11	11	2	2	1		104.4	0.2	(3)	114.7	1.3	682	100.6	0.6
eil23_66	15	7	2	2	1		112.7	0.2	(3)	127.1	1.2	649	102.9	0.5
eil23_80	18	4	2	2	2		113.7	0.2		111.5	1.3	625	102.3	0.4
eil30_50	15	14	2	2	2	(3)	115.9	0.2		146.2	1.8	501	100.6	0.7
eil30_66	20	9	3	3	1		120.7	0.2		163.0	1.7	542	110.2	1.2
eil30_80	24	5	3	3	1		119.8	0.2	(4)	159.1	1.8	519	106.1	1.2
eil33_50	16	16	3	3	2		123.7	0.2		131.2	1.9	764	106.0	1.6
eil33_66	22	10	3	3	1		123.7	0.2	(4)	163.4	1.8	763	107.8	1.8
eil33_80	26	6	3	3	1	(4)	136.0	0.3	(4)	158.7	1.9	761	111.7	1.7
eil51_50	25	25	3	3	3	(4)	121.4	0.7		126.3	2.1	561	101.8	4.1
eil51_66	34	16	4	4	2		119.6	0.7		122.4	2.3	551	102.8	5.6
eil51_80	40	10	4	4	1	(5)	119.1	1.9	(5)	127.8	2.1	584	106.2	6.2
eilA76_50	37	38	6	6	5		117.0	2.6		128.4	2.8	756	105.3	18.9
eilA76_66	50	25	7	7	4		124.4	2.7		145.0	2.7	776	106.4	16.5
eilA76_80	60	15	8	8	2	(9)	130.2	2.7	(9)	138.7	2.9	839	119.7	26.6
eilB76_50	37	38	8	8	7	(9)	123.5	2.8	(9)	149.9	3.1	836	107.9	24.7
eilB76_66	50	25	10	10	5		124.7	2.7	(11)	155.6	2.8	897	113.7	27.9
eilB76_80	60	15	12	12	3		132.0	3.0		147.7	3.5	951	121.6	36.1
eilC76_50	37	38	5	5	4		123.5	2.7		134.0	2.6	714	102.4	13.6
eilC76_66	50	25	6	6	3		119.6	3.0		135.5	3.2	748	104.9	17.9
eilC76_80	60	15	7	7	2		122.7	2.8		140.0	3.2	757	110.8	23.1
eilD76_50	37	38	4	4	3		118.0	2.8		125.9	3.9	704	103.1	10.2
eilD76_66	50	25	5	5	2		119.1	2.7		128.1	3.0	730	104.1	10.8
eilD76_80	60	15	6	6	2		118.9	2.7		131.7	3.7	715	106.6	19.8
eilA101_50	50	50	4	4	4	(5)	113.3	6.8	(5)	126.7	5.1	849	105.3	22.3
eilA101_66	67	33	6	6	3		114.2	6.4		134.6	5.6	879	105.1	34.6
eilA101_80	80	20	6	6	2	(7)	117.4	6.6	(7)	135.6	5.0	908	111.5	32.0
eilB101_50	50	50	7	7	7	(8)	120.3	7.0	(8)	138.1	5.6	954	108.5	44.5
eilB101_66	67	33	9	9	5	(10)	129.0	6.6	(10)	147.9	6.1	1056	118.5	45.8
eilB101_80	80	20	11	11	3	(12)	130.2	7.6	(12)	147.7	5.8	1076	119.3	62.8

HYBRID GENETIC ALGORITHMS FOR BUS DRIVER SCHEDULING

Raymond S. K. Kwan

Lecturer, School of Computer Studies, University of Leeds, Leeds LS2 9JT, UK.

Anthony Wren

Professor of Scheduling and Constraint Management, School of Computer Studies, University of Leeds, Leeds LS2 9JT, UK.

A hybrid approach for bus driver scheduling is described, which incorporates Genetic Algorithms (GAs), a rule-based driver duties Estimator, and an integer linear programming method called IMPACS. IMPACS produces near-optimal schedules. However, a scheduling run might take hours, and larger problems may have to be decomposed into sub-problems. Earlier research indicates that GAs for directly producing driver schedules are quick, but so far they tend to converge at sub-optimal solutions. In the hybrid approach, GAs take on an indirect role. They do not form schedules, but instead would yield a population of vastly condensed versions of the problem. IMPACS is then applied using the elite population members to yield the optimal schedule almost trivially. The Estimator plays an essential role in the evaluation of population members during the evolution process. The resulting scheduling system captures the complementary strengths of its components and has potential for solving large problems optimally and quickly.

1. Introduction

Given a set of work for a fleet of buses, driver scheduling is the follow-on process of arranging the bus work into suitable duties for drivers. A state-of-the-art review of research into this problem is given by [Wren and Rousseau 1994]. The most widely used systems at present are probably those of Teleride-Sage, UMA Trapeze, HASTUS

[Rousseau et al 1985], IMPACS [Smith and Wren 1988] and HOT [Daduna and Mojsilovic 1988]. Despite their success, practical problems often require hours of computer time for generating an efficient driver schedule. No system is yet able to solve the larger problems entirely satisfactorily: problems are either over simplified or decomposed into sub-problems, which often result in using more duties than necessary.

At Leeds, along with continuing research into improving IMPACS, which is based on a set covering integer linear programming model, other complementary and innovative methods are being pursued. Among these are the use of Genetic Algorithms [Clement and Wren 1994, Wren and Wren 1995] and applications of an intelligent rule-based Estimator [Kwan et al 1992, Zhao et al 1994]. Experimental results indicate that each of the methods investigated has its own inherent disadvantages. However, the strengths of these individual approaches may be combined to form a powerful solution method. This paper describes such a hybrid approach.

Instead of deriving complete driver schedules, the role of Genetic Algorithms (GA) is to evolve a population of data sets representing greatly compressed versions of the original problem. Because of the much reduced sizes, IMPACS would be able to derive optimal solutions using one or more of the elite members in the evolved population very easily and quickly. All GAs must have the ability to evaluate population members. This function is provided by the Estimator, which safeguards the GA from going off-course for compact but bad data sets: ones that will not yield the optimal schedule eventually.

2. Relief Opportunities as Genes

Whereas a bus may be used continuously throughout the day, a driver has to be given a break off the bus after working for a length of time. In some places, especially less developed countries, there may be exceptions; but we shall only consider the normal common practices. When a driver leaves for a break, another driver has to take over the bus unless there is no more work for that bus. This usually cannot just take place anytime anywhere the bus happens to be. Nevertheless, there is usually quite a number of suitable opportunities for each bus; which are called *relief opportunities*. After a break, the driver resumes work either from the start or at a relief opportunity on usually a different bus. Duties forming a driver schedule must fit seamlessly at the relief opportunities used.

Figure 1 shows part of a bus schedule with drivers' duties superimposed. We assume a local rule permitting unbroken stretches of work up to five and a half hours (allowed in some relatively slack operating agreements). Duty D1 works on bus 1 until 0937, then on bus 2 from 1007 until 1407. Duty D2 works on bus 2 until 1007, then on bus 3 from 1037 until 1437. Duty D4 takes over bus 1 from D1 at 0937, working until 1337, taking over bus 2 from 1407 until 1807. Duty D6 works on bus 3 from 1437 until 1740, and on bus 1 from 1840 until 2345. Other parts of duties D3, D5, D7 and D8 would be found on other buses.

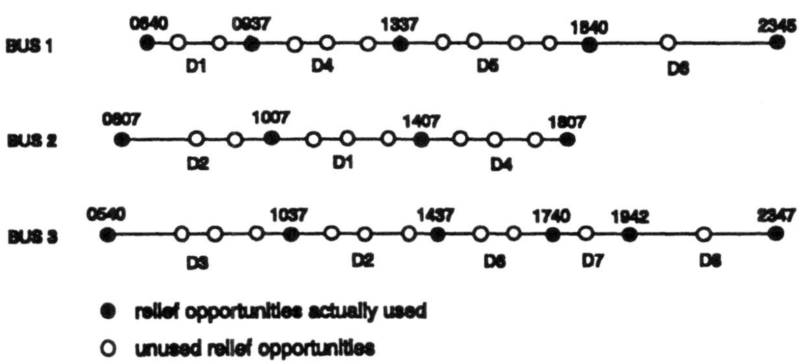

Figure 1

This is obviously a combinatorial problem the hardness of which increases exponentially with the number of relief opportunities available.

In an optimal schedule, only a small percentage of the total relief opportunities will actually be used. For example, suppose every duty has only one break. Each duty uses at most four relief opportunities corresponding to the beginning and end of duty, and the beginning and end of break. But since any of these four relief opportunities which are not at the beginning or end of a bus are also used by other duties, a 100 duty schedule might only involve just over 200 relief opportunities. The original problem might consist of well over 1000 relief opportunities.

Most bus driver scheduling methods naturally include mechanisms for reducing the number of relief opportunities to be considered. At the simplest level, manual on/off switches enable the human scheduler to exercise judgment through experience. Other mechanisms are automatic and more sophisticated. The ideal is to be able to suspend, right at the start, all but the relief opportunities going to be used in the optimal solutions. No existing methods are close to this ideal yet.

In IMPACS, there is an optional module that makes use of a few parameters, such as the average length of a piece of driver's work, and some heuristic rules to identify relief opportunities that are unlikely to be used. For example, a relief opportunity at 30 minutes after the bus has started work will unlikely be used. This process does not consider interactions between duties in the overall schedule, and is only meant for small scale suspension of relief opportunities. Sometimes manual adjustments may also be needed to yield satisfactory results.

HASTUS constructs duties after a set of relief opportunities to be used have been determined, the process of which heuristically approximates the relief opportunity times to round figures, for example half hours. The solutions would be refined by revising the set of relief opportunities to be used iteratively.

The hybrid GA approach makes analogy between relief opportunities and genes in a chromosome. Each chromosome has as many genes as the number of relief opportunities available, and each gene has a binary value: "1" for the active ones and "0" for those suspended. The chromosome represents the full set of relief opportunities on all buses, bus 2 following directly after bus 1, etc.

The full original problem is therefore represented by a bit string (chromosome) of all "1"s, and a compressed version of the problem would have many "0"s in the corresponding bit string. The GA works on the principle that there is a fast and accurate process for predicting how many duties the compressed problem represented by a chromosome is going to need. Chromosomes implying the least number of duties in the prospective schedules will be the fittest, and amongst them those with the least number of "1"s are fitter than others. The law of evolution namely survival of the fittest operates, resulting in highly compressed versions of the original problem after some generations. A modified version of the Estimator already mentioned is used to assess the number of duties implied by a given chromosome.

3. The Estimator

There is no fail-safe way of predicting the right number of duties apart from actually constructing the optimal schedule. The best an estimator could achieve is a very tight lower bound, and to guarantee that it will not over-estimate. The solution to the integer relaxation phase of IMPACS is an excellent estimate [Parker et al 1994], but being a linear programming model the process is too time consuming for this application. Instead, the Estimator used here originated from an optional module of IMPACS for decomposing problems that are too large to be solved in one go [Wren and Smith 1988]. The estimates it has yielded in experiments have accurately matched the optimal solutions [Zhao et al 1994].

The high accuracy of the Estimator is attributable to the vast and refined experience in bus driver scheduling the research team at Leeds has accumulated since the 1960s. Such know-how is embedded in heuristic rules, details of which are discussed in [Zhao et al 1995], forming the Estimator. For example, there are rules for identifying situations where extra duties might be needed, and there are rules to make rational judgment of whether such problems could be resolved.

To illustrate, Figure 2 shows the early part of a small bus schedule. We assume a maximum permitted stretch length of four and a half hours (typical in British and other northern European urban operations).

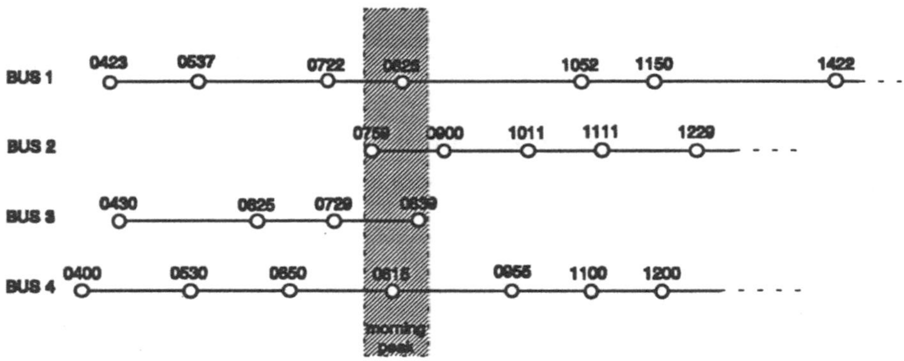

Figure 2

Four buses operate over the morning peak, and we therefore need four duties over this period. Buses 1 and 4 must be relieved at 0825 and 0815 at the latest, and we apparently need two further duties starting at these times. However, if we relieve the driver on bus 4 at 0650 to have a meal and then work on bus 1 from 0722 until 1150, we eliminate the need for the 0825 start. If we then allow the driver who is relieved at 0722 to take over bus 2 after a break from 0759 until 1229, we have covered the peak with only four duties (see Figure 3), starting at 0400 on bus 4, 0423 on bus 1, 0430 on bus 3 and 0650 on bus 4 (until 1100). Depending on local rules, the duty on bus 3 may be the first part of a split duty taking over bus 4 at 1100. We then need other duties starting at 1150 and 1229, but these can contribute to the afternoon peak, after having suitable breaks.

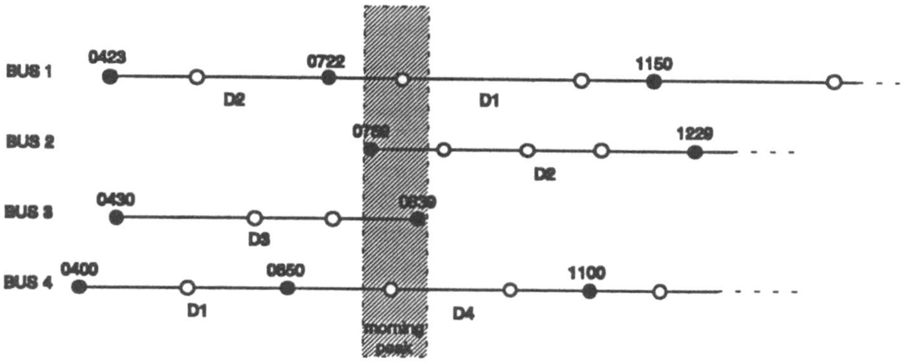

Figure 3

The combinations which may be used are much more complex on larger schedules. It is a function of the Estimator to seek such opportunities and to determine whether an apparent need for additional duties above the peak bus requirement can be resolved.

The Estimator is computationally very fast. Although the heuristic rules used are complex, only one straightforward pass without any major iterative loops is required. This is important in a GA application because there would be numerous times when population members have to be evaluated.

As pointed out above, the Estimator will not always give perfect predictions. This is well acceptable. IMPACS would only have difficulty in finding the optimal solutions quickly for relatively large problems. Compression to the absolutely minimal set of relief opportunities would have scaled a given problem down manyfold. In fact, the power of IMPACS makes it not significantly important whether a problem is perfectly compressed or just near perfectly compressed; it would still return solutions at desirable promptness. To compensate for imperfection, when in fact some other relief opportunities will be needed to achieve the estimate, the elite members in the evolved population are actually merged to boost the chance of finally yielding the optimal schedule. The elite members would be very similar to each other and only differ in a few critical relief opportunities.

4. Genetic Algorithms

The approach being described was inspired from some pilot research into the application of GAs in bus driver scheduling [Wren and Wren 1995, Clement and Wren 1994]. In the earlier research, duties were made analogous to genes. All "reasonable" potential duties are generated, and the chromosome consists of a list of duties in a solution. A non-standard form of recombination is used to form offspring. The objective of the GA is therefore to obtain the optimal schedule directly, in contrast with the indirect approach described here, through evolution. Results showed that GAs have one big advantage and that is they converge very quickly. However, they may level-off at sub-optimal solutions. Although the sub-optimal solutions were only up to a few duties more than the optimal solutions, the current method would not be acceptable to most operators. Perhaps it may be more competitive with other existing methods for very large problems; which if solved by existing methods would require decomposition or approximation and might incur extra duties.

The previous approach faced a number of difficulties. The set of "reasonable" potential duties is usually large, which is about 5,000 for average problems and 10,000 for the large ones. Only a tiny fraction of the potential duties will be used in an optimal solution. Just to minimise the number of "1"s in a chromosome bit string would have been insufficient: the evaluation function would have had to check how the bus work is covered; each piece of bus work must be uniquely covered in a final operational schedule. If a straightforward mating and crossover process were used and we threw away offspring that covered the bus work infeasibly, viable offspring would be produced too infrequently. In a bid to overcome these problems, more and more complex heuristics were added to the mating and crossover processes. There had been improvements with each adaptation, but the method will not yet usually yield the optimal solutions.

In the current approach, a first estimate (*firstEstimate*) is obtained (see “The Estimator” in Section 3) for the full problem, i.e. a chromosome with a bit string (representing relief opportunities) of all "1"s. The system proceeds as if the first estimate were a perfect prediction as indicated by experiments [Zhao et al 1994]. A fixed-size initial population is generated with random gene values. The evaluation function for a chromosome X is:

$$worth(X) \begin{cases} = K - \left(estimate(X) \times W + \sum Xgenes\right) & \text{if } estimate(X) \geq firstEstimate \\ = K - \left(firstEstimate \times W + \sum Xgenes\right) & \text{otherwise} \end{cases}$$

where W is a large constant and *Xgenes* are the values of genes in X, K is a constant much larger than W so that a fitter chromosome will have a higher *worth*. Estimates less than the first estimate are regarded as imperfect lower bounds and therefore raised to *firstEstimate* in the evaluation function. Survival of the fittest means that an elite chromosome will yield the least estimated number of duties with the least number of relief opportunities.

Simple standard strategies for reproduction and evolution, e.g. single point crossover, roulette-wheel selection of mating partners [see Davis 1991], seem to be adequate.

5. Schedule Production

IMPACS is used for the final production of driver schedules. It first generates a set of "reasonable" potential duties based on the selected relief opportunities, and then uses specialised integer linear programming techniques to select from the set the least cost subset that covers all the bus work. IMPACS is being used successfully by over 30 bus operators. The biggest drawback of IMPACS is that it often takes hours to run, and the computer time required tends to increase dramatically with the number of relief opportunities. In this approach, the GA component acts as a superior replacement of the existing crude optional module for selecting relief opportunities. With this approach, the run time for IMPACS will decrease dramatically. It may even be practical to run IMPACS several times using different compressed versions of the problem.

6. Results and Conclusions

Experiments to date have concentrated on compressing the relief opportunities in the morning bus work only, which can easily be achieved by keeping the afternoon portion of the chromosome bit string values at "1"s. The same data sets used in experiments as described in [Zhao et al 1994] are used. Over 60% reduction in the number of relief opportunities are consistently achieved, and consistently good schedules have been produced by applying IMPACS to these, with substantial reduction in processing time. It is hoped that further detailed comparison and analysis of results and computational performance will be reported in the near future.

An interesting application of this approach will be to adapt it for scheduling train drivers. One differing problem characteristic in train operations is that relief opportunities may occur in time windows of significant duration, instead of at definite times for bus operations. If each time window is treated as containing as many relief opportunities as there are minutes in it, the problem will be beyond the capabilities of existing bus driver scheduling systems. The hybrid genetic approach described hopefully will overcome such a problem.

The approach described may also be used as a generic approach for problem reduction in other domains. The development may start with a crude estimator. When the estimate for an elite member of the evolved population cannot be achieved, analysis would lead to new rules for enhancing the estimator. Repeating this process, a crude estimator may evolve and mature rapidly approaching ever closer to perfection.

References

[1] R. Clement, A. Wren, "Greedy genetic algorithms, optimising mutations and bus driver scheduling", to appear *in* J.R. Daduna and J. Paixao (eds.) *Computer-aided transit scheduling*. Springer-Verlag, Berlin (also available as University of Leeds School of Computer Studies report 93.28, 1993), 1994.

[2] J.R. Daduna, M. Mojsilovi, "Computer-aided vehicle and duty scheduling using the HOT programme system", In J.R. Daduna and A. Wren (eds.) *Computer-aided transit scheduling*. Springer-Verlag, Berlin, pp.133-146, 1988.

[3] L. Davis (ed.). *"Handbook of genetic algorithms", Van Nostrand Reinhold*, New York, 1991.

[4] R.S.K. Kwan, A. Wren, L. Zhao, "Driver scheduling using intelligent estimation techniques", In M. Desrochers and J-M. Rousseau (eds.) *Computer-aided transit scheduling*, Springer-Verlag, Berlin, pp.379-394, 1992.

[5] M.E. Parker, A. Wren, R.S.K. Kwan, "Modelling the scheduling of train drivers", To appear in J.R. Daduna and J. Paixao (eds.) *Computer-aided transit scheduling. Springer-Verlag, Berlin* (also available as University of Leeds School of Computer Studies report 93.27, 1993), 1994.

[6] J-M. Rousseau, R. Lessard, J-Y Blais, "Enhancements to the HASTUS crew scheduling algorithm", In J-M Rousseau (ed.) *Computer scheduling of public transport* - 2. North-Holland, Amsterdam, pp.295-310, 1985.

[7] B.M. Smith, A. Wren, *"A bus crew scheduling system using a set covering formulation"*, *Transportation Science*, 22A, pp.97-108, 1988.

[8] A. Wren, J-M. Rousseau, "Bus driver scheduling - an overview", To appear in J.R. Daduna and J. Paixao (eds.) *Computer aided transit scheduling* - 3, Springer-Verlag, Berlin (also available as University of Leeds School of Computer Studies report 93.31, 1993), 1994.

[9] A. Wren, B.M. Smith, "Experiences with a crew scheduling system based on set covering", In J.R. Daduna and A. Wren (eds.) *Computer-aided transit scheduling*. Springer-Verlag, Berlin, pp.104-118, 1988.

[10] A. Wren, D.O. Wren, "A genetic algorithm for public transport driver scheduling", *Computers and Operations Research*, Vol.22, No.1, pp.101-110, 1995.

[11] L. Zhao, A. Wren, R.S.K. Kwan, "Enriching rules in a driver duty estimator", To appear in J.R. Daduna and J. Paixao (eds.) *Computer aided transit scheduling*, Springer-Verlag, Berlin (also available as University of Leeds School of Computer Studies report 93.22, 1993), 1994.

[12] L. Zhao, A. Wren, R.S.K. Kwan, "Development of a driver duty estimator", To appear in *Journal of Operational Research Society*, 1995.

MIX
Papier aus verantwortungsvollen Quellen
Paper from responsible sources
FSC® C105338

If you have any concerns about our products,
you can contact us on
ProductSafety@springernature.com

In case Publisher is established outside the EU,
the EU authorized representative is:
Springer Nature Customer Service Center GmbH
Europaplatz 3, 69115 Heidelberg, Germany

Printed by Libri Plureos GmbH
in Hamburg, Germany